Lecture Notes in Networks and Systems

Volume 197

The series "Lecture Notes in Networks and Systems" publishes the latest developments in Networks and Systems—quickly, informally and with high quality. Original research reported in proceedings and post-proceedings represents the core of LNNS.

Volumes published in LNNS embrace all aspects and subfields of, as well as new challenges in, Networks and Systems.

The series contains proceedings and edited volumes in systems and networks, spanning the areas of Cyber-Physical Systems, Autonomous Systems, Sensor Networks, Control Systems, Energy Systems, Automotive Systems, Biological Systems, Vehicular Networking and Connected Vehicles, Aerospace Systems, Automation, Manufacturing, Smart Grids, Nonlinear Systems, Power Systems, Robotics, Social Systems, Economic Systems and other. Of particular value to both the contributors and the readership are the short publication timeframe and the world-wide distribution and exposure which enable both a wide and rapid dissemination of research output.

The series covers the theory, applications, and perspectives on the state of the art and future developments relevant to systems and networks, decision making, control, complex processes and related areas, as embedded in the fields of interdisciplinary and applied sciences, engineering, computer science, physics, economics, social, and life sciences, as well as the paradigms and methodologies behind them.

Indexed by SCOPUS, INSPEC, WTI Frankfurt eG, zbMATH, SCImago.

All books published in the series are submitted for consideration in Web of Science.

More information about this series at http://www.springer.com/series/15179

Vikrant Bhateja · Suresh Chandra Satapathy ·
Carlos M. Travieso-Gonzalez ·
Wendy Flores-Fuentes
Editors

Computer Communication, Networking and IoT

Proceedings of ICICC 2020

Springer

Editors
Vikrant Bhateja
Department of Electronics
and Communication Engineering
Shri Ramswaroop Memorial Group
of Professional Colleges (SRMGPC)
Lucknow, Uttar Pradesh, India

Dr. A.P.J. Abdul Kalam Technical
University, Lucknow, Uttar Pradesh, India

Carlos M. Travieso-Gonzalez
Department of Signals and Communication
Institute for Technological Development
Las Palmas de Gran Canaria, Spain

Suresh Chandra Satapathy
School of Computer Engineering
Kalinga Institute of Industrial Technology
Bhubaneswar, India

Wendy Flores-Fuentes
Faculty of Engineering
Universidad Autónoma de Baja California
Mexicali, Baja California, Mexico

ISSN 2367-3370 ISSN 2367-3389 (electronic)
Lecture Notes in Networks and Systems
ISBN 978-981-16-0979-4 ISBN 978-981-16-0980-0 (eBook)
https://doi.org/10.1007/978-981-16-0980-0

This Springer imprint is published by the registered company Springer Nature Singapore Pte Ltd.
The registered company address is: 152 Beach Road, #21-01/04 Gateway East, Singapore 189721, Singapore

Preface

This book is a collection of high-quality peer-reviewed research papers presented at the '4th International Conference on Intelligent Computing and Communication (ICICC 2020)' held at the School of Engineering (SoE), Dayananda Sagar University (DSU), Bengaluru, Karnataka, India, during September 18–20, 2020.

After the success of the past three editions of ICICC held in 2016 at the University of Kalyani, West Bengal, India; in 2017 at MAEER's MIT College of Engineering, Pune, India; and in 2019 at SoE, DSU, Bengaluru, India, the 4th edition of ICICC is again organized by SoE, DSU, Bengaluru, India. All papers of the past ICICC editions are published by Springer AISC Series. Presently, ICICC 2020 provided a platform for academicians, researchers, scientists, professionals and students to share their knowledge and expertise in the diverse domain of intelligent computing and communication.

ICICC 2020 had received a number of submissions from the fields of ICT, computer communication, networking, IoT and its prospective applications in different spheres of engineering. The papers received have undergone a rigorous peer-review process with the help of the technical program committee members of the conference from various parts of the country as well as abroad. The review process has been crucial with minimum two–three reviews each along with due checks on similarity and content overlap. This conference has featured seven exclusive theme-based special sessions as enlisted below along with the main track and hence witnessed more than 400 submissions in total.

S1: Application of Artificial Intelligence in Mobile Apps: A Machine and Deep Learning Techniques Perspective
S2: Deep Learning for Intelligent Systems
S3: Computer Vision for Multimedia Security and Forensics
S4: Emerging Trends in Machine Learning Methodologies and Applications in Artificial Intelligence
S5: Artificial Intelligence and Nature-Inspired Algorithms
S6: Recent Innovations in Artificial Intelligence Research and Computing Technologies
S7: Artificial Intelligence and Machine Learning Applications (AIML)

Out of the above pool, only quality papers were finally presented in 10 parallel tracks/day (of virtual presentation sessions chaired by esteemed professors from premier institutes of the country) and compiled in this volume for publication. This volume consists of these selected papers focussing on the theme of 'Data Engineering and Intelligent Computing.'

The conference featured many distinguished keynote addresses by eminent speakers like: Mr. Upendra M Kulkarni (Vice President and G M, Firmware & Software (Non-volatile Memory Solutions Group) at Intel Corporation, California, USA) delivered a lecture on Role of High Performance and High-Density Storage in Intelligent Computing; Mr. Andrew Haschka (APAC Technology Practice Lead, Application Modernization, Google Cloud at Google, Australia) delivered a lecture on Building on Application Platform; Dr. Dinakar Sitaram (Chairperson of the Cloud Computing Innovation Council of India Working Group and Ex-CTO, HP System Technology, Software Division, India), delivered a lecture on Speech Recognition for Indian and Low Resource Languages; Mr. Swaminathan Venkataraman (World-wide Product and Solution Architect, Communications and Media Solutions (CMS), Hewlett Packard Enterprise India) delivered a lecture on Cloud Native 5G.

These keynote lectures/talks embraced a huge toll of the audience of students, faculties, budding researchers as well as delegates. The editors thank the general chair, TPC chair and the organizing chair of the conference for providing valuable guidelines and inspirations to overcome various difficulties in the process of organizing this conference. The editors also thank SoE, DSU, Bengaluru, for their whole-hearted support in organizing this conference.

The editorial board takes this opportunity to thank the authors of all the submitted papers for their hard work, adherence to the deadlines and patience during the review process. The quality of a refereed volume depends mainly on the expertise and dedication of the reviewers. We are indebted to the TPC members who not only produced excellent reviews but also did these in short time frames.

Uttar Pradesh, India — Dr. Vikrant Bhateja
Bhubaneswar, India — Dr. Suresh Chandra Satapathy
Las Palmas de Gran Canaria, Spain — Dr. Carlos M. Travieso-Gonzalez
Mysuru, India — Dr. V. N. Manjunath Aradhya

Conference Organization Committees

Chief Patrons
Dr. D. Hemachandra Sagar, Chancellor, DSU, and Chairman, DSI
Dr. D. Premachandra Sagar, Pro-Chancellor, DSU, and Chairman, DSI

Patrons
Shri Galiswamy, Secretary, DSI
Dr. K. N. B. Murthy, Vice Chancellor, DSU
Shri. R. Janardhan, Pro-Vice Chancellor, DSU
Dr. Puttamadappa C., Registrar, DSU

Honorary Chairs
H. P. Khincha, Evangelist, DSU
K. Jairaj, Advisor, DSU

Honorary General Chair
Dr. Hao Ying, Wayne State University, Michigan, USA

General Chair
Dr. S. C. Satapathy, KIIT, Bhubaneshwar, Odisha, India

Publication Chairs
Dr. Vikrant Bhateja, SRMGPC, Lucknow, U.P., India
Dr. Carlos M. Travieso-Gonzalez, University of Las Palmas de Gran Canaria (ULPGC), Spain

Organizing Chairs
Dr. M. K. Banga, Dean Research, DSU
Dr. A Srinivas, Dean, SoE, DSU
Dr. Sanjay Chitnis, Chairman, CSE, DSU

Technical Program Chairs
Dr. V. N. Manjunatha Aradhya, SJCE, Mysore, Karnataka, India
Dr. Wendy Flores-Fuentes, Autonomous University of Baja California, Mexico

Program Chairs
Dr. Rajesh T. M., Dept. of CSE, DSU
Dr. Jasma Balasangameshwara, Dept. of CST, DSU

Organizing Secretaries
Dr. Shaila S. G., Dept. of CSE, DSU
Dr. Girisha G. S., Dept. of CSE, DSU
Dr. Mallanagouda Patil, Dept. of CSE, DSU

Advisory Committee
Dr. Aime' Lay-Ekuakille, University of Salento, Lecce, Italy
Dr. Amira Ashour, Tanta University, Egypt
Dr. Aynur Unal, Stanford University, USA
Dr. Bansidhar Majhi, IIIT Kancheepuram, Tamil Nadu, India
Dr. Dilip Kumar Sharma, Vice-Chairman, IEEE UP Section
Dr. Ganpati Panda, IIT Bhubaneswar, Odisha, India
Dr. Jagdish Chand Bansal, South Asian University, New Delhi, India
Dr. João Manuel R. S. Tavares, Universidade do Porto (FEUP), Porto, Portugal
Dr. Jyotsana Kumar Mandal, University of Kalyani, West Bengal, India
Dr. K.C. Santosh, University of South Dakota, USA
Dr. Le Hoang Son, Vietnam National University, Hanoi, Vietnam
Dr. Naeem Hanoon, Multimedia University, Cyberjaya, Malaysia
Dr. Nilanjan Dey, TIET, Kolkata, India
Dr. Noor Zaman, Universiti Teknologi PETRONAS, Malaysia
Dr. Roman Senkerik, Tomas Bata University, Zlin, Czech Republic
Dr. Swagatam Das, Indian Statistical Institute, Kolkata, India
Dr. Siba K. Udgata, University of Hyderabad, Telangana, India
Dr. Shri Nivas Singh, MMMUT, Gorakhpur, U.P., India
Dr. Steven L. Fernandez, The University of Alabama at Birmingham, USA
Dr. Tai Kang, Nanyang Technological University, Singapore
Dr. Valentina Balas, Aurel Vlaicu University of Arad, Romania

Technical Program Committee
Prof. H R Vishwakarma, VIT Vellore, India
Prof. Dan Boneh, Stanford University, California, USA
Prof. Alexander Christea, University of Warwick, London, UK
Prof. Ahmad Al- Khasawneh, The Hashemite University, Jordan
Dr. Bharat Singh Deora, JRNRV University, India
Prof. Jean Michel Bruel, Departement Informatique IUT de Blagnac, Blagnac, France
Prof. Ngai-Man Cheung, Singapore University of Technology and Design, Singapore
Prof. Yun-Bae Kim, Sungkyunkwan University, South Korea
Prof. Ting-Peng Liang, National Chengchi University, Taipei, Taiwan
Prof. Sami Mnasri, IRIT Laboratory, Toulouse, France
Prof. Lorne Olfman, Claremont, California, USA
Prof. Brent Waters, University of Texas, Austin, Texas, USA
Prof. Philip Yang, PricewaterhouseCoopers, Beijing, China

Contents

Editors and Contributors

About the Editors

Vikrant Bhateja is Associate Professor, Department of Electronics and Communication Engineering, Shri Ramswaroop Memorial Group of Professional Colleges (SRMGPC), Lucknow, and also Head (Academics and Quality Control) in the same college. He completed his doctorate in Bio-Medical Imaging and Signal Processing and has a total academic teaching experience of 16 years with around 125 publications in reputed international conferences, journals and online book chapter contributions. His areas of research include digital image and video processing, computer vision, medical imaging, machine learning, pattern analysis and recognition. Dr. Vikrant has edited 15 proceeding books/editorial volumes with Springer Nature. He is Editor-in-Chief of IGI Global—*International Journal of Natural Computing and Research* (IJNCR). He is also Associate Editor in *International Journal of Synthetic Emotions* (IJSE) and *International Journal of Ambient Computing and Intelligence* (IJACI) under IGI Global press. He is Guest Editor of Special Issues in reputed Scopus/SCIE indexed journals under Springer Nature: "*Evolutionary Intelligence*" and "*Arabian Journal of Science and Engineering*".

Suresh Chandra Satapathy is a Ph.D. in Computer Science, currently working as Professor and at KIIT (Deemed to be University), Bhubaneshwar, Odisha, India. He held the position of the National Chairman Div-V (Educational and Research) of Computer Society of India and is also Senior Member of IEEE. He has been instrumental in organizing more than 20 International Conferences in India as Organizing Chair and edited more than 30 Book Volumes from Springer LNCS, AISC, LNEE and SIST Series as Corresponding Editor. He is quite active in research in the areas of swarm intelligence, machine learning and data mining. He has developed a new optimization algorithm known as Social Group Optimization (SGO) published in Springer Journal. He has delivered number of keynote address and tutorials in his areas of expertise in various events in India. He has more than 100 publications in reputed journals and conference proceedings. Dr. Suresh is in Editorial Board of IGI

Global, Inderscience, Growing Science journals and also Guest Editor for *Arabian Journal of Science and Engineering* published by Springer.

Carlos M. Travieso-Gonzalez received the M.Sc. degree in 1997 in Telecommunication Engineering at Polytechnic University of Catalonia (UPC), Spain, and Ph.D. degree in 2002 at the University of Las Palmas de Gran Canaria (ULPGC-Spain). He is Full Professor on Signal Processing and Pattern Recognition and Head of Signals and Communications Department at ULPGC; teaching from 2001 on subjects on signal processing and learning theory. His research lines are biometrics, biomedical signals and images, data mining, classification system, signal and image processing, machine learning, and environmental intelligence. He has researched in 50 International and Spanish Research Projects, some of them as Head Researcher. He has 440 papers published in international journals and conferences. He has published 7 patents in Spanish Patent and Trademark Office. He is Evaluator of project proposals for European Union (H2020), Medical Research Council (MRC—UK), Spanish Government (ANECA—Spain), Research National Agency (ANR—France), DAAD (Germany), and other Institutions. He has been General Chair in 16 international conferences, mainly sponsored by IEEE and ACM. He is Associate Editor in CIN and Entropy (Q2-WoS journals). He has been awarded in "Catedra Telefonica" Awards in editions 2017, 2018 and 2019 on Knowledge Transfer Modality.

Wendy Flores-Fuentes received the bachelor's degree in Electronic Engineering from the Autonomous University of Baja California in 2001, the master's degree in Engineering from Technological Institute of Mexicali in 2006, and the Ph.D. degree in science, applied physics, with emphasis on Optoelectronic Scanning Systems for SHM, from Autonomous University of Baja California in June 2014. By now she is the author of 30 journal articles in Elsevier, IEEE Emerald and Springer, 13 book chapters and 5 books in Springer, Intech, IGI Global Lambert Academic and Springer, 35 proceedings articles in IEEE ISIE 2014–2019, IECON 2014, 2018, 2019, the World Congress on Engineering and Computer Science (IAENG 2013), IEEE Section Mexico IEEE ROCC2011 and the VII International Conference on Industrial Engineering ARGOS 2014. Recently, she has organized and participated as Chair of Special Session on "Machine Vision, Control and Navigation" at IEEE ISIE 2015, 2016, 2017, 2019 and IECON 2018, 2019. She holds 1 patent of Mexico and 1 patent of Ukraine. She has been a reviewer of several articles in Taylor and Francis, IEEE, Elsevier, and EEMJ. Currently, she is Full-time Professor at Universidad Autónoma de Baja California, at the Faculty of Engineering, and is the coordinator of the physics area in the basic science department. She has been incorporated into CONACYT National Research System in 2015. She did receive the award of "Best session presentation" in WSECS2013 in San-Francisco, USA. She did receive as the co-author the award of "Outstanding Paper in the 2017 Emerald Literati Network Awards for Excellence".

Contributors

Swati Adhikari Department of Computer Science, The University of Burdwan, Purba Bardhaman, West Bengal, India

Sarita P. Ambadekar K. J. Somiaya Institute of Engineering and Information Technology, Mumbai, India

Blessy Antony SCMS Engineering and Technology, Ernakulam, Kerala, India

B. Durga Anuja Govt. Degree College for Womens Srikalahasti, SriKalahasti, Andhra Pradesh, India

V. N. Manjunath Aradhya Department of Computer Applications, JSS Science and Technology University, Mysuru, India

Chirag Arora KIET Group of Institutions, Delhi-NCR, Ghaziabad, UP, India

E. Arul Department of Information Technology, Coimbatore Institute of Technology, Coimbatore, Tamil Nadu, India

A. R. Aswatha Dayananda Sagar College of Engineering, Bangalore, Affiliated to VTU, Belgaum, Karnataka, India

Prashant A. Athavale Electrical and Electronics Engineering Department, BMS Institute of Technology and Management, Bengaluru, India

Bijitha Balakrishnan SCMS Engineering and Technology, Ernakulam, Kerala, India

Akansha Baral Dayananda Sagar College of Engineering, Bangalore, India

Mukul Anand Bathija Dayananda Sagar College of Engineering, Bangalore, India

Jorge A. Benitez Departamento de Ciencias Juridicas and Departamento de Jurisprudencia, Universidad Técnica Particular de Loja, Loja, Ecuador

Ashwin R. Bharadwaj Department of Computer Science and Engineering, People's Education Society University, Bangalore, India

Seema S. Bhat Department of ISE, Dayananda Sagar College of Engineering, Bengaluru, India

Vikrant Bhateja Department of Electronics and Communication Engineering, Shri Ramswaroop Memorial Group of Professional Colleges (SRMGPC), Lucknow, Uttar Pradesh, India;
Dr. A.P.J. Abdul, Kalam Technical University, Lucknow, Uttar Pradesh, India

P. V. Bhavyasree SCMS Engineering and Technology, Ernakulam, Kerala, India

V. R. Bindu School of Computer Sciences, Mahatma Gandhi University, Kottayam, Kerala, India

C. V. Chaithra SCMS Engineering and Technology, Ernakulam, Kerala, India

Suhith Shekar Chandra Department of Computer Science and Engineering, SJB Institute of Technology, Bangalore, Karnataka, India

Dharamendra Chouhan Department of Computer Science and Engineering, University Visvesvaraya College of Engineering, Bengaluru, India

Haritha Darisi Department of CSE, Aditya College of Engineering and Technology, Surampalem, India

S. Deepthi Dayananda Sagar College of Engineering, Bangalore, India

K. Devender Reddy Institute of Aeronautical Engineering, Hyderabad, Telangana, India

K. Anjani Devi Department of Physics, Computer Science and Engineering, Aditya Engineering College Surampalem, Kakinada, Andhra Pradesh, India

Thotakura Sri Satya Dhanesh Department of Mechanical Engineering, Aditya College of Engineering and Technology Surampalem, Kakinada, Andhra Pradesh, India

S. M. Dilip Kumar Department of Computer Science and Engineering, University Visvesvaraya College of Engineering, Bengaluru, India

Nitla Stanley Ebenezer Department of Mechanical Engineering, Aditya College of Engineering and Technology Surampalem, Kakinada, Andhra Pradesh, India

T. Edwin Fabricio Lozada Carrera de Software, Universidad Regional Autónoma de Los Andes—UNIANDES, Ambato, Ecuador

Rahul Elagoudanavar School of Computer Science and Engineering, KLE Technological University, Hubballi, Karnataka, India

N. Freddy Patricio Baño Carrera de Software, Universidad Regional Autónoma de Los Andes—UNIANDES, Ambato, Ecuador

Narendra K. Garg Department of ECE, Amity School of Engineering & Technology, Gwalior, MP, India

Shelly Garg Greater Noida Institute of Technology, Greater Noida, India

R. B. Geeta Department of CSE, KLE Institute of Technology, Hubballi, Karnataka, India

G. S. Girisha Department of CSE, Dayananda Sagar University, Bengaluru, India

Hardik Gourisaria Department of Computer Science and Engineering, People's Education Society University, Bangalore, India

Leena Govekar Mumbai University, Mumbai, India

M. S. Greeshma School of Computer Sciences, Mahatma Gandhi University, Kottayam, Kerala, India

Swati Gupta National Institute of Technology Kurukshetra, Kurukshetra, India; P.I.E.T., Panipat, India

Nguyen Ha Huy Cuong Software Development Centre—University of Dang Nang, Da Nang, Viet Nam

M. C. Hanumantharaju Department of Electronics and Communication, BMS Institute of Technology and Management, Bengaluru, India

Prasanna J. Hiremath School of Computer Science and Engineering, KLE Technological University, Hubballi, Karnataka, India

Rohini S. Hongal Electronics and Communication Department, KLE Technological University, Hubli, India

Sanjay K. Hoysala Department of Computer Science and Engineering, SJB Institute of Technology, Bangalore, Karnataka, India

Kausar Jahan Department of Electronics and Communication Engineering, Koneru Lakshmaiah Education Foundation, Vaddeswaram, Guntur, Andhra Pradesh, India

Vishal Jain Panipat Institute of Engineering and Technology, Samalkha, Panipat, India

Paramananda Jena Department of Telecommunication, Dayananda Sagar Collage of Engineering, Bangalore, India

V. R. Jyothish School of Computer Sciences, Mahatma Gandhi University, Kottayam, Kerala, India

B. Kalivaraprasad Department of Electronics and Communications Engineering, K.L. University, Vaddeswaram, Andhra Pradesh, India

Meghana Kambali Electronics and Communication Department, KLE Technological University, Hubli, India

Mahesh Kankar National Institute of Technology Karnataka, Surathkal, India

Rajib Kar Department of ECE, NIT Durgapur, Durgapur, West-Bengal, India

K. G. Karibasappa School of Computer Science and Engineering, KLE Technological University, Hubballi, Karnataka, India

K. Kavitha Department of CSE, Gokaraju Rangaraju Institute of Engineering and Technology, Hyderabad, India

Jayashree Khanapuri K. J. Somiaya Institute of Engineering and Information Technology, Mumbai, India

Pallavi Khare ECE Department, Matrusri Engineering College, Hyderabad, India

Abdul khurshid Department of Mechanical Engineering, JNTUK University College of Engineering, Vizianagaram, Andhra Pradesh, India

A. N. Krishna Department of Computer Science and Engineering, SJB Institute of Technology, Bangalore, Karnataka, India

Sunil Kumar Department of Computer Science, Central University of Rajasthan, Ajmer, India

A. Lakshmi HCL Technologies, Bangalore, India

R. V. S. Lalitha Department of CSE, Aditya College of Engineering and Technology, Surampalem, India

Vikas Maheshwari Department of ECE, Guru Nanak Institutions Technical Campus, Hyderabad, India

Somashekhar Malipatil Department of ECE, Malla Reddy Engineering College & Management Sciences, Medchal, Hyderabad, India

Nisith Manika Electronics and Telecommunications Department, Dr. D. Y. Patil Institute of Technology, Pune, India

Penke Pragnana Manipal Department of Mechanical Engineering, Aditya College of Engineering and Technology Surampalem, Kakinada, Andhra Pradesh, India

B. S. Manohar Department of ECE, JSS Academy of Technical Education, Bangalore, Karnataka, India

Aishwarya Mathapati Electronics and Telecommunications Department, Dr. D. Y. Patil Institute of Technology, Pune, India

Rashmi Mishra Delhi Technological University, Delhi, India

Shreya Mishra JSS Academy of Technical Education, Noida, India

Lini Mohan Electronics and Telecommunications Department, Dr. D. Y. Patil Institute of Technology, Pune, India

S. Mohan Department of Computer Science Engineering, University of South Pacific, Suva, Fiji

Krishna Kumar Mohbey Department of Computer Science, Central University of Rajasthan, Ajmer, India

Safikureshi Mondal Department of Computer Science and Engineering, Narula Institute of Technology, Kolkata, India

Nandini Mukherjee Department of Computer Science and Engineering, Jadavpur University, Kolkata, India

E. Nagarjun Department of Computer Science and Engineering, University Visvesvaraya College of Engineering, Bengaluru, India

Surya Kala Nagireddi Department of CSE, Aditya College of Engineering and Technology, Surampalem, India

K. Nandini Department of CSE, Dayananda Sagar University, Bengaluru, India

A. Naresh Kumar Institute of Aeronautical Engineering, Hyderabad, Telangana, India

Le Ngoc Khanh The University of Da Nang, Da Nang, Viet Nam

T. Nithish School of Computer Science and Engineering, KLE Technological University, Hubballi, Karnataka, India

B. G. Padmageetha Department of Computer Science, Surana College, Bengaluru, India

Mallanagouda Patil Department of CSE, Dayananda Sagar University, Bengaluru, India

B. D. K. Patro Rajkiya Engineering College, Kannauj, India

A. K. Paul Department of Electronics and Communication Engineering, Desh Bhagat University, Punjab, India

Digambar Pawar School of Computer and Information Sciences, University of Hyderabad, Hyderabad, Telangana, India

M. V. D. Prasad Department of Electronics and Communications Engineering, K.L. University, Vaddeswaram, Andhra Pradesh, India

A. Punidha Department of Computer Science and Engineering, Coimbatore Institute of Technology, Coimbatore, Tamil Nadu, India

Madas Pushpalatha SSSUTMS, Sehore, Bhopal, Madhya Pradesh, India

P. S. Puttaswamy Electrical and Electronics Engineering Department, GSSS Institute of Engineering and Technology for Women, Mysore, India

R. Vadivel Department of Information Technology, Bharathiar University, Tamil Nadu, Coimbatore, India

Deepthi Raj Dayananda Sagar College of Engineering, Bangalore, India

Rajeev Panipat Institute of Engineering and Technology, Samalkha, Panipat, India

K. Rajesh Babu Department of Electronics and Communications Engineering, K.L. University, Vaddeswaram, Andhra Pradesh, India

T. Rama Reddy Aditya Engineering College, Surampalem, Andhra Pradesh, India

P. Ramadevi Vardhaman College of Engineering, Hyderabad, Telangana, India

K. Ramanjaneyulu Electronics and Communication Engineering Department, Prasad V. Potluri Siddhartha Institute of Technology, Vijayawada, India

Nisha Rao School of Computer Science and Engineering, KLE Technological University, Hubballi, Karnataka, India

S. Koteswara Rao Department of Electronics and Communication Engineering, Koneru Lakshmaiah Education Foundation, Vaddeswaram, Guntur, Andhra Pradesh, India

Y. S. Rao Sardar Patel Institute of Technology, Mumbai, India

Ch. Rathna Jyothi Computer Science and Engineering Department, Andhra Loyola Institute of Engineering and Technology, Vijayawada, India

Christina Roseline SCMS Engineering and Technology, Ernakulam, Kerala, India

Abhijit Roy Dr. Bhupendra Nath Dutta Smriti Mahavidyalaya, Burdwan, West Bengal, India

Parthajit Roy Department of Computer Science, The University of Burdwan, Purba Bardhaman, West Bengal, India

Bhawana Rudra National Institute of Technology Karnataka, Surathkal, India

C. Sahaya Kingsley Department of Computer Science and Engineering, Amrita College of Engineering and Technology, Kanyakumari, India

Shaikh Sahil Ahmed National Institute of Technology Karnataka, Surathkal, India

Gayatri Sakya JSS Academy of Technical Education, Noida, India

Freddy G. Salazar Carrillo Centro de Posgrados, Universidad Técnica de Ambato, Ambato, Ecuador

Chodisetti Naga Sandeep Department of Mechanical Engineering, Aditya College of Engineering and Technology Surampalem, Kakinada, Andhra Pradesh, India

K. S. Santhosh Department of Computer Science and Engineering, SJB Institute of Technology, Bangalore, Karnataka, India

Syed Shameem Department of Electronics and Communications Engineering, K.L. University, Vaddeswaram, Andhra Pradesh, India

Jailaxmi Shankar Department of Telecommunication, Dayananda Sagar Collage of Engineering, Bangalore, India

Spandana Shashikumar Department of ECE, JSS Academy of Technical Education, Bangalore, Karnataka, India

Rahul R. Shetty Dayananda Sagar College of Engineering, Bangalore, India

P. Shiva Kumar Institute of Aeronautical Engineering, Hyderabad, Telangana, India

J. Shreyas Department of Computer Science and Engineering, University Visvesvaraya College of Engineering, Bengaluru, India

J. Shwetha Dayananda Sagar College of Engineering, Bangalore, Affiliated to VTU, Belgaum, Karnataka, India

Anuj Kumar Singh Amity University Haryana, Gurugram, India

Harinder Pal Singh Department of Technical Education and Industrial Training, Govt. of Punjab, Punjab, India

Harpreet Singh Department of Electrical and Computer Engineering, Wayne State University, Detroit, USA

Monalisa Singh Department of Electronics and Communication Engineering, Shri Ramswaroop Memorial Group of Professional Colleges (SRMGPC), Lucknow, Uttar Pradesh, India;
Dr. A.P.J. Abdul, Kalam Technical University, Lucknow, Uttar Pradesh, India

Niraj Pratap Singh National Institute of Technology Kurukshetra, Kurukshetra, India

Shubhanshu Singh JSS Academy of Technical Education, Noida, India

P. M. Siva Raja Department of Computer Science and Engineering, Amrita College of Engineering and Technology, Kanyakumari, India

D. R. Srihari Department of Computer Science and Engineering, SJB Institute of Technology, Bangalore, Karnataka, India

S. Srikanth Institute of Aeronautical Engineering, Hyderabad, Telangana, India

N. N. Srinidhi Department of Computer Science and Engineering, Sri Krishna Institute of Technology, Bengaluru, India

Rayudu Srinivas Aditya Engineering College, Surampalem, Andhra Pradesh, India

M. Srinivasan SSSUTMS, Sehore, Bhopal, Madhya Pradesh, India

B. V. Srividya Department of Telecommunication, Dayananda Sagar Collage of Engineering, Bangalore, India

Tasleem Sulthana School of Computer and Information Sciences, University of Hyderabad, Hyderabad, Telangana, India

Shruti Suman ECE Department, Koneru Lakshmaiah Education Foundation (K L University), Guntur, Andhra Pradesh, India

Wladimir L. Tenecota Facultad de Ciencias Humanas de la Educación y Desarrollo Social, Universidad Indoamérica, Ambato, Ecuador

Shashank Mani Tiwari JSS Academy of Technical Education, Noida, India

Shashikumar G. Totad School of Computer Science and Engineering, KLE Technological University, Hubballi, Karnataka, India

Nguyen Trong Trung Dong A University, Da Nang, Viet Nam

Konstantinos G. Tsiknas Department of Electrical and Computer Engineering, Democritus University of Thrace, Xanthi, Greece

Shaik Vahida Aditya College of Engineering and Technology, Surampalem, India

Nguyen Van Loi Vietnam-Korea University of Infomation and Communication Technology, Da Nang, Viet Nam

Vasudevan Vani Department of Computer Science Engineering, University of South Pacific, Suva, Fiji

Mithra Venkatesan Electronics and Telecommunications Department, Dr. D. Y. Patil Institute of Technology, Pune, India

Ashok D. Vidhate ECE Department, Koneru Lakshmaiah Education Foundation (K L University), Guntur, Andhra Pradesh, India

Gorthi Siva Vijay Department of Mechanical Engineering, Aditya College of Engineering and Technology Surampalem, Kakinada, Andhra Pradesh, India

Hrishikesh Viswanath Department of Computer Science and Engineering, People's Education Society University, Bangalore, India

W. Rose Varuna Department of Information Technology, Bharathiar University, Tamil Nadu, Coimbatore, India

Mohit Kumar Yadav Department of Electronics and Communication Engineering, Shri Ramswaroop Memorial Group of Professional Colleges (SRMGPC), Lucknow, Uttar Pradesh, India;
Dr. A.P.J. Abdul, Kalam Technical University, Lucknow, Uttar Pradesh, India

R. K. Yadav Delhi Technological University, Delhi, India

Kyriakos E. Zoiros Department of Electrical and Computer Engineering, Democritus University of Thrace, Xanthi, Greece

BPSK Demodulation Technique Using Xilinx Vivado

Akansha Baral, S. Deepthi, Mukul Anand Bathija, Rahul R. Shetty, and Deepthi Raj

Abstract Demodulation of commands for effective communication between systems (irrespective of the location of system: be it present on ground or in space) is essential. The main motive behind demodulation of commands in telecommand is to ensure the transmission of errorless command signals to the satellite. After the verification of demodulated commands to be devoid of any errors, the errorless commands are transmitted to the satellite, else the command set is discarded. The telecommand encoder is responsible for generating uplink satellite command with the required subcarriers and carrier modulation by interfacing with telemetry, tracking and command (TTC) station computer through Ethernet. Earlier, peripheral component interconnect direct digital synthesis-based integrated command encoder (PCDICE) which is a PC add-on card was used to get the information like modulation scheme, subcarrier frequency, voltage level, command mode, mode of operation, etc. However, the proposed model to carry out effective demodulation is designed and simulated using Xilinx VIVADO. After successful simulation, the required demodulation system is implemented on FPGA Artix-7 board.

Keywords Demodulation · Vivado · FPGA · PSK · Modulation

1 Introduction

Carrier modulation enables frequency transmission of modulating signals as a medium without using a transmission wire. However, the information must be retrieved at the receiving site in its original form for the correspondence process to be completed or to be useful. Almost any type of modulation is different, and different techniques are required to recover (demodulate) the original information. The circuit where recovery is accomplished is called the detector or demodulator. Depending on the design of the demodulator, the distortion may be in amplitude,

---

A. Baral (✉) · S. Deepthi · M. A. Bathija · R. R. Shetty · D. Raj
Dayananda Sagar College of Engineering, Bangalore, India

D. Raj
e-mail: deepthi-tce@dayanandasagar.edu

V. Bhateja et al. (eds.), *Computer Communication, Networking and IoT*, Lecture Notes in Networks and Systems 197, https://doi.org/10.1007/978-981-16-0980-0_1

frequency or phase. Demodulation of commands is important for communication between systems (be it ground or space systems). The key purpose behind demodulation of telecommand commands is to ensure that the errorless command signals are transmitted to the satellite. The coding to be implemented on hardware and the software simulation is done using VHDL.

VHSIC-HDL, Very High-Speed Integrated Circuit Hardware Description Language (VHDL) is a hardware description language used to define digital and mixed-signal systems such as field-programmable gate arrays and integrated circuits in electronic design automation. VHDL could also be used as a parallel programming language of general use. A test bench or workbench test is an area used to validate a concept or model's correctness or soundness. A test bench is an area in which the product being produced is tested using software and hardware instruments. For some instances, the program can need to be slightly changed to fit with the test bench but proper coding can ensure that the modifications can be quickly reversed and without bugs being introduced. In Vivado software, a test bench is a file that is written as an HDL file (VHDL, Verilog) usually providing stimuli (inputs, clocks) to a unit under test (UUT).

Xilinx developed Vivado software for the development, synthesizing and implementation of HDL projects for their FPGA devices. It is an IDE that compiles the HDL code to FPGA-programmable bitstreams. The Vivado Simulator is an important part of Vivado software. It is a simulator compiled language that supports mixed language, TCL scripts, encrypted IP and enhanced tests. The Vivado IP Integrator enables engineers to integrate and customize the network from the large Xilinx IP library easily. The Xilinx Vivado Product Suite offers a centric product flow of intellectual property (IP) that allows you to connect IP modules from different IP sources to your application. A scalable IP library that includes Xilinx-delivered Plug-and-Play IP is key to the system.

2 Proposed Method

PSK demodulation is a method of demodulation by which the input signal that is a binary PCM waveform is obtained from the output waveform whose phase has been transferred to one of a fixed number of states [1]. BPSK demodulator block diagram is composed of a mixer with local oscillator circuit, a low-pass filter, a detector circuit. The first stage of demodulation is completed by retrieving the band-limited message signal, with the aid of the mixer circuit and the low-pass filter. The baseband signal is obtained which is a band restricted and this signal is used to recreate the binary message bitstream. If the bit rate is a submultiple of the frequency of the carrier, then the regeneration of the bit clock is simplified [2]. A decision-making circuit can also be implemented at the second stage of detection to make the circuit easily comprehensible.

The coding to be implemented on hardware and the software simulation is done using VHDL. Generally, VHDL is used for writing text models representing a logic

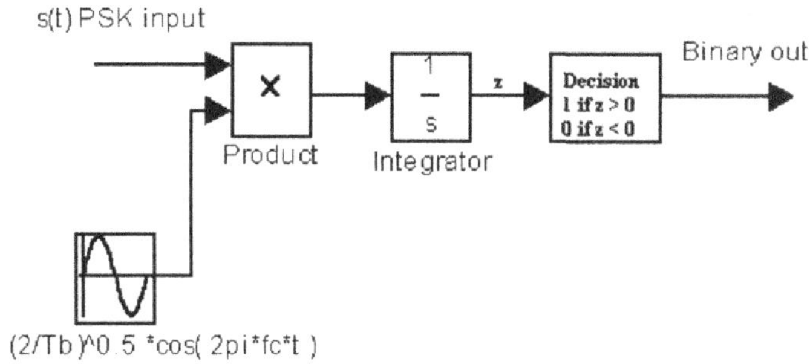

Fig. 1 Block diagram of BPSK demodulator

circuit. A synthesis software processes such a pattern, only if it is part of the logic design. To check the logic design using simulation models, a simulation system is used to represent the logic circuits that interface to the design. This simulation model set is generally called a test bench.

The Vivado Design Suite IP Integrator tool helps you to make complicated device designs by instantiating and interlinking IP cores and Vivado IP catalog module references onto a design canvas. You may generate output products after developing a specification, or delay generation until later. If the output products are not present in project mode, the Vivado tools automatically generate the necessary output products prior to synthesis or simulation. The output products for synthesis are generated out of context (OOC) by design. The Vivado IP packager helps you to build plug-and-play IPs to connect to the Vivado IP extensible catalog. IP packager wizard is based on IEEE IP-XACT standard (IEEE Std 1685). After you have created a user interface for the Vivado Product Suite, the IP packager helps you to convert the interface into a reusable IP module that you can then add to the Vivado IP catalog so that others can use it to create. Within a project design, you can use packaged IP [3] (Fig. 1).

3 Methodology

The Xilinx Vivado Product Suite offers a centric product flow of intellectual property (IP) that allows you to connect IP modules from different IP sources to your application. A scalable IP library that includes Xilinx-delivered Plug-and-Play IP is key to the system.

In order to use an IP specification in a system, you need to instantiate the IP in your top-level application HDL code. The IP output products have instantiation templates in both VHDL and Verilog that are generated automatically. The compiler core of Xilinx LogiCORE IP Direct Digital Synthesizer (DDS) incorporates high-performance, automated phase generation and sinusoidal phase circuits. In many applications, the main sources of sinusoidal waveforms are for use. A DDS comprises

Fig. 2 DDS compiler IP

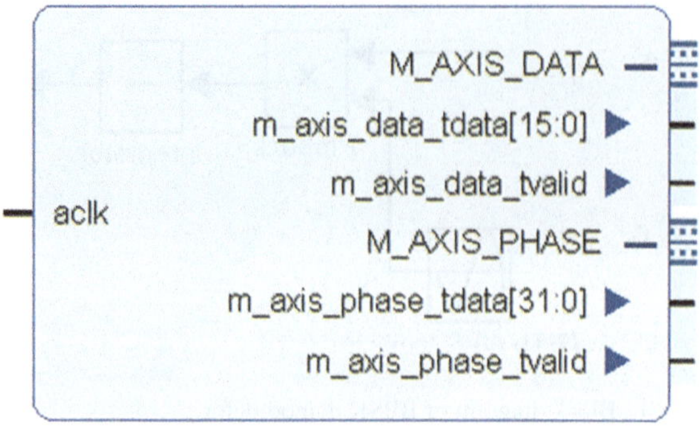

Fig. 3 Multiplier IP

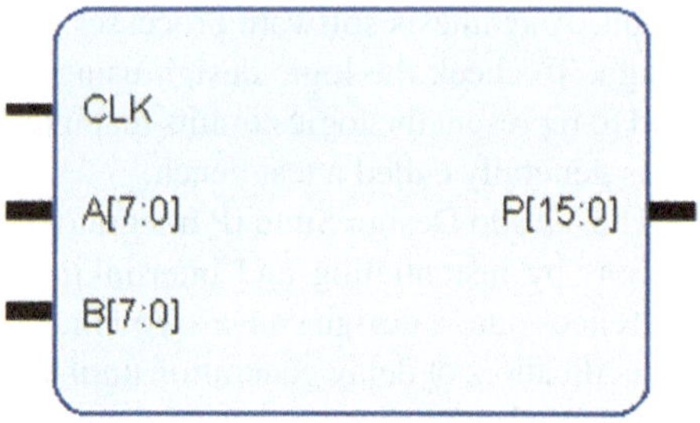

of a phase generator and a SIN/COS Lookup Table (conversion phase to sinusoid). With this core, certain pieces are available individually or in combination (Fig. 2).

The Xilinx LogiCORE IP multiplier implements customized, high-performance multipliers. Trade-off solutions for resources and output are available to adapt the core to a specific application (Fig. 3).

One of the most common and basic building blocks in DSP systems is the finite impulse response (FIR) filter. The FIR compiler cuts filter implementation time by pressing a button, while also offering users with the ability to trade-offs between different FIR filter specification hardware architectures [4]. The core of the Xilinx LogiCORE IP FIR compiler provides a standard interface to create highly parameterizable, high-performance FIR filters that are effective at the field. In the Vivado Integrated Design Environment (IDE), a broad variety of filter types can be incorporated: single-rate, polyphase decimators and interpolators, and half-band decimators and interpolators. The framework in the set of coefficients is used to produce FPGA implementations which are efficient in area [5]. Here, a few assumptions are made: The message signal used in the process is band limited. The bit rate of recreated binary message bite stream is assumed to be a submultiple of the frequency of the carrier signal for the ease of simplification of regeneration of bit clock. The approach of out-of-context (OOC) designing used for designing the proposed model is sufficient for the synthesis of output product (Fig. 4).

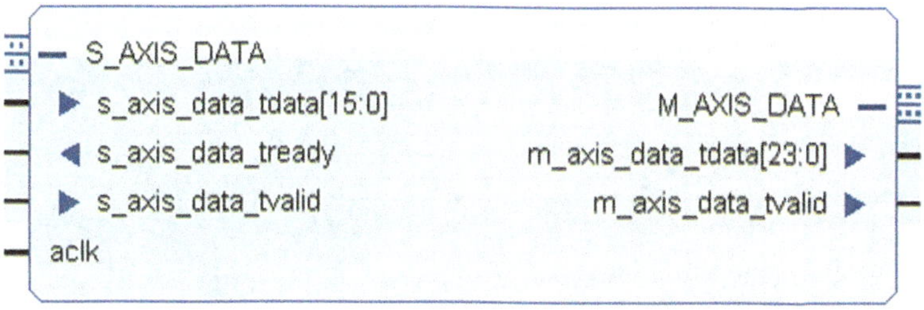

Fig. 4 FIR compiler IP

4 Results and Discussions

In the figure below, we take two sin inputs which are 180° out of phase from each other, and the multiplexer selects one of the sin signals bases on the input signal (whether it is 1 or 0) and gives the output as a BPSK modulated waveform as shown (Figs. 5, 6 and 7).

The BPSK demodulator which was explained in the proposed method chapter was simulated in Vivado, and its schematic diagram is shown above. It takes a BPSK modulated wave as input, and each block of the block diagram was executed sequentially. The output for each block is shown in the waveforms below. The 'data_out' represents the modulated wave, 'sin_out' is the carrier wave given to the multiplier, 'prod' shows the output of the multiplier after multiplying the modulated wave and carrier, 'fil_out' shows the output of the low-pass filter (integrator) after filtering the product and finally 'comp' gives the output of the comparator by setting a threshold value. 'comp' is nothing but the original message signal (Fig. 8).

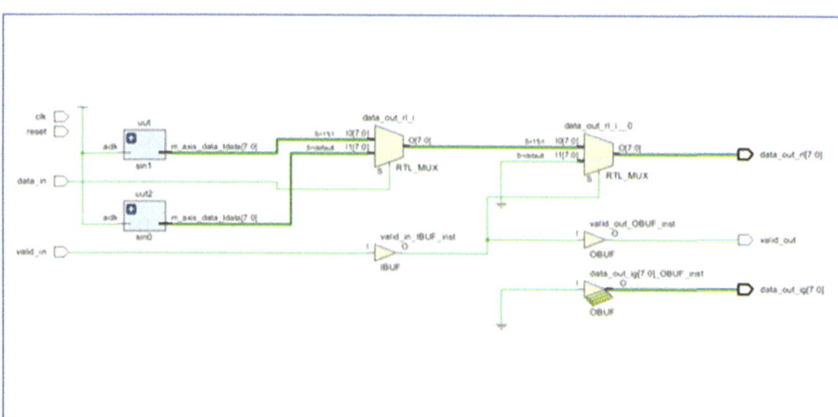

Fig. 5 Schematic block diagram of BPSK modulation simulated in Vivado

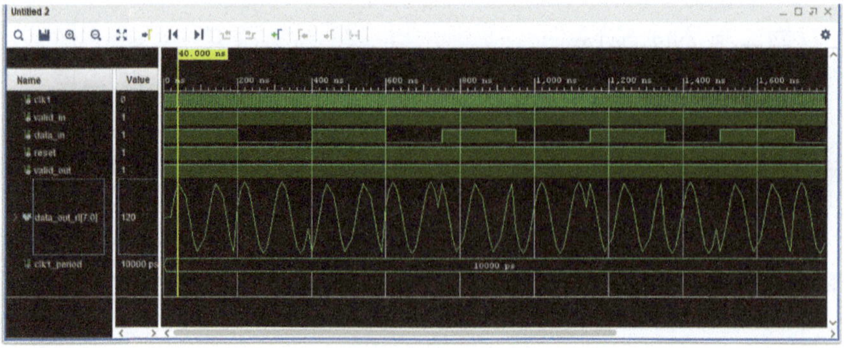

Fig. 6 BPSK modulated waveform simulated on Vivado

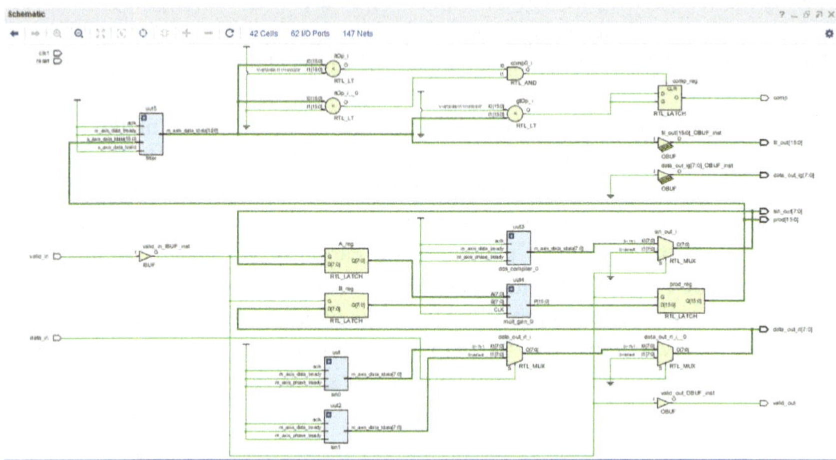

Fig. 7 Schematic block diagram of BPSK demodulator simulated in Vivado

5 Conclusion

In this paper, demodulation of commands using various digital modulation techniques such as PSK, FSK and FSK/FM is investigated. Utilization of Xilinx Vivado software to simulate the proposed model for demodulation is carried out successfully with the software implementation of the proposed demodulation system which is presented with the satisfactory simulation results. The need for demodulation of commands is to check for the presence of any unforeseen error in the time period between commands are modulated and before further transmission is instigated. It is thus seen to have provided a technique to ensure only error less command transmission to the satellite. It can be concluded that Xilinx Vivado is reliable and an efficient tool to implement a demodulation system for various purposes including satellite communication [6].

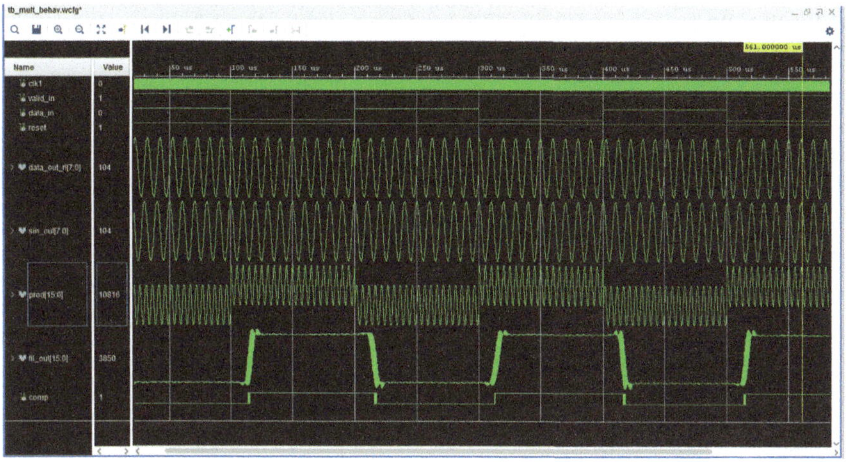

Fig. 8 BPSK demodulator waveforms simulated in Vivado

6 Future Scope

There are numerous approaches and procedures for demodulation including coherent and non-coherent techniques [7]. The other strategies have to be acknowledged to contrast between them with the need of deciding which methods perform better and the strategy reasonable for specific application has to be catered to. Also, there are a few limitations of designing block designs in Xilinx Vivado being: It is not possible to add pre-customized IP(s) to block design. The customized IP can only be used by the option of 'Add Sources' only and not the IP module customized in an IP repository. And while the entire block design can be set as out of context, it is not possible to do the same for a specific single IP. In addition to this, IP(s) seperately inside block design cannot be set as out of context. IP location of the IP created by user cannot be changed for the ease of use. This service is reserved only for the IP personalized in IP catalog only. All of these limitations are yet to be overcome.

Demodulation in systems can be extended to be used in systems similar or fairly different than satellite communication for errorless command signal transmission [8]. The inspiration driving the investigation of demodulation procedure on a software platform like Xilinx Vivado is that the quantity of various command signals can be taken care of without utilizing additional hardware that is reusing identical equipment with the adjustment which, in turn, can aid in decreasing the time to advertise the final system and put a cap on updating the framework while curbing the improvement costs [9].

Acknowledgements Indebted to the supportive faculty at Department of Telecommunication Engineering, Dayananda Sagar College of Engineering, Bangalore, India, and authors or various papers cited for a smooth and easier outcome.

Declaration We have taken permission from competent authorities to use the images/data as given in the paper. In case of any dispute in the future, we shall be wholly responsible.

References

1. Kumar, M., Tripathi, S., Vidyarathi, A., Mohanty, S., Gowri, R.: Various PSK modulation schemes for wireless communication. In: 2011 2nd International Conference on Computer and Communication Technology (ICCCT-2011), Allahabad, pp. 545–549 (2011). https://doi.org/10.1109/ICCCT.2011.6075129
2. Devi, M.S., Saketh, M., Muthumeenakshi: Design of costas loop for carrier recovery mechanism of BPSK demodulation. In: 2017 International Conference on Intelligent Computing and Control Systems (ICICCS), Madurai, pp. 424–429 (2017). https://doi.org/10.1109/ICCONS.2017.8250758
3. Townsend, T., Nelson, B.: Vivado design interface: an export/import capability for Vivado FPGA designs. In: 2017 27th International Conference on Field Programmable Logic and Applications (FPL), Ghent, pp. 1–7 (2017). https://doi.org/10.23919/FPL.2017.8056809
4. Owen, J., Manus, H.: 384 TMAC/s FIR filtering on an Artix-7 FPGA Using Prism signal processing (2018). https://doi.org/10.1109/IECON.2018.8591619
5. Goswami, A., Agarwal, M., Rawat, T.K., Singh, K.: FPGA implementation of reconfigurable architecture for half-band FIR filters. In: 4th IEEE International Conference On Signal Processing, Computing and Controlling ICSPCC (2017)
6. Kumar, M., Tank, V.K., Ram, T.V.S.: Design and FPGA implementation of a programmable data rate PSK digital demodulator for onboard and ground applications. In: 2013 Third International Conference on Advances in Computing and Communications, Cochin, pp. 437–440 (2013). https://doi.org/10.1109/ICACC.2013.93
7. Wei, C., Dian-ren, C., Xing-guang, L.: A novel wide dynamic range digital PSK demodulation. In: 2009 ISECS International Colloquium on Computing, Communication, Control, and Management, Sanya, pp. 136–138 (2009). https://doi.org/10.1109/CCCM.2009.5267808
8. Nivin, R., Sheeba Rani, J., Vidhya, P.: Design and hardware implementation of reconfigurable nano satellite communication system using FPGA based SDR for FM/FSK demodulation and BPSK modulation. In: 2016 International Conference on Communication Systems and Networks (ComNet)|21–23 July 2016|Trivandrum (2016)
9. Lin, J., Zhou, Y., Pang, Y., Li, Z., Zhao, Z., Wang, H.: A low-power circuit for BPSK and QPSK demodulation for body area networks applications. In: International Symposium on Bioelectronics and Bioinformations 2011, Suzhou, pp. 240–243 (2011). https://doi.org/10.1109/ISBB.2011.6107691

Dynamic ICT Modeling for Handling Student Data Using Big Data Technology and Hybrid Cloud Computing

Harinder Pal Singh, Harpreet Singh, and A. K. Paul

Abstract The Indian State of Punjab is a prosperous state. It has three-tier technical education systems consisting of industrial training institutes, polytechnics and engineering colleges. Due to faster Internet technology available at cheaper costs and the students using IoT like mobile phones and other devices, they are searching for courses and colleges of their choice along with options of scholarships to fund their education. Also, a large number of underprivileged students apply for scholarships every intake and generate a large amount of student data consisting of enrollments, scholarships, academic content, literature search, testing results and finally placement tracking. Such a high volume of data requires integrity security and scalability. Therefore, dynamic modeling for handling high volumes of student data is needed. In this paper, a new ICT model for providing need-based scholarships and fee waivers to large number of students has been suggested using high-end database technology and hybrid cloud computing. A case study for scholarships to under-represented students is discussed. This model is useful for other states looking for similar interests in a cost-effective and reliable manner.

Keywords Database model · ICT model · Hybrid cloud computing · Student scholarships

---

H. P. Singh (✉)
Department of Technical Education and Industrial Training, Govt. of Punjab, Punjab, India
e-mail: harinderpal_s@yahoo.com

H. Singh
Department of Electrical and Computer Engineering, Wayne State University, Detroit, USA
e-mail: hsingh1@wayne.edu

A. K. Paul
Department of Electronics and Communication Engineering, Desh Bhagat University, Punjab, India
e-mail: Akpaul53@gmail.com

V. Bhateja et al. (eds.), *Computer Communication, Networking and IoT*, Lecture Notes in Networks and Systems 197, https://doi.org/10.1007/978-981-16-0980-0_2

1 Introduction

The model is a cloud-based platform using virtualized cluster of servers over data centers over SLA [1] and dynamic resource provisioning of the servers' storage and the networks. There may be different servers doing their main job but they can share need-based data queries on the information stored in them or if they are all connected to common hybrid cloud, algorithms can be developed for hybrid cluster computation. Data flow from such servers shall be dynamic for instant decision-making process. Starting from basic algorithms, the application software can be developed for users. For example, if we have four servers connected to hybrid cloud through a dynamic document-oriented database engine like MongoDB, the objective of students' data processing as per the particular requirements is possible. In this example, P1 is NIC scholarship server which is responsible for collecting scholarship data from large number of students through application software connected to their mobile phones or laptops. This server also has the application control opportunities. Another server P2 is UID server which hosts data of personal id numbers of all the people living in a country but here we need its services only to verify personal identification of students applying to scholarship server P1. Similarly, third server in the network loop which is P3 is university/college server. Although this server is doing services like hosting and processing academic and attendance performance of all the students, it can also supply information related to students applying for scholarship to the scholarship server P1 dynamically when asked for. Fourth server in the parallel loop may be Internet banking server(s) so that as and when decision is made to award scholarship to the student, it will go to the bank account of that particular student through Internet banking as the bank account is linked to UID number of the student. All their servers are doing their jobs with a common objective of awarding scholarships to students well within the specified time frame. Student data through database engine is stored in a hired memory space in a hybrid cloud for need-based decision making and analytics. Figure 1 shows the development of data flow methodology for developing dynamic models for hybrid cluster computation as it is showing parallel processing layout.

2 Requirement of Dynamic Modeling

2.1 Detection of Data Security Risks

Security of data and integrity of information are big challenges as far as data of educational institutions is concerned. There is a risk of personal data and classified information of students can be stolen online by hackers. Examples of such data are national security number or unique identification number (UID) of the applicant or bank account numbers may be stolen by hackers which can result in financial loss to the applicants. Also, unauthorized access to such a classified data can result in

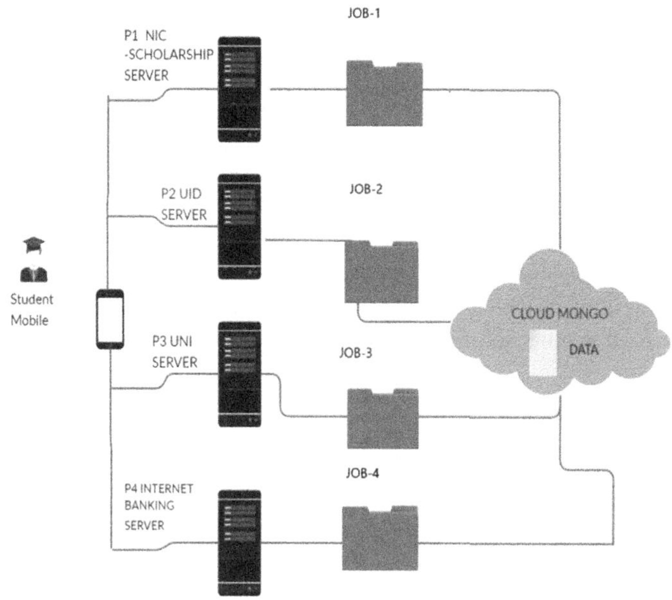

Fig. 1 Parallel processing layout of servers and their jobs

financial scams. So detection of risks and analysis algorithms using modern security techniques and encryption algorithms are suggested to be inbuilt in the data mining system [2].

2.2 *Data Visualization*

Data of the students become more complex as its size increases. Visualization of such a data is possible using data visualization techniques and to easily identify the trends and relations of data just by just generating the reports.

2.3 *Prediction of Performance*

The real-time prediction of performance of pupils whether he/she is continuing in his studies after obtaining financial assistance needs to be ascertained before granting the financial assistance application for the next term. His board/university scores need to be linked using various data tools to the database management system. If he fails to appear in examination or does not pass any of the subjects, his application is liable to be rejected till he clearly passes his subjects/core of the requisite numbers and resubmits his application for next term. In the proposed study, data alert has been incorporated. Dropout rates can be seen while doing analysis of the data, so finally, the decision making can be improved for further award of similar financial assistance applications.

2.4 Feedback Intelligence

Educational systems may give intelligent and immediate feedback to the pupils in response to their inputs which would lead to improvements of student interaction and performances. It is proposed to implement a new feedback system in the database.

3 Development of Algorithm and Implementation Strategy

3.1 Halstead Complexity Measures

These are software matrices introduced by Maurice Howard Halstead in the year 1977 as part of his treatise on establishing an empirical science of software development. Halstead made the observation that matrices of the software should reflect the implementation or expression of algorithms in different languages but be independent of their execution on a specific platform [3]. These metrics are therefore computed statically from the code. Halstead's goal was to identify measurable properties of software and the relations between them. This is similar to the identification of measurable properties of matter—(like the volume, mass, and pressure of a gas)—and the relationship between them is analogous to the gas equation. Thus, his metrics are actually not just complexity metrics.

For a given problem,

Let

$\eta 1$ = the number of distinct operators
$\eta 2$ = the number of distinct operands
N1 = the total number of operators
N2 = the total number of operands.

From these numbers, several measures can be calculated:

$$\text{Program vocabulary: } \eta = \eta 1 + \eta 2$$

$$\text{Program length: } N = N1 + N2$$

Calculated estimated program length: $\hat{N} = \eta 1 \log_2 \eta 1 + \eta 2 \log_2 \eta 2$

$$\text{Volume: } \mathbf{V} = \mathrm{N} \times \mathbf{log_2}\eta$$

$$\text{Difficulty: D} = \frac{\eta 1}{2} \times \frac{N2}{\mathrm{n2}}$$

$$\text{Effort: } \mathbf{E} = \mathbf{D} \times \mathbf{V}$$

The difficulty measure is related to the difficulty of the program to write or understand.

The effort measure translates into actual coding time using the following relation:

Time required to program:

$$\mathbf{T} = \frac{\boldsymbol{E}}{18} \text{ seconds}$$

Halstead's delivered bugs (B) are an estimate for the number of errors in the implementation.

$$\text{Number of delivered bugs: } \frac{E2/3}{3000}$$

$$\text{Or more recently is accepted } B = \frac{V}{3000}$$

3.2 Case Study and Development of Algorithm

Let us take the case study of state department of social justice awarding scholarships designated as implementing department. Let there be four line departments, first technical education catering to students going to technical streams like ITI, polytechnics and engineering colleges with total students 60,000 every year. Second line department is medical education and research catering to 50,000 students going to medical streams like nursing, pharmacy and MBBS, third line department is higher education catering to about 70,000 students of arts, commerce and mathematics streams and fourth line department is school education catering to high school and secondary school students numbering 90,000 students every year. So, the total students applying for scholarships every year will be around 270,000. If we count the returning students to next year, the number will increase to many folds. There are total of more than 30 Indian states and union territories, which means if there are 270,000 average students per state, a total of 8,100,000 students apply for scholarships across the country according to a rough estimate, and millions are being spent every year to award scholarships.

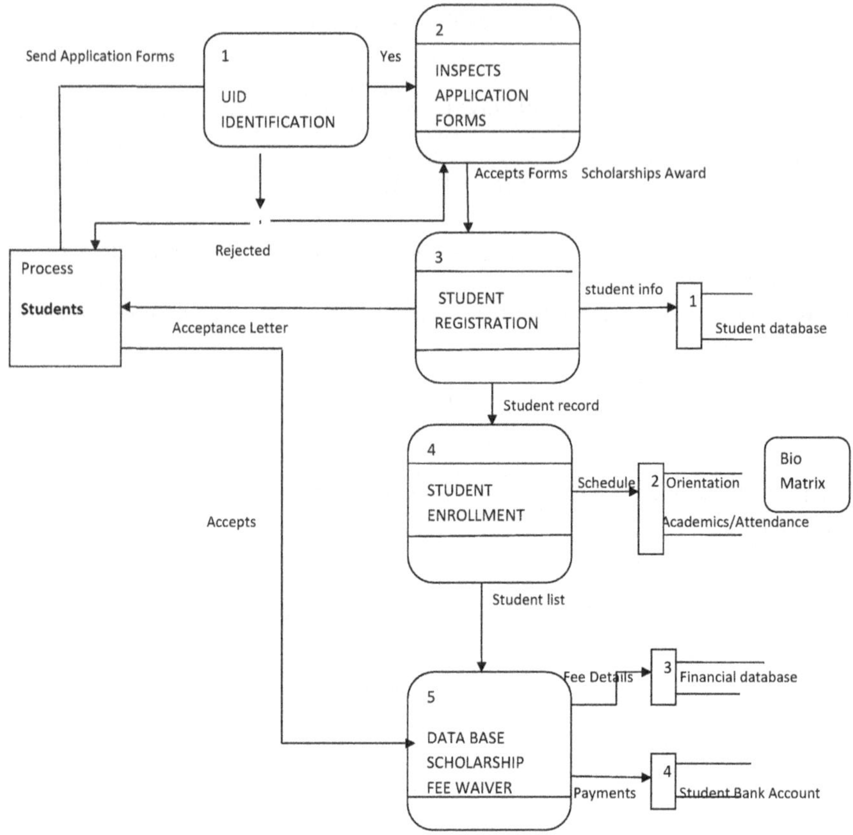

Fig. 2 Processing of student scholarships

To process applications of such a large number of students, dynamically ICT modeling is required for parallel processing of their claims as shown in Fig. 2 throughout the year. The parallel processing can be achieved in following three different ways, the following different ways:

(a) All the line departments have their own processors and technical education P1 serving database of 60,000 students, P2 department of medical education catering to 50,000 students, P3 is processor for higher education department catering to 70,000 students and P4 is processor of school education catering to students of school education. All the four processors processing data and sending to implementation department of social justice for making payments to bank accounts of students. These four processors are connected to UID processor and processors of all the universities/colleges and Internet banking server.

(b) Another way of doing it is placing 22 processors (P1–P22) at all the district headquarters of state and students of all the line departments (270,000) together applying to each district of their jurisdiction. In this way, averages of 12,273 students are processed by each processor at district level. All the 22 processors are connected to all the university servers, UID server and Internet banking server.

(c) Suggested way of parallel processing will be only one processor hosting application controls at state level, and through application software and Internet, it is connected to database hybrid cloud server where database of all the students processed after fetching need-based useful information from university/college servers/processors and other useful information like unique identification check for duplicity check and also connected to Internet banking server through payment gateway.

Dynamic ICT model with hybrid cloud computing can cater to structured data like rows and columns of student application, student grades for performance check from university/college server also for biometric attendance check which is unstructured data needed to be accessed dynamically. The UID identification happens through UID server and is finally connected to Internet banking through payment gateway for transferring money to student accounts. Figure 3 shows cloud depiction. Gordon Bell, Jim Gray and Alex Szalay [4] pointed out that "computational science is changing to be data-intensive. Supercomputers must be balanced systems, not only CPU farms but also petascale I/O and networking arrays." Therefore, working with large-scale datasets will mean sending the application programs to the data, rather than copying the data to workstations. Computation may be moving from desktops to large data centers where there is on-demand provision of hardware, software and data as a service. This data explosion has promoted the idea of cloud computing.

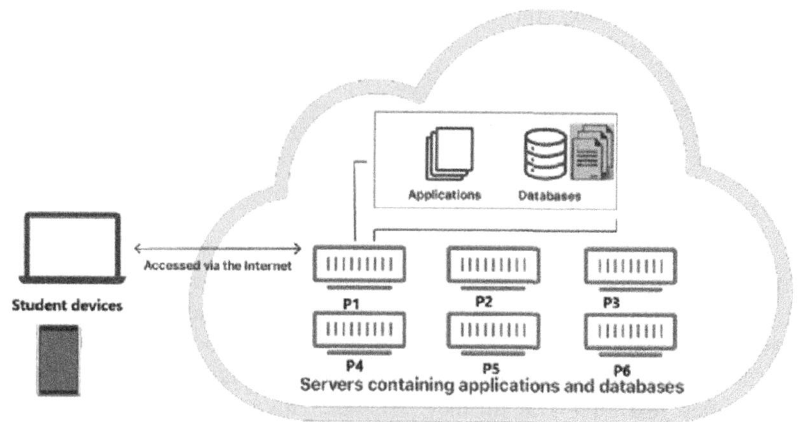

P1 - Scholarship Server P2 - UID Server P3 –P4 -University /College Servers P5-P6- Internet /Gateway Servers

Fig. 3 Cloud depiction of servers connected through internet

3.3 Ford-Fulkerson Algorithm

The following Ford-Fulkerson algorithm [5] for dynamic ICT model describes it as

input: Applications form from students
output: send checks f to banks for awards
for each application (u, v) in *Database* do
implement clustering approach to distribute the applications
while there exists appropriate application to scholarship.
return f

It is based on the following example:

Here follows a longer example of mathematical-style pseudocode, for the Ford-Fulkerson algorithm:

Algorithm Ford-Fulkerson is
input: Graph G with flow capacity c,
  source node s,
  sink node t
output: Flow f such that f is maximal from s to t
(Note that $f_{(u,v)}$ is the flow from node u to node v, and $c_{(u,v)}$ is the flow capacity from node u to node v)
for each edge (u, v) in G_E do
  $f_{(u,v)} \leftarrow 0$
  $f_{(v,u)} \leftarrow 0$
while there exists a path p from s to t in the residual network G_f do
  let c_f be the flow capacity of the residual network G_f
  $c_f(p) \leftarrow \min\{c_f(u, v) \mid (u, v) \text{ in } p\}$
  for each edge (u, v) in p do
  $f_{(u,v)} \leftarrow f_{(u,v)} + c_f(p)$
  $f_{(v,u)} \leftarrow -f_{(u,v)}$

3.4 Implementation Strategy

The dynamic ICT model using hybrid mongo cloud is shown in Fig. 4. The department of social justice sends the money to UID linked bank account of the student through Internet banking as the account number and IFS code data already available in the student application form. Many students are doing multiple times the same activities year after year till they pass out. This kind of big data generated ranging from admission, claiming financial assistance, attendance, performance, etc. is stored in the mongo cloud and available for decision making and analysis by the line department and the department of social justice for arranging funds, estimations, budgeting and other decision-making analytics. Lot of server memory and some manual processing of re-verifying the performance of students is now not

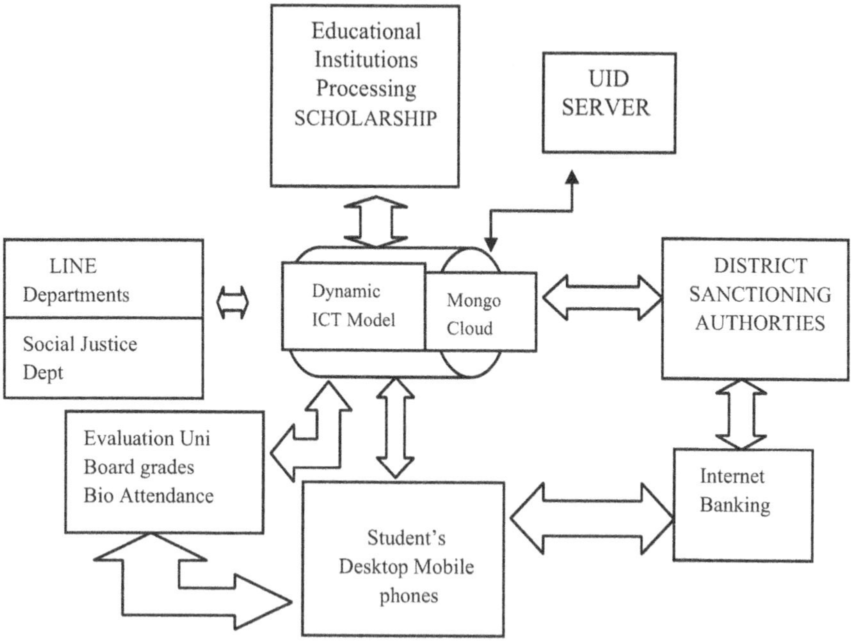

Fig. 4 Dynamic ICT model for providing scholarships

needed. Also, manual deleting fake/duplicate claims take time, and there were delays in releasing the scholarship/financial assistance amount to the students account [6].

3.5 *Hardware and Software Specifications*

The new data model requires only a server and the application software giving access to hybrid mongo cloud platform, which can be hired for need-based memory requirements. The server hosts the software application controls. There is no need for adding more hard disk memory which may become expensive year after year. The application software shall be connecting all the existing servers like mongo cloud, university/board server for student performance query, Internet banking and (unique identification authority) UID server to the student mobile phones, laptops or tabs. For this software app, the student can install on his mobile phone or use laptop to access the app from Internet using normal browser. All the data can be added and processed simultaneously using the app interface. In the following sections for developing a computer and mobile application, the basic algorithms for importing existing data to MongoDB are given in next section.

3.6 Importing Existing Data to the New Database System

Importing CSV file into MongoDB Create a folder on disk C—c:\importMongo then Download the file "ImportDataMongo.rar" from the Google drive then Extract file ImportDataMongo.exe from "ImportDataMongo.rar" in the folder: c:\importMongo then copy here the CSV file which you want to import to Mongo in the folder—c:\importMongo then launch command prompt and change folder to—c:\importMongo. Run the file c:\importMongo\ImportDataMongo.exe.

Note: the CSV file after import will be move to the folder c:\importMongo\Archive.

Checking the imported data in MongoDB: Launch the application Compass from MongoDB. Click on Sample_StdRec Click on stdRecords Click on Table.

3.7 Comparison of Big Data Technology Dynamic Model with Other Database Systems

The area of education to prospective students is gaining importance as large volumes and variety of real-time data are being produced. Universities and colleges are producing high volumes of data, from grades or examination scores to number of admissions made and number of scholarships awarded [6]. The big data systems are required for modern world to support data mining approaches for increasing the efficacy of educational institutions [2]. MongoDB platform uses for data storage and analyzing educational data is being proposed. Present-day open-source tools like MongoDB along with cloud technology are being used to test real data samples on real-time basis for analysis of students' scholarships data and comparisons thereof with other already used old technologies are carried out. Such databases support systems based on MapReduce for processing large amounts of data in parallel fashion. The MapReduce framework deals with data mapped on distributed file systems, with intermediate data being stored on local disks and can be retrieved remotely by reducers [7]. MapReduce which is a Google's proprietary reads and writes to Google file system (GFS). But modern platforms of the type MongoDB, Apache Hadoop HDFS, Hive, Big Table, and HBASE, etc. have emerged to cater to the needs of high storage requirements of data. Best suited may be MongoDB for storing data of students as this is NoSQL. Also, this is an open-source document-oriented database system of the 10Gen Company. It can store structured data as JSON-like heterogeneous documents with dynamic schemas and it can scale horizontally. This database has capability of handling unstructured data as well and also has capability of querying database quickly and suitable for storing student data due to its scalability and flexibility in the structural format for storing solutions. The database is worthy of using content delivery and all management because of features as below:

(a) Storage of data in the form of JSON style documents and using JavaScript engine which is simplified.
(b) Index on any attribute possible and fast in-place updates on data possible as well.
(c) GridFS is supported for storage of data.
(d) This is a document database system as one collection (i.e., data store) can hold different type of documents. Field numbers, size and content of the document can be different from one another other document.
(e) There is no need of conversion of application objects to structural format of database objects.
(f) There are no complex joins, as it is present in old database systems.
(g) Dynamic queries are being supported by mongo on documents using a document-based query language.
(h) Internal memory is being used for storage of the (windowed) working set; faster access of data is enabled
(i) Very easy to scale.
(j) Replication is easy with this database, shredding and the availability is high for it.

4 Results and Discussion

Public funding schemes like post-matric scholarship enable free education for scheduled caste students and other backward class students. The students can apply for scholarship as per eligibility every year. The payment is being made directly to the UID linked bank accounts of students/institutes by the department of social justice and empowerment of minorities. The data for eligible students is processed by line departments. Dropouts and duplicate are also of considerable numbers. The new database model is very important in the sense that duplicates and dropouts are weeded out at the first instance using UID identification and using online university/college server being connected to hybrid mongo cloud. If any student drops out or his attendance is not continuous, the system will not clear such students for transferring scholarship money for next semester until the student completes requisite attendance and his/her performance requirements. The new database model suggested to address the risk of duplicate claims as same student may be applying with other line departments or claiming through multiple schemes. Also, ease of operation and saving of time and disk space happens. Earlier conventional data management models were not that robust as these were stand-alone scheme-specific models, and the financial implications were involved and there were data security risks involved as UID numbers and bank accounts apart from personal information of the students are part of student personal data. The new database model would help eliminating duplicity and dropouts and shall provide inter se data analysis opportunities to the states with other line departments and other scholarship schemes including with the other states of the country offering similar scholarship schemes query times can also be achieved. Management of bulk data through practical approaches by building a new data base model using MongoDB, an open-source tool which is a cross-platform document-oriented database tool is possible; also it is simultaneously possible for analyze the data using real-time data sets of students. The new database model is flexible enough to cater to any number of additions of new scholarship schemes state and national level like persons with disabilities-based scholarship can be operated through same model. The model can be useful for all Indian states, processing similar scholarships for millions of students every year. This may save them cost and time apart from avoiding possible frauds and scams. In the future, all the students may use smartphones for all kinds of educational activity, so this database model will come handy for them. Also, in the future, the usage of combinations various parallel programming models based on MapReduce, Hadoop, PACT and similar for different types of data analytics could be explored to improve the educational data analysis for social justice objectives as these technologies even can cater to the unstructured student data from social media networks, thereby accelerating the economic growth and also help in designing scalable models in the educational field and may provide a better scope of improvement in the field of educational analytics as the unstructured data from social media networks can also be taken into account to know the student's future interests as using platforms like MongoDB may also help in even building applications on social media networks thereby handling unstructured and structured students data with more flexibility. Real-time linking of performance and attendance of the students would help them remain serious in their studies and the performance and required attendance before availing next semester/year scholarship amount. This will lead to better decision making for uplifting the average.

5 Conclusion

It is successfully concluded that the new ICT model based on big data technology for providing scholarships to underprivileged students will help in weeding out duplicity of claims with other scholarship schemes of the state and the country as the UID server authenticates the student id before application form opens up. It will also help in asserting the student performance like checking attendance, grades from university/server simultaneously without any delays whatsoever. It also helps in transparency in processing the student claims as no manual interface with students and authorities. Main findings of the model are decreasing processing time, saving memory space, achieving increased data security and reducing cost of the new hardware and software. It also helps achieving authorized access to important data due to capabilities of using MongoDB-based tool including saving lots of hardware memory space as the hybrid cloud technology is used and need-based cloud server can be hired on rent. So, it will help in reducing costs related to adding hardware like memory disks and associated software along with considerable improvements in the data analytics.

Declaration The permission has been taken by us from competent authorities to use the images/data as given in the paper. In case of any dispute in the future, we shall be wholly responsible.

References

1. Kai, H., Fox, G.C., Dongarra, J.J.: Distributed and cloud computing from parallel processing to the internet of things. JCSI **11**(5), 35–48 (2014)
2. Kim, Y.H., Hoseo, J.-H.A.: A study on the application of big data to the korean college education system. In: Conference Paper (Information Technology and Quantitative Management (ITQM 2016). University of Korea (2016)
3. Halstead, M.H.: Elements of software science, vol. 7. Elsevier, New York (1977)
4. Bell, G., Gray, J., Szalay, A.: Petascale computational systems (2006). https://ieeexplore.ieee.org/abstract/document/1580395
5. Ford, L.R., Fulkerson, D.R.: Maximal flow through a network. Can. J. Math. **8**, 399–404 (1956)
6. Singh, H.P., Singh, H.: Big data technologies in scholarship management of technical education system. Int. J. Emerg. Technol. Innov. Res. **6**(6), 492–496 (2019). www.jetir.org ISSN: 2349-5162, Available at https://www.jetir.org/papers/JETIR1906H50.pdf
7. Cuzzocrea, A.: Warehousing and protecting big data: state-of-the-art-analysis, methodologies, future challenges. In: ICC'16: Proceedings of the International Conference on Internet of things and Cloud Computing, Article number 14 (2016). https://doi.org/10.1145/2896387.2900335

Gesture Recognition of Indian Sign Language

K. S. Santhosh, Sanjay K. Hoysala, D. R. Srihari, Suhith Shekar Chandra, and A. N. Krishna

Abstract In today's world, many people are not as fortunate as us. They suffer from speaking and hearing disabilities that make daily tasks very difficult to perform. This paper mainly deals with people suffering from the loss of hearing and cannot speak. In this study, we have introduced a recognition and classification technique which can detect hand gestures used by the deaf and dumb community, which is then translated to an equivalent English word. The vision-based system is built using two machine learning algorithms, namely convolution neural network (CNN) and recurrent neural network (RNN). The video sequences are not directly fed into the model instead those videos are first converted into frames and then fed into the model. Further, each frame is directed into CNN, the exact features are extracted, trained on those features and stored in a model file, these model files are fed into RNN for further feature extraction, and the model is trained on those features. The model can convert the sign language to text with an accuracy of 96%. To make the system as user-friendly as possible, we have provided a feature where the general public can record the video of the person doing the hand gesture and use it to make the equivalent English word prediction.

Keywords Sign language · Convolutional neural network · Recurrent neural network · Long-term short memory · Spatial features · Temporal features

1 Introduction

Gesture recognition is a concept where hand movements are performed in front of a camera, which is then taken as input to an image/video processing algorithm. This algorithm is usually a machine learning algorithm. Input can either be a single image or a video. If the input is a video, then it is converted into frames and then it is further processed. Video processing is intensive but given the proper hardware and tools, a fast outcome can be achieved. People with impaired hearing and speech use

K. S. Santhosh (✉) · S. K. Hoysala · D. R. Srihari · S. S. Chandra · A. N. Krishna
Department of Computer Science and Engineering, SJB Institute of Technology, Bangalore, Karnataka, India

V. Bhateja et al. (eds.), *Computer Communication, Networking and IoT*, Lecture Notes in Networks and Systems 197, https://doi.org/10.1007/978-981-16-0980-0_3

sign language for communication. Sign language gestures are used as a means of non-verbal communication to express their thoughts and emotions. People who do not know the gestures find it extremely hard to comprehend what the deaf or dumb person is trying to convey. In the past five years, the need for interpreting services has increased by many folds to meet this demand, and alternatives, such as video remote human interpreting using high-speed Internet connections, have been brought to light. This alternative though helpful and effective has its limitations such as access to a high-speed Internet connection and a capable device.

To solve this, we are using two models to recognize the gestures. We are using our custom recorded Indian Sign Language dataset. The dataset is simple and has only 25 distinct words. This dataset has been created by using multiple people doing so gives variation in context and video conditions. We propose to use a convolutional neural network (CNN) named inception to extract spatial features from the video stream for sign language recognition (SLR). Then, by using an long short-term memory (LSTM) and recurrent neural network (RNN) model, we can extract temporal features from the video sequences.

2 Literature Survey

To classify certain alphabets and words of the American Sign Language, Aarti and Vijaylakshmi have prepared a model [1], which uses a microcontroller Arduino Uno where the predefined threshold values were used to compare readings of hand rotation and each finer bending, which is then used to generate text with the help of hidden Markov model-based text to a speech synthesizer. Chandra et al. [2] present a prototype that makes use of gyroscope, accelerometers and flex sensors embedded gloves. The data was then collected with the help of a microcontroller Arduino Nano, and the data was then sent to the PC with the help of Bluetooth. The prototype classified the gestures with the help of support vector machine (SVM) algorithm. As an improvement to the previous model, Bhujbal and Warhade [3] presented a prototype that made use of a wearable device that includes five flex sensors, one Bluetooth module, a 16 * 2 l;iquid crystal display and a three-axis accelerometer. Then, the data collected from the sensors were compared with the saved data. If a match was found, then the LCD screen was able to display a particular gesture. Badhe and Kulkarni have proposed an idea that converts alphabets of Indian Sign Language into English [4]. A combination of YCbCr model and Canny edge detection technique was used for processing step, and then, Fourier descriptors were used to extract the features. The euclidian distance was used for sign recognition. Further, for Arabic Sign Language Recognition, hidden Markov model and spatiotemporal, feature extraction-based model was presented by Assaleh et al. [5], where 80 words were used to create a dataset of 40 sentences. This model had an accuracy of around 75 and 94% for sentence and word recognition rates, respectively. Ekbote and Joshi [6] presented a prototype that helped in automatically recognizing the numerals of Indian Sign Language. A self-created database that consisted of 100 images for each of the numerals, resulting in a total database of 1000 images was used. The prototype used various techniques such as histogram of oriented gradients (HOG) and scale invariant feature transform (SIFT) for desired feature extraction. The classification was done with the help of support vector machines and artificial neural networks. But the major limitation of this prototype is the limited number of gestures that consisted of only numerals. Lahoti et al. [7] present a prototype, which uses an android application that converts ASL to text, which makes use of a YCbCr system for skin segmentation of the image. A HOG was used for extracting various features from the image, followed by the classification of extracted images done with the help of support vector machine technique. Kolkur et al. [8] propose an algorithm for improving the skin pixel recognition in the input images. This method considers a combination of colour parameters instead of restricting to only three-colour parameters.

3 Proposed Methodology

Neural networks are similar to biological nervous systems that consist of largely interconnected processing elements. These elements work together to solve a specific problem. It is made up of numerous algorithms that recognize the relationship in a set of data similar to the way the human brain works. They are capable of adapting to the changes in the input, so that the network can generate the best possible output, without having to redesign the output criteria.

Convolutional neural network is a machine learning algorithm where images are fed as input, and further, it assigns learnable weights and biases to various objects present in the image so that it can differentiate one image from another. CNN can briefly work in three steps or layers: convolution, pooling layer and fully connected layer.

Figure 1 shows the layers of a ConvNet. A convolutional layer serves as a major layer of CNN. In the convolution layer, a filter is applied to the input image, which

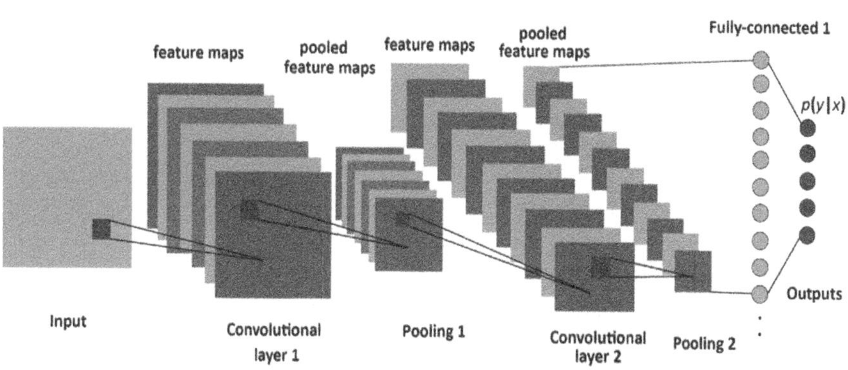

Fig. 1 Architecture of ConvNet

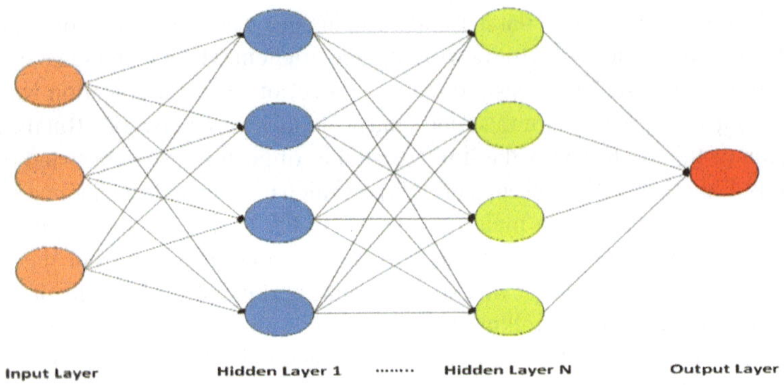

Fig. 2 Fully connected layer

results in activation. Repeated application of these filters results in a feature map, indicating the locations and strength of a detected feature in an image. The next layer in CNN is the pooling layer which is used to reduce the spatial size of the image, and it also reduces the parameters required which in turn reduces the computation that undergoes in the network. The next layer is the fully connected layer as shown in Fig. 2 which accepts the results of the convolution layer and the pooling layer and finally classifies them into labels. The convolution layer and the pooling layer play a very important role in prediction, whereas a fully connected layer is used for classification.

Since learning from a video means the computer has to learn from a series of images that make up the video, an employed neural network must be able to learn from a series of input. Remembering the past decisions is one of the characteristics of the recurrent neural network and this characteristic influences the RNN to make decisions based on the learned decisions of the past. The similarity in learning between training the various input vectors and also remembering the things learned from prior inputs to predict an appropriate output can be found in RNN, which is a part of a network. We can feed multiple input vectors to the RNNs to produce multiple output vectors. Weights applied to the inputs are not the only parameters that influence an RNN as it does in a regular neural network, instead, it also depends on the hidden state which is used to indicate the prior inputs or outputs. So, if in the series, the previous input change, then there are chances that a different output can be obtained for the same input as shown in Fig. 3.

$$\boldsymbol{h}_t = \varphi(\boldsymbol{W}\boldsymbol{x}_t + \boldsymbol{U}\boldsymbol{h}_{t-1}) \tag{1}$$

The function (1) h_t represents a time step t in the hidden state h. At a given time t, in which t is a function of the input, x_t is modified by a weight matrix W which is present in the hidden state, and this is then added to the hidden state of the previous time step h_{t-1} which is then multiplied to the matrix U which is a hidden-state-to-hidden-state transition matrix. These weight matrices act as filters which aide in

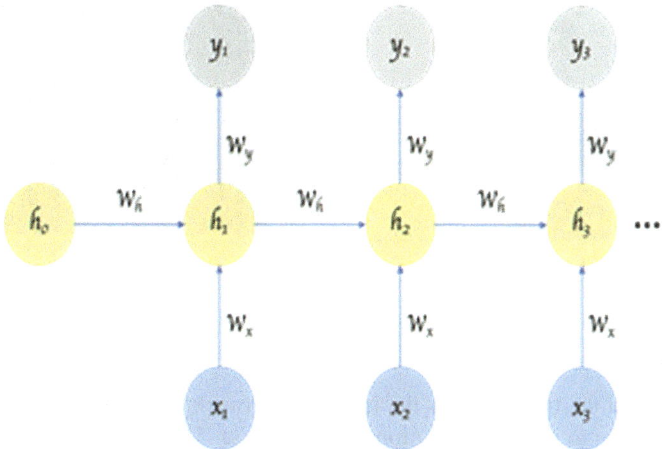

Fig. 3 Representation of an RNN that consists of a hidden state that is used for passing the data from one input to the other in the series

deciding the level of importance to be given to the present input and the past hidden layer. The function φ which is either a logistic sigmoid function or tanh function squashes the sum of the weight input and hidden state.

4 Result and Discussion

The model is implemented using a software tool named Google Colab, and the programming language used was Python 3.6. We have created 25 categories of Indian sign words which consist of 50 videos for each category and around 1500 videos in a whole dataset. Among 50 videos, 40 videos are used for training the video, and the rest is used for testing. We used the concept of spatial features and temporal features which can be extracted from the video [9]. To extract spatial features, we used the inception model and a recurrent neural network consisting of 256 LSTM's to extract temporal features. The model first converts videos into 201 snapshots called as frames, since we are just concentrating on the hand gesture, we just require the position of the hand gesture concerning time and space. So, all the noises including the background will be removed with the help of segmentation. In our model, the segmentation is applied for red colour so whenever the model identifies the red colour in the video, it removes everything and only enhances the red part. Below shows the frame of a video dataset (Fig. 4).

Each frame is segmented as shown above, and the sequence of segmented frames is fed into inception model to train on spatial features. We have used the Inception v3 model of the tensor flow library. Inception v3 is an image classification model that consists of millions of parameters that can differentiate a large category of images.

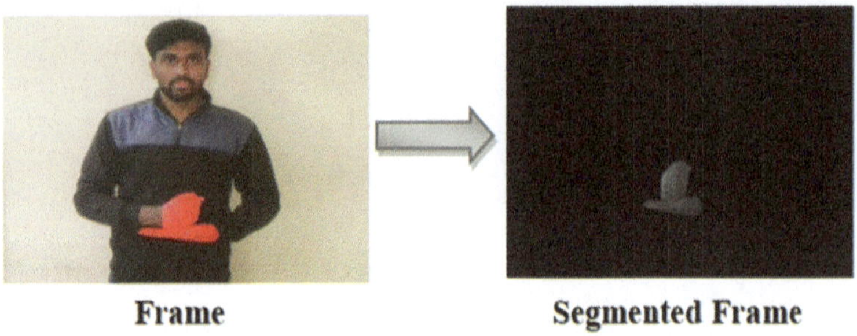

Fig. 4 Conversion of the frame to a segmented frame

We just trained the final layer of the inception network. Inception v3 is trained for the ImageNet large visual recognition challenge (Fig. 5).

We have used transfer learning on the inception model. Transfer learning on the inception model allows retraining the final layer for the existing model. The result of retraining only the final layer does not only reduce the training time but also the dataset required will be reduced. Once a prediction is done, all those predictions will be wrapped in a model file. Then that model file is fed into RNN.

There is a sequence of information in a model file, this information can be used to perform tasks, and it can only be performed by recurrent neural nets but not be feedforward nets. These recurrent neural nets are way different from feedforward nets by the fact that they have a feedback loop, looping back their output after feeding input (Fig. 6).

The first layer in the RNN is an input layer, which is built in order to feed the input to the next consecutive layers. Its size is insisted by the amount of the input

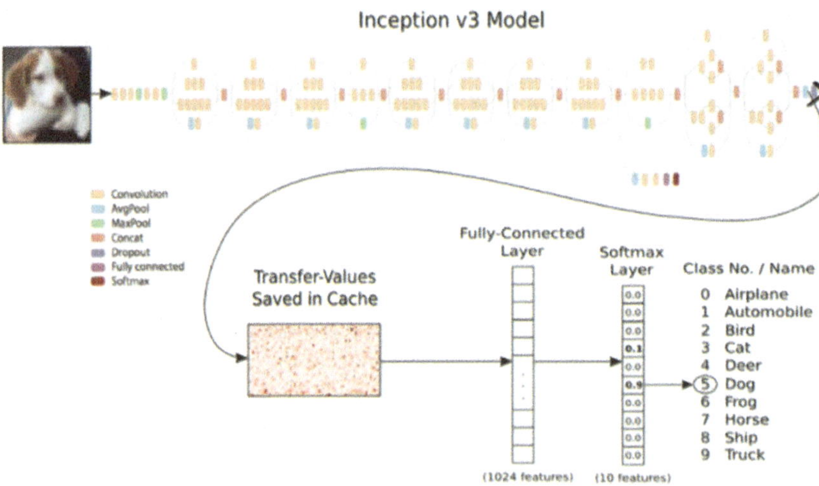

Fig. 5 Example of how the inception model predicts an image as a dog

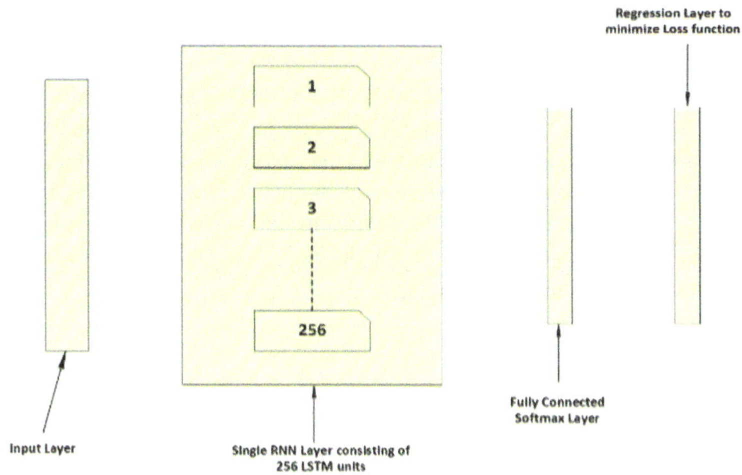

Fig. 6 Our RNN model

being fed. The model consists of a single RNN layer with 256 LSTM units. The next layer is the fully connected layer with an activation function called SoftMax activation. In a fully connected layer, all neurons are connected to every other neuron of the previous layer. The fully connected layer comprises as many neurons as there are categories/classes. The final layer is the regression layer, which is used to apply regression (linear/regression) to the input provided. We have used Adam [10] (Adaptive Moment Estimation) which is a stochastic optimizer, as a gradient descent optimizer to minimize the provided loss function "categorical_crossentropy" (which calculate the errors). Finally, the predictions will be stored in a model file, and those predictions are made available to the user. The model can predict 25 categories for words with an accuracy of 96%.

Keeping in mind the time and processing speed required to train the model, the model is trained with 25 distinct words given gesture as an input and can identify the word among those 25 gestures. The words that the model can identify are (Table 1):

Table 1 List of words in the dataset used

Hello	Pray	Puja	Temple	Disturb
Priest	Hanuman	Ghost	Fruits	Railway
Ship	Aeroplane	Guard	Ticket	Bandh
Flag	Government	Father	Mother	Friend
Job	Measurement	Weight	Gold	Farewell

5 Conclusion

The goal of this study is to recognize the Indian Sign Language using machine learning. The model we have proposed is efficient, fast and accurate and has a high prediction rate. It is mainly aimed to solve the communication problem faced by the deaf and dumb people with the general public. It is composed of two major components, analysing the gestures from images and classifying images. A nation truly grows when all parts of its economy develop at an equal rate. It is our job as responsible citizens to make this happen.

References

1. Aarthi, M., Vijayalakshmi, P.: Sign language to speech conversion. In: Fifth International Conference on Recent Trends in Information Technology, India (2016)
2. Chandra, M.M., Rajkumar, S., Kumar, L.S.: Sign languages to speech conversion prototype using the SVM classifier. In: TENCON 2019—2019 IEEE Region 10 Conference (TENCON), Kochi, India, pp. 1803–1807 (2019)
3. Bhujbal, V.P., Warhade, K.K.: Hand sign recognition based communication system for speech disable people. In: 2018 Second International Conference on Intelligent Computing and Control Systems (ICICCS), Madurai, India, pp. 348–352 (2018)
4. Badhe, P.C., Kulkarni, V.: Indian sign language translator using gesture recognition algorithm. In: 2015 IEEE International Conference on Computer Graphics, Vision and Information Security (CGVIS), Bhubaneshwar, pp. 195–200 (2015)
5. Assaleh, K., Shanableh, T., Fanaswala, M., Amin, F., Bajaj, H.: Continous Arabic sign language recognition in user dependent mode. J. Intell. Learning Syst. Appl. **2**(1), 19–27 (2010)
6. Ekbote, J., Joshi, M.: Indian sign language recognition using ANN and SVM classifiers. In: 2017 International Conference on Innovations in Information, Embedded and Communication Systems (ICIIECS), Coimbatore, pp. 1–5 (2017)
7. Lahoti, S., Kayal, S., Kumbhare, S., Suradkar, I., Pawar, V.: Android based american sign language recognition system with skin segmentation and SVM. In: 2018 9th International Conference on Computing, Communication and Networking Technologies (ICCCNT), Bangalore, pp. 1–6 (2018)
8. Kolkur, S., Kalbande, D., Shimpi, P., Bapat, C., Jatakia, J.: Human skin detection using RGB. HSV and YCbCr Color Models (2017). https://doi.org/10.2991/iccasp-16.2017.51
9. Tang, X., Luo, B., Gao, X., Pissaloux, E., Zhan, H.: Video text extraction using temporal feature vectors. In: Proceedings. IEEE International Conference on Multimedia and Expo (n.d.). https://doi.org/10.1109/icme.2002.1035724
10. Kingma, D., Ba, J.: Adam: A Method for Stochastic Optimization. arXiv preprint arXiv:1412.6980 (2014)
11. Pang, J., Bai, Z.-Y., Lai, J.-C., Li, S.-K.: Automatic Segmentation of Crop Leaf Spot Disease Images by Integrating Local Threshold and Seeded Region Growing. IEEE (2011)
12. Li, H.-N., Feng, J., Yang, W.-P., Wu, X.-S., Li, Z.-D., Liu, W.: Spectrum-Based Method for Quantitatively Detecting Diseases on Cucumber Leaf. IEEE (2011)
13. Hashim, H., Haron, M.A., Osman, F.N., Al Junid, S.A.M.: Classification of Rubber Tree Leaf Disease Using Spectrometer. IEEE (2010)

Energy Efficient Geo-distributed Data Center Resizing Markov Chain Model

P. M. Siva Raja and C. Sahaya Kingsley

Abstract Big data contains large-volume, complex and growing data sets with multiple, autonomous sources. Big data processing is the explosive growth of demands on computation, storage and communication in data centers, which hence incurs considerable operational expenditure to data center providers. Therefore, to minimize the cost is one of the issues for the upcoming big data era. Using these three factors, i.e., task assignment, data placement and data routing, deeply influenced by the operational expenditure of geo-distributed data centers. In this paper, we are ambitious to study the cost minimization problem via a joint optimization of these three factors for big data processing in geo-distributed data centers. Proposed using n-dimensional Markov chain and procure average task completion time.

Keywords Bid data · Distributed data center · Minimize cost

1 Introduction

Big data usually includes data sets with sizes beyond the ability of commonly used software tools to capture, manage and process data within a tolerable elapsed time. Big data is a set of techniques and technologies that require new forms of integration to uncover large hidden values from large data sets. First, the proposed data center resizing (DCR) to reduce the computation cost by adjusting the number of activated servers via task placement. They distribute data in entire geographical data centers to lower the electricity cost. They distribute databases on no of user and distribution of industries.

Second, the links in networks vary on the transmission rates and costs according to their unique features, e.g., the distances and physical optical fiber facilities between data centers. However, the existing routing strategy among data centers fails to exploit the link diversity of data center networks. Due to the storage and computation capacity

P. M. Siva Raja (✉) · C. Sahaya Kingsley
Department of Computer Science and Engineering, Amrita College of Engineering and Technology, Kanyakumari, India

V. Bhateja et al. (eds.), *Computer Communication, Networking and IoT*, Lecture Notes in Networks and Systems 197, https://doi.org/10.1007/978-981-16-0980-0_4

constraints, not all tasks can be placed onto the same server, on which their corresponding data reside. Third, the quality-of-service (QoS) of big data tasks has not been considered in existing work. Similar to conventional cloud services, big data applications also exhibit service-level agreement (SLA) between a service provider and the requesters. Existing studies, on general cloud computing tasks mainly focus on the computation capacity constraints, while ignoring the constraints of transmission rate. We are the first to consider the cost minimization problem of big data processing with joint consideration of data placement, task assignment and data routing. To reduce the communication cost, a few recent studies make efforts to improve data cost minimization for big data processing. Big data is one of the emerging hot research topic because it is mostly used in data center application in human society, such as government, climate, finance and science.

Currently, most research work on big data falls in data mining, machine learning and data analysis [1]. The name itself contains the meaning of data will be—so big || in large volume of both structured and unstructured data present. The challenges include capture, curation, storage, search, sharing, transfer, analysis and visualization. The trend to larger data sets is due to the additional information derivable from analysis of a single large set of related data, as compared to separate smaller sets with the same total amount of data, allowing correlations to be found to spot business trends, prevent diseases, combat crime and so on.

Big data is difficult to work with using most relational database management systems and desktop statistics and visualization packages, requiring instead "massively parallel software running on tens, hundreds, or even thousands of servers." What is considered "big data" varies depending on the capabilities of the organization managing the set, and on the capabilities of the applications that are traditionally used to process and analyze the data set in its domain. Big data is a moving target; what is considered to be "Big" today will not be so years ahead. For some organizations, facing hundreds of gigabytes of data for the first time may trigger a need to reconsider data management options. For others, it may take tens or hundreds of terabytes before data size becomes a significant consideration. Figure 1 shows the definition of the big data and the important characteristics of big data.

Everyday, 2.5 quintillion bytes of data are created, and 90% of the data in the world today were produced within the past two years [2]. Another example is our capability for data generation has never been so powerful and enormous ever since the invention of the information technology in the early nineteenth century. Example for big data, on October 4, 2012, the first presidential debate between President Barack Obama and Governor Mitt Romney is triggered more than 10 million tweets within 2 h [3].

Among all these tweets, the specific moments that generated the most discussions actually revealed the public interests, such as the discussions about Medicare and vouchers. Such online discussions provide a new means to sense the public interests and generate feedback in real time and are mostly appealing compared to generic media, such as radio or TV broadcasting. The most fundamental challenge for big data applications is to explore the large volumes of data and extract useful information or knowledge for future actions [4]. Currently, big data processing mainly

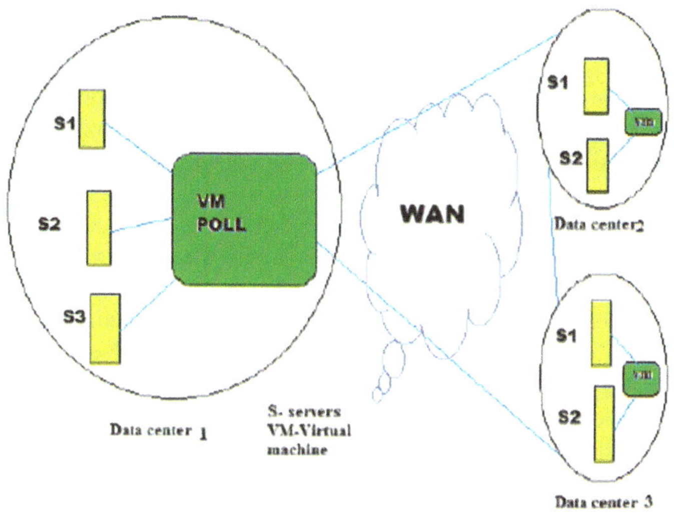

Fig. 1 One geo-distributed data centers

depends on parallel programming models like MapReduce, as well as providing a cloud computing platform of big data services for the public. MapReduce is a batch-oriented parallel computing model. There is still a certain gap in performance with relational databases. Improving the performance of MapReduce and enhancing the real-time nature of large-scale data processing have received a significant amount of attention, with MapReduce parallel programming being applied to many machine learning and data mining algorithms. Data mining algorithms usually need to scan through the training data for obtaining the statistics to solve or optimize model parameters. It calls for intensive computing to access the large-scale data frequently. To improve the efficiency of algorithms [5], geo-distributed data centers are operated by many organizations such as Google and Amazon are the powerhouses behind many Internet-scale services. They are deployed across the Internet to provide better latency and redundancy. These data centers run hundreds of or thousands of servers, so it consumes megawatts of power with massive carbon footprint and also incurs electricity bills of millions of dollars [6]. Data explosion is one of the rising demand for big data processing in recent years and modern data centers that are usually distributed at different geographic regions, e.g., in worldwide Google's, 13 data centers over 8 countries in 4 continents [7]. Gartner predicts that by in 2015, 71% of worldwide data centers will be spending hardware from the big data processing, which will surpass $126.2 billion. These are the reasons behind for we study the cost minimization problem for big data processing in geo-distributed data centers. Cost of big data processing has following issues. First, data locality may result in a waste of resources. For example, most computation resource of a server with less popular data may stay idle. Second, data center resizing. The low resource utility further causes more servers to be activated and hence higher operating cost.

2 Related Works

Sivasubramanian et al. proposed the data center power consumption which has one of a significant impact on both its recurring electricity bill (Op-ex) and one-time construction costs (Cap-ex). They develop peak reduction algorithms that combine the UPS battery knob with existing throttling-based techniques for minimizing power costs in data center. MapReduce runs on a large cluster of commodity machines and is highly scalable and its support to programmers for the system easy to use. Zheng et al. proposed the data center power optimization has recently received a great deal of research attention. Traffic consolidation has one to recently propose to save energy for data center networks (DCNs). We propose PowerNetS, a power optimization strategy that leverages workload correlation analysis to jointly minimize the total power consumption of servers. Dynamically adjusting server capacity and performing load shifting in different time scales, we propose three different loads shifting and joint capacity allocation schemes with different complexity and performance. Our schemes leverage both stochastic multiplexing gain and electricity-price diversity. Liu and Lin, energy expenditure has become a significant fraction of data center operating costs.

Recently, geographical load balancing has been suggested to reduce cost by exploiting the electricity price differences across regions. However, this reduction of cost can paradoxically increase total energy use. This paper explores whether the geographical diversity of Internet-scale systems can additionally be used to provide environmental gains. Geographical load balancing can encourage use of green renewable energy and reduce use of brown fossil fuel energy. They using two factors for reducing the energy cost in data centers. The factors are energy-gobbling cooling and location independent. Temperature diversity can be used to reduce the overall cooling energy overhead. Cost minimization data center resizing (DCR) has been proposed to reduce the computation cost by adjusting the number of activated servers via task placement. To describe the rate-constrained computation and transmission in big data processing process, a two-dimensional Markov chain derives the expected task completion time in closed form that has been proposed. To deal with the high computational complexity of solving MINLP, a mixed-integer linear programming (MILP) problem is linearized, which can be solved using commercial solver. DCR and task placement are usually jointly considered to match the computing requirement.

Agarwal et al. [8] proposed the Now a days services grow to span more and more globally distributed data centers, so we need urgent automated mechanisms to place application data across these data centers. We present the Volley algorithm to a system that addresses these challenges Dean et al. [9] proposed the MapReduce is a programming model and its associated with implementation for processing and to generating large data sets. MapReduce runs on a large cluster of commodity machines and is highly scalable and its support to Programmers for the system easy to use.

Zheng et al. [10] proposed the data center power optimization has recently received a great deal of research attention. Traffic consolidation has one to recently proposed to save energy for data center networks (DCNs). we propose PowerNetS, a power optimization strategy that leverages workload correlation analysis to jointly minimize the total power consumption of servers. Xu et al. [11] the goal is to achieve an optimal tradeoff between energy efficiency and service performance over a set of distributed IDCs with dynamic demand. Dynamically adjusting server capacity and performing load shifting in different time scales. We propose three different load-shifting and joint capacity allocation schemes with different complexity and performance. Our schemes leverage both stochastic multiplexing gain and electricity-price diversity.

Liu and Lin [12] energy expenditure has become a significant fraction of data center operating costs. Recently—geographical load balancing|| has been suggested to reduce energy cost by exploiting the electricity price differences across regions. However, this reduction of cost can paradoxically increase total energy use. This paper explores whether the geographical diversity of Internet-scale systems can additionally be used to provide environmental gains. Geographical load balancing can encourage use of green renewable energy and reduce use of brown fossil fuel energy.

3 System Model

3.1 Geo-distributed Data center

Geo-distributed data center means many data centers are geo-graphically distributed and connected through the WAN environment. In recently, many organizations move to this geo-distributed data center. Because they stored large or massive volume of data. If they are using our own data center means, only limited storage will be there so only many of them used this geo-distributed data centers.

Google's and Amazon using this geo-distributed data centers for storing and managing their data.

Markov chain. A Markov is a mathematical system that undergoes transitions from one state to another on a state space. It is a random process usually characterized as memoryless: The next state depends only on the current state and not on the sequence of events that preceded it. Memorylessness is called the Markov property. Markov chains have many applications as statistical models of real-world processes. A Markov chain is a sequence of random variables $X1, X2, X3$, with the Markov property, namely that, given the present state, the future and past states are independent [13].

3.2 Data Placement

WAN bandwidth between data centers, and over-provisioning data center capacity to tolerate highly skewed data center utilization. In this paper, we show that a more sophisticated approach can both dramatically reduce these costs and still

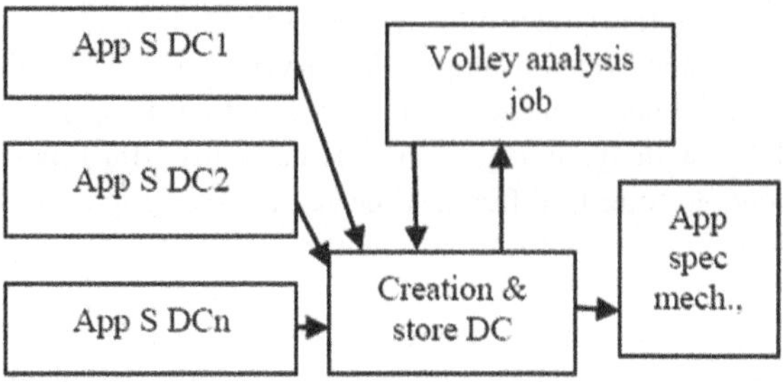

Fig. 2 Volley algorithm

further reduce user latency [8]. Proposed using volley algorithm for automated data placement in geo-distributed data centers.

Recursive Step. Volley algorithm contains three phases such as compute initial placement, iteratively move data reduce latency and iteratively collapse data to data centers (Fig. 2).

3.3 Electricity Cost

All the hardware's work without electricity. proposed a novel, data-centric algorithm used to reduce energy costs and with the guarantee of thermal reliability of the servers in geo-distributed data centers [12]. And also using the n-dimensional Markov chain algorithm to reduce the electricity cost.

3.4 Server Cost

In geo-distributed data centers hundreds of servers used. Because of this automatically the server cost will be increases [14]. How to reduce the server cost means using communications and data placement and task assignment approach. Number of sever will be reduced means at a meantime the energy cost also decreases [15]. Server cost reduced using the joint optimization of these three factors such as task assignment, data placement and data routing via a *n*-dimensional Markov chain.

4 System Implementation

This system output is a simulation-based studies. It will be evaluated using NS2 tool. The evaluation base on the effect of the number of servers, on the effect of task arrival rate, on the effect of data size, on the effect of expected task completion delay and on the effect of number replicas. Based on these, we using joint optimization of task assignment, data placement and data routing via an n-dimensional Markov chain algorithm.

5 Results

See Figs. 3 and 4.

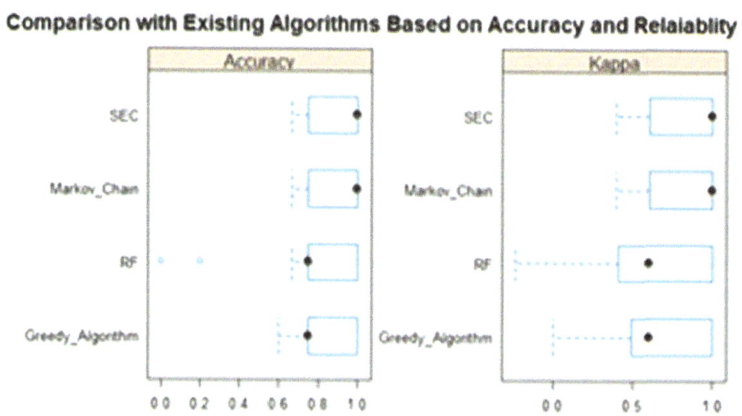

Fig. 3 Comparison with existing algorithms based on accuracy and reliability

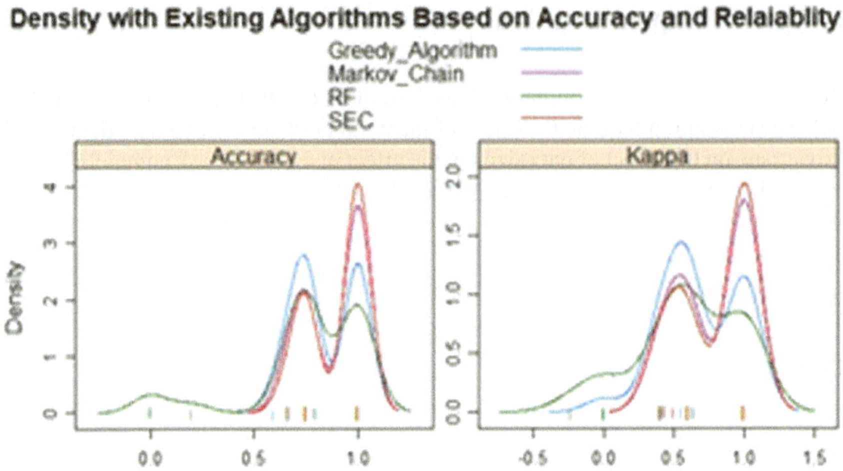

Fig. 4 Density with existing algorithms based on accuracy and reliability

6 Conclusion and Future Work

In this paper, we study the geo-distributed data centers issues. We jointly study the data placement, data center resizing and data routing to reduce the operational cost in geo-distributed data centers for big data processing. And we characterize the data processing and task assignment using n-dimensional Markov chain based on joint optimization of that three factors.

References

1. Agarwal, S., Dunagan, J., Jain, N., Saroiu, S., Wolman, A., Bhogan, H.: Volley: automated data placement for geo-distributed cloud services. In: 7th USENIX Symposium on Networked Systems Design and Implementation (NSDI), pp. 17–32 (2010)
2. Czajkowski, K., Fitzgerald, S., Foster, I., Kesselman, C.: Grid information services for distributed resource sharing. In: 10th IEEE International Symposium on High Performance Distributed Computing, pp. 181–184. IEEE Press, New York (2001)
3. Foster, I., Kesselman, C., Nick, J., Tuecke, S.: The Physiology of the Grid: an Open Grid Services Architecture for Distributed Systems Integration. Technical report, Global Grid Forum (2002)
4. Gao, P.X., Curtis, A.R., Wong, B., Andheshav, S.: It's not easy being green. In: Proceedings of ACMSIGCOMM (2012)
5. Rao, L., Liu, X., Xie, L., Liu, W.: Minimizing electricity cost: optimization of distributed internet data centers in a multi-electricity-market environment. In: Proceedings of the 29th International Conference on Computer Communications (INFOCOM). IEEE (2010)
6. Foster, I., Kesselman, C.: The Grid: Blueprint for a New Computing Infrastructure. Morgan Kaufmann, San Francisco (1999)
7. Smith, T.F., Waterman, M.S.: Identification of common molecular subsequences. J. Mol. Biol. **147**, 195–197 (1981)

8. May, P., Ehrlich, H.C., Steinke, T.: ZIB structure prediction pipeline: composing a complex biological workflow through web services. In: Nagel, W.E., Walter, W.V., Lehner, W. (eds.) Euro-Par 2006. LNCS, vol. 4128, pp. 1148–1158. Springer, Heidelberg (2006)
9. INTEL INC. Reducing Data Center Cost with an Air Economizer (2008)
10. Sathiamoorthy, M., Asteris, M., Papailiopoulos, B., Dimakis, A.G., Vadali, R., Chen, S., Borthakur, D.: Xoring elephants: novel erasure codes for big data. In: Proceedings of the 39th International Conference on Very Large Data Bases, ser. PVLDB'13. VLDB Endowment, pp. 325–336 (2013)
11. Dean, J., Ghemawat, S.: MapReduce: Simplified Data Processing on Large Clusters (2004)
12. Joint Power Optimization of Data Center Network and Servers with Correlation Analysis
13. Govindan, S., Siva Subramaniam, A., Urgaonkar, B.: Benefits and limitations of tapping into stored energy for datacenter. In: Proceedings of the 38th Annual International Symposium on Computer Architecture (ISCA). ACM, pp. 341–352 (2011)
14. Twitter Blog, Dispatch from the Denver Debate, https://denvedebate.html
15. Gu, L., Zeng, D., Li, P., Guo, S.: Cost minimization for big data processing in geo-distributed data centers. IEEE Trans. Emerg. Top. Comput. (2014. https://doi.org/10.1109/TETC.2014.2310456

Voice-Controlled Intelligent Wheelchair for Quadriplegics

Mithra Venkatesan, Lini Mohan, Nisith Manika, and Aishwarya Mathapati

Abstract The major challenge faced by paralyzed people is that they depend on other people for their movement. Further, because of their failing health, there is a growing need to maintain track of the status of their health too. Also, they require assistance in normal activities in daily life. Toward addressing the above challenges, electric wheelchairs catering needs of paraplegics have been developed. However, these wheelchairs are not suitable for quadriplegics as their needs and problems differ. Apart from providing mobility, continuous monitoring of health condition will be an additional feature vital to the patient. The major objective of this paper is to design a fully automated wheelchair for quadriplegics with features to continuously monitor health of the patient and reduce the burden of performing various activities of routine life. The proposed method uses the voice-controlled movement of the wheelchair along with the inbuilt manual operation for the movement. Health monitoring unit in the system is set up to govern the health status and emergency measures and precautions. The main motive behind this paper is to design a well-equipped wheelchair which is based on current technology and can get integrated into today's world for better application and ease of living.

Keywords Intelligent wheelchair · Health monitoring system · Home automation · Obstacle detection · Automatic umbrella

1 Introduction

A special type of paralysis is called quadriplegia which is the illness caused due to partial or complete loss of functioning of four limbs and torso. This is more severe compared to paralysis which has no effect on arms. Further, this disorder is also associated with loss of sensation where the patient losses sensory and motor skills control. It has been established through study that in a population of 1 million in a year, 18.5% of the population suffer from quadriplegia. Also, it is seen that

M. Venkatesan (✉) · L. Mohan · N. Manika · A. Mathapati
Electronics and Telecommunications Department, Dr. D. Y. Patil Institute of Technology, Pune 411018, India

V. Bhateja et al. (eds.), *Computer Communication, Networking and IoT*, Lecture Notes in Networks and Systems 197, https://doi.org/10.1007/978-981-16-0980-0_5

patients who suffer from quadriplegia also demonstrate cardiovascular sickness. Any heart-related illness proves fatal when left unattended, and elderly people with co-morbidities are more vulnerable than youngsters. Hence, to ease the burdens and challenges in the life of quadriplegic patients, there is a growing need for an exclusive wheelchair having feature also to monitor health condition of the patient. Since the patient suffering from quadriplegics is unable to use their arms, a voice-controlled wheelchair would be the requirement toward enabling their mobility. Also, pulse rate tracking continuously should be integrated to the voice-controlled intelligent wheelchair to measure heart rate of the patient whose record could be maintained. In the voice-controlled intelligent wheelchair, the heart rate of patients is recorded by sensors and the records are maintained.

2 Literature Survey

Singh et al. [1] have proposed awareness about different technologies that can be implemented in wheelchair module including artificial intelligence. Using those technologies, he defined how a patient can be monitored using smart wireless system. The paper delivered the technical approaches in wheelchair designing. Megalingam et al. [2] have described creation of system with general functionality of microcontrollers and upgrading it with GPS technology, that indeed pictured the need of positioning in designing a wheelchair. Dalsaniya et al. [3] have elucidated the formulation of an intelligent wheelchair which is capable of navigation with help of inputs received from many sources. This makes it possible for users with varied kinds of disabilities to utilize the same system. Ruajiz et al. [4] in their work on design and implementation of low-cost intelligent wheelchair has safety mechanism for avoiding collision with obstacles in the event of mistakes occurring in voice commands. The authors have used of GPS and GSM technology that could be accessible in case of emergencies. Zhang et al. [5] in their work described the necessary control system for development of voice-controlled wheelchair having feature of providing different postures. The system is capable of avoiding obstacles and can navigate automatically according to requirements. In [6], the authors describe human-centric models and how efficiently they learn their environment from natural language descriptions.

This paper tries to combine features in the above-stated works and proposes the development of a voice-controlled intelligent wheelchair for quadriplegics. The proposed work will be entirely built up using hardware and software interface and try to accommodate all the required features. The developed model will be capable of operating in real life beyond laboratory trials and also will be made intelligent by having voice prompting feature to help and enable users give voice command.

3 Research Method and System Architecture

3.1 System Overview

Figure 1 presented the block diagram of the proposed system which clearly symbolizes all the units that have been collaborated in the intelligent wheelchair.

The controlling of the wheelchair can be done through voice control or manually using switches. Either a manual joystick or an android application is being connected to the hands of the wheelchair.

The android application which is connecting via Bluetooth module or a manual joystick is fixed on the hands of the chair. The usage of both of these features will decrease the dependency of user on another person. The wheelchair has an ultrasonic sensor which detects the presence of obstacles and thus helping in avoiding accidents. Health monitoring system is also integrated to constantly monitor health of the patient and send the updates to the application developed. An interactive unit is also combined with the developed system.

Fingerprint for unlocking of wheelchair, automatic umbrella which will automatically open when it detects the rain, and controlling the lights of the room are the additional features which have also been augmented in the system.

Figure 1 describes the block diagram of the system which comprehensively covers the various functionalities and modules which have been discussed below.

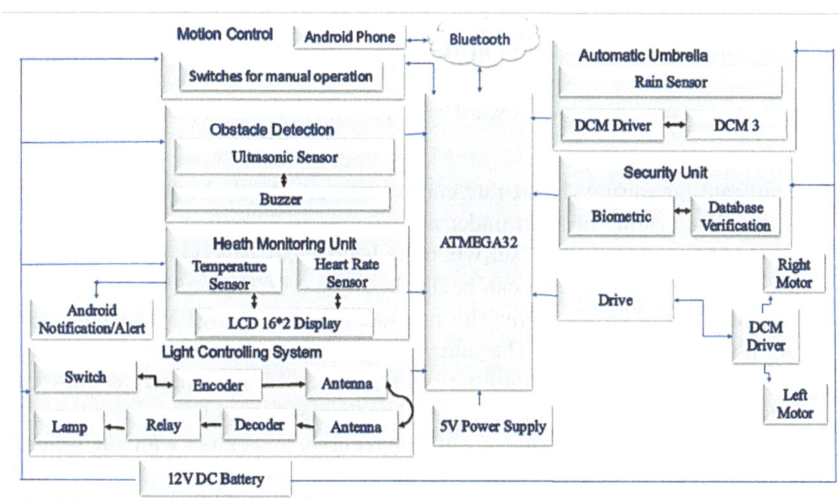

Fig. 1 Block diagram overview of the voice-controlled intelligent wheelchair

3.1.1 Sensing and Measuring Division

There are different features and quantities that are sensed and accordingly divided into different divisions. They are categorized as environmental parameters, physiological parameters, and voice-related parameters. All the data which are being sensed and being utilized toward making decisions on the movement to the wheelchair as well as creation of alert signal if there are some issues.

3.1.2 Environment Related Parameters

Three types of sensors are used under environment-related parameter. These sensors take information necessary from the environment and use it for movement of the wheelchair. The following were the sensors which are to be used.

1. Sensing of obstacles: The presence of obstacle can be detected by use of ultrasonic sensor. The measurements from the sensor are given to microcontroller and aid in stopping of the wheelchair.
2. Sensing of Rain: Rain sensor senses the moisture and sends the signal to microcontroller to act accordingly by opening the umbrella.
3. Fingerprint Sensing: The biometric in the wheelchair gives the security to lock and unlock the wheelchair and then proceed with processing of data from other sensors. This sensing ensures security to the whole wheelchair.

3.1.3 Sensing of Physiological Parameters

Different sensors are being used toward sensing of physiological parameters as given below:

1. Heart rate measuring: Heart rate can be measured with the help of heart rate sensor. The beating of heart under normal case is 72 beats for a minute while threshold for alarm can be set when the heart beats beyond 80 every minute. The measurements of heart can be displayed on a LCD.
2. Measurement of Temperature: This measurement is done by LM35 having accuracy up to tenth of a degree. The values being measured are converted into digital. In the measurement temperature of 40 and beyond is considered as threshold. This value which will be measured will be displayed on LCD.
3. Voice: Through detailed analysis, the physiology associated with any user can be understood based on the voice recording.

All the above-mentioned parameters are measured, and it gives insights about physiological state of the patient. These help in making the navigation of the intelligent wheelchair.

3.1.4 Light Controlling Module

In this module, the switches which are attached to the wheelchair can be used to control the lights. The output of switches is sent to encoder which then passes it to the transmitting antenna. The signal is received in the receiving antenna and then passed on to decoder which decodes the value. This value is given to the relay which makes the lamp off and on.

3.1.5 Division for User Command

Since the wheelchair is voice controlled, voice commands can be given by patient which can in turn control the movement of the wheelchair. The commands given through the android phone are in turn given to the microcontroller which takes decisions on the actions to be taken based on the commands.

3.1.6 Home Automation Unit

This module helps the system to operate the lights and fans using the voice commands given through the voice. The home automation module inbuilt in the wheelchair work with correspondence with different relays and Bluetooth which incase makes it operate in the defined range.

3.1.7 Automatic Umbrella Unit

With the help of rain sensor, the system detects the moisture and opens the flap as per the requirement when taking the chair outdoors.

3.1.8 Buzzer Alert

Whenever an emergency is encountered or the system gets out of contact, at every instant, a buzzer is provided. This ensures the system to be up to date and so in case makes it more reliable.

3.2 Methodology

Figures 2, 3, 4, and 5 individually describe the methodology to be followed in the separate sub-modules.

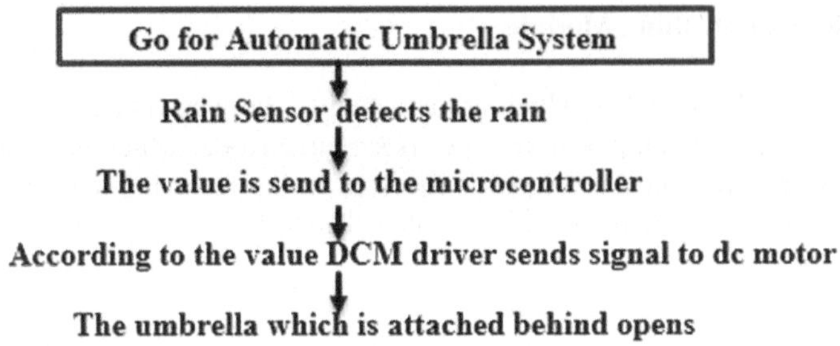

Fig. 2 Methodology of automatic umbrella system

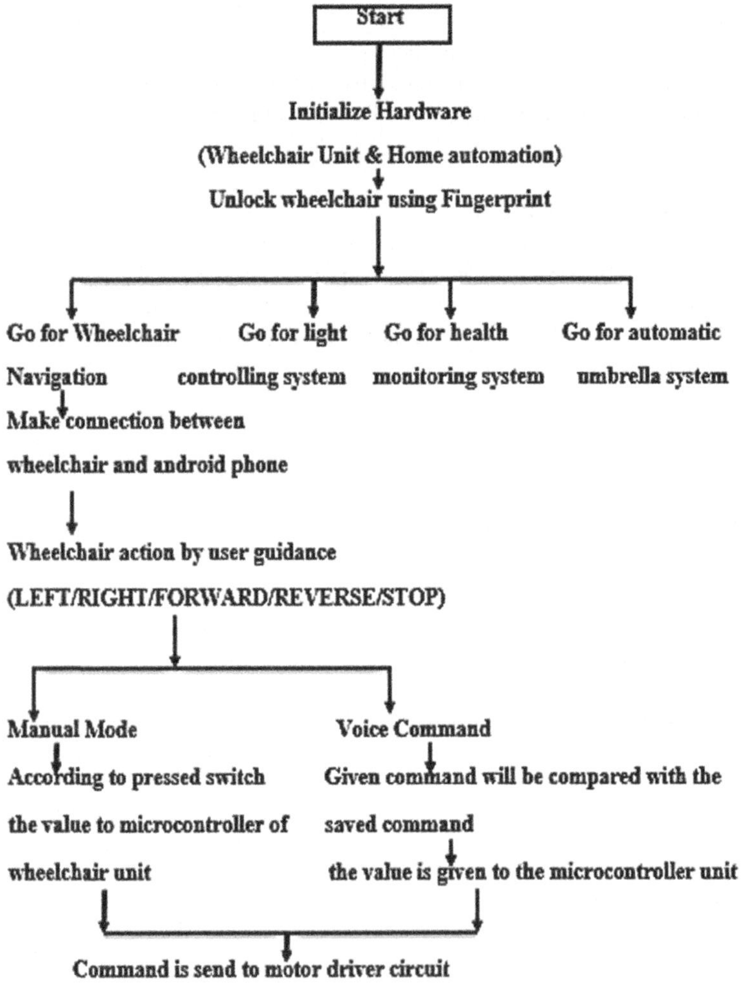

Fig. 3 Methodology for navigation system

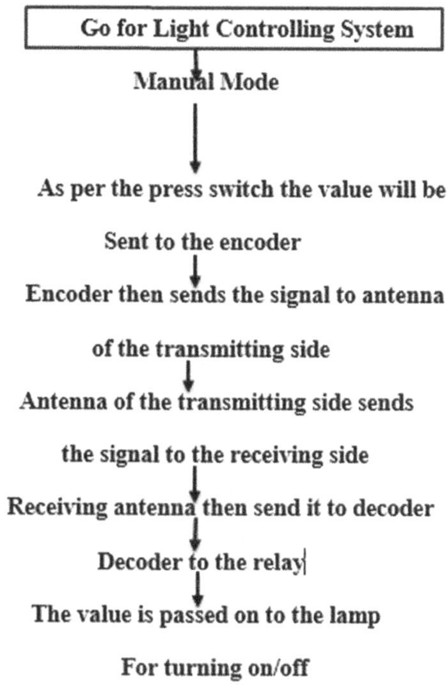

Fig. 4 Methodology for light controlling system

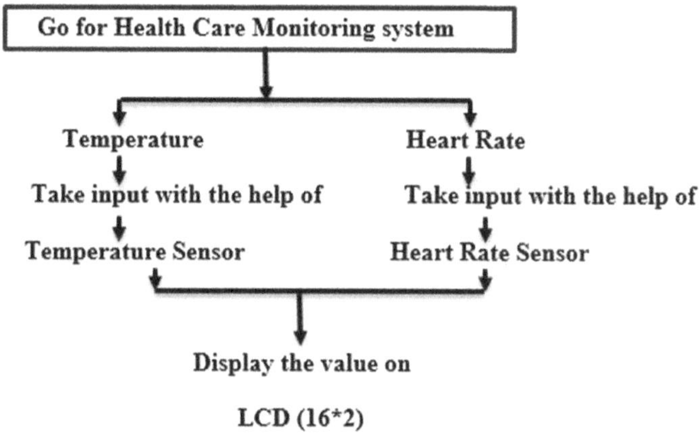

Fig. 5 Methodology for healthcare monitoring system

4 Flowchart

Figure 6 depicts the flowchart of the system. A threshold is being set with maximum and minimum values at every sensor. Whenever measurements go beyond thresholds,

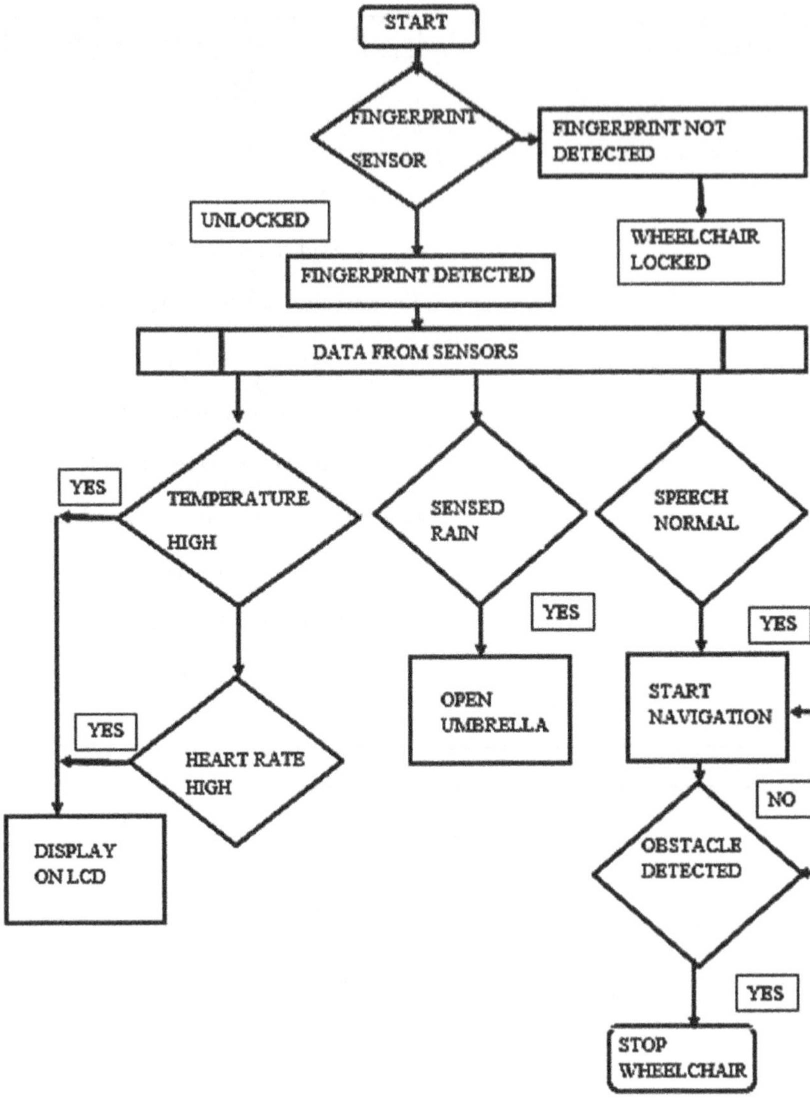

Fig. 6 Flowchart

then movement of the wheelchair and decision of alarm is to be taken specific to a particular situation.

1. Finding the state associated with disease: The physiological factor of temperature and heart rate measurement gives indication about the health of the patient. High temperature and high heart rate indicate a health issue and should alert the alarm system and can be useful in predicting illness with respect to the patient.

2. Finding changes in environmental conditions: The changes in environmental conditions are being sensed by the corresponding sensors, and when the values go beyond the threshold limits, it alerts the alarm system or the navigation system through decisions by the microcontroller.

The inputs given by both the above-mentioned sensor sets to the microcontrollers enable the microcontroller to take decision. The sensors not only predict disease but it also enables the system to detect any abnormality which will enable the system take further decision regarding the mobility of the system and alarm system activation. The flowchart clearly depicts that the need of another person toward assistance is completely eliminated in the current scenario, and the patient can independently navigate using the current setup.

5 Simulation and Results

5.1 Working

Microcontroller ATMEGA32 (40 pins) which has 28 digital I/O pins and 8 analog pins and belongs to AVR family is being used. Manual Reset pin is provided. Upon reset, microcontroller will execute process from beginning. Firstly, input is given from fingerprint sensor which is connected via UART to activate system.

Temperature sensor (LM35) and heartbeat sensor are to be connected to microcontroller whose reading is to be displayed on LCD. Next, we need to take voice command from android smart phone with some suitable app which needs to be connected via Bluetooth module. Motor Driver L293D is provided which moves wheelchair F, B, L, R, and Stop, respectively.

For obstacle avoidance, ultrasonic sensor (HC-SR04) is used to sense the obstacles on a given path which is intimated by a buzzer. Four switches are needed to operate the wheelchair manually by moving F, B, L, R, and Stop.

There is a provision for lamp ON/OFF. A switch is assigned to input to encoder which is further connected to antenna followed by decode connected to relay and lamp. It also senses rain. It needs to cover the wheelchair with flaps. For that, rain sensor needs to be connected to separate DC motor of 12 V, 150 mA. Figure 7 depicts the detailed modules which are brought together in the circuit.

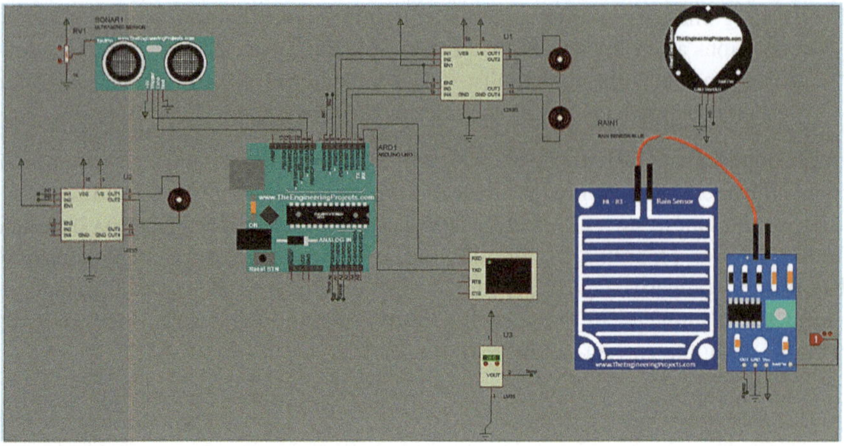

Fig. 7 Detailed modules on the circuit

6 Conclusion and Future Scope

6.1 Conclusion

This paper accomplishes in the design and fabrication of a voice-controlled wheelchair for quadriplegics which would greatly enable in the movement of elderly through voice commands. The various commands such as forward and backward through voice recognition are recognized, and hence, in navigation of the wheelchair, a joystick is also fitted on the handle of the wheelchair if the patient has the ability to operate it if required.

The system also has provision for monitoring the health of the patient through temperature and pulse rate and sends the data wirelessly for suitable alarm activation system. Further, the automatic umbrella which will open only when it detects the rain is also included. The security system is enhanced by making use of fingerprint to lock and unlock the system. The notification of pulse rate and temperature will be sent to the user as well as the caretaker.

6.2 Future Scope

The future scope of this paper is to indulge in various interactive activities like music box and games. Alexa can also be included to enhance the interactive with the user and wheelchair. The GPS tracking can be added to the features of wheelchair for getting the location of the wheelchair. The power consumption can be reduced more, and solar panel can be used for charging purpose of wheelchair.

References

1. Singh, A.K., Gaharwar, M.: Artificial intelligence based wireless smart wheelchair. Int. J. Emerg. Technol. Eng. Res. (IJETER) **6**(3) (2018)
2. Megalingam, R.K., Pocklassery, G., Jayakrishnan, V., Mourya, G., Thulasi, A.A.: Smartphone based continuous monitoring system for home-bound elders and patients. In: 2014 International Conference on Communication and Signal Processing, pp. 1173–1177. IEEE (2014)
3. Dalsaniya, A.: Smart Phone Based Wheelchair Navigation and Home Automation for Disabled, pp. 1–5 (2016). https://doi.org/10.1109/ISCO.2016.7727033
4. Ruzaij, M.F., Poonguzhali, S.: Design and implementation of low cost intelligent wheelchair. In: 2012 International Conference on Recent Trends in Information Technology, Chennai, Tamil Nadu, pp. 468–471 2012. https://doi.org/10.1109/ICRTIT.2012.6206809
5. Zhang, R., Li, Y., Yan, Y., Zhang, H., Wu, S.: An intelligent wheelchair based on automated navigation and BCI techniques. In: 36th Annual IEEE International Conference of the Engineering in Medicine and Biology society (EMBC), Chicago, IL, pp. 1302–1305 (2014)
6. Nguyen, J.S.: A Smart Wheelchair System Using a Combination of Stereoscopic and Spherical Vision Cameras, Ph.D. dissertation, Department of Engineering and Information technology, Sydney University, Sydney (2012)

Digital Forensic Investigator for Cloud Computing Environment

Tasleem Sulthana and Digambar Pawar

Abstract Cloud computing has got a dominant position in the field of computing with its roots in virtualization techniques and its characteristics such as scalability, elasticity, availability and many more. As cloud computing has been adopted by many companies and businesses, security becomes a main challenging aspect. Though cloud computing comes with secured deployment models, criminals still find a way to break it and commit crimes; therefore, cloud computing must always be ready to support the forensics investigation to deal with such security issues and the acts of criminals. Cloud forensics is a field in cyber forensics domain using which investigators can collect the digital evidence of a crime which took place in a cloud environment and present the same as a proof of evidence. This paper intends to provide a prototype model with implementation that makes the forensic task easier by collecting, preserving and presenting the data to the investigator by simplifying the task of data collection phase of the forensic investigation.

Keywords Digital forensics · Cyber-crime · Cloud computing · Cloud crime · Cloud forensics

1 Introduction

Cloud computing got popularized with the development of Elastic Compute Cloud (EC2) by Amazon in 2006, but from then, it has broadened its roots in the field of computer technology. According to Gartner, Inc. Feb, 2019 report, the cloud service market to rise to 266.4 billion dollars in 2020 from 224.8 billion dollars in 2019 making a rise of 17%. Multinational companies have invested almost 30% of their budget amount on cloud computing in 2018 according to Forbes. As the numbers show how the cloud has become the most trending business model, it is mandatory to

T. Sulthana · D. Pawar (✉)
School of Computer and Information Sciences, University of Hyderabad, Hyderabad Central University Road, Gachibowli, Hyderabad 500046, Telangana, India
e-mail: dpr@uohyd.ac.in

V. Bhateja et al. (eds.), *Computer Communication, Networking and IoT*, Lecture Notes in Networks and Systems 197, https://doi.org/10.1007/978-981-16-0980-0_6

deploy the cloud in a secured and also in such a way that it helps forensic investigation [1].

1.1 Cloud Computing

Cloud computing is a business model that provides various computing, storing and processing services over the Internet when required in order to provide scalability, elasticity and high availability by using virtualization techniques.

A widely cited "The NIST Definition of Cloud Computing" defined in 2011 as "Cloud computing is a model for enabling ubiquitous, convenient, on-demand network access to a shared pool of configurable computing resources that can be rapidly provisioned and released with minimal management effort or service provider interaction" [2]. It is composed of three service models.

- ✓ Software-As-A-Service [SAAS]: Software applications running on the cloud infrastructure are provided to consumers.
- ✓ Platform-As-A-Service [PAAS]: Cloud components are provided to consumers as a platform for deploying the software or other developments but have no control on the cloud infrastructure including network, operating system and storage.
- ✓ Infrastructure-As-A-Service [IAAS]: The consumer is provided with virtualized computing resources such as operating system, network and storage that are maintained by the consumer to deploy and run any application or software.

On-demand self-service, broad network access, resource pooling, rapid elasticity and measured service are the five characteristics of cloud computing, whereas private cloud, public cloud, community cloud and hybrid cloud are the four deployment models according to the final version published by NIST [2].

1.2 Forensics: Digital Forensics and Cloud Forensics

With an increase in technology and digitalization, the crime rate has also increased gradually in the digital world. To uncover a crime, the investigation is carried out using the forensic tools and techniques, and evidence is procured from the site of crime. This evidence can be used in the court of law only if it has not been tampered. The process of forensics consists of scientific tests, techniques and tools used to trace out the crime and collects the untampered evidence by acquiring, preserving and examining. Forensics is a post-event of crime. Digital forensics deals with the collection and analysis of digital evidence from digital devices such as mobiles, laptops, desktops, router and so on through or on which crime has been carried out. Cloud forensics is a set of methods to uncover the crime in the cloud using forensic

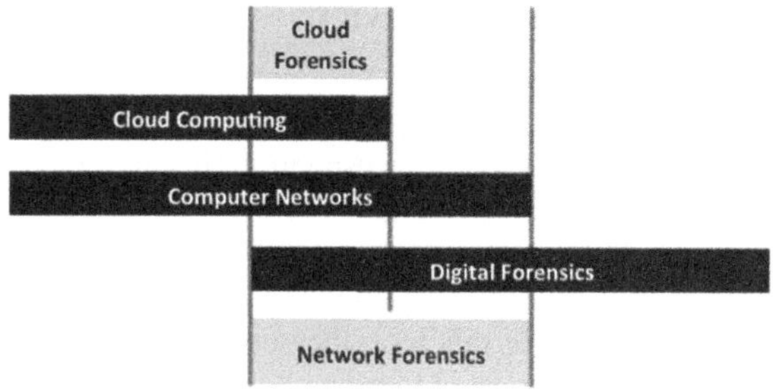

Fig. 1 Where cloud forensics fits? [3]

tools by extracting the required evidence from the cloud environment. Cloud computing requires huge access to networks to provide the connection between various virtual resources deployed at various locations. So, cloud forensics is considered as a subset of network forensics as the network forensics involves forensics of various types of private and public networks. Cloud forensics is a cross-discipline of cloud computing and digital forensics which means applying the digital forensics techniques in every phase in the cloud [4] (Fig. 1).

1.3 Requirements and Challenges of Cloud Forensics

Cloud forensics has different phases and every phase has diverse challenges. The traditional forensics can deal with the areas where the data and devices are physically located, whereas in cloud forensics is required to acquire the evidence of a crime committed in the cloud. It becomes a more complex task to acquire evidence when the physical location of the data is not known [5, 6]. The NIST has identified 65 major challenges involved in cloud computing.

Event logs play a vital role in resolving a crime incident, but the diverse formats of the logs make the forensic task challenging. Thus, a standard, simple log structure is needed. This challenge further increases due to the large volume of log data, as it is difficult to trace crime-related log from huge volume of logs [4]. Other challenges involving logs are inaccessibility of logs, absence of logs and incompatible format of logs. The accessibility of logs also depends on the service model deployed and cloud service provider [7]. Table 1 summarizes the log-related challenges for forensics in cloud environment.

The remaining part of the paper consists of a literature review followed by the cloud forensic analysis and storage section where we provide a detailed description of the proposed model, its setup, implementation details and results. The final section has a brief discussion and conclusion.

2 The Literature Review

There is a lot of work done till now to support forensics of cloud environments in order to facilitate the investigation. In time, there are many frameworks, tools, applications built to serve cloud forensic purposes [9–11].

In 2013, Sherman and Dykstra [12] developed a toolkit called Forensic OpenStack Tools [FROST] to support digital forensics in cloud computing on the OpenStack's Folsom version. In this tool, they enhanced the existing OpenStack dashboard by including Authenticated Logging Service by making slight changes in the code that implements the Horizon (OpenStack dashboard). This service stores the logs of each OpenStack user in the merkle tree data structure by making them the child of the owner's root. A cryptographic hash algorithm is used in this model for segregating the users' data and obtaining integrity. This model fails to address the problem of preservation of the data and as a tool it fails to support the versions after pike as the devstack is run as systemd unit files with all its services.

A generalized prototype model was proposed in 2017 [13]. This paper lists all the possible challenges for each and every phase of cloud forensics and suggests the possible solutions. According to this prototype, the intrusion detection systems are used to identify odd activities and the related logs are thus collected and stored by the cloud service provider.

In our model, we try to collect all the logs from the CSP as any other models and build a separate application for the forensic purpose which is available over the cloud. We also reduced the overhead of searching the logs by providing the filtering methods and present the result in a tabular form.

3 Cloud Log Storage and Analysis

To enable cloud forensics, it is required to aggregate all the logs from various locations, segregate and preserve them for future investigations. As the cloud environment has no physical location, one should either make use of tools available or use a tech-

Table 1 Summary of log-related challenges in cloud forensics [8]

Cloud forensic challenge	Exists in the cloud model		
	SaaS	PaaS	IaaS
Decentralization of logs	Yes	Yes	Yes
Volatility of logs	No	No	Yes
Logs in multiple tiers and layers	Yes	Yes	Yes
Log dependents on CSP	Yes	Yes	Yes
Accessibility of logs	Yes	Yes	Yes
Absence of critical information in logs	Yes	Yes	Yes
Absence of standard log format	Yes	Yes	Yes

nique for collecting the log data. Apart from log collection, log storage is also a difficult task. There are various cloud-based log monitoring tools available, which pushes the data on to cloud from the hosts.

In this section, we provide the proposed work with a block diagram presenting the abstract architecture of the model, environment setup details, stepwise implementation procedure and the results obtained.

3.1 Proposed Work

Architecture: The proposed model collects the logs from the host machine where the cloud environment is deployed and stores those logs on the cloud, so that a forensic investigator can further analyze or search for evidence. Figure 2 shows the workflow of the proposed model.

Implementation details: The model is implemented on the OpenStack cloud environment (Ussuri version). OpenStack cloud environment can be set up by using either manual installation or devstack. OpenStack comes with a set of modular components which can be either installed as whole or in components as required on single or multiple nodes. Initially, a set of virtual machines, routers and networks are created on a single node deployment of OpenStack in order to generate log files.

The following are the four phases of the implementation of the proposed model (Fig. 3).

- Phase-1: Data collection
  The log data stored in the host machine is collected in this phase by using a Unix-based job scheduler called cron, which runs a task periodically. The log data is

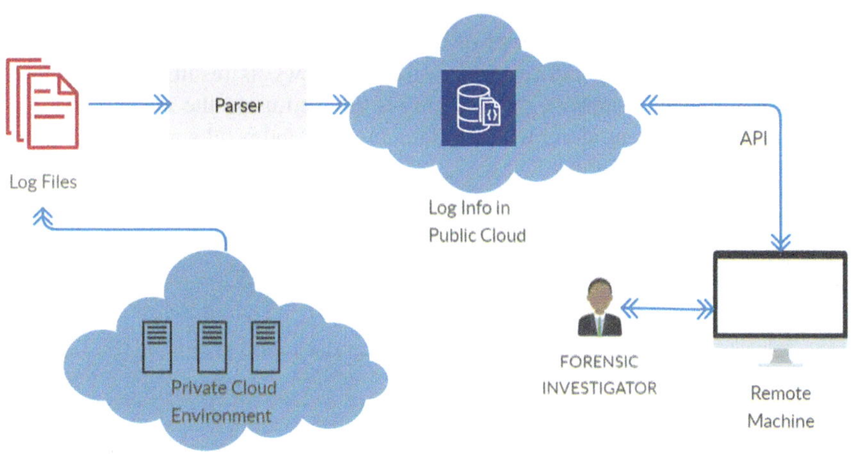

Fig. 2 Architectural diagram for the proposed model

stored in journald for the current version of OpenStack, as all the services are run as systemd unit files. The cron job collects the data of every component using journalctl command and stores it in text formats.

- Phase-2: Parsing and Segregation
  The log format of every component is understood and parsed into fields according to the requirement and then converted into JSON format.
- Phase-3: Storing
  In this phase, the log data in JSON format is stored in MongoDB, each component's log data is considered as a collection. The data in the MongoDB is then stored in the cloud (MongoDB Atlas's AWS cloud) to provide remote access from anywhere, to avoid tampering and facilitate storage for large data. As log data is stored away from the host machine, the chances of manipulating the log data are minimized. Eventually, there will be an improvement in fault tolerance because of data replication.
- Phase-4: Visualization
  An application is built to query the event logs based on date, component ID or based on the log filters. This application fetches the log data stored in the cloud by applying the given filters and visualizes the event log matching the filters. Only an authenticated person with valid login credentials will be able to operate this application.

3.2 *Results*

The results are presented to an authorized user through the dashboard of an API, which allows the user to query the logs by applying available filters. This dashboard provides filter forms for every component of the OpenStack cloud environment, and the results are displayed in an organized tabular form. Figure 4 shows the form for filtering the neutron (network) logs of the OpenStack environment along with the statistics of available log information and Fig. 5 shows its results after querying. For this component (neutron), the querying is allowed using the network-ID, log-level and network-agent along with the date duration and results are presented in a well-organized tabular form.

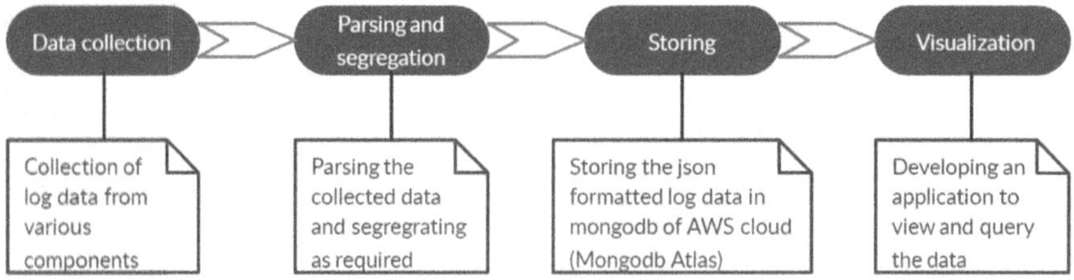

Fig. 3 Phases of implementation

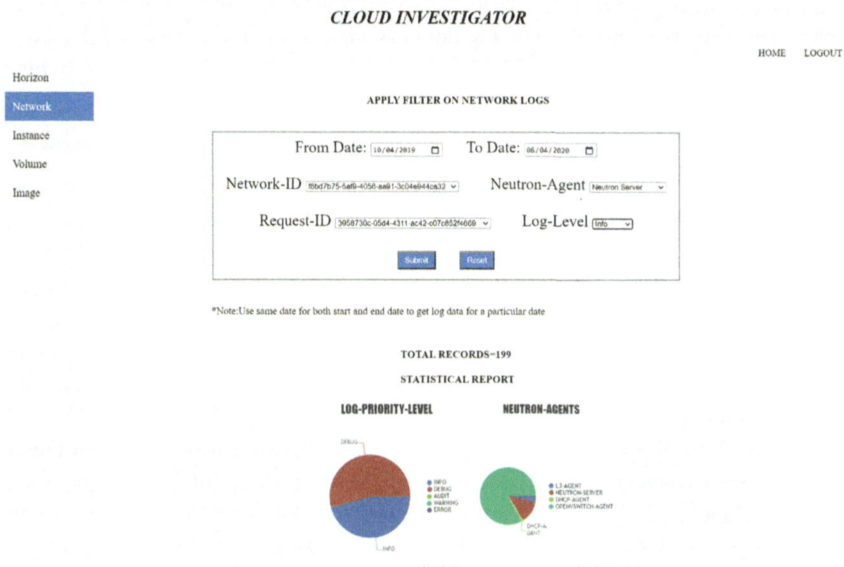

Fig. 4 Interface to filter neutron logs

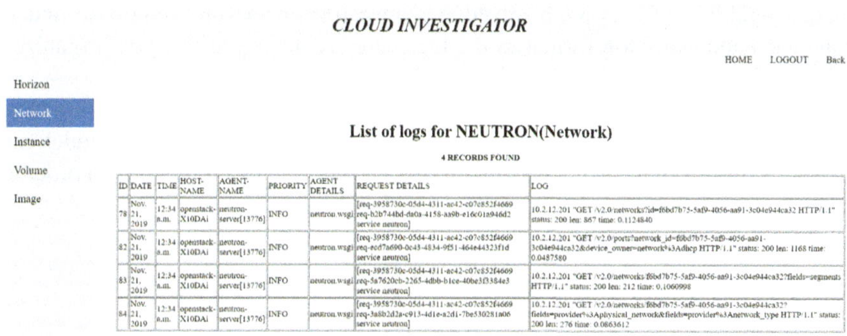

ID	DATE	TIME	HOST-NAME	AGENT-NAME	PRIORITY	AGENT DETAILS	REQUEST DETAILS	LOG
78	Nov. 21, 2019	12:34 a.m.	openstack-X10DAi	neutron-server[13776]	INFO	neutron.wsgi	[req-3958730c-05d4-4311-ac42-c07c852f4669 req-b2b744bd-da0a-4158-aa9b-e16c01a946d2 service neutron]	10.2.12.201 "GET /v2.0/networks?id=f6bd7b75-5af9-4056-aa91-3c04e944ca32 HTTP/1.1" status: 200 len: 867 time: 0.1124840
82	Nov. 21, 2019	12:34 a.m.	openstack-X10DAi	neutron-server[13776]	INFO	neutron.wsgi	[req-3958730c-05d4-4311-ac42-c07c852f4669 req-ecd7a690-0c43-4834-9f51-464e44323f1d service neutron]	10.2.12.201 "GET /v2.0/ports?network_id=f6bd7b75-5af9-4056-aa91-3c04e944ca32&device_owner=network%3Adhcp HTTP/1.1" status: 200 len: 1168 time: 0.0487580
83	Nov. 21, 2019	12:34 a.m.	openstack-X10DAi	neutron-server[13776]	INFO	neutron.wsgi	[req-3958730c-05d4-4311-ac42-c07c852f4669 req-5a7620cb-2265-4dbb-b1ce-40be3f3384e3 service neutron]	10.2.12.201 "GET /v2.0/networks/f6bd7b75-5af9-4056-aa91-3c04e944ca32?fields=segments HTTP/1.1" status: 200 len: 212 time: 0.1060998
84	Nov. 21, 2019	12:34 a.m.	openstack-X10DAi	neutron-server[13776]	INFO	neutron.wsgi	[req-3958730c-05d4-4311-ac42-c07c852f4669 req-3a8b2d2a-c913-4d1e-a2d1-7be530281a06 service neutron]	10.2.12.201 "GET /v2.0/networks/f6bd7b75-5af9-4056-aa91-3c04e944ca32?fields=provider%3Aphysical_network&fields=provider%3Anetwork_type HTTP/1.1" status: 200 len: 276 time: 0.0863612

Fig. 5 Results after submitting the form as shown in Fig. 4

Similarly, the logs of Horizon, Nova, Glance, Cinder are also collected and parsed in the same way as neutron's. The log data can also be filtered based on ID such as image-ID, instance-ID, volume-ID. One of the focuses of this model is to provide a user-friendly graphical user interface (GUI).

4 Conclusion and Future Work

The requirement of cloud forensics has rapidly increased with an increase in malicious events in cloud environments. In this scenario, building a forensic-enabled cloud would enable the investigators to speed up the investigation. The presented model helps an investigator by providing a dashboard to find the evidence in the log data of a cloud environment.

The model we proposed is implemented on a single node deployment of OpenStack cloud environment, similarly, it can be implemented for multi-node deployment and other cloud services by applying few changes. This model can be easily set up and involves very less complexity in its implementation. The cloud crime investigator can directly search for the evidence of the logs information from any remote machine instead of examining the host machine physically. In this way, the task of the investigator is minimized, and the challenges such as decentralization of logs and understandability of logs are resolved. The investigator does not require to understand each and every log format as the logs here are displayed in a well-organized format.

In future, we intend to extend this model to support other cloud service models, i.e., PaaS and SaaS as the current model that support IaaS only. This model can further be improved in such a way that it can also be used by cloud users in order to debug and trace the errors by using the event log data.

References

1. Alenezi, A., Atlam, H.F., Wills, G.B.: Experts reviews of a cloud forensic readiness framework for organizations. J. Cloud Comp. **8**, 11 (2019)
2. Mell, P., Grance, T.: The NIST Definition of Cloud Computing, NIST, SP 800–145 [online] (2011)
3. Ruan, K., Carthy, J., Kechadi T., Crosbie, M.: Cloud forensics: an overview (2011)
4. Ruan, K., Carthy, J., Kechadi, T., Crosbie, M.: Cloud forensics. In: Peterson, G., Shenoi, S. (eds) Advances in Digital Forensics VII. DigitalForensics 2011. IFIP Advances in Information and Communication Technology, Vol. 361. Springer, Berlin, Heidelberg (2011)
5. Dillon, T., Wu, C., Chang, E.: Cloud computing: issues and challenges. In: 2010 24th IEEE International Conference on Advanced Information Networking and Applications, pp. 27–33 Perth, WA (2010)
6. Jain, S., Rana, I.I.T.: A review of cloud forensics issues and challenges. Int. J. Adv. Res. Comput. Sci. Technol. **2**(3). ISSN 2347–8446 (2014)

7. Simou, S., Kalloniatis, C., Kavakli, E., Gritzalis, S.: Cloud forensics: identifying the major issues and challenges. In: Jarke, M., et al. (eds) Advanced Information Systems Engineering. CAiSE 2014. Lecture Notes in Computer Science, vol. 8484. Springer, Cham(2014)
8. Shams, Z., Hasan, R.: Cloud Forensics: A Meta-Study of Challenges, Approaches, and Open Problems. arXiv preprint arXiv:1302.6312 (2013)
9. Liu, C., Singhal, A., Wijesekera, D.: Identifying evidence for implementing a cloud forensic analysis framework. In: Thirteenth IFIP WG 11.3 International Conference on Digital Forensics 2017, NIST (2017)
10. Sibiya, G., Venter, H.S., Fogwill, T.: Digital forensic framework for a cloud environment. In: IST-Africa 2012 Conference Proceedings, Tanzania, 9 May 2012
11. Agrawal, V., et al.: Log-based cloud monitoring system for openstack. In: 2018 IEEE Fourth International Conference on Big Data Computing Service and Applications (BigDataService), pp. 276–281 (2018)
12. Dykstra, Josiah: Design and implementation of FROST: digital forensic tools for the OpenStack cloud computing platform. Digit. Investig. **10**, S87–S95 (2013). https://doi.org/10.1016/j.diin.2013.06.010
13. Yasir Arafat, Md., Mondal, B., Rani, S.: Technical challenges of cloud forensics and suggested solutions. Int. J. Sci. Eng. Res. **8**(8) (2017)

Extended Instruction Set Design for Transcoding Application Using RISC-V ISA

Rohini S. Hongal and Meghana Kambali

Abstract RISC-V is a coeval, refined and flexible open-source instruction set architecture (ISA) basically designed for computer architecture research. It presents a conventional and open base ISA and competitive reference hardware designs that anybody can freely use as a base for further innovation. This makes RISC-V unambiguously suitable for the later phase of development in microprocessor hardware architectures. Transcoding plays a crucial role in a multimedia streaming service. Content owners and publishers require transcoders to modify their data to different bitrates, formats and qualities before streaming them to end users with the best quality of service. In this paper, improving the performance of transcoding application using RISC–V ISA is explained by adding an instruction to RISC-V instruction set architecture. The three stages of a pipeline architecture, i.e., fetch, decode, and execute are verified. Each stage except multiplication operations taking one clock cycle used in DCT operation and the designed system is synthesized using Spartan 6 XC6SLX45 fgg484 device.

Keywords Transcoding · Instruction set architecture (ISA) · Bitrate · Multimedia streaming · RISC-V

1 Introduction

In this present epoch of communication, multimedia plays a very important role. One of the most essential task in reinforcing multimedia systems, like education, space endeavors, telemedicine, crisis management, marketing, military and transportation, is to provide smooth and seamless transmission of audio-visual information, all over the place. But, the properties of media like bitrate and format will be different at both the base station and receiving station. To match these, media processing is the most important requirement. In all of the multimedia processing, discrete cosine transform (DCT) is the most common operation which can transform the signals from

R. S. Hongal (✉) · M. Kambali
Electronics and Communication Department, KLE Technological University, Hubli, India
e-mail: rohini_sh@kletech.ac.in

V. Bhateja et al. (eds.), *Computer Communication, Networking and IoT*, Lecture Notes in Networks and Systems 197, https://doi.org/10.1007/978-981-16-0980-0_7

any domain to frequency domain. This can be implemented either through hardware or through software approach. But, as DCT is used in almost all of the multimedia processing, the operating speed has to be increased to boost the performance of the system.

This paper presents how the operating speed can be improved by adding an instruction, which can perform DCT function. In the current advanced technology, there are many numbers of instruction set architecture (ISA) but most of the ISAs are proprietary, complex and are only popular in certain market domain and are not designed for extensibility. The two most ISAs which can overcome the above drawbacks are Open RISC and RISC-V but, the Open RISC does not support IEEE-754 standard. Hence, RISC-V is the most suitable ISA for this topic. RISC-V has four base ISA, namely RV-32I, RV-32E, RV-64I and RV-128I. RV-32I is used in this project. The next section describes about the literature survey undertaken for the current work.

System-on-chip (SOC) and field programmable gate arrays (FPGAs) may be used to speed up the development steps and prototyping for mission-critical security products, which frequently need the functions like supply chain assurance, trusted intellectual property (IP), secure communications, inspection and certification of register transfer level (RTL) source code and more. FPGA with a specific processor is not at all the best choice to fulfill these challenges, and hence, a new RISC-V open ISA can be used for this. The combination of secure FPGA with IP core for RISC-V open ISA processors with a complete ecosystem permits the architects to enhance the development, accelerate innovations, safeguard software assets using a trusted processor and address all of the rigorous safety and other requirements of latest generation strategic defense system [1–3]. The instruction set architecture (ISA) is a standard, which explains how a computing machine would appear to a compiler or machine language programmer. The ISA defines the: (1) instruction format, forms and ways, (2) memory model and (3) registers, types and data addressing [4].

Multimedia transcoding is chosen as one application which uses operations, like bitrate and format conversions, to change one form of media stream to another. Transcoding allows multimedia devices of different proficiencies and formats to exchange the multimedia data on diverse network platforms like Internet [5, 6]. The widely used multimedia processing algorithm is the discrete cosine transform (DCT), employed in many standards such as MPEG and JPEG. As it shows better energy compaction capability, it is used in most of the speech and image compression. However, for VLSI design, computational requirement of DCT is a substantial consignment for the real-time mobile applications. To enrich the trade of between computation time and hardware requirement, different DCT architectures are recommended to apply signal proprieties. Among these, number of multiplication operations required by signal processing is reduced by Loeffler [7]. The next section describes about the organization of the paper.

The paper is organized as follows: The second segment presents the detailed note on instruction cycle and instruction formats. Section 3 details about transcoder application, the need for transcoding. Section 4 details on the operation used for instruction design, simulation, and synthesis results are presented in Sect. 5. Finally, an inference is presented in last section.

2 Methodology

This section briefly describes about the pipelining architecture, the instruction templates used in RISC-V architecture, opcode map—number of opcodes available for ISA extension and the brief description about the instruction format used in the proposed work.

A. **Instruction Cycle**
   The operation of all the processors is characterized by an instruction cycle, i.e., fetch, decode and execute cycle. As pipelining is a standard feature in reduced instruction set computing, the classic RV-32I pipeline comprises of fetch, decode, execute, memory read and write back. The next section explains about the instruction design and format used in RV-32I.
B. **Instruction format**
   The base ISA contains four basic instruction formats (R/I/S/U) and two variants. SB and UJ type are the two variants which are similar to S and U type as shown in Fig. 1. Each instruction frame format is of 32-bit length and should be lined up on a four-byte boundary in memory. Suppose intended address is not four-byte aligned, an instruction address misaligned exception will be generated on a taken/unconditional branch. A significant characteristic of this encoding format is that these register specifiers when exist, always conquer the same location in the instruction. This allows register fetch and instruction decoding to proceed in parallel.

For the current DCT operation register, R type of format is used as shown in Fig. 2. The funct7 and funct3 fields opt the type of operation; for DCT, it requires three operations to be performed. Hence, the function is defined, and the format is stored in memory.

1. **Instruction fetch:**
   The RISC machines instruction has one cycle latency which means that if the instruction is in the cache, it will be available on the next succeeding clock cycle. A 32-bit instruction will be fetched from the cache during the instruction fetch stage. The program counter (PC) retains the address of the present instruction,

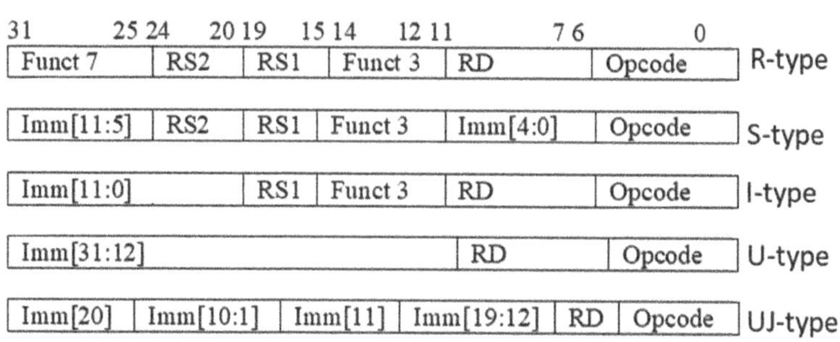

Fig. 1 Instruction format of RV-32I

31 25	24 20	19 15	14 12	11 7	6 0
Funct 7	Rs2	RS1	Funct 3	RD	Op code
7	5	5	3	5	7
0000000	src2	src1	ADD/SLT/SLTU	dest	op
0000000	src2	src1	AND/OR/XOR	dest	op
0000000	src2	src1	SLL/SRL	dest	op
0100000	src2	src1	SUB/SRA	dest	op

Fig. 2 *R* type of instruction format

31 25	24 20	19 15	14 12	11 7	6 0
Funct 7	RS2	RS1	Funct 3	RD	Opcode
0100000	xxxxx	00001	010	010000	1111111

Fig. 3 Extended instruction format

it stores the present instruction into the PC predictor and then forwards the pc to the instruction cache to read out the current instruction; meanwhile, the pc predictor anticipates the address of next instruction by augmenting the PC by 4 (except during branch instruction fetch). The instructions are stored in the binary form, and the designed instruction for the particular operation is stored in the memory and shown in Fig. 3 [8].

2. **Instruction decode:**
   The process of decoding enables the CPU to decide which instruction is to be executed so that the CPU ascertains how many operands it has to fetch in order to execute the instruction. The opcode read from the memory is decrypted to perform the next step.
3. **Instruction execution:**
   The actual computation takes place in execute stage. This stage usually contains a bit shifter, logic and arithmetic unit. In this paper, the execution cycle is performing the discrete cosine transform (DCT) operation. This operation is explained in detail in the next section.
4. **Memory access:**
   The data access at any particular step is performed at memory access stage.
5. **Register write back:**
   During this stage, instructions write their results into register file.

3 Transcoding

The most crucial task in reinforcing multimedia systems, like education, space endeavors, telemedicine, marketing, crisis management, military and transportation, is to provide smooth and seamless transmission of audio-visual data, all over the place. A multimedia system may include several devices (laptops, PCs, smart phones, PDAs, etc.) interconnected via heterogeneous wireless and wireline networks. In such a systems, streaming information initially compressed and conceptualized with a certain format may require some format conversion and bitrate modification so as to permit access to receiving devices of different aptitudes (memory, display and decoder). Hence, a transcoding mechanism is essential to make the data flexible to the networks and client devices of different capabilities. When all the multimedia applications are viewed, DCT is the most common operation performed in all the transcoding applications. Hence, DCT operation is chosen in this particular paper. The next section describes about DCT [9].

3.1 Discrete Cosine Transform (DCT)

Discrete cosine transform (DCT) is the most commonly applied multimedia algorithm, which is employed in many standards. Because of its better energy compaction, it extensively finds application in image and speech processing. However, the computational requirement of DCT is a huge load in the real-time mobile applications for VLSI implementation. Various DCT designs have been contemplated to retain signal properties to improve the trade of between computation time and hardware requirement. Among these, the DCT algorithm suggested by Loeffler has given a new direction to digital signal processing. The proposed method of compression translates the image to be compressed in many lines of 8 pixels and then applies an optimized 1D-DCT algorithm for compression. By lowering the use of number of multiplication units, DCT design is optimized while calculating coefficients [10].

The Loeffler architecture computes DCT results in four steps as shown in Fig. 4. Every step consists of some arithmetic operations. The initial step is carried out by evaluating 4 subtracters and 4 adders while the second one consists of 2 subtracters 2 adders and 2 Mult-Add-Sub (Multiplier, Adder and Subtracter) blocks, and the third

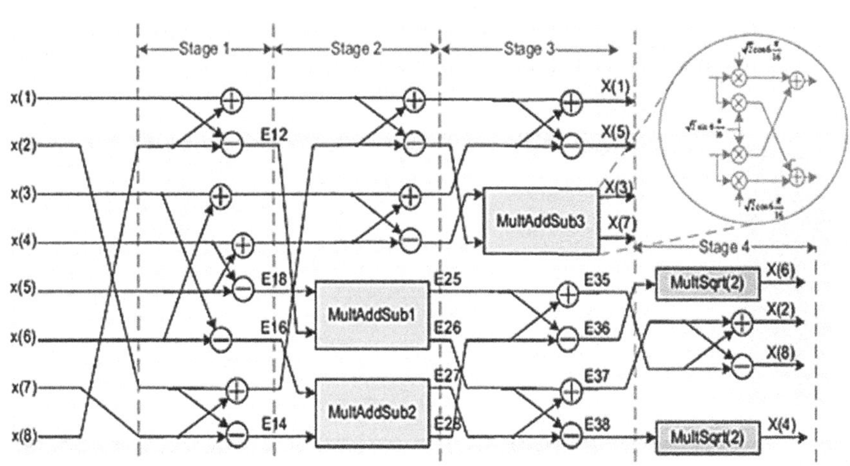

Fig. 4 Loeffler architecture

stage is calculated by evaluating 3 adders, 3 subtracters and 1 Mult-Add-Sub-block, and the final stage uses a adder, a subtracter and 2 MultSqrt(2) blocks to compute the multiplication by sqrt(2). The n point DCT can be calculated using the formula given below

$$X(n) = \sqrt{\frac{2}{N}} C(n) \sum_{K=0}^{N-1} x(k) \cos \frac{(2k+1)n\pi}{2N} \tag{1}$$

where $C(0) = 1/\text{sqrt } 2$ and $C(n) = 1$ if n not equal to 0. Basically, DCT optimization is centered on minimizing the number of arithmetic operators required.

4 Results and Discussion

Register to register instruction frame format is used to perform the DCT operation, the inputs to be given to get the results of 8-point DCT are clock, wr_en and 8-bit input. The clock has to be externally specified as it is not calculated for each stages in this work, and wr_en is used for storing and retrieving the designed opcode to and from the memory, and only one source operand is used to get the 8-bit input from the user, and the second operand is left unused, and the wr_en should have high to low transition (high for storing the opcode to memory and low for retrieving the opcode from the memory) for the proper working of the designed system. The simulated and the synthesized results are shown in Figs. 5 and 6. Figure 5 indicates the simulation results, and Fig. 6 indicates synthesized results of the sub-modules in the main module.

By observing at the simulated result, we can identify that all fetch and decode operations have used one clock cycle to output the result. wr_en control signal is used to supervise the memory, during the logic high of wr_en, the designed opcode is written into memory, and at logic low, the opcode is read from the memory. And

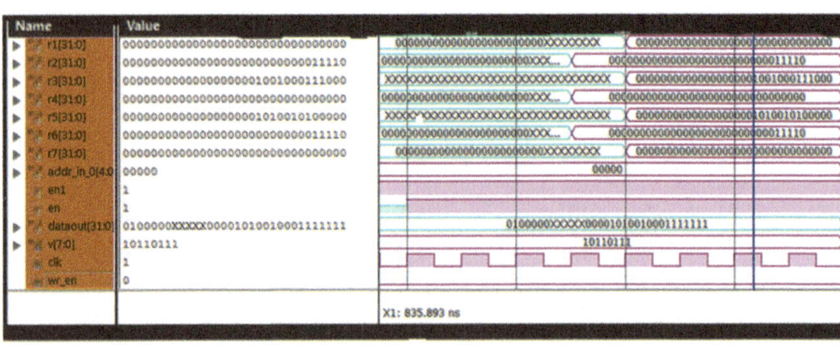

Fig. 5 Simulation result

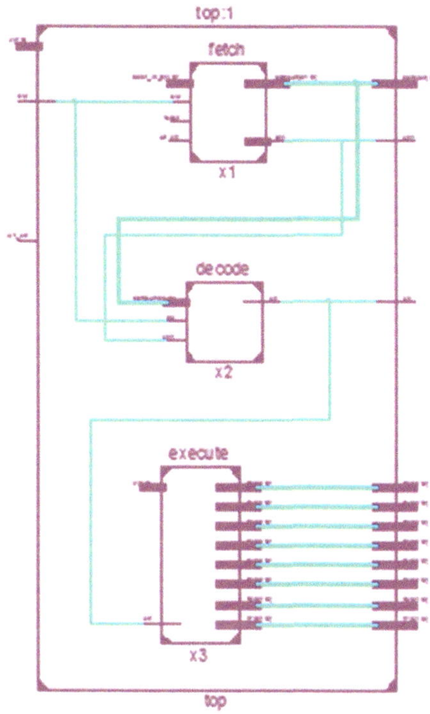

Fig. 6 Sub-modules of top module

all the three stages are executed one after another by passing control variables for each from its previous stage; en1 at first clock cycle indicates the output from fetch stage, and it acts as a controller for the decode stage; en at the second clock cycle indicates the output from the decode stage which acts as a control variable to the next phase, and in the execute stage, the final output which performs the DCT operation is obtained, and here, in this phase, 11 multiplication operation are used, and the result is written back to registers at positive edges of the clock. Execute phase has consumed many clock cycles due to many number of multiplication units, and the outputs are of 32-bits wide because instruction is added to RV-32 ISA which has all the registers of 32-bit wide.

The designed system is synthesized, and it is as shown in Fig. 6. All the pipelining stages are interconnected and act serially as the subsequent modules are controlled by previous models (Fig. 7).

The total power consumed during the instruction execution is 37 mw in that, 1 mw is the dynamic power, and 36 mw is the quiescent power. Timing analysis of proposed work is done and shown in Fig. 8.

The total time taken by the instruction to reach the destination from source is 6.785 ns. The ratio of total number of paths from source to destination ports is 15628:208.

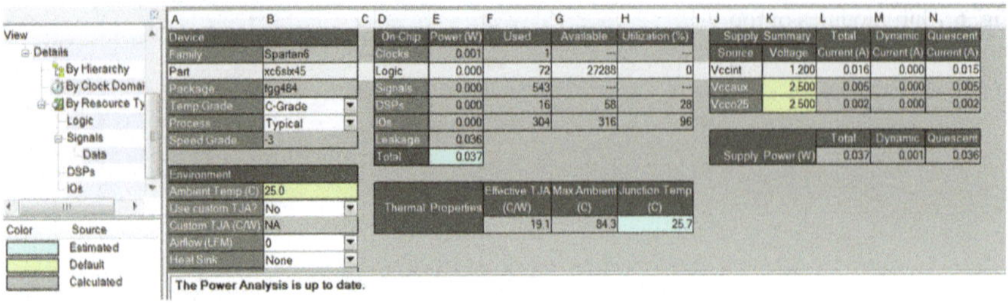

Fig. 7 Power report

```
Clock to Setup on destination clock clk
---------------+---------+---------+---------+---------+
               | Src:Rise| Src:Fall| Src:Rise| Src:Fall|
Source Clock   |Dest:Rise|Dest:Rise|Dest:Fall|Dest:Fall|
---------------+---------+---------+---------+---------+
clk            |    6.785|         |         |         |
---------------+---------+---------+---------+---------+

=========================================================================


Total REAL time to Xst completion: 5.00 secs
Total CPU time to Xst completion: 4.90 secs
```

Fig. 8 Timing report

5 Conclusion

The three stages of a pipelined 32-bit instruction which performs the 8-point DCT and which is compatible with RISC-V ISA have been simulated and synthesized. As RISC-V ISA has a special feature of ISA extensibility, many complex operations of any domain can be designed as an instruction and can be added to architecture which helps the user and increases the speed of calculation which is proved in this work. The results of DCT are performed through instruction designed and match with the manual calculation and are checked on spartan6 FPGA XC6SLX9-CSG324 board.

6 Future Scope

It is possible to design the instructions for the other blocks/operations in multimedia transcoding, so that the computational speed can be increased, and as RISC-V is not over architected, this ISA can be used for the applications of other domain also.

References

1. Using RISC-V in FPGAs for strategic defense system by Ted Marena
2. Waterman, A., Lee, Y., Avizienis, R., Cook, H., Patterson, D., Asanovic, K.: Design of the RISC-V Instruction Set Architecture, spring 2016
3. Wang, Y., Wen, M., Zhang, C., Lin, J.: RVNet: A Fast and High Energy Efficiency Network Packet Processing System on RISC-V. 978-1–5090–4825–0/17/2017. IEEE (2017)
4. Schmalz, M.S.: Organization of Computer Systems: 2: ISA, Machine Language, and Number Systems. Department of Electrical Engineering, National Cheng Kung University, University Road, Tainan, Taiwan 701, ROC, available in https://www.cise.ufl.edu/~mssz/CompOrg/CDA-lang.html#:~:text=The%20instruction%20set%20architecture%20(ISA,%2C%20types%2C%20and%20data%20addressing.
5. Ahmad, I., Wei, X., Sun, Y., Zhang, Y.-Q.: Video transcoding: an overview of various techniques and research issues. IEEE Trans. Multim. **7** (2005)
6. Ma, D.-S., Yang, J.-F., Lee, J.-Y.: Programmable and parallel variable-length decoder for video systems. IEEE Trans. Consum. Electron. **39**(3) (1993)
7. Sun, C., Yang, E.-H.: An efficient DCT-based image compression system based on Laplacian transparent composite model. IEEE Trans. Image Process. **24**(3), 886–900 (2015)
8. Waterman, A., Asanovic, K.: The RISC-V Instruction Set Manual Volume I: User-Level ISA Document Version 2.2. California, University of Berkley, Department of Electrical Engineering and Computer Sciences (2014)
9. Coelho, F.G., Nimmalapalli, S., Dimitrov, V.S., Madanayake, A., Cintra, R.J., Tisserand, A.: Computation of 2D 8×8 DCT Based on the Loeffler Factorization Using Algebraic Integer Encoding. https://hal.archives-ouvertes.fr/hal-01797957
10. Jridi, M., Alfalou, A.: A low-power, high-speed DCT architecture for image compression: principle and implementation. In: Conference: VLSI System on Chip Conference (VLSI-SoC), 2010 18th IEEE/IFIP (2010). https://doi.org/10.1109/VLSISOC.2010.5642678

A Novel Image Functioned Framework for Steganography Using PS-DCT Approach

Pallavi Khare

Abstract This research paper presents a unique approach of image primarily based steganography method, wherever pseudo-random sequence generators operate is employed to randomize the mapping sequence in an exceedingly pseudo-random fashion. The embedding method of the key message is completed per the pseudo-random sequence, and extraction is completed exploitation identical sequence at the opposite side. At the bottom level, 2D-DCT is employed to cover the bits of secret message within the stego image by exchanging the LSBs of DCT coefficients. Since in previous paper, the LSB substitution technique is already used; here during this paper, the previous concept is extended and used DCT rather than LSB technique. A replacement text randomization algorithmic rule (4-2 swap text randomization algorithm) is additionally accustomed to modify the key message before embedding.

Keywords Steganography · Security · DCT · Pseudo-random

1 Introduction

Steganography is the art as well as the science of concealing info into another covering media in such the simplest way that no one except the supposed recipient is aware of concerning the key message and retrieving it. Steganography (means—covered writing || in Greek) is associated with nursing previous art that has been used since the Golden Age of Balkan country wherever some practices were recorded like writing a message on a wood table then covering it with wax. Different techniques use invisible ink, microdots, changing channels, and character arrangement. Steganography may be a form of cryptography within which the key message is hidden in a very digital image. Steganography differs from cryptography within the sense that wherever cryptography focuses on keeping the contents of a message secret, whereas steganography focuses on keeping the terribly existence of the message secret [1–3].

Various image primarily based steganography methodology specifically least significant bit (LSB), pixel-value differencing (PVD), gray level modification (GLE),

P. Khare (✉)
ECE Department, Matrusri Engineering College, Hyderabad, India

V. Bhateja et al. (eds.), *Computer Communication, Networking and IoT*, Lecture Notes in Networks and Systems 197, https://doi.org/10.1007/978-981-16-0980-0_8

and also the methodology projected by Ahmed et al. has been in brief mentioned within the following section.

1.1 Data Hiding by Least Significant Bit (LSB)

The most standard and common techniques are predicated on manipulating the least significant bit (LSB) [4–6] planes by directly replacing the LSBs of the cover image with the message bits. LSB methods generally come through high capability; however, unfortunately, LSB insertion is susceptible to slight image manipulation like cropping and compression.

1.2 Data Hiding by Pixel-Value Differencing Method

Another method proposed by Wu and Tsai, the pixel-value differencing (PVD) method can successfully provide both high embedding capacity and outstanding imperceptibility for the stego image. Based on PVD method, various approaches have also been proposed. Among them, Chang et al. propose a new method using tri-way pixel-value differencing.

1.3 Data Hiding by GLM

In 2004, Potdar et al. [6] propose gray level modification (GLM) technique which is used to map data by modifying the gray level of the image pixels. Gray level modification steganography is a technique to map data (not embed or hide it) by modifying the gray level values of the image pixels.

1.4 Data Hiding by the Method Proposed by Ahmad et al.

In this work, a novel steganography technique for concealing data among the abstraction domain of the grayscale image has been planned. The planned approach works by dividing the quilt into blocks of equal size and so embeds the message within the fringe of the block reckoning on the quantity of ones in left four bits of the pixel. The paper has been organized in some sections. Section 1 discussed a short introduction and a discussion on the related works in the relevant field. Section 2 outlined concerning some steganographic strategies. Consequently, Sect. 3 delineates the planned procedure. Section 4 delineates the general procedure via associate degree example. Next, Sect. 5 of the paper is concerning and analyzing the experimental result. The last Sect. 6 concluded on the research paper.

2 Steganographic Methods

There are mainly three ways to hide a digital message in a digital cover [4, 7, 8]. **(A) Injection, (B) Substitution, (C) Generation of New files**. A cover is generated for the sole purpose of concealing a secret message. A sender creates image of one thing original that may be passed to receiver; the innocent picture is that the cover that has the mechanism for transferring the message. Digital images will be dealt in two totally different domains. One is the abstraction domain, wherever the image is taken into account because the two-dimensional array of pixels is oriented in an exceedingly table format. On the opposite hand, within the frequency domain, images are thought of as the digital signal transmitted through the physical media. Another method proposed by Wu and Tsai, the pixel-value differencing (PVD) method can successfully provide both high embedding capacity and outstanding imperceptibility for the stego image. During this research paper, 2D discrete cosine transform (2D-DCT) technique [7] is employed rather than least significant bit (LSB) substitution technique as the basic hiding technique [7].

3 Proposed Scheme of Steganography

In this paper, a replacement image primarily based steganography scheme is presented within which both the key message yet because the cover image is split into a number of small but equal sized segments [7]. Once for that segmentation, separate segments of the key message are embedded by using LSB technique at the DCT coefficients of those image segments; however, a selected section of the cover image for a selected section of text is chosen in an exceedingly pseudo-random fashion. This technique makes the embedding or insertion method tough, however, imposes additional security [7]. Analysis might facilitate to discover that steganography has been silent on the cover image; however, attackers cannot be able to extract the key message from the cover image without knowing the parameters employed in this method.

A. ***Embedding Method***

This new steganography technique is block mapping technique. The complete steganography method has two components. One half is to insert or hide the key text message within a canopy image, desirable colored image. This method is completed at the sending end. Another half is employed at the receiving end to extract the first secret text message from the stego image.

STEP 1:

The secret message is taken as the primary input. This is the secret message to be hidden inside the cover image. The total number of characters (Ch) of the secret message is counted.

STEP 2:

A unique integer number means thresold value which is kept unique for every stegnography process. It takes a one-time password, a set key (For advanced security purpose), and the secret message as input looking on the total range of characters, and therefore, the strength of password and threshold is calculated.

STEP 3:

This steganography algorithmic rule segments the entire text into some equal sized blocks known as cells containing equal range of characters. The size of each cell (denoted as C) depends on the amount of characters within the secret message and, therefore, the value of threshold. The dimensions are calculated by adding the digits of the full range of characters (Ch) within the text and threshold (T); i.e., the amount of character per cell or size of cell is denoted by

$$C = \text{sum of digits of Ch} + T \quad (1)$$

STEP 4:

To calculate the entire range of cell(s) (denoted by C_T), divide the entire no. of characters (Ch) by the size of each cell (denoted as C). The size of every cell (C) might not always be an element of the entire range of characters (Ch). Thus, take the ceiling value of the result and add some redundant bit(s) when the last character of the last cell to equalize it with others. Total numbers of cells are calculated by

$$C_T = \left[{}^{\text{Ch}}\!/_{C} \right] \quad (2)$$

STEP 5:

This calculation depends on stowing away of content into measure up to estimated square pieces of cover picture. That implies, the cover picture must be partitioned or portioned into various equivalent estimated consistent square pieces. Each intelligent piece will shroud a specific cell of content. This number gives the measurement of each sensible piece of picture that conceals a unit of content, i.e., right roof of foundation of b:

$$D = \left[\sqrt{b} \right] \quad (3)$$

STEP 6:

The measurement is then ascertained as $D \times D$. This is the measure of each square formed legitimate piece of the cover picture, and 2D-DCT is connected on each of those pieces. To make the concealing procedure fruitful, $D \times D$ must be more noteworthy than or equivalent to the quantity of bits in a unit (b), i.e.,

$$D \times D \geq b \text{ or } D2 \geq b \quad (4)$$

STEP 7:

D is the measurement of each legitimate square piece. Subsequent to isolating tallness and width of the cover picture by D, and overlooking the partial piece of the outcomes, figure $M = H/D$ and $N = W/D$. M and N are the quantity of intelligent divisions alongside tallness and width of the photo and length of every division is D. The aggregate number of consistent squares of picture is indicated by $B = M \times N$, where B is add up to number of Cells. B must be more noteworthy than CT, i.e., $B >$ CT

$$M = H/D \text{ and } N = W/D \text{ and } B = M/N \quad (5)$$

In this progression, the whole cover picture is fragmented into B number of legitimate square pieces. On each of those legitimate $D \times D$ square pieces, 2D-DCT is connected, and coefficients are ascertained.

Mapping is done in pseudo-arbitrary mold so that amid extraction, a similar pseudo-irregular grouping can be created, yet it will be hard to foresee any example of mapping. The mapping capacity relies upon size of cell (C), threshold (T), and aggregate number of blocks (B). To create a pseudo-arbitrary succession that will indicate the specific locales for every unit of content, area recognizable proof numbers or district ID number, likewise called as piece numbers must be remarkable, whatever the quantity of squares ought to be. Another confinement must be kept up here which is the upper bound of the produced number. The qualities must not surpass the aggregate number of squares (B). This suggests that the capacity must be a modulus capacity of B or ($B + 1$). To produce the pseudo-arbitrary grouping having adequate spaces between the numbers with the goal that likelihood of getting covered or to maintain a strategic distance from crashes between those numbers, a base separating is presented by including energy of C and threshold esteem. Exploratory outcome demonstrates that the mapping capacity works best if the articulation will be as per the following:

$$B_i = \left[\left(T^i \times C^i\right) \bmod (B - i)\right] + i \quad (6)$$

where $i = 1, 2, 3\ldots C_{\mathrm{T}}$. B_i is a pseudo-arbitrary number created for a particular estimation of C_{Ti}. Bi is the square number of unit number C_{Ti}.

STEP 8:

Every unit of content is changed over to bit stream, and afterward, the bit stream is randomized utilizing another randomization system called 4-2 swap content randomization method. Utilizing this randomization strategy, the randomized grouping can be changed over again into the first arrangement. The capacity likewise takes threshold and passkey as contribution for doing randomization. This makes the procedure one time and solid.

STEP 9:

Subsequent to doing this randomization of the bit stream, inserting of the mystery message is finished by supplanting the LSB of each DCT coefficient of each chose intelligent piece of cover picture.

STEP 10:

The stego-picture in the change area is then changed over into the spatial space by applying IDCT. The stego-picture got is like the cover picture, and the contrast between the two pictures is not recognizable by the human eye [7].

B. ***Retrieval Method***

The recovery of the mystery message from the stego-picture is finished by following a similar calculation connected for installing the mystery message. Every one of the means depicted before is rehashed for the stego-picture less than desirable end with the exception of the IDCT technique, on the grounds that there is no compelling reason to change over the picture from change area to spatial space. After the division of the picture, 2D-DCT is connected on the stego-picture [7]. At that point, LSBs are separated from the DCT coefficients of each intelligent square. After extraction of the whole piece stream of the mystery message, the 4-2 swap randomization work is connected on it. As this capacity is a reversible capacity, it re-organizes the bit stream into its unique grouping. Presently from this bit stream, the mystery message is produced.

C. ***Utilization of 2D-DCT on Cover Image***

The essential thought to shroud data in the recurrence area is to modify the size of the majority of the DCT coefficients of cover picture. The 2-D DCT change over the picture hinders from spatial space to recurrence area. The schematic piece chart of the entire procedure is given in Fig. 1.

Let $I(x,y)$ denote an 8-bit grayscale cover image with $x = 1, 2, \ldots, M$ and $y = 1, 2, \ldots, N$. This $M \times N$ cover image is divided into $D \times D$ blocks, and two-dimensional (2-D) DCT is performed on each of $L = M \times N/D^2$ blocks.

The mathematical definition of DCT is:

Forward DCT:

$$F(u, v) = \frac{1}{4}C(u)C(v)\sum_{x=0}^{7}\sum_{y=0}^{7} f(x, y)\mathrm{Cos}\left[\frac{\pi(2x+1)u}{16}\right]\mathrm{Cos}\left[\frac{\pi(2y+1)v}{16}\right] \tag{7}$$

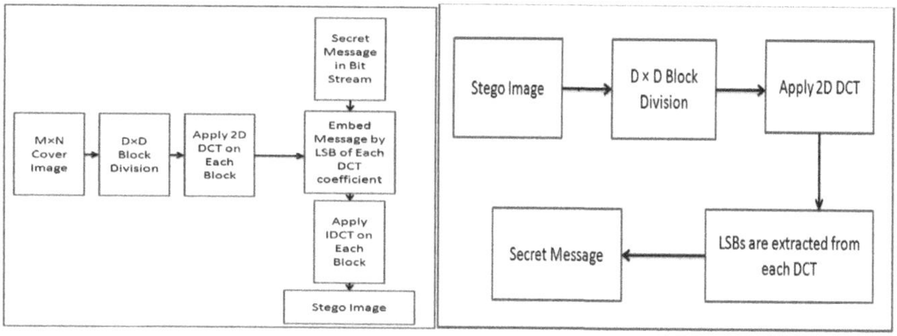

Fig. 1 Insertion and retrieval of a secret message into a cover image

For $u = 0, \ldots, 7$ and $v = 0, \ldots, 7$ Where $C(k) = \begin{cases} \frac{1}{\sqrt{2}}, & \text{for } k = 0 \\ 1, & \text{otherwise} \end{cases}$

Inverse DCT:

$$f(x, y) = \frac{1}{4} \sum_{u=0}^{7} \sum_{v=0}^{7} \begin{matrix} C(u)C(v)F(u, v)* \\ \text{Cos}\left[\frac{\pi(2x+1)u}{16}\right]\text{Cos}\left[\frac{\pi(2y+1)v}{16}\right] \end{matrix} \quad (8)$$

D. ***4-2 Swap Text Randomization Method***

The secret message is randomized before the embedding process to impart more security to the entire steganography process. The randomization algorithm is described here.

The secret message is taken as input, and the length of the message is calculated. The entire message is segmented into some equal sized cells and padding space if required. Number of cells depends on the total no. of characters (Ch) and size of cell (C).

Evaluate the ASCII values for all characters of each part for the secret message.

Modify the ASCII values by using the passkey and threshold value.

Convert each ASCII value into its corresponding 8bits binary value for each character.

Swap left most 4 bits with the right most 4 bits binary value.

Swap first two bits with last two bits for efficient randomization for binary image.

Convert each 8 bits binary value into its corresponding hexadecimal value.

4 Implementation of an Example

A. ***Illustration of 4-2 swap Text Randomization Method***

The content randomization strategy is delineated with a case in this area. For instance, consider the mystery message as—the college foundation of innovation at burdwan.

Here, add up to number of characters (Ch) = 49, assume the computed threshold (T) = 7. The measure of cell (C) = 4 + 9 + 7 = 20, add up to number of cell required = roof (49/20) = 3, min number of character must contain every cell = roof (49/3) = 17. To level every cell with same number of character requires ((17 * 3) – 49) = 2 underscores (_) or spaces ( ) to add after the last character of the mystery message. Presently, every one of the cells contains measure up to number of characters.

Cell 1 = "the university in" = 17
Cell 2 = "statute of techno" = 17
Cell 3 = "logy at burdwan_" = 17.
Presently, for cell 1, the message is "the university in."

The accompanying table demonstrates that after alteration of each character, the comparing ASCII esteems are (Table 1).

In this mold, the changed decimal likeness ASCII benefits of comparing characters will produce. Thus, another two tables for Cell2 and Cell3 individually will likewise create. So, the mystery content is randomized utilizing an interesting capacity which called as 4-2 swap content randomization strategy. In the wake of getting the decimal equal estimations of the considerable number of characters, a bit stream is created from each of them and that bit stream is installed by supplanting the LSBs of all DCT coefficients of the cover picture.

Table 1 Result of 4-2 swap text randomization method

Character	Modified ASCII	8 bits swapped binary equivalent	4 bits swapped binary equivalent	First 2 bits swap with last 2 bits	Decimal equivalents
T	129	10000001	00011000	00011000	024
h	117	01110101	01010111	11010101	213
e	114	01110010	00100111	11100100	228
	45	00101101	11010010	10010011	147
u	130	10000010	00101000	00101000	040
n	123	01111011	10110111	11110110	246
i	118	01110110	01100111	11100101	229
v	131	10000011	00111000	00111000	056
e	114	01110010	00100111	11100100	228
r	127	01111111	11110111	11110111	247
s	128	10000000	00001000	00001000	008
i	118	01110110	01100111	11100101	229
t	129	10000001	00011000	00011000	024
y	134	10000110	01101000	00101001	041
	45	00101101	11010010	10010011	147
i	118	01110110	01100111	11100101	229
n	123	01111011	10110111	11110110	246

Fig. 2 Secret text

B. Algorithm Used for Embedding the Secret Message
The algorithm used for embedding process is illustrated in this section. Here, the calculation of the threshold value is also shown.

STEP 1 Take a string as input and calculate the length. Calculate the threshold T $T \leftarrow C\%\ 9$, $T \leftarrow T + P_{\text{L}}$, $T \leftarrow T\%\ 9$, $T \leftarrow T + [(9 - X)((P_{\text{L}} - P_{\text{M}})/P_{\text{L}})]$ (Where P_{L} = original password length and P_{M} = minimum password length/passkey).

STEP 2 Ascertain the measure of unit (u) by including the digit(s) of the string length and the threshold (T). Ascertain the aggregate number of unit (U) $U \leftarrow [C/u]$.

STEP 3 Ascertain bits per unit (B), $B \leftarrow u \times 7$ (Each character is of 7 bits)

STEP 4. Ascertain the measurement of each legitimate piece (D), $D \leftarrow \sqrt{b}$

STEP 5. Figure the total dimension (D^2) *and* $D^2 \geq B$. Take a picture as information and compute number of consistent division alongside height (H) and width (W) $M = H/D$. Figure the total no. of logical pieces (B), $B \leftarrow M \times N$. Create the piece numbers by $B_i \leftarrow \left[\left(F^i \times u^i\right) \bmod (B - i)\right] + i$, where $i = 1$ ton.
Modify the ASCII values by using the passkey and threshold value. Evaluate the ASCII values for all characters of each part for the secret message.

STEP 6 Convert each ASCII value into its corresponding 8 bits binary value for each character. Swap left most 4 bits with the right most 4 bits binary value. Swap first two bits with last two bits for efficient randomization for each binary value.

STEP 7 Convert each 8 bits binary value into its corresponding hexadecimal value. Get DCT coefficients of all the $D \times D$ logical segments of the cover image.

STEP 8 Replace the LSBs of DCT coefficients by the bits of the modified bit stream of secret message generated by the 4-2 swap randomization method (Figs. 2, 3 and 4).

This algorithm illustrates the whole embedding process at the sending end. After embedding the secret message, IDCT is applied [7] on the image to convert it from transform domain to spatial domain, and the output stego image is generated.

5 Result Analysis

Exploratory outcome demonstrates that 2D-DCT grants more unwavering quality and security as opposed to the LSB substitution method. Some assessment parameters are

Fig. 3 Cover image

Fig. 4 Stego image

utilized to gage the execution of the calculation. The regular assessment parameters are [4, 7] MSE, PSNR, capacity, and so on.

Mean Square Error (MSE): It is defined as the square of error between cover image and stego image. The distortion in the image can be measured using MSE and is calculated using the following equation

Table 2 Experimental result

Image block size	Technique	MHC (%)	PSNR (dB)	MSE
150 × 150	LSB	55	39.56	32.45
	DCT	57	37.82	32.38
200 × 200	LSB	46	38.58	32.28
	DCT	48	32.62	32.14
400 × 400	LSB	34	35.90	28.98
	DCT	34	32.46	26.87

$$\mathrm{MSE} = \left[\frac{1}{N \times N}\right]^2 \sum_{i=1}^{N} \sum_{j=1}^{N} \left(X_{ij} - \overline{X_{ij}}\right)^2 \tag{9}$$

where

X_{ij} = The intensity value of the pixel in the cover image
$\overline{X_{ij}}$ = The intensity value of the pixel in the stego image
N = Size of the image.

Peak Signal to Noise Ratio (PSNR): It is the measure of quality of the image by comparing the cover image with the stego image; i.e., it measures the statistical difference between the cover and stego image and is calculated using the following equation.

$$\mathrm{PSNR} = 10 \log_{10} \frac{255^2}{\mathrm{MSE}}\ \mathrm{db} \tag{10}$$

Capacity: It is the size of the data in a cover image that can be modified without deteriorating the integrity of the cover image. The steganography embedding operation needs to preserve the statistical properties of the cover image in addition to its perceptual quality. Capacity is represented by bits per pixel (bpp) and the maximum hiding capacity (MHC) in terms of percentage [7].

Experimental result shows that the new approach using DCT coefficient is better than that of the earlier method using LSB substitution technique. Taking the picture of Lena and converting it into a 800 × 600 dimensional image, the following table shows the MSE values for three different block sizes (Table 2).

6 Conclusion

This steganography calculation infers a multi-level security in light of the fact that before installing the characters of the mystery instant message into a cover picture utilizing 2D-DCT, the whole content is divided first into different cells, and each fragment of mystery message is randomized at bit level. This randomization or change is finished utilizing a one of a kind capacity which is a reversible capacity, i.e., the capacity can re-create the first succession from the randomized yield. Again, every one of the fragments of content is arbitrarily embedded into various areas of the cover picture. A bit-stuffing technique is utilized to make the quantity of characters of every unit measured up to. Another pseudo-irregular succession generator work is utilized to produce a pseudo-arbitrary arrangement to install each of the unit of mystery message into the sensible square areas or pieces of the cover picture in a pseudo-arbitrary mold. Utilizing a similar pseudo-arbitrary grouping, extraction should be possible to get the first message. As specified before, the information esteem named threshold (*T*) of the pseudo-irregular arrangement generator work relies upon the mystery secret key and a passkey known to the two expected gatherings (sender and beneficiary) as it were. As these are mystery esteems, another level of security is likewise forced here. This technique can be enhanced again by expanding the quality of edge esteem era work, altering the strategy for ascertaining the aggregate number of characters in every unit of message and furthermore by forcing the randomization work (4-2 swap randomization technique) at the character level and in addition at the bit level. Here, 2D-DCT strategy is inferred rather than LSB substitution method in this broadened rendition of the past paper.

References

1. Attabya, A.A., Mursi Ahmedb, M.F.M., Alsammak, A.K.: Data hiding inside JPEG images with high resistance to steganalysis using a novel technique: DCT-M3. Ain Shams Eng. J. (2017)
2. Fatnassi, A., Gharsellaoui, H., Bouamama, S.: A novel approach for embedding information in image cover media with security. In: World Symposium on Computer Applications and Research (WSCAR), added in IEEE explore: 22 Dec 2016 (2016)
3. Shelke, S.G., Jagtap, S.K.: A novel approach: pixel matching based image steganography. In: International Conference on Pervasive Computing (ICPC), 2015 added in IEEE Explorer: Sept 2015
4. Pujari, S., Mukhopadhyay, S.: An image based steganography scheme implying pseudo-random mapping of text segments to logical region of cover image using a new block mapping function and randomization technique. Int. J. Comput. Appl. (0975–8887) **50**(2) (2012)
5. Yang, C., Wang, S.: Transforming LSB substitution for image-based steganography in matching algorithms. J. Inf. Sci. Eng. **26**, 1199–1212 (2010)
6. Shiva Kumar, K.B., Raja, K.B., Chhotaray, R.K., Pattanaik, S.: Bit length replacement steganography based on DCT coefficients. Int. J. Eng. Sci. Technol. **2**(8), 3561–3570 (2010)
7. Bhardwaj, R., Sharma, V.: Image steganography based on complemented message and inverted bit LSB substitution. Elsevier Procedia Comput. Sci. **93**, 832–838 (2016)
8. Huang, W., Zhao, Y., Rong-Rong, N.: Block-based adaptive image steganography using LSB matching revisited. J. Electron. Sci. Technol. **9**(4) (2011)

A Novel Approach to Find Minimum Cost Spanning Tree (MST) of a Graph

Rayudu Srinivas, T. Rama Reddy, R. V. S. Lalitha, and Shaik Vahida

Abstract Finding minimum cost spanning tree plays a major role in many applications like transportation, communication, network, design of computer, etc. There are a number of algorithms proposed in literature but popularly used algorithms are Prim's and Kruskal's algorithms. The time complexity of these algorithms is not linear, so still finding MST in linear time is open to do research. These algorithms require the edges of the graph in sorted order. In this paper, a novel approach is proposed to find the MST for given graph which processes the edges based on levels of the graph. This proposed method requires level-wise sorted order of edges. This proposed a level-wise sorting of edges minimizes the total execution time required to execute the overall procedure. Another advantage with this approach is, cycle checking is not required in phase 1 of algorithm.

Keywords Cost · Graph · Kruskal · Minimum cost spanning tree · MST · Prims · Tree

1 Introduction

A spanning tree of a graph is a tree that contains all the vertices (V) of the given graph and minimum number of edges E (if n is the number of vertices in the graph, then a number of edges to connect these E are $n - 1$) to connect these vertices. The number of spanning trees can be constructed for a given graph based on number of edges and vertices. There are a number of methods proposed to construct minimum and maximum cost spanning trees. The commonly used methods are Prim's [1–3] developed in 1930 by Jarnik and Kruskal's [4–6] developed by Joseph Kruskal in

R. Srinivas (✉) · T. Rama Reddy
Aditya Engineering College, Surampalem, Andhra Pradesh, India
e-mail: srinivas.rayudu@aec.edu.in

R. V. S. Lalitha · S. Vahida
Aditya College of Engineering and Technology, Surampalem, India
e-mail: rvslalitha@acet.ac.in

V. Bhateja et al. (eds.), *Computer Communication, Networking and IoT*, Lecture Notes in Networks and Systems 197, https://doi.org/10.1007/978-981-16-0980-0_9

1956 [7]. These methods require the edges in nondecreasing order. Prim's algorithm constructs the spanning tree by initially considering all the vertices and by adding edge by edge based on the procedure of the algorithm until a spanning tree is constructed. In Kruskal's algorithm the least-cost edge is considered first and keeps on adding the next least-cost edge which does not violate the property of spanning tree until $n - 1$ edges are added. These algorithms execute the same procedure at least $n - 1$ times, where n is the number of vertices in the graph. The main drawback with these methods is, if n is very large and a number of edges are almost equivalent to a number of vertices, then also the algorithms proceed in the same manner. Another classic algorithm is called Boruvka's algorithm. It was first published by Otakar Boruvka in 1926 [8]. This algorithm was again discovered in 1938 by Choquet, and then it rediscovered in 1951 by Florek, Lukasiewicz, Perkal, Steinhaus, and Zubrzycki, and then again in 1965 discovered by Sollin [2, 7, 9]. This algorithm was improved by Yao [10]. A similar but slightly more practical algorithm was proposed by Cheriton and Tarjan [11]. Fredman and Tarjan [12] proposed new algorithm with the invention of an efficient form of heap, called the Fibonacci heap (F-heap). In addition to classical methods, a number of algorithms were proposed in literature. The detailed surveys are given by Pierce, Maffioli, and Graham and Hell [13–15]. Recently, Karger et al. [16] proposed an algorithm. Theoretical expected time of this algorithm is linear [17]. The contributions to this paper are: (i) proposed a new method to pruning costly edges when number of vertices are almost equivalent to number of edges, (ii) in both Prim's and Kruskal's algorithms, consider the cycle checking step only after adding two edges to edge list of a spanning tree, (iii) proposed novel algorithm and improvement to this algorithm to find MST. These contributions are explained in Sect. 2.

The rest of the paper is organized as follows. The proposed novel algorithm and contribution are explained in Sect. 2. Application of the proposed algorithm on an example graph is presented in Sect. 3. Concluding remarks are given in Sect. 4.

2 Proposed Algorithm

2.1 Algorithm to Prune Costly Edges

When given graph has number of edges almost equivalent to number of vertices, then this method can be used. This method prunes the costly edges until the tree has $n - 1$ edges without violating the spanning tree properties.

Algorithm

Step 1: d = 0 // consider the edges in non decreasing order.
Step 2: calculate number of vertices |V|= N, and number of edges |E|= e.
Step 3: calculate D = e − N.
Step 4: while d < D + 1.

i Select next highest cost edge from edge list
ii. If removal of this edge creates disconnected components, then add this edge to list of spanning tree edges, otherwise discard this edge and update d value as d + 1.

Step 5: add all other edges to list of spanning tree edges which are not considered for decision making.

2.2 *Loop Testing*

In a graph, if self-loops are not considered, then there must be at least three edges are required to form a loop. So, while applying the procedure to construct spanning tree using both Prim's and Kruskal's algorithms directly, add the first two least-cost edges to solution list because no cycle is formed by using two edges. Third edge onwards, loop should be tested.

2.3 *Novel Algorithm to Find MST*

The proposed algorithm to find a minimum cost spanning tree is given below.

Step 1: Start.
Step 2: Initially, order all the edges based on the cost of the edges in each level starting from a vertex. (Suggestion: consider the vertex as source node which has incident edge that has least cost.)
Step 3: Consider level-wise ordering of edges using heap sort.
Step 4: Select one least-cost edge from each level and add to E and update edge count value.
Step 5: Repeat step 5 until spanning tree edge list E has n − 1 edges.

i. Select unselected least-cost edge from each level and add to E if a cycle is not formed and update edge count, otherwise discard.

2.4 *Time Complexity of the Algorithm*

The proposed algorithm time complexity is not linear but overall time complexity will be less than the time required to execute Prim's and Kruskal algorithms. In each level, sort the edges using heap sort [18–21]. In general, the best sorting has time complexity $O(n\log n)$. If a graph contains n edge and if there are m levels, then there will be on average n/m edges in each level. The complexity of sorting edges in each level is $O((n/m)\log(n/m))$. So, sorting of all the edges in all the levels is $O(m \cdot (n/m)\log(n/m)) = O(n\log(n/m))$.

The proposed algorithm adds at most *m* edges in phase 1 (step 4), and this phase does not require any method to check for cycle. In the second phase (step 5), it considers at most m edges but this requires for checking cycles.

2.5 Proposal to Improve the Performance Novel Algorithm

The performance of the proposed algorithm can be improved by considering minheap for considering least-cost edge. Initially, edge lists will be sorted level-wise like proposed algorithm.

In phase 1, select one least-cost edge from one level and create minheap. Select root of minheap and add to spanning tree edge list if addition of this edge does not form a cycle. Add new least-cost edge to minheap from the level where root node is formed in the previous stage. Repeat this process until spanning tree edge list has n-1 edges, where n is the number of vertices.

3 Results

The proposed algorithm is applied on the graph shown in Fig. 1. The number of vertices in the graph is $|V| = 12$ and the number of edges considered is $|E| = 17$ (Fig. 2).

From level L0 to L1, the least-cost edge is (1, 2) so it is added to the edge list *E*.

From L1 to L2, L2 to L3, L3 to L4, the least-cost edges in each stage are (2, 4), (4, 10), (8, 11), and in level L4, there are edges, and the edges are also considered further and select least-cost edge (11, 10).

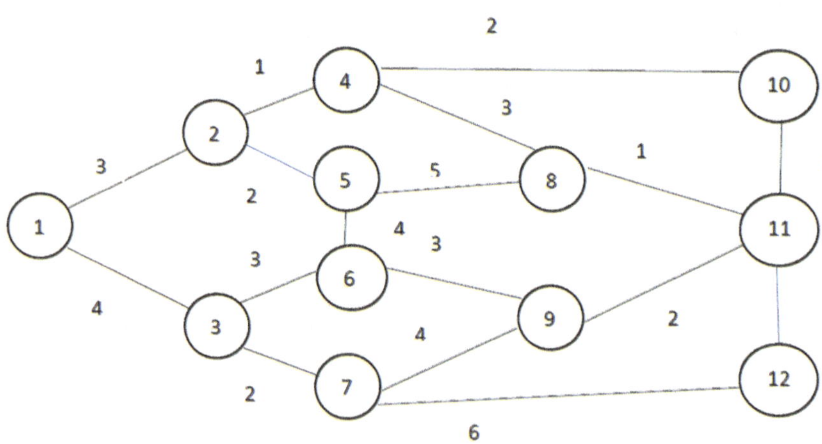

Fig. 1 Example graph with $|V| = 12$ and $|E| = 17$

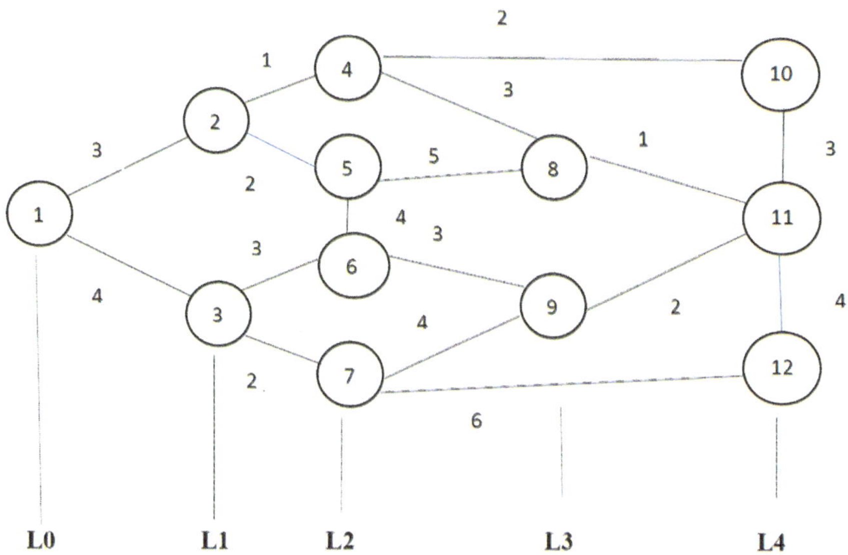

Fig. 2 Example graph with level numbers

After adding these edges, the edge list has edge $E = \{(1, 2), (2, 4), (4, 10), (8, 11), (11, 10)\}$ and graph looks like Fig. 3.

The advantage of the method is, in first phase, no need to check for cycles. Only from the second phase onward, addition of edge will form a cycle or not to be checked.

Phase 2 procedure is repeated until E has $n - 1$ edges. In first iteration of phase 2, the second least-cost edge of each level will be selected, and this edge is added to edge list if a cycle is not formed as shown in Fig. 4.

$L0$–$L1$ no edges are left no need to consider. $L1$–$L2$, edge (3, 6) and (3, 7) are left. The least-cost edge is (3, 7) so add this edge to E. Even the cost of (3, 6) that edge will not be added because it will form a cycle. After adding edge (3, 7) to E, the E will have $n - 1$ i.e. 11 edges. So, the procedure is stopped.

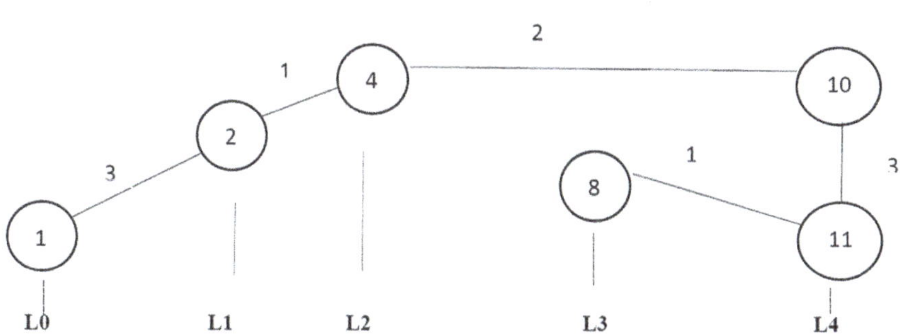

Fig. 3 Spanning tree after phase 1

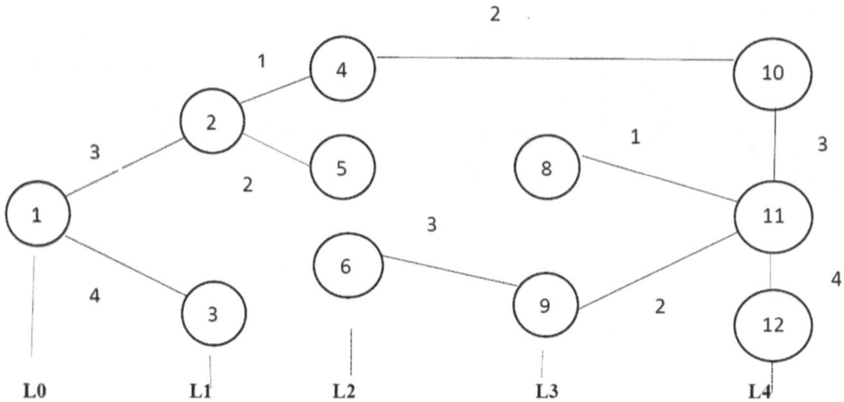

Fig. 4 Spanning tree after phase 2 first iteration

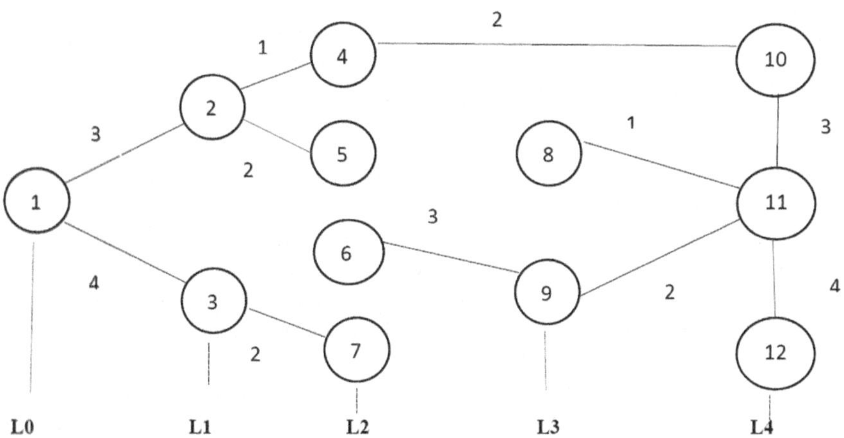

Fig. 5 Spanning tree after phase 2 second iteration

The result will be shown in Fig. 5.

The cost of the tree is $3 + 4 + 1 + 2 + 2 + 2 + 3 + 1 + 2 + 3 + 4 = 27$.

4 Conclusion

In this paper, a novel algorithm is proposed to construct MST by processing edge of the graph in level-wise. Further, the performance of this algorithm is improved by using minheap. The proposed algorithm time complexity is good compared to existing algorithm but it is not guaranteed that the optimal minimum cost spanning tree is constructed. The proposed algorithm is easy to understand and requires less time to generate code and execute. In future, the proposed algorithm can be improved to get optimal MST that has linear time complexity. For certain graphs which have |V|-1 levels, the complexity may be linear. The proposed method is used to construct minimum spanning tree and maximum cost spanning tree.

References

1. Dagar, S.: Modified Prim's algorithm. IJCIT **03**(02) (2012). ISSN 2218-5224
2. Eisner, J.: State of the art algorithms for minimum spanning trees a tutorial discussion (1997).
3. Karger, D.R., Tarjan, R.E., Klein, P.N.: A randomized linear-time algorithm to find minimum spanning trees. J. ACM **42**(2), 321–328 (1995)
4. Cormen, T.H., Leiserson, C.E., Rivest, R.L., Stein, C.: Introduction to Algorithms (3rd edn.). PHI Learning Private Ltd., pp. 624–642 (2009)
5. Mehlhorn, K., Sanders, P.: Algorithms and Data Structures: The Basic Toolbox. Springer, Berlin (2008)
6. Pettie, S., Ramachandran, V.: An optimal minimum spanning tree algorithm. In: Automata, Languages and Programming. Springer, Berlin, pp. 49–60 (2000)
7. Patel, N., Patel, K.M.: A survey on: enhancement of minimum spanning tree. Int. J. Eng. Res. Appl. **5**(1)(Part 3), 06–10. ISSN: 2248-9622, Jan 2015
8. Boruvka, O.: O jistem problemu mininalnim. Praca Moravske Prirodovedecke Spolecnosti **3**, 37–58 (1926). In czech
9. Nesetril, J., Nesetrilova, H., Milkovs, E.: Otakar Boruvka on minimum spanning tree problem translation of both the 1926 papers, comments, history. Discret. Math. **233**(1), 3–36 (2001)
10. Yao, A.: An o(|e|loglog|v|) algorithm for finding minimum spanning trees. Inf. Process Lett. **4**, 21–23 (1975)
11. Cheriton, D., Tarjan, R.E.: Finding minimum spanning trees. SIAM J. Comput. **5**, 724–742 (1976)
12. Fredman, M.L., Tarjan, R.E.: Fibonacci heaps and their uses in improved network optimisation algorithms J. . Assoc. Comput. Mach. **34**(3), 596–615 (1987)
13. Pierce, A.R.: Bibliography on algorithms for shortest path, shortest spanning tree and related circuit routing problems (1956–1974). Networks **5**, 129–149 (1975)
14. Maffioli, F.: Complexity of optimum undirected tree problems: a survey of recent results. In: Ausiello, G., Lucertini, M. (eds.) Analysis and Design of Algorithms in Combinatorial Optimization. International Center for Mechanical Sciences. CISM Courses and Lectures, vol. 266, pp. 107–28. Springer, New York (1981).
15. Graham, R.L., Hell, P.: On the history of the minimum spanning tree problem. Ann. History Comput. **7**, 43–57 (1985)
16. Karger, D.R., Klein, P.N., Tarjan, R.E.: A randomized linear-time algorithm to find minimum spanning trees. J. Assoc. Comput. Mach. **42**(2), 321–328 (1995)
17. Bazlamacci, C.F., Hindi, K.S.: Minimum-weight spanning tree algorithms: a survey and empirical study. Comput. Oper. Res. **28**, 767–785 (2001)
18. Schaffer, R., Sedgewick, R.: The analysis of heapsort. J. Algorithms 15(1), 76–100 (1993)
19. Marszałek, Z.: Performance Test on Triple Heap Sort Algorithm. Technical Sciences/University of Warmia and Mazury in Olsztyn (2017)
20. Meng, S., Zhu, Q., Xia, F.: Improvement of the dynamic priority scheduling algorithm based on a heapsort. IEEE Access **7**, 68503–68510 (2019)
21. Hirata, H., Nunome, A.: A modified parallel heapsort algorithm. Int. J. Software Innov. (IJSI) **8**(3), 1–18 (2020)

Application of AUV to Track a Maneuvering Target

S. Koteswara Rao and Kausar Jahan

Abstract In this research work, a target maneuvering in its course is tracked using Aerial Unmanned Vehicle (AUV) in three-dimensional space making use of bearing angle, range, and elevation angle measurements. An Extended Kalman filtering algorithm is considered for processing noise altered measurements. An algorithm that uses chi-square distribution is proposed for the detection of any maneuver in target parameters. The statistics about the estimated target parameters are provided to armament administration with the help of a communication arrangement like a global positioning system. Details of mathematical modeling for simulating and implementation of the target and observer paths and outcomes are presented in this work.

Keywords Extended Kalman filter · Estimation · Motion analysis of maneuvering target · Three-dimensional tracking · Aerial unmanned vehicle

1 Introduction

Aerial unmanned vehicle (AUV) is a harmless inflight warfare vehicle present in recent times. AUV is an automaton system hovering in the air mostly used for tracking a target. Parameters that are used to track the target, like bearing, range, and elevation are detected by sending radio waves. AUV these days are furnished with global positioning systems so that the armament administration system of AUV keeps track of it. The armament administration system may be an aircraft in the air or a ship on the surface. Data observed from AUV is directed to the armament administration system with the help of a global positioning system so that the armament administration system will get to know the target's location and path and to release armament in that way. The most common Extended Kalman filter (EKF) algorithm is used for tracking the target. Parameters representing target motion at prolonged ranges are mostly nonlinear. Consequently, EKF is thought based on balancing and speedily converging filter difficulties arising in the Kalman filter [1, 2].

S. K. Rao · K. Jahan (✉)
Department of Electronics and Communication Engineering, Koneru Lakshmaiah Education Foundation, Vaddeswaram, Guntur, Andhra Pradesh, India
e-mail: kausar465@ieee.org

V. Bhateja et al. (eds.), *Computer Communication, Networking and IoT*, Lecture Notes in Networks and Systems 197, https://doi.org/10.1007/978-981-16-0980-0_10

Target tracing is conducted utilizing EKF [3–7]. In this research, the major involvement is tracing a target that is maneuvering in course, as advised in [4, 5]. Observing the residual plot of the bearing target alone cannot visualize. So, the target maneuver is detected by the residuals obtained from a random sequence with zero-mean using chi-square distribution in the gliding window. An innovation that is square and normalized is utilized to detect if the target is under maneuver or not. For obtaining the finest result for the period of target maneuver, an ample quantity of plant noise is tallied to the plant noise covariance matrix. Once the maneuver is concluded, plant noise is dropped back.

The relation of target state elements is nonlinear to the observations (bearing and elevation), which makes the process more nonlinear in nature. So, the Kalman filter that is optimal for the linear process is not applicable for 3-D tracking of the target. For minimalism of the complexity in method, the target moving with steady speed and maneuvering only in its course angle is presumed. The system noise measured is white Gaussian noise delivered because of ruckus in the target's velocity.

Section 2 comprises of precise modeling of the filter and process execution. It also provides the process of detecting the target maneuver execution. Section 3 illustrate the implementation process and the outcomes attained. This paper is concluded in Sect. 4.

2 Mathematical Modeling

2.1 System Model

Contemplate the state vector as follows:

$$X_S(\kappa t) = \left[\, \dot{x}(\kappa t)\ \dot{y}(\kappa t)\ \dot{z}(\kappa t)\ R_x(\kappa t)\ R_y(\kappa t)\ R_z(\kappa t) \,\right]^T \tag{1}$$

Here $\dot{x}(\kappa t), \dot{y}(\kappa t), \dot{z}(\kappa t)$. denotes the speed components of target, and $R_x(\kappa t)$, $R_y(\kappa t)$, $R_z(\kappa t)$ are its range components in x, y, and z directions correspondingly. The state vector for subsequent time is calculated using the following equation.

$$X_s(\kappa t + 1) = \emptyset X_s(\kappa t) + b(\kappa t + 1) + \Gamma w(\kappa t). \tag{2}$$

$\emptyset$ is given by

$$\emptyset = \begin{bmatrix} 1\,0\,0\,0\,0\,0 \\ 0\,1\,0\,0\,0\,0 \\ 0\,0\,1\,0\,0\,0 \\ t\,0\,0\,1\,0\,0 \\ 0\,t\,0\,0\,1\,0 \\ 0\,0\,t\,0\,0\,1 \end{bmatrix} \tag{3}$$

Here t is the time frame at which observation is acquired. $b(\kappa t+1)$ is a deterministic control matrix and is provided by

$$b(\kappa t+1) = \begin{bmatrix} 0 \\ 0 \\ 0 \\ -(x_0(\kappa t+1)+x_0(\kappa t)) \\ -(y_0(\kappa t+1)+y_0(\kappa t)) \\ -(z_0(\kappa t+1)+z_0(\kappa t)) \end{bmatrix}^{\mathrm{T}} \tag{4}$$

Here x_0, y_0, z_0 denotes the observer location in x, y, and z directions. To lessen the mathematical complication, Y-axis is a reference for computing all the bearing angles and Z-axis for elevation angles. Let $w(\kappa t)$ represent Gaussian process noise.

$$w(\kappa t) = \begin{bmatrix} w_x \; w_y \; w_z \end{bmatrix}^{\mathrm{T}} \tag{5}$$

The variance of $w(\kappa t)$ is given by

$$E\left[\Gamma(\kappa t)w(\kappa t)w^T(\kappa t)\Gamma^{\mathrm{T}}(\kappa t)\right] = Q\delta_{ij}. \tag{6}$$

where

$$\begin{aligned} \delta_{ij} &= \sigma_w^2 \; (i = \kappa t) \\ &= 0 \quad \text{otherwise} \end{aligned} \tag{7}$$

$$Q = \begin{bmatrix} ts^2 & 0 & 0 & ts^3/2 & 0 & 0 \\ 0 & ts^2 & 0 & 0 & ts^3/2 & 0 \\ 0 & 0 & ts^2 & 0 & 0 & ts^3/2 \\ ts^3/2 & 0 & 0 & ts^3/4 & 0 & 0 \\ 0 & ts^2/2 & 0 & 0 & ts^3/4 & 0 \\ 0 & 0 & ts^2/2 & 0 & 0 & ts^3/4 \end{bmatrix} \tag{8}$$

$$\Gamma(\kappa t) = \begin{bmatrix} t & 0 & 0 \\ 0 & t & 0 \\ 0 & 0 & t \\ t^2/2 & 0 & 0 \\ 0 & t^2/2 & 0 \\ 0 & 0 & t^2/2 \end{bmatrix} \quad (9)$$

$Z(\kappa t)$ denotes the matrix of all observations and is represented as:

$$Z(\kappa t) = \left[\, R_m(\kappa t)\ B_m(\kappa t)\ \theta_m(\kappa t) \,\right]^{\mathrm{T}}. \quad (10)$$

Here $R_m(\kappa t)$, $B_m(\kappa t)$ and $\theta_m(\kappa t)$ are measured range, bearing, and elevation.

$$R_m(\kappa t) = R(\kappa t) + \xi_R(\kappa t). \quad (11)$$

$$B_m(\kappa t) = B(\kappa t) + \xi_B(\kappa t). \quad (12)$$

$$Z(\kappa t) = \left[\, R_m(\kappa t)\ B_m(\kappa t)\ \theta_m(\kappa t) \,\right]^{T}. \quad (13)$$

Here $R(\kappa t)$, $B(\kappa t)$, and $\theta(\kappa t)$ are the simulated values of the range, bearing angle, and elevation angle.

$$R(\kappa t) = \sqrt{R_x^2(\kappa t) + R_y^2(\kappa t) + R_Z^2(\kappa t)}. \quad (14)$$

$$B(\kappa t) = \tan^{-1}\left(R_x(\kappa t)/R_y(\kappa t)\right).. \quad (15)$$

$$\theta(\kappa t) = \tan^{-1}\left(R_{xy}(\kappa t)/R_Z(\kappa t)\right). \quad (16)$$

where

$$R_{xy} = \sqrt{R_x^2 + R_y^2}. \quad (17)$$

The measurement vector is given by

$$Z(\kappa t) = H(\kappa t) X_s(\kappa t) + \xi(\kappa t). \quad (18)$$

$$H(\kappa t) = \begin{bmatrix} 0\,0\,0 & \sin(B)\sin(\theta) & \sin(\theta)\cos(B) & \cos(\theta) \\ 0\,0\,0 & \frac{\cos(B)}{R_{xy}} & \frac{-\sin(B)}{R_{xy}} & 0 \\ 0\,0\,0 & \frac{\sin(B)\cos(\theta)}{R} & \frac{\cos(\theta)\cos(B)}{R} & \frac{-\sin(\theta)}{R} \end{bmatrix}. \quad (19)$$

And

$$\xi(\kappa t) = \left[\xi_R \ \xi_B \ \xi_\theta \right]^{\mathrm{T}}. \tag{20}$$

2.2 EKF Algorithm

All EKF implementation is as follows.

i. The initial state vector's estimate and its covariance matrix estimate be as of $X(0|0)$ and $P(0|0)$.
ii. For the subsequent time, the state vector is $X_s(\kappa t + 1)$:

$$X_s(\kappa t + 1) = \emptyset(\kappa t + 1|\kappa t) X_{\kappa t}(\kappa t) + b(\kappa t + 1) + \omega(\kappa t). \tag{21}$$

iii. The state vector's covariance matrix for the subsequent time is as follows.

$$P(\kappa t + 1|\kappa t) = \emptyset(\kappa t + 1|\kappa t) P(\kappa t) \emptyset^{\mathrm{T}}(\kappa t + 1|\kappa t) + Q(\kappa t + 1). \tag{22}$$

iv. The gain of the EKF is as follows:

$$\begin{aligned} G(\kappa t + 1) = {} & P(\kappa t + 1|\kappa t) \emptyset^{\mathrm{T}}(\kappa t + 1|\kappa t) \\ & \left[H(\kappa t + 1) P(\kappa t + 1|\kappa t) H^{\mathrm{T}}(\kappa t + 1) + R \right]^{-1} \end{aligned} \tag{23}$$

v. The state estimation and its error covariance:

$$X_s(\kappa t + 1|(\kappa t + 1) = X_s(\kappa t + 1|\kappa t) + G(\kappa t + 1) \left[Z(\kappa t + 1) - \hat{Z}(\kappa t + 1) \right] \tag{24}$$

$$P(\kappa t + 1|\kappa t + 1) = [1 - G(\kappa t + 1) H(\kappa t + 1) P(\kappa t + 1|\kappa t)] \tag{25}$$

vi. For the next iteration

$$X_s(\kappa t|\kappa t) = X(\kappa t + 1|\kappa t + 1) \tag{26}$$

$$P(\kappa t|\kappa t) = P(\kappa t + 1|\kappa t + 1) \tag{27}$$

2.3 Target Maneuver Detection

At the time of the target's movement at constant speed and course, the plant noise is a smaller amount. But, as the target starts its maneuver, the plant noise is gradually risen [8, 9]. To increase the plant noise the plant covariance matrix is increased by multiplying it with a fledge factor of 10 till the target executes maneuvering. When the target maneuver is complete i.e., it attains the required course, the plant noise is brought back to its reduced value. The regulated squared innovation, $\gamma_{\varphi}(\kappa t)$, is calculated as follows.

$$\gamma_{\varphi}(\kappa t) = \varphi^{\mathrm{T}}(\kappa t)S^{-1}(\kappa t + 1)\varphi(\kappa t + 1) \tag{28}$$

where $\varphi(\kappa + 1)$ is

$$\varphi(\kappa t + 1) = Z(\kappa t + 1) - h(\kappa t + 1, X(\kappa t + 1/\kappa t)) \tag{29}$$

Let $S(\kappa t)$ is

$$S(\kappa t + 1) = H(\kappa t + 1)P(\kappa t + 1/\kappa t)H^{\mathrm{T}}(\kappa t + 1) + \sigma^2 \tag{30}$$

$$\mathrm{d}(\xi) = \gamma^{\mathrm{T}} S^{-1} \gamma \geq c \tag{31}$$

where S is diag$\{S(\kappa t)\}$ and

$$\gamma = \varphi 1 \varphi(2) \ldots \varphi \kappa t \quad \gamma = [\varphi(1)\varphi(2) \ldots \varphi(\kappa t)]^{\mathrm{T}} \tag{32}$$

Here c is the threshold with constant value and d is the statistical value of the chi-square distribution. This gliding window of size five samples is chosen for this application.

3 Simulation and Results

Assuming that research is steered at satisfactory ecological circumstances, simulation is conducted on a workstation using Matlab. The trajectories that are followed by target and observer are chosen in Table 1, for performance validation of the process. For instance, scenario 1 defines a target at an opening distance of 2 km away from the spectator, moving with an initial course angle and speed of 170° and 300 m/s correspondingly. The preliminary bearing observed is 0°. The observations, bearing angle, and distance are assumed to be tarnished having a standard deviation in the error of 0.33° (1σ) and 0.01 km (1σ) correspondingly. From 300 s onwards, the target starts its maneuver in course to 295° with a rotating frequency of 3° for every second. The target's initial elevation angle is 0° for ease. The observer is presumed to move with a continual pace of 25 m/s and with 90° course.

The endless accessibility of observations for every time sample is presumed. The actual values of the target position and observer position are generated using Matlab software. Hence, the estimated parameters are authenticated based on the actual modeled parameter values built on specific acceptable standards. The acceptance

Table 1 Scenarios of target and observer positions

Parameters	Scenarios	
	1	2
Opening range of target (m)	2000	3000
The opening bearing of a target (°)	0	0
The opening course of the target (°)	135	170
The course of the target after 300 s (°)	235	295
Speed of target (m/s)	300	400
Elevation of the target (°)	0	0
Speed of observer (m/s)	25	20
The course of the observer (°)	90	90
Bearing angle noise (1σ) (°)	0.33	0.33
Range noise (1σ) (m)	10	10
Elevation angle noise (1σ) (°)	0.33	0.33

measure is preferred based on armament control necessity. The solution is acknowledged or believed to be obtained if inaccuracy in the estimated course is less than 3° and inaccuracy in the estimated speed of the target is less than or equal to 1 m/s.

For scenario 2, the approximations and factual tracks of the target along with that of observer trajectory are shown in Fig. 1. For precision of the notions, Figs. 2 and

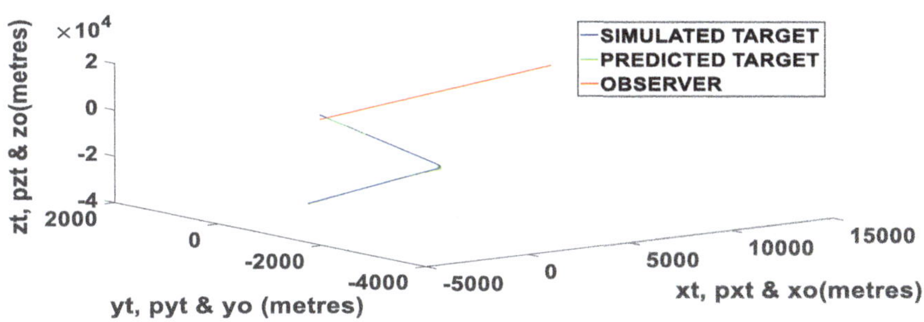

Fig. 1 Simulated and true trajectories of target and observer of scenario 2

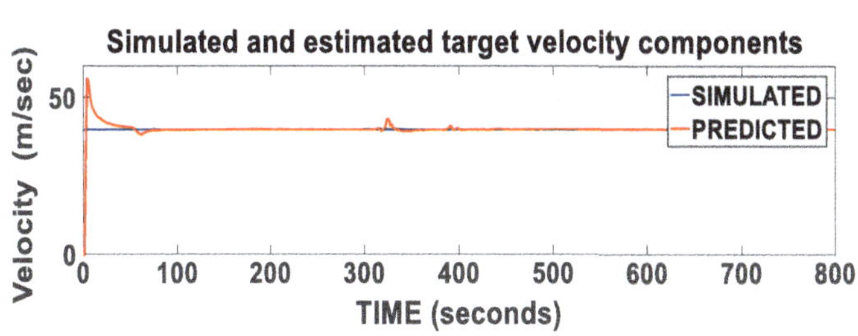

Fig. 2 Factual velocity versus projected velocity of a target for scenario 2

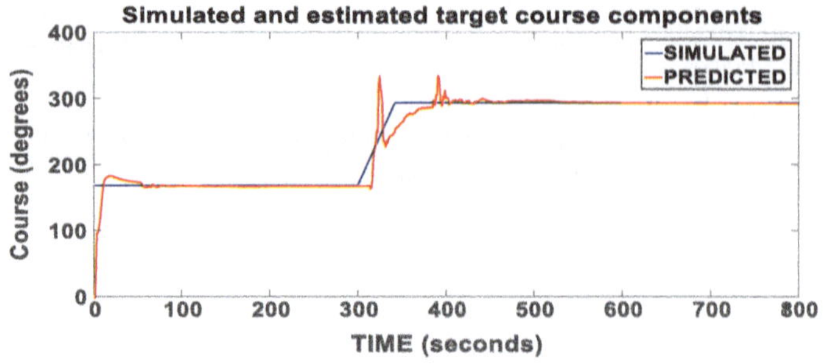

Fig. 3 Factual course versus projected course of a target for scenario 2

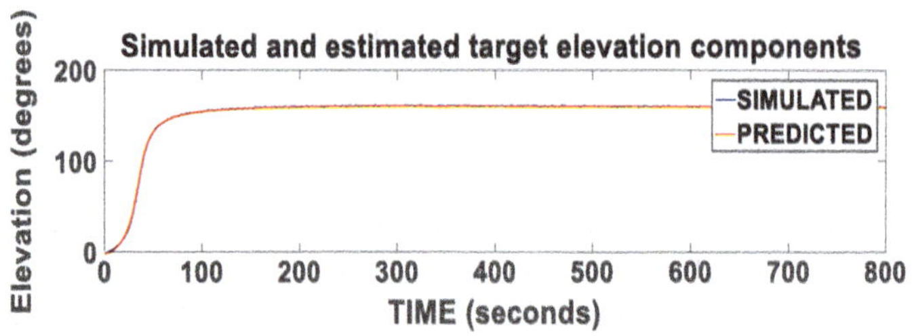

Fig. 4 Factual elevation versus projected elevation of a target for scenario 2

3 shows the actual and estimated course angle and speed of the target for scenario 2 correspondingly. Likewise, the target's actual and estimated elevation angle for the same scenario is depicted in Fig. 4. The result is acknowledged or encountered if the inaccuracies in the estimated course and estimated speed of the target are in the interior of the receiving conditions. Table 2 provides the solution convergence time samples in seconds for the situations provided as in Table 1.

Let us consider scenario 2 to evaluate the algorithm, where the target is maneuvering in its course. The convergence time of outcomes within acceptance criteria

Table 2 Convergence time sample of the solution in seconds

Parameter converged		Scenarios	
		1	2
Earlier to target maneuver	Course	54	67
	Speed	84	43
	Elevation	8	2
	Solution convergence	84	67
Post target maneuver	Course	385	460
	Speed	84	43
	Elevation	8	2
	Solution convergence	385	460

for the estimated course once the target completes the maneuver, is at the 460th-time sample and at the 67th time sample before target maneuvers. As the speed of the target is unchanged, there is only one convergence time, i.e., 43 s, before and after the target maneuver.

4 Conclusion

In this research, an attempt is made to develop an algorithm to track a maneuvering target in a three-dimensional plane. Grounded on the outcomes attained in simulation, EKF is suggested to approximate target parameters in a dynamic model of tracking using AUV systems.

References

1. Edwin Westerfield, E., Dennis Duven, J., Warnke, L.L.: Development of a global positioning system/Sonobuoy system for determining Ballistic missile impact points. John Hopkins APL Techn. Digest **5**, 335—340 (1984)
2. Baker, G.J., Bonin, Y.R.M.: GPS equipped Sonobuoy. https://www.sokkia.com.tw/NOVATEL/Documents/Waypoint/Reports/sonobuoy.pdf. In: Blackman S., Popolli, R. (eds.) Design and analysis of modern tracking systems, Norwood, MA, Artech House (1999)
3. Koteswara Rao, S.: Algorithm for detection of maneuvering targets in bearings only passive target tracking. IEE Proc. Sonar Navig. **146**(3), 141—146 (1999)
4. Koteswara Rao, S.: Modified gain extended Kalman filter with application to bearings only passive maneuvering target tracking. IEE Proc. Radar Sonar Navig. **152**(4), 239—244 (2005)
5. Kavitha Lakshmi, M., Koteswara Rao, S., Subramanyam, K.: Passive Object Tracking Using MGEKF Algorithm. Advances in Intelligent Systems and Computing, vol. 701, pp. 277–287. Springer Nature Singapore Pte Ltd (2018)
6. Jing, D., Han, J., Zhang, J.: A method to track targets in three-dimensional space using an imaging sonar. Sensors (Basel) **18**(7) (2018). https://doi.org/10.3390/s18071992
7. Isbitiren, G., Akan, O.B.: Three-dimensional underwater target tracking with acoustic sensor networks. IEEE Trans. Veh. Technol. **60**(8) (2011)
8. Clark, D.E., Bell, J., de Saint-Pern, Y., Petillot, Y.: PHD filter multi-target tracking in 3D sonar. Europe Oceans 2005, Brest, France, 2005, pp. 265–270. https://doi.org/10.1109/OCEANSE.2005.1511723
9. De Ruiter, H., Benhabib, B.: On-line modeling for real-time 3D target tracking. Mach. Vis. Appl. **21**(17) (2009). https://doi.org/10.1007/s00138-008-0138-y

of the estimated course once the target completes the maneuver, as at the 160th time sample and at the 170th time sample before target maneuvers. As the speed of the target is unchanged, there is only one maneuvering time ... before and after the target maneuver.

4 Conclusion

References

Analysis of LTE MIMO Antenna Parameters Based on Modulation Techniques and Interference

Leena Govekar and Y. S. Rao

Abstract OFDM united with multiple input multiple output (MIMO) system is a very important technology to improve the transmission performance for next-generation wireless communications, because of the need of reliable high-speed wireless communication. An attempt is made to study the characteristics of LTE MIMO 1×1 and 2×2 antenna parameters such as EVM peak, EVM rms, and BER in this paper. The above parameters are studied for QPSK and QAM modulation techniques. We have used Zedboard for carrying out the simulations on MATLAB software. The bit error rate (BER) is calculated for all the methods and is compared with each other. We found that BER for 16 QAM is 0.04588, for 64 QAM BER is 0, for 256 QAM BER is 0.00064. For 64 QAM with interference, it is 0.03445.

Keywords MIMO-OFDM · QPSK · QAM · BER · Channel estimation · Harmonic distortion

1 Introduction

For 4G and 5G broadband wireless communications, MIMO-OFDM is the leading air interface. It combines orthogonal frequency division multiplexing (OFDM), which splits a radio channel into a large number or closely spaced subchannels, and multiple input multiple output (MIMO) technology, which boosts capacity by transmitting different signals over various antennas, to give more respectable communications at high speeds. It was shown in research conducted during the middle of 1990s that while MIMO can be used with other prominent air interfaces such as CDMA and TDMA, the fusion of OFDM and MIMO is most feasible at greater data rates. For most progressive wireless local area network which we call WLAN and mobile broadband network standards, MIMO-OFDM is the foundation because the maximum spectral

L. Govekar (✉)
Mumbai University, Mumbai, India

Y. S. Rao
Sardar Patel Institute of Technology, Mumbai, India
e-mail: ysrao@spit.ac.in

V. Bhateja et al. (eds.), *Computer Communication, Networking and IoT*, Lecture Notes in Networks and Systems 197, https://doi.org/10.1007/978-981-16-0980-0_11

efficiency is achieved by it and, thus, it delivers the highest data throughput and capacity. MIMO was invented by Greg Raleigh in 1996 when he demonstrated that various data streams could be transmitted at the same time on the same frequency because the signals transmitted through space bounce off objects, e.g., ground and follow multiple paths to the receiver. That is, different data streams could be sent over different paths by employing multiple antennas and precoding the data. Raleigh suggested and later proved that because OFDM transforms a high-speed data channel into a number of parallel lower-speed channels, the processing required by MIMO at higher speeds would be most feasible using OFDM modulation [1].

2 Implementation

2.1 Hardware Description

ZedBoard is an economical development board for the Xilinx Zynq7000 SoC, and it provides key features such as Zynq 7000 sustem on chip (SoC), 512 MB random access memory (RAM), 4 GB SD card for data storage, Ethernet port supporting 10 Gigabit, 100 Gigabit, 1000 Gigabit speeds, onboard USB JTAG Programming, and USB port. The following shows the block diagram of Zynq XC7Z020-1CLG484. The Zynq-7000 family of SoCs is primarily used for advanced embedded system applications, such as automotive driver assistance, video supervision, factory automation, and advanced wireless applications. Zynq-7000 integrates an ARM Cortex A9 MP Core processor based on 28 nm system. The architecture of Zynq is slightly different from previous marriages of embedded processors and programmable logic by moving to a processor centric model from an FPGA centric model. The Zynq 7100 model was developed in the year 2013 by Xilinx. It interfaces digital signal processing (DSP) to meet developing programmable systems integration requirements of broadcast, wireless, military, and medical applications. The Zynq 7000 product family had some limitations on its usage. One of the reasons was that the design software of ISE was not completely developed for handling the designing complexity and capacity with FPGA and ARM core. To overcome this disadvantage, we now use new Vivado Design Suite of Xilinx as it was developed for higher capacity FPGAs. Also, it has high-level synthesis (HLS) feature [2].

We have interfaced MIMO LTE antennas with the Xilinx Zedboard. Using MATLAB software, the antenna parameters such as gain, SNR, and BER have been studied. The block diagram of Zynq XC7Z020-ICLG484 is shown in Fig. 1.

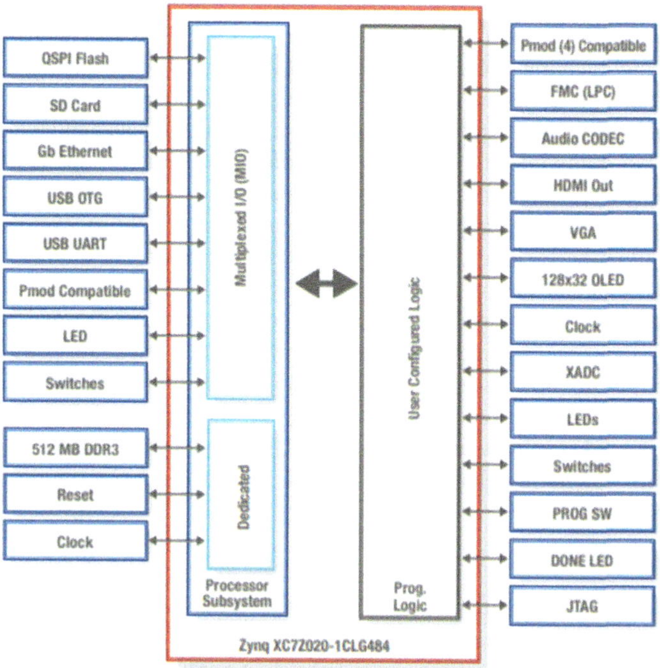

Fig. 1 Block diagram of Zynq XC7Z020-1CLG484

2.2 *Antenna Parameters*

Using Zedboard hardware, we simulated 1 * 1 and 2 * 2 LTE MIMO antennas and studied the following antenna parameters using MATLAB software.

Constellation Diagram

An illustration of a signal modulated by digital modulation scheme like phase shift keying (PSK) or quadrature amplitude modulation (QAM) is called a constellation diagram. It demonstrates the signal as a 2D XY plane scatter diagram in the complex plane at symbol sampling instants. The phase shift of the carrier wave from a reference phase is defined by the angle of a point measured anticlockwise from horizontal axis. A measure of the amplitude or power of the signal is represented by the distance of a point from the origin. With feature extraction of the signal constellation, the recognition method of M-QAM signals including 16QAM, 32QAM, and 64QAM is researched. The past knowledge about carry frequency and baud rate of the received signals is not required by this scheme, but these parameters are directly estimated by it, measuring the number of the signal constellation points by blind clustering algorithm. Therefore, the modulation type is determined in this way [3].

Channel Estimation

During communication, the signal propagates through a medium or channel and it gets distorted, i.e., noise gets added to the signal when it goes through the channel. This transmitted signal must be accurately decoded at the receiver end without much errors [4]. For doing this, the initial step is finding the characteristics of the channel through which the signal has propagated. Channel estimation is the name of this technique to characterize the channel [5]. In such large mobility system standard channel estimation methods that apply a block-type pilot placement or consider the wireless channels to be proportional in one OFDM symbol cannot be used. The estimation of the time domain channel coefficients is performed by making use of pilot tones and by linear interpolation of time domain channel estimates. In one method, minimum mean square error (MMSE) and least square (LS) channel estimation methods are explored along with various linear interpolation methods like linear, second order, spline-cubic, lowpass, and time domain interpolation. In another technique for estimating channel coefficients, a basis expansion model (BEM) is recommended [6].

Harmonic Distortion

Nonlinear devices which are present in the power system cause harmonic distortion. A device in which current is not proportional to the applied voltage is a nonlinear device. When a sinusoidal voltage is enforced to a nonlinear load, it causes current to be distorted. The current may be doubled and take on a different wave shape even by increasing the voltage by a few percent. In a power system, this is the cause of most harmonic distortion [7]. To compute the level of harmonics in voltage or current waveforms, total harmonic distortion (THD) is an important figure of merit that is used. In literature, two distinct definitions of THD can be found. One of the definition states that the harmonic content of a waveform is correlated to the waveform's rms vale and the second definition states that the harmonic content of a waveform is compared to its fundamental. To discriminate between the two, the former is occasionally expressed by THD_R and the second by THD_F [8].

Intermodal Distortion

The amplitude modulation of signals containing two or more distinct frequencies caused by time variance or nonlinearities in a system is called intermodal distortion. The intermodulation between frequency components gives rise to additional components at frequencies which are not just at harmonic frequencies (integral multiples) of either, like harmonic distortion, but also at the sum and difference frequencies of the original frequencies and at sums and differences of multiples of those frequencies [9].

Occupied Bandwidth

The ITU-R originally defines it as a peak bandwidth, outside of which emissions do not outpace a certain percentage of the total emissions. The occupied bandwidth is for LTE equal to the channel bandwidth, outside of which a maximum of 1% of the emissions is allowed (0.5% on each side). The occupied bandwidth measures up to the aggregated channel bandwidth, in the case of carrier aggregation.

2.3 Test Setup

As shown in Fig. 2, we have used Zedboard along with transmitting and receiving antennas (LTE MIMO). The Zedboard is connected to the laptop with a gigabit Ethernet cable. To simulate the results, we have used MATLAB 2013b software. Based on the antenna cases, we have successfully designed our code for each case to carry out our research. This research has been carried out for both with and without interference. Here, we have used a mobile phone to generate interference. As soon as the mobile phone receives a call, we see an interference in our results. The research is carried out with the test setup as shown in Fig. 2 under normal atmospheric conditions.

3 Results

We have successfully obtained results by using the hardware shown in Fig. 1 and simulated the parameters such as 'Constellation Diagram', 'Channel Estimation', 'Harmonic Distortion', 'Intermodal Distortion' and 'Occupied Bandwidth'. The practical results are almost equal to theoretical values. We have categorized our results based on the antenna type and modulation type.

Fig. 2 Test setup

3.1 For 1 * 1 Antenna and Modulation Type 16 QAM

See Figs. 3, 4, 5, 6, and 7.

We have calculated following antenna parameters using MATLAB software and are shown in Tables 1 and 2.

We have calculated following antenna parameters using MATLAB software and are shown in Tables 3 and 4.

Fig. 3 Transmitted and received image

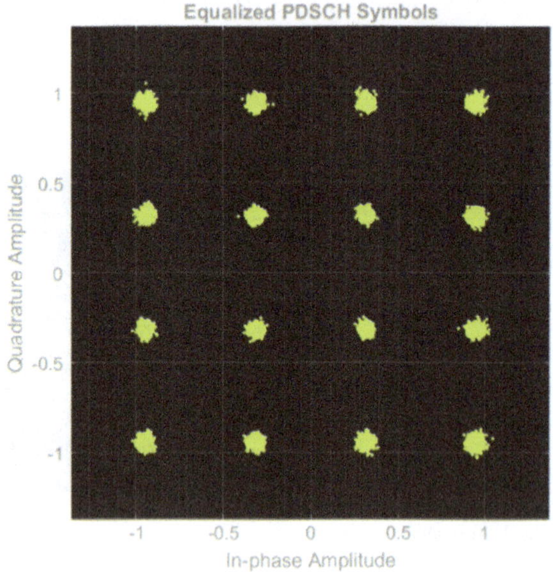

Fig. 4 Constellation diagram

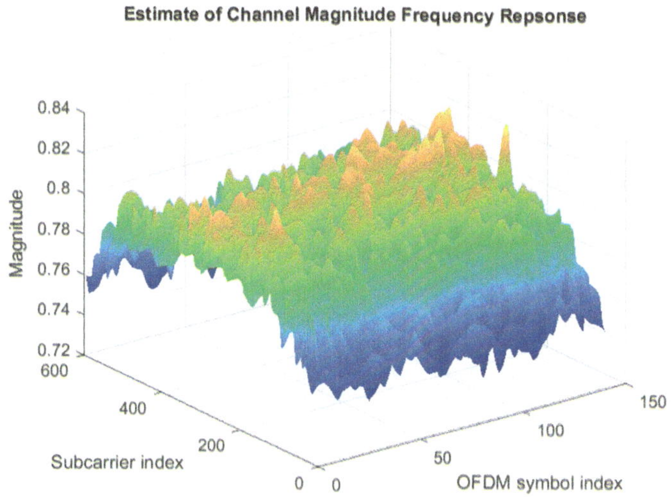

Fig. 5 Channel estimation

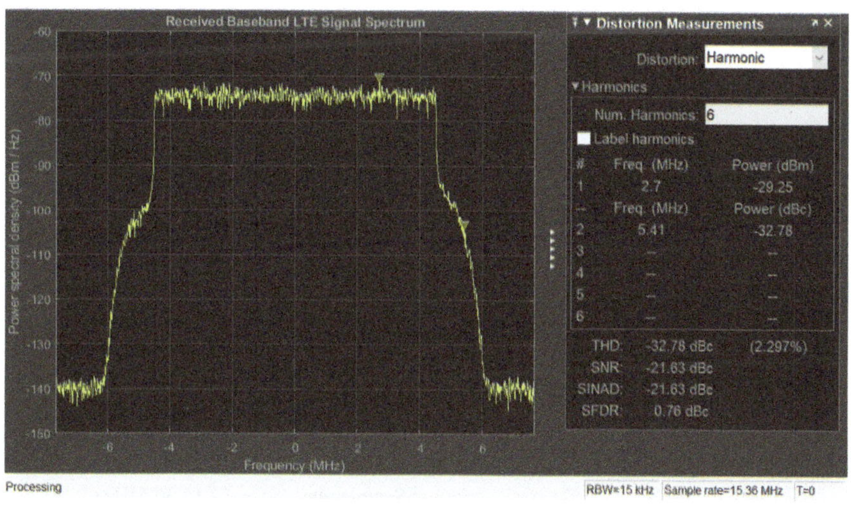

Fig. 6 Harmonic distortion

3.2 *For 1 * 1 Antenna and Modulation Type 64 QAM*

We have calculated following antenna parameters using MATLAB software and are as follows. Here, the number of harmonics is 6 (Figs. 8, 9, and 10; Tables 5, 6, and 7).

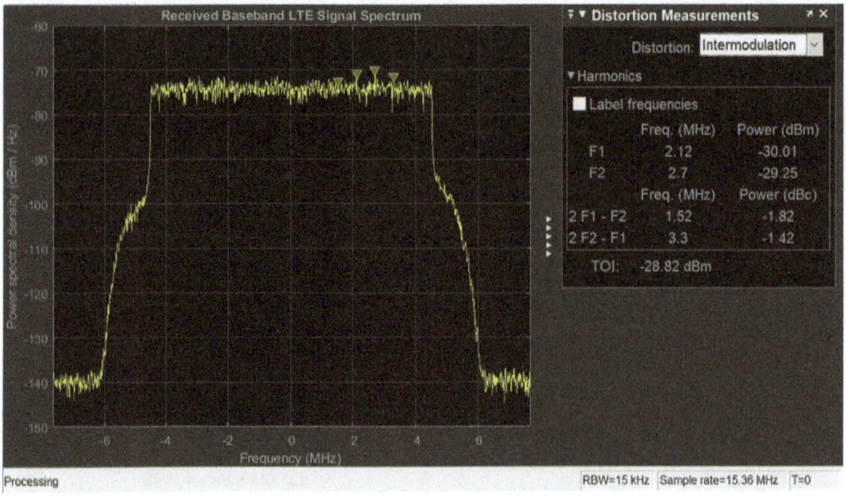

Fig. 7 Intermodal distortion

Table 1 Antenna parameters and its value

S. no.	Parameter	Value
1	RBW	15 kHz
2	Sample rate	15.36 MHz
3	SNR	−21.63 dBc
4	SINAD	−21.63 dBc
5	SFDR	0.76 dBc

Table 2 Antenna parameters and its value

S. no.	Frequency (Hz)	Value
1	2.7	−29.25
2	5.41	−32.78

Table 3 Antenna parameters and its value

Sr. no.	Parameter	Value
1	RBW	15 kHz
2	Sample rate	15.36 MHz
3	TOI	−28.82 dBm

Table 4 Antenna parameters and its value

S. no.	Parameter	Value
1	2.12	−30.01
2	2.7	−29.25

Fig. 8 Transmitted and recieved image

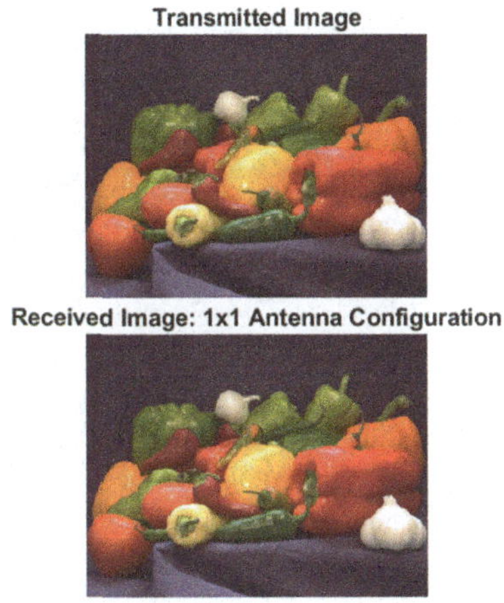

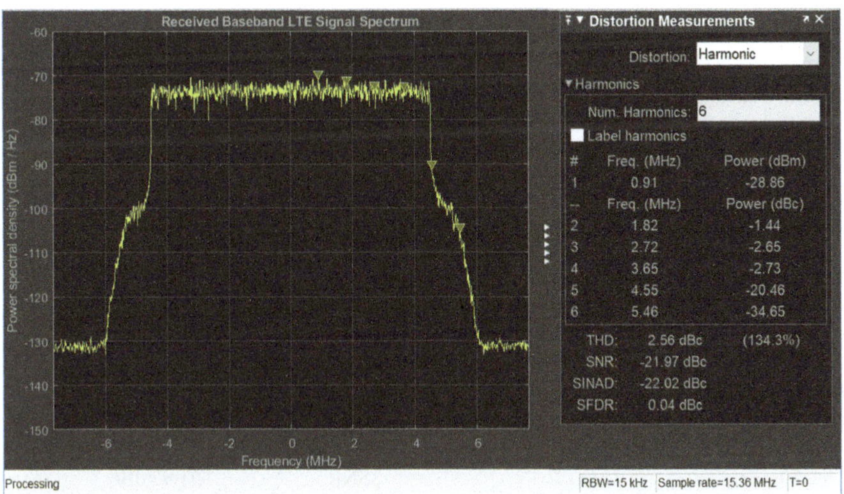

Fig. 9 Harmonic distortion

We have calculated following antenna parameters using MATLAB software and are as follows. Here, the number of harmonics is 6.

3.3 For 1 * 1 Antenna with Interference and Modulation Type 64 QAM

See Figs. 11, 12, and 13.

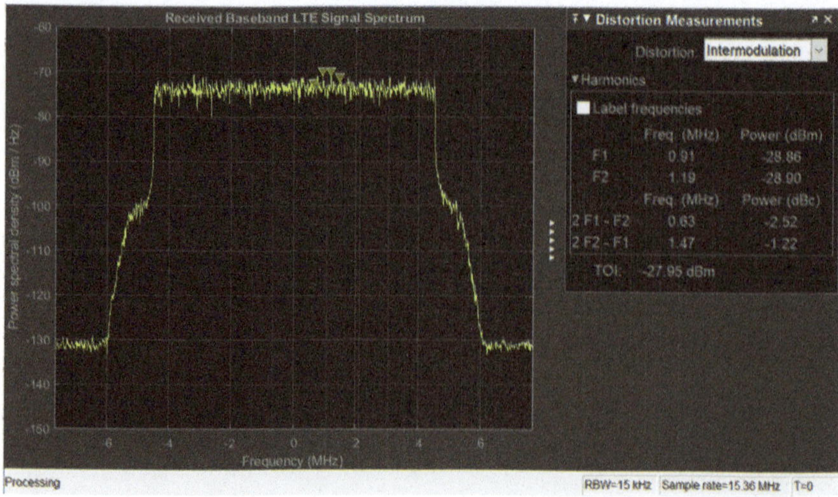

Fig. 10 Intermodal distortion

Table 5 Antenna parameters and its value

S. no.	Frequency (MHz)	Power
1	0.91	−28.86 dBm
2	1.82	−1.44 dBc
3	2.72	−2.65 dBc
4	3.65	−2.73 dBc
5	4.55	−20.46 dBc
6	5.46	−34.65 dBc

Table 6 Antenna parameters and its value

S. no.	Parameter	Value
1	THD	2.56 dBc (134.3%)
2	SNR	−21.97 dBc
3	SINAD	−21.02 dBc
4	SFDR	0.04 dBc

Table 7 Antenna parameters and its value

S. no.	Freq	Frequency (MHz)	Power
1	$F1$	0.91	−28.86 dBm
2	$F2$	1.19	−28.90 dBm
3	$2F1 - F2$	0.63	−2.52 dBc
4	$2F2 - F1$	1.47	−1.22 dBc

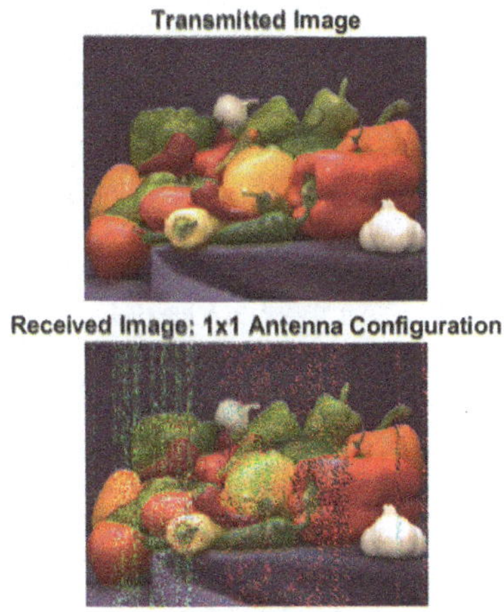

Fig. 11 Transmitted and recieved image

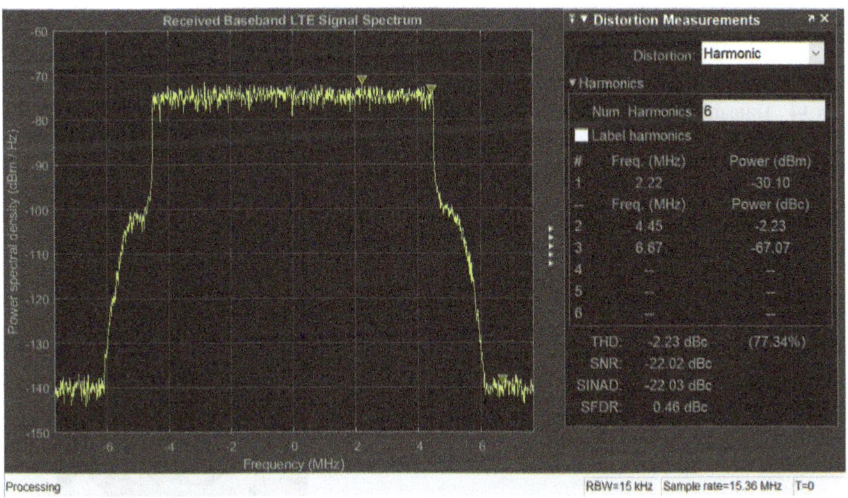

Fig. 12 Harmonic distortion

*3.4 For 2 * 2 Antenna with Interference and Modulation Type 64 QAM*

See Figs. 14, 15, and 16.

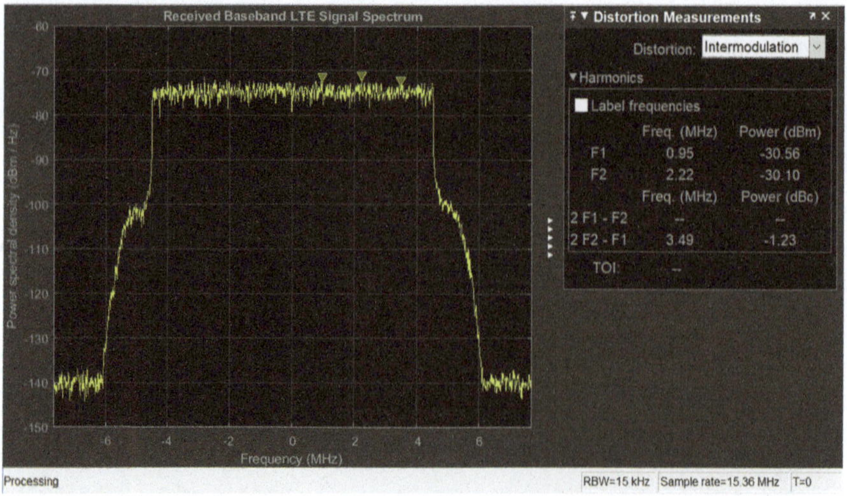

Fig. 13 Intermodal distortion

Fig. 14 Transmitted and recieved image

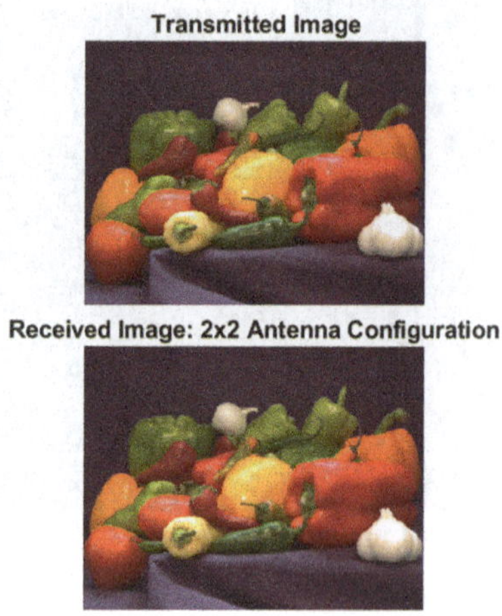

4 Conclusion

MIMO-OFDM achieves better performance than SISO-OFDM and is the result given by the experiment conducted in real environment by using 'Zedboard' hardware with SDR platform. For MIMO, the BER is 0 when 64 QAM technique is employed. We can also conclude from the above comparison that 64-QAM technique is more efficient than QPSK technique. Among the QAM technique, 64 QAM has possessed lower BER which is 0. The 64 QAM technique is also efficient during interference and has BER equal to 0 for 2 * 2 antenna and 0.04588 for 1 * 1 antenna. Also, we should employ 2 * 2 antenna with 64 QAM modulation technique to reduce interference

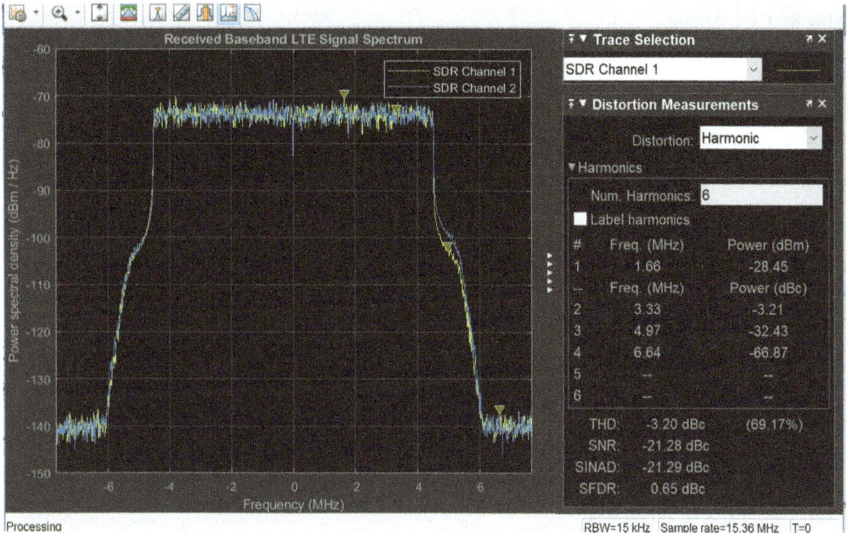

Fig. 15 Harmonic distortion

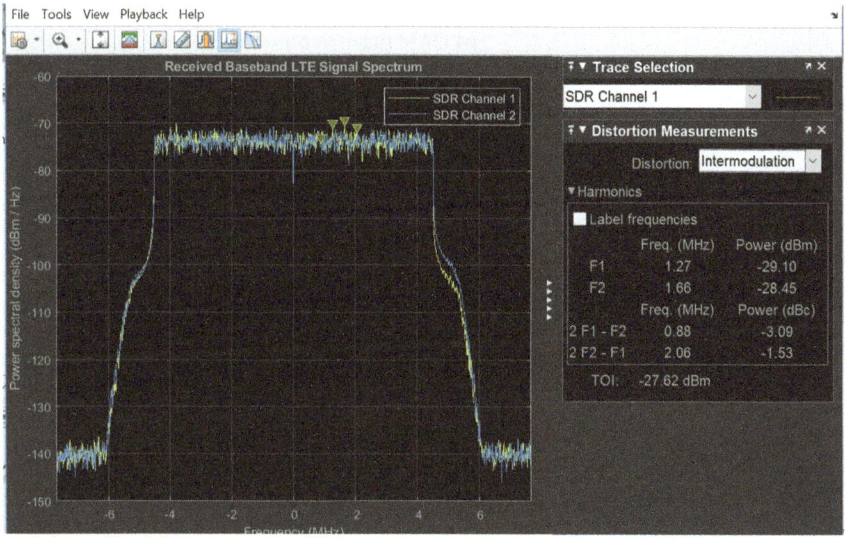

Fig. 16 Intermodal distortion

and loss of transmitted data. Therefore, we can conclude that by employing more number of antennas at both transmitting and receiving end we can obtain signal with 0 errors. Also, the loss of information is prevented. Our conclusion is based on the comparison of antenna parameters shown in Tables 8 and 9.

Table 8 1 * 1 Antenna parameters and its value

S. no.	Modulation type	Parameter	1 * 1 antenna
1	16 QAM	EVM peak	273.594%
2	16 QAM	EVM rms	12.284%
3	16 QAM	BER	0.04588
4	64 QAM	EVM peak	13.535%
5	64 QAM	EVM rms	3.204%
6	64 QAM	BER	0
7	256 QAM	EVM peak	303.816%
8	256 QAM	EVM rms	8.959%
9	256 QAM	BER	0.00064
10	64 QAM (interference)	EVM peak	598.193%
11	64 QAM (interference)	EVM rms	38.639%
12	64 QAM (interference)	BER	0.03445

Table 9 2 * 2 antenna parameters and its value

S. no.	Modulation type	Parameter	2 * 2 antenna
1	64 QAM (interference)	EVM peak	3.168%
2	64 QAM (interference)	EVM rms	0.763%
3	64 QAM (interference)	BER	0

References

1. MIMO-OFDM. Retrieved from https://en.wikipedia.org/wiki/MIMO-OFDM
2. Zedboard. Retrieved from https://www.zedboard.org/product/zedboard
3. Zhendong, C., Weining, J., Changbo, X., Min, L.: Modulation recognition based on constellation diagram for M-Qam signals. In: ICEMI, pp. 70–74 (2013)
4. Hardjawana, W., Li, R., Vucetic, B., Li, Y.: A new iterative channel estimation for high mobility MIMO-OFDM systems. IEEE (2010)
5. Communication—Channel Estimation. Retrieved from https://www.sharetechnote.com/html/Communication_ChannelEstimation.html
6. Zijian, T., Cannizaro, R.C., Leus, G., Banelli, P.: Pilot assisted time-varying channel estimation for OFDM systems. IEEE Trans. Signal Process. **55**, 2226–2238 (2007)
7. Periyaazhagar, D., Irrusapparajan, S.: A survey of harmonic distortion and reduction techniques. IJPT **08**, 23581–23589 (2016)
8. Shmilovitz, D.: On the definition of total harmonic distortion and its effect on measurement interpretation. IEEE Trans. Power Deliv. **20**, 526–528 (2005)
9. Intermodulation. Retrieved from https://en.wikipedia.org/wiki/Intermodulation

Infrastructure as a Service (IaaS) for Smart Education

Nguyen Ha Huy Cuong, Nguyen Trong Trung, Le Ngoc Khanh, and Nguyen Van Loi

Abstract The Fourth Industrial Revolution puts a great impact in technology in all aspects such as artificial intelligence, big data, robotics, and IoT. The prospect of the development of education in the context of the Industrial Revolution for the fourth time, Vietnam will certainly face many difficulties in education. Sometimes, transmission of knowledge is not sufficient without looking into the quality and capacity of the learners; quality teaching staff, and management staff. With the influx of new learning models and the development of science and technology, the methods of traditional education will certainly be challenging, and each student needs different learning abilities. Advances in technology allowed the school to design individual learning pathways which is suitable for each particular case. Within the scope of the study, this paper deals with the technology infrastructure solutions to solve the whole distribution of resources, avoid deadlock, and concurrent access to the system for the learners.

Keywords Industrial · Cloud computing · Infrastructure · Cluster focus · Smart education

---

N. Ha Huy Cuong
Software Development Centre—University of Dang Nang, Da Nang, Viet Nam
e-mail: nhhcuong@sdc.edu.vn

N. Trong Trung (✉)
Dong A University, Da Nang, Viet Nam

L. Ngoc Khanh
The University of Da Nang, Da Nang, Viet Nam
e-mail: lnkhanh@ac.udn.vn

N. Van Loi
Vietnam-Korea University of Infomation and Communication Technology, Da Nang, Viet Nam
e-mail: nvloi@vku.udn.vn

V. Bhateja et al. (eds.), *Computer Communication, Networking and IoT*, Lecture Notes in Networks and Systems 197, https://doi.org/10.1007/978-981-16-0980-0_12

1 Introduction

In fact, in recent years, cloud computing has always been preferred by newly established businesses or deploying new software systems. However, the transition from an application running on the physical server to the cloud is still slow. There are three main factors leading to this problem.

- Firstly, the system running stably on a physical server moving to the cloud will require the process of moving data, impacting on the system ... and when implemented, it does not eliminate the possibility of appearing risks. Therefore, some people in charge of information technology (IT) of universities and organizations are psychologically afraid, not supportive of moving existing applications on physical infrastructure to "cloud."
- Second, many universities have invested in physical server systems and continue to use this system instead of switching to cloud. If you convert immediately, the number of machines invested will do what, where to arrange ... are difficult questions.
- Third, the deployment costs include human training as well as the services provided from abroad leading to very high prices.

For these three reasons, most organizations and universities only use cloud for new applications. Migrating legacy applications to the cloud is not much of a school unless they invest in new or their servers have started to run sluggishly. These are also the reasons why the transition to using cloud in Vietnam is slower than other countries in the world. In recent years, the university expanded its role to strengthen scientific research and teaching. Promote computerization in management, building IT infrastructure through connectivity between universities in the country and globally. Therefore, it is required to develop a cost-effective solution in providing resources such as digital libraries, electronic lectures, registered credits, and online learning (Fig. 1).

In this paper, we study the appropriate cloud types in the educational environment, solutions to support virtual server services, and provide effective resources that meet the requirements of students, faculty, and user. The solution meets the following specific requirements:

- Assurance against resource conflicts, flexible solution to set up virtual server resources to respond more quickly to changes of infrastructure.
- Technology to prevent congestion when simultaneous traffic from students, faculty, and users.
- Technical proposals ensure appropriate cost for users.
- Solution applications conflict when simultaneously running on an operating system platform, if one application goes down, it can affect other applications running on the same server.

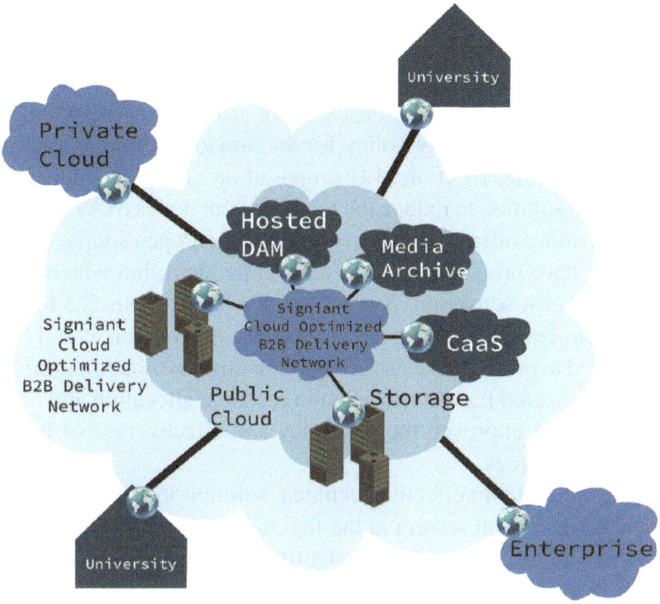

Fig. 1 Physical machine server system at university

The article consists of several parts and is presented as follows: Part 1 provides an overview. Part 2 presents basic research. Part 3 presents the proposed algorithms. Part 4 presents the above test setup and compares the test results. Part 5 concludes and recommends.

2 Related Works

This section deals with the background study of the paper. Abderrahim El Mhouti et al. [1] have discussed on the concepts of e-learning and cloud computing with their structure, benefits, and challenges. They have presented the impact of using cloud computing for e-learning [1]. Germana Manca et al. [2] discussed how an integrated virtual system based on cloud computing can be developed to enhance GIS online courses [3]. Maria et al. [4] applied various optimization techniques to minimize the user's cost with more scenarios. Goldberg [5] and Zhan [6] have explained some ways of computing shortest paths over a network. While Nan [7] proposed the optimal VM allocation schemes for single-site cloud and multi-site cloud. Yan [8] has discussed on the optimal routing strategy and the resource allocation strategy for VMs in the media cloud. Elsharkawey [9] has discussed a novel load balancing algorithm which has been proposed over a novel architectural model in the fog computing environment. Cuong [10, 11] has provided an algorithmic approach to detect deadlock and resource allocation issues in the virtualization of heterogeneous platform. Walsh's et al. [12] have suggested different classes like automatic and autonomous scenes; Yazir et al. [13] propose solutions to use adaptive controllers to tailor virtual resources to meet client access needs. They also propose a suitable algorithm to provide resources with priority policy for the previous resource request registration process. Meanwhile, Li et al. [14] proposed an automatic multi-dimensional resource allocation solution to reduce the load on agent nodes (NA). Stillwell [3] has discussed on providing infrastructure services based on open source cloud computing system. They have proposed optimal scheduling algorithm which considers the ability to allocate spirit activation. Xiao et al. [15] have proposed a load balancing policy with a smart resource management mechanism. Nguyen et al. [10, 11] have discussed a method to predict the lease completion time distribution that is applicable to making a sophisticated tradeoff decision in resource allocation and rescheduling. The authors have also improved the efficiency and effectiveness of heterogeneous cloud computing resources.

In the paper, we build an effective technical solution in providing virtualization resources, which are virtual servers at the first layer in the cloud computing model that is infrastructure as a service. Two algorithms provide optimal resources, and resources are released.

To perform simulation scenarios, we experimented with running on the OpenStack virtual machine platform. In addition, the maths we advanced into experiments on the CloudSim simulation tool. Then proceed to analyze and evaluate the experimental results obtained. Currently, the university research centers have applied data integration centers, so it will be very effective for us to conduct experiments on private clouds.

3 Proposed Resource Allocation Algorithm and Solution of Technological Infrastructure for Smart Education

Resource allocation of cloud computing resources has become a topic of great interest in recent times. An effective resource supply policy brings economic benefits for business point of view. Organizations also want to develop information technology infrastructure in the context of technological development. Equity in resource allocation applies from the concept of dominant resource fairness (DRF) developed by A. Ghodsi and colleagues in a multi-resource environment [9]. In addition, Xu et al. [16] devised a model for equitable allocation of resources while ensuring efficient use of resources in other words minimize waste of resources. Accordingly, the equity of the decision to allocate resources is calculated by the following formula:

$$v(A) = \left(\sum j \sum k\left(\frac{x_{jk}}{c_j} - \beta * d_{jk}\right)^{y-1}\right)^{\frac{1}{y}}\Big) \, (y \in R) \qquad (1)$$

X_{jk}: is the number of resources k allocated for request j.

$\beta = \frac{1}{\max \sum jd_{jk}}$: is the largest percentage of any given resource allocated to users.

$d_{jk} = \frac{y_{jk}}{\max y_{jk}}$: is a standardized request.

In the formula below, to estimate the energy consumption of physical servers, we choose the approximate energy consumption of physical server *j* as a linear function (), interpreting energy according to wake (1).

$$(u_i) = (P - P_{\text{idle}}) * u_j + P_{\text{idle}}, \forall j \tag{2}$$

Among them, P_{max} and P_{idle} are the power of physical machine *j* in the maximum used utilities state and idle state, respectively. u_j is the total used utilities of all resources on the physical machines *j*, $u_j \in [0, 1]$, and it is calculated by a Formula (2).

$$u_j = \sum_{k=1}^{D} \frac{U_{jk}}{C_{jk}}, \forall_j \tag{3}$$

In which, U_{jk} is the resources *k* of physical servers *j* provided for virtual services and C_{jk} is the capacity of resources *k* on physical servers *j*. The unused physical server in the process automatically shuts down. Therefore, the energy consumption of *M* physical servers in phase is set as Formula (3).

$$E(t) = \{t \times \sum_{j=1}^{M} P(u_j)\tag{4}$$

At the data center, we proceed to use the algorithm to provide resources based on a physical server and monitor the use of hard public resources. From the decentralized virtual infrastructure layer, there is at least one processing platform related to distributed virtual infrastructure. The processor platform consists of at least one processor device with a processor paired with memory. The processing platform operates to allocate at least one of the information technology resources to users that meet the corresponding requirements, to track the use of information technology resources assigned to users, and to create one or more messages if the usage of IT resources specified by the user is below the specified threshold. Distributed virtual infrastructure can include an enterprise's own cloud infrastructure.

The distributed computing system has changed from the mainframe model to the client–server model. Today, users exploit computing applications with computational resources placed in a cloud environment. Infrastructure service (IaaS) in cloud computing will provide virtual server resources and a setup based on data centers (DC). Data centers are built from one or more distributed physical servers. At that time, telecommunications networks acted as a means of exchanging information, sharing resources to access remote data and information.

Our technical solution is to provide a solution to optimize data transfer between mobile devices and cloud storage providers. In particular, patent options may select one or more files to transfer from a mobile device to a cloud storage device based on the analysis of a set of (at least one) factors such as ability link between mobile

Algorithm 1 Resource Allocation for Tecnological Infrastructure

Input: $W[c_i; m_i; n_i; d_i]$
Output: Resource Allocation t_i
BEGIN
Do
Work.Insert t_i
$k \leftarrow 0$; $\max_k = input$
While (Read.isfree()=true do
Ready.Insert (Work.delete(t_i));
J-cur←Readly.delete(ti);
J-cur.Status = Assigned;
for each $j \in R_i$ do;
If(usedbyi[j]=0) send request (ti) to pj end for;
Else if respone = Succes;
Return result
Else wait 2^{k++} if (Timeout Or $k = \max\_k$;
Return fail
While (True)

END.

Algorithm 2 Resource Allocation for Tecnological Infrastructure

Input:Number of virtual services N, type of resources D and additional factor Q_{ij}.
Number of physical machines *M*

Output: list of used physical machines, allocation t_i

BEGIN
for $i := 1$ to N do
int $j \leftarrow random(M)$;
for $k := 1$ to D do
$Sum_{ik}{}^{e} \leftarrow r_{ik}{}^{e} + Q_{ij} \times f_{ik}{}^{e}$
$Sum_{ik}{}^{a} \leftarrow r_{ik}{}^{a} + Q_{ij} \times f_{ik}{}^{a}$
if $(C^e_{jk} \geq Sum^e_{ik} and C^a_{jk} \geq Sum^a_{ik})$
$C^a_{ik} \leftarrow C^a_{jk} - Sum^a_{jk}$
end if
end for $k := 1$ to D do
$allo[i] \leftarrow j$;
end for $i := 1$ to N
return allocation;

END.

device and cloud storage provider; last access date for files stored on mobile devices; and/or remaining battery power in mobile devices (if applicable).

The mission of a telecommunication network is to transmit information quickly and accurately. The current trend is to build a layer of infrastructure services with outstanding features like flexible scalability, security, and smooth network transmission. The resource provisioning method uses virtualization technology, which allows the aggregation and exploitation of physical resources in a flexible manner.

4 Experimental Results of Resource Allocation Algorithm

After running the test, we have the results Table 1. We have created eight virtual machines with the results obtained as shown below.

Table 2 shows the result of creating eight required virtual machines from cloudlet, represented by VM1 to VM8 of 04 physical servers, start time is 0.1 ms, finish is 85 ms for VM1 and 613 ms virtual machines for VM8, and virtual machines have many different scenarios, such as we increase the number of virtual machines needed, as well as allocate physical servers on different data centers D.

In Fig. 2, with the result of test case 2, after four virtual machines were created at the second point, besides test point 3, 2 virtual machines were created. In the third test, point 2 also created 8, then point 3 created 2, point 4 created 1. With limited

Table 1 Resource characteristics and power consumption of physical machine

Type of PM	CPU (Mhz)	RAM	BW	DISK	Pidle	Pmax
HP proliant G4	2core*1860	16	1	40	86	110
HP proliant G5	2core*2660	16	1	40	93.7	115
IBM proliant x3250	2core*1860	16	1	40	46.1	113
IBM proliant x3250	2core*1860	16	1	40	58.4	122

Table 2 CPU resource and RAM resource characteristics of virtual machine

Type of PM	f^e_{CPU}	r^e_{CPU}	Num	Total	f^e_{RAM}	r^e_{RAM}	Num	Total
VM1	250	500	1	500	0.4	0.44	1	0.85
VM2	250	500	1	500	1.1	2.74	1	3.5
VM3	250	500	1	500	0.7	1.4	1	1.95
VM4	250	500	1	500	0.112	0.5	1	0.615
VM5	250	500	1	500	0.4	0.45	1	0.85
VM6	250	500	1	500	1.0	2.75	1	3.85
VM7	250	500	1	500	0.7	1.0	1	1.85
VM8	250	500	1	500	0.113	0.5	1	0.615

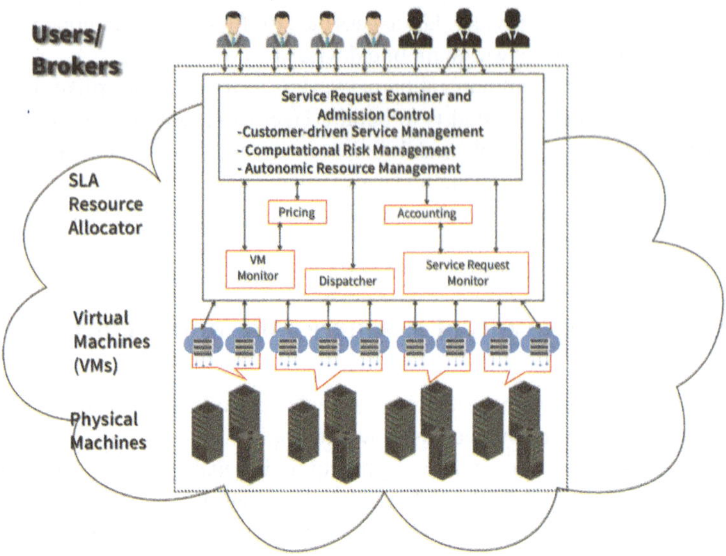

Fig. 2 Infrastructure architecture of cloud computing services

Fig. 3 Computer system for research and practice at university

capabilities of the tested laptop configuration, it could create up to eight machines. Virtual with 4 data points (Figs. 3 and 4).

Based on the results in Table 3, we find that the power source increases proportionally and linearly with the number of physical servers used; the value of the physical servers used and the energy consumption of the FF-ECMDRA-SA algorithm are the best. However, the execution time of ECMDRA-SA is larger than FFD algorithm. One reason for this is affected by the initial temperature parameter and the number of iterations in the ECMDRA-SA algorithm. Besides, the implementation time of the algorithm is small and can be implemented in practice.

It can be seen that the number of nodes requested is increasing. The application of technical solutions in algorithm 2 has brought very positive research results.

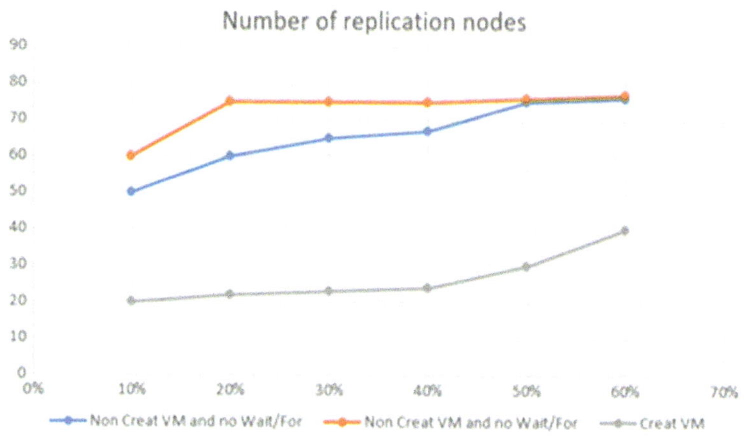

Fig. 4 Infrastructure architecture of cloud computing services

Table 3 BW resource and DISK resource characteristics of virtual machine

Type of PM	f_{BW}^{e}	r_{BW}^{e}	Num	Total	f_{DISK}^{e}	r_{DISK}^{e}	Num	Total
VM1	0.2	0.25	1	0.95	0.5	0.44	1	0.85
VM2	0.2	0.25	1	0.75	2.5	2.74	1	3.5
VM3	0.2	0.25	1	0.95	3.0	1.4	1	1.95
VM4	0.2	0.25	1	0.15	5.0	0.5	1	0.615
VM5	0.2	0.25	1	0.95	0.5	0.45	1	0.85
VM6	0.2	0.25	1	0.75	2.0	2.75	1	3.85
VM7	0.2	0.25	1	0.95	2.7	1.0	1	1.85
VM8	0.2	0.25	1	0.15	5.0	0.5	1	0.615

5 Conclusion

In this paper, we discussed infrastructure services for the smart education system in the context of Vietnam. Analysis of several factors leading to the slow implementation of cloud computing systems in universities. In the general context, currently in the world, the data centers in CSP use virtual server technology to increase both in research and application deployment. In this paper, we proposed an algorithm to suggest additional precautions and prevent deadlock, based on the results of running physical servers and creating eight virtual machines. However, computers that perform simulations have a limited configuration, which affects the efficiency of the algorithm. This article will also help universities to refer and select solutions to build cloud infrastructure to provide smart education systems. In the next study, we conduct research on solutions based on resource allocation algorithms to improve efficiency such as ant algorithm and genetic algorithm.

References

1. El Mhouti, A., Erradi, M., Nasseh, A.: Using cloud computing services in e-learning process: benefits and challenges. (2017). Print ISSN1360-2357
2. Ghodsi, A., Zaharia, M., Hindman, B., Konwinski, A., Shenker, S., Stoica, I.: Dominant Resource Fairness: Fair Allocation of Multiple Resource Types, p. 24 (2011)
3. Stillwell, M., Schanzenbach, D., Vivien, F., Casanova, H.: Resource allocation algorithms for virtualized survive hosting platforms. Int. J. Parallel Distib. Comput. (JPDC) **70**(9), 962–974 (2013)
4. Maria Alejandra, R.S.: Resource Provisioning and Scheduling Algorithms for Scientific Workflows in Cloud Computing Environments. Department of Computing and Information Systems University of Melbourne, Australia (2016)
5. Goldberg, A.V., Harrelson, C.: Computing the shortest path: a search meets graph theory. IN: Proceedings of the Sixteenth Annual ACM-SIAM Symposium on Discrete Algorithms. Society for Industrial and Applied Mathematics, 156–165 (2005)
6. Zhan, F.B.: Three fastest shortest path algorithms on real road networks: data structures and procedures. J. Geogr. Inf. Dec. Anal. **1**(1), 69–82 (1997)
7. Nan, X., He, Y., Guan, L.: Optimal allocation of virtual machines for cloudbased multimedia applications, 3rd ed. In: 2012 IEEE 14th International Workshop on Multimedia Signal Processing (MMSP). IEEE (2012)
8. Ya, J.: Optimal routing and resource allocation for multimedia cloud computing, **2**(8) (2016)
9. Mohamed, A.: Elsharkawey, MLRTS: multi-level real-time scheduling algorithm for load balancing in fog computing environment. Mod. Educ. Comput. Sci. **2**, 1–15 (2018)
10. Nguyen, H.H.C., Solanki, V.K., Van Thang, D., Nguyen, T.T.: Resource allocation for heterogeneous cloud computing. Resource **9**(1–2) (2017) ISSN 1943-3581
11. Nguyen, H.H.C., Le, V.S., Nguyen, T.T., Algorithmic Approach to Deadlock Detection for Resource Allocation in Heterogeneous Platforms. Inc. IEEE (2014)
12. Walsh, W.E., Tesauro, G., Kephart, J.O., Das, R.: Utility functions in autonomic systems. Proceeding International Conference on Autonomic Computing, pp. 70–77. 17-18 May 2004 (2010)
13. Yazir, Y.O., Matthews, C., Farahbod, R.: Neville, dynamic resource allocation in computing clouds using distributed multiple criteria decision analysis. In: Proceeding of the 3ht Cloud Computing (CLOUD) pp. 91–98, 5–10 July 2010, Miami FL. IEEE (2010)
14. Li, Q., Hao, Q., Xiao, L., Li, Z.: Adaptive management of virtualized resources in Cloud Computing using feeback control. In: Proceedings of the 1th Information Science and Engineering (ICISE), pp. 99–102 26–28 Dec. 2009. IEEE (2009)
15. Xiao, Zhen, Song, Weijia, Chen, Qi: Dynamic resource allocation using virtual machines for cloud computing environment. Int. J. IEEE Trans. Parallel Distrib. Syst. **24**(6), 1107–1117 (2013)
16. Xu, X., Yu, H.: A game theory approach to fair and efficient resource allocation in cloud computing. Math. Probl. Eng. (2014)
17. Manca, G., Waters, N.W., Sandi, G.: Using cloud computing to develop an integrated virtual system for online GIScience programs. Knowl. Manage. E- Learn. **8**(4), 514–527 (2016)

Intrusion Detection Techniques for Detection of Cyber Attacks

Shaikh Sahil Ahmed, Mahesh Kankar, and Bhawana Rudra

Abstract Intrusion detection system (IDS) is a software-related application where we can detect the system or network activities and notice if any suspicious task happens. Excellent broadening and the use of the Internet lift examine the communication and save the digital information securely. Nowadays, attackers use variety of attacks for fetching private data. Most of the IDS techniques, algorithms, and methods assist to find those various attacks. The central aim of the project is to come up with an overall study about the intrusion detection mechanism, various types of attacks, various tools and techniques, and challenges. We used various machine learning algorithms and found performance metrics like accuracy, recall, and F-measure and compared with the existing work. After this research, we got good results that can help to detect the cyber attacks being performed in the network.

Keywords Intrusion detection · Cybersecurity · Machine learning · Computer networks

1 Introduction

Intrusion detection systems (IDSs) are a range of cybersecurity-based technology at first developed to find the exploits and vulnerabilities against a destination host. The individual use of the IDS is to detect threats. Algorithms of machine learning have played a necessary part in the field of cybersecurity. The fundamental way is to use machine learning algorithms to build a model of honest and reliable activity, and then, it is compared to the new behavior against this model.

We evaluated our project based on various performance metrics such as accuracy, recall, precision, f-measure, and entropy. Accuracy is defined as the ratio of the number of correct assumptions to the total number of input samples, precision (also called positive predictive value) is defined as the fraction of necessary samples among the retrieved samples, and recall (also known sensitivity of data) is the part of the total

S. Sahil Ahmed (✉) · M. Kankar · B. Rudra
National Institute of Technology Karnataka, Surathkal, India
e-mail: bhawanarudra@nitk.edu.in

V. Bhateja et al. (eds.), *Computer Communication, Networking and IoT*, Lecture Notes in Networks and Systems 197, https://doi.org/10.1007/978-981-16-0980-0_13

number of necessary samples that were actually retrieved. The F-measure (F1 score) is a computation of a test's precision, and it is defined with the weighted harmonic mean of the accuracy and recall of the test. Entropy is a measure of the randomness in the information being processed.

2 Literature Review

Khraisat [1] discussed various techniques of intrusion detection system such as statistics-based techniques, knowledge-based techniques, supervised learning, unsupervised learning, semi-supervised learning, and hybrid-based techniques. Attacks are classified into four categories into denial-of-service (DoS), probing, user-to-root (U2R), and remote-to-local (R2L). They have compared various datasets such as NSL-KDD, ADFA-WD, and ADFA-LD. In their paper, they have presented, in detail, a survey of intrusion detection system methodologies, types, and technologies with their advantages and limitations.

Tiwari [2] discussed various techniques such as an anomaly-based detection system and signature-based intrusion detection. They also discussed two types of IDS as client-based and network-based. Two models described are local (one system to monitor) and client/server for centralized analysis.

Kulhare [3] discussed the detection and prevention of cyber attacks and also about various types of IDS. They surveyed different approaches such as pattern matching and fuzzy clustering. They concluded that the study of IDS is great for network security.

Raghunath [4] discussed the supervised anomaly detection that shows how anomalous the network is an association pattern analysis indicating the network as highly anomalous. Further, they are given a very huge amount of connections observed per given unit time which is more useful for the detection of attacks.

Zhou [5] discussed propose a heuristic algorithm called correlation-based feature selection bat algorithm (CFS-BA) to detect the cyber attacks. They performed experiments using KDDCup 99, NSL-KDD, and CIC-IDS2017 datasets. They got good accuracy and examined various performance metrics. They proposed a novel machine learning-based IDS. Firstly, they propose a CFS-BA algorithm with the aim of discarding irrelevant features. Then, the ensemble classifier based on C4.5, random forest, and forest by penalizing attributes with the average of probabilities rule is used to construct the classification model. The proposed IDS is evaluated on KDDCup'99, NSL-KDD, and CIC-IDS2017 datasets.

Table 1 Category and types of attack

Category of attacks	Types of attacks
Denial-of-service-attack (DoS)	Back, land, neptune, pod, smurf, teardrop
User-to-root-attack (U2R)	Buffer_overflow, loadmodule, perl, rootkit
Remote-to-local-attack (R2L)	Ftp write, guess passwd, multihop, spy, imap, phf, warezclient, warezmaster
Probing-attack	Ipsweep, nmap, portsweep, satan

3 Methodology

Till now, we worked on KDD Cup 99 dataset. This dataset is grouped together by almost 4,900,000 individual connections which include a feature count of 41. The simulated attacks were categorized as given below:

1. **Denial-of-Service-Attack (DoS)**: Host becomes inaccessible due to flooding, and hence, the host is overloaded.
2. **User-to-Root-Attack (U2R)**: unauthorized access from a remote machine.
3. **Remote-to-Local-Attack (R2L)**: unauthorized access to local superuser (root) privileges.
4. **Probing-Attack:** surveillance and other probing.

The dataset contains 42 features like duration, protocol_type, service, src_bytes, and dst_bytes. Various attacks are labeled under the label attribute present in the 42nd index of the dataset. Based on the dataset of attacks collected in the past, we build and train our model so that the machine can detect future attacks (suspicious activity) going into the network. The dataset contains the attacks under each of four cases as shown in Table 1.

Various machine learning algorithms used are deep neural network (with one and five layer), artificial neural network (ANN), naive Bayes, logistic regression, decision tree, and random forest.

4 Implementation

All the experiments are performed over the dataset "kddcup.data_10_percent.gz" as mentioned in [6]. The dataset contains 494,020 rows. We used 70% of the dataset for training purposes and the remaining 30% for testing purposes. We go through various algorithms and get the performance metrics of the work done as shown in Table 2.

Table 2 Performance metrics for all algorithms

Algorithms	Accuracy	Recall	Precision	F-measure	Entropy
DNN (1 layer)	0.99937	0.99792	0.99894	0.99843	0.00105
DNN (5 layer)	0.99908	0.99863	0.99673	0.99768	0.00326
Artificial neural network	0.99931	0.99877	0.99775	0.99826	0.00224
Naive Bayes	0.94803	0.79267	0.99911	0.88400	0.00088
Logistic regression	0.99862	0.99602	0.99703	0.99652	0.00295
Decision tree	0.99975	**0.99942**	0.99931	0.99937	0.00068
Random forest	**0.99985**	0.99938	**0.99986**	**0.99962**	**0.00013**

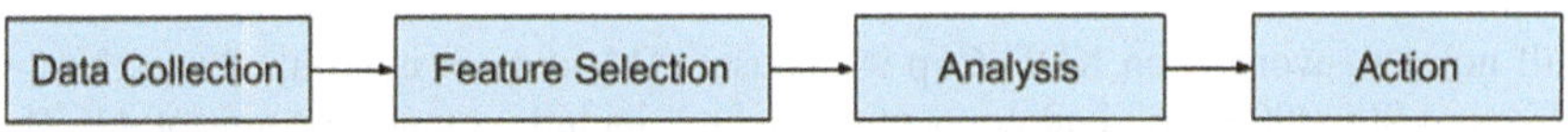

Fig. 1 Phases of intrusion detection system

Intrusion detection system consists of four main parts as data collection, feature selection, analysis, and action as shown in Fig. 1.

- Data Collection : The dataset is given as input to ID System.
- Feature Selection : Various features as duration, protocol_type, service, src bytes, and dst_bytes all are extracted, processed, and given to the second stage.
- Analysis : The data is analyzed for its correctness.
- Action : It states the reaction of attack over the system.

4.1 Deep Neural Network (DNN)

A deep neural network (DNN) is an artificial neural network (ANN) with numerous layers between the input and output layers. DNN helps in finding the correct mathematical calculations to convert the input into the output, as input may be a linear or a nonlinear relationship.

Working Connections between neurons are associated with a weight, dictating the importance of the input value. All 42 attributes will be given as input to the first layer of DNN; then, on the second layer for each hidden unit, we will compute its value based on previous layer output and so on. Neurons apply an activation function on the data to "standardize" the output coming out of the neuron. Each attribute in dataset has its own importance in finding whether this will lead to attack or not so in training phase deep neural model is trained in such a way that if any data packet comes with its out of threshold value, then it will be classified as attack otherwise not an attack, weights of neuron will be adjusted with consideration of bias and activation function. Value of any neuron is shown by $f(x)$.

$$f(x) = W_1X_1 + W_2X_2 + W_3X_3 + \ldots + W_{42}X_42 + B \quad (1)$$

where W_n= weight of each unit, X_n= attribute, and B is the bias.

4.2 Artificial Neural Network (ANN)

ANN or neural networks are computational algorithms. It is intended to mimic the behavior of biological systems composed of the "neurons." ANN is initialized with a classifier as sequential. There is one input layer, three hidden layers and output layer where "relu" is used as an activation function for the input and hidden layer and "sigmoid" for the output layer. The ANN compiled using an optimizer used as "adam." The model is trained, a confusion matrix is observed, and various performance metrics are calculated.

Working The arrow between the two layers represents connectivity joining the two neurons. It is used to signify the channel for the flow of data. It was observed that each relation has a load, an integer number that is owned to rule the signal connecting the two neurons. If the output is fine that was produced by the network, then we do not need to adjust the loads. Also if the bad output is produced, then firmly the system will modify the loads to upgrade the results.

4.3 Naive Bayes

It is a classification method based on the popular Bayes theorem with an assumption of independence among predictors. Here, the classifier is used as GaussianNB, i.e., fitting naive Bayes to the training set. After that, the test set results are predicted, and a confusion matrix is observed.

Working Naive Bayes is Bayes theorem-based algorithm in which it will classify into four categories based on probability $P(A/B) = [P(B/A).P(A)]/P(B)$, where $P(A/B)$ finds what is the probability of being classified into one of the attack category given all 42 attributes. For example, let any packet that enters the system through a LAN port, we will extract all info about the packet (length, checksum, header length, etc.). In the naive Bayes algorithm, all of these 42 attributes will be treated as prior to determining posterior info (classification).

4.4 Logistic Regression

Logistic regression is a machine learning algorithm which is used for classification problems, it is a predictive analysis algorithm, and based on the concept of probability, the classifier is used as logistic regression. The confusion matrix is observed, and various performance metrics are calculated.

Working This is a simple algorithm in which we will find y (output) based on multiple input variables (all 42 attributes).

$$y = M_0 + M_1A_1 + M_2A_2 + M_3A_3 + \ldots + B \tag{2}$$

where M is coefficient for various attributes, B is the bias, and A_n are the attributes that are multiple variables. It will give the best fit line among the dataset data samples will be divided by line, distance between the data point and line is considered as error it uses maximum likelihood function to find the best fit line among the datapoints in dataset.

4.5 Decision Tree

Decision tree is a non-invariable supervised learning algorithm that is owned for both classifications as well as regression of the tasks. The aim is to build a model that forecasts the value of a target variable by gaining simple decision-based rules deduced from the data quality. The classifier is used as a decision tree classifier. The confusion matrix is observed, and various performance metrics are calculated.

Working Decision tree will take input at the root of all the data points, and on each node, partial decision of being classified into one of four categories will be determined, and leaf node will work as a final destination where four nodes will represent that. For each row in training dataset, prediction will be there, and we will check whether our model is classified correctly or not based on the testing dataset.

4.6 Random Forest

Random forest algorithm builds decision trees based on data trials and then forecasts from each of them and ultimately selects the finest solution by state of voting. The classifier is used as a random forest classifier. The confusion matrix is observed, and various performance metrics are calculated.

Working Firstly, random data samples are opted from a KDD training dataset. Next, this algorithm will build a decision tree for each and every sample. Then, it will acquire the forecasted result from every decision tree that where it has to be classified in the attack category. Final voting says the overall attack pattern from all given sample dataset. At last, opt the most voted forecasted result as the ending prediction result as mentioned earlier.

Algorithm:
1. Load the dataset.
2. Apply pre-processing technique with normalizer.
2.1 Normalize samples individually to unit norm.
3. Make a partition of train and test data.
4. Dataset is given input to random forest.
5. The test dataset is then provided to random forest for classification.
6. Calculate accuracy, recall, precision, F-measure, and entropy.

5 Experiment and Results

All the model with their performance metrics is shown in Table 2. Among all the models, random forest gives better accuracy, precision, F-measure, and least entropy as compared to others, and decision tree shows best over recall in comparison with other models.

So, we can observe that among all the algorithms, random forest shows the best accuracy, precision, and F-measure. The randomness measured by random forest is least among all. In case of recall, decision tree gives better results. So, overall we can say that decision tree algorithm has performed well over other algorithms.

We also compared our work to the existing papers as described in Table 3. Zhou [5] proposed a heuristic algorithm called correlation-based feature selection bat algorithm (CFS-BA) and concluded results over various datasets such as KDD CUP 99, NSL-KDD, and CIC-IDS2017. By their method, they achieved an accuracy of 0.976, precision of 0.998 as well as F-measure of 0.998 using KDD CUP 99 dataset, whereas in our work, we got an accuracy of 0.99985, precision of 0.99986, and F-measure of 0.99962 using KDD CUP 99 dataset. As well as for the other dataset for the existing work, our accuracy for detecting the network intrusion is much better.

Table 3 Results compared with existing work

Authors	Performance metrics (KDD CUP 99)
Zhou [5]	0.976 (Accuracy) 0.998 (Precision) 0.998 (F-measure)
Proposed work	
Random forest	0.99985 (Accuracy), 0.99986 (Precision), 0.99962 (F-measure)

6 Conclusion and Future Work

IDS is fetching the head role for many institutions after establishing the firewall technology across the network circumference. IDS can supply protection from outside users and inner attackers, where traffic does not go behind the firewall. All the following points are should always be kept in mind. In the future, we will see a various algorithm that can give better results, and also we will make one add-on which will be added to the browser.

Our proposed model works only on host side, this can further be extended to server side as well, and KDD-cup 99 dataset is used for all proposed algorithms but newer version of dataset can be used with new deep learning model as well. Since latest version of KDD dataset contains more features, it will give more accuracy over old dataset.

References

1. Khraisat, A., Gondal, I., Vamplew, P., Kamruzzaman, J.: Survey of intrusion detection systems: techniques, datasets and challenges, cybersecurity (2019)
2. Tiwari, M., Kumar, R., Bharti, A., Kishan, J.: Intrusion detection system. Int. J. Techn. Res. Appl. **5**(2), 2320–8163 (2017)
3. Kulhare, R., Singh, D.: Survey paper on intrusion detection techniques. Int. J. Comput, Technol (2013)
4. Raghunath, B.R., Nitin Mahadeo, S.: Network intrusion detection system (NIDS). In: First International Conference on Emerging Trends in Engineering and Technology (2008)
5. Zhou, Y., Cheng, G., Senior Member, IEEE, Jiang, S., Dai, M.: An efficient intrusion detection system based on feature selection and ensemble classifier. J. Latex Class Files **14**(8) (2015)
6. http://kdd.ics.uci.edu/databases/kddcup99/kddcup.data_10_percent.gz
7. Vijayarani, S., Maria Sylviaa, S.: Intrusion detection system—a study. Int. J. Secur. Privacy Trust Manag. (IJSPTM) **4**(1) (2015)
8. Park, K., Song, Y., Cheong, Y.-G.: Classification of attack types for intrusion detection systems using a machine learning algorithm. In: Fourth International Conference on Big Data Computing Service and Applications (2018)
9. Nur Cahyo, A., Hidayat, R., Adhipta, D.: Performance comparison of intrusion detection system based anomaly detection using artificial neural network and support vector machine. In: AIP Conference Proceedings 1755, 070011 (2016)
10. Arul, A., Subburathinam, K., Sivakumari, S.: Classification techniques for intrusion detection an overview. Int. J. Comput. Appl. **76**(16), 33–40 (2013)
11. Wankhede, R., Chole, V.: Intrusion detection system using classification technique. Int. J. Comput. **139**(11), 0975–8887 (2016)
12. Ravindra Chaudhari, R., Pramod Patil, S.: Intrusion detection system: classification, techniques and datasets to implement. Int. Res. J. Eng. Technol. (IRJET) **04**(02) (2017)

Efficient NoSQL Graph Database for Storage and Access of Health Data

Safikureshi Mondal and Nandini Mukherjee

Abstract Graph data models are mostly used for storage and access of large, unstructured real-life data generated from real-life applications. The graph data model can be implemented by NoSQL graph databases. In this paper, eight well-known NoSQL graph databases are compared to study their properties. After rigorous review of different research works which are focused on different parametric measures of storage and access, only the best three NoSQL graph databases, Neo4j, OrientDB and ArangoDB, are chosen. The efficiency of these three graph databases is compared based on searching or traversing and querying operations on the databases for storage and access. A particular type of graph, i.e., disease-symptom graph database has been used for this purpose.

Keywords Health data · Disease-symptom graph · NoSQL graph database

1 Introduction

To design any emerging real-life applications, the major problem is how to manage the heterogeneous and unstructured data. It is a challenging task for the researchers and practitioners to choose appropriate platform to handle the huge data in real-life applications [1]. As a solution to this challenge and to process big data quickly, NoSQL databases are used. The main advantages of using NoSQL databases over relational databases are as follows:

1. Storing large volume of structured, semi-structured and unstructured data.
2. Supporting dynamic schema.
3. Less time required to access the large heterogeneous data.

S. Mondal (✉)
Department of Computer Science and Engineering, Narula Institute of Technology, Kolkata, India

N. Mukherjee
Department of Computer Science and Engineering, Jadavpur University, Kolkata, India
e-mail: nmukherjee@cse.jdvu.ac.in

V. Bhateja et al. (eds.), *Computer Communication, Networking and IoT*, Lecture Notes in Networks and Systems 197, https://doi.org/10.1007/978-981-16-0980-0_14

For heterogeneous data, NoSQL databases are more preferred than relational databases [2]. NoSQL databases are categorized into four classes based on their organization of data, as shown in Table 1 along with examples of each.

After rigorous review of different research works studying efficiency of storage and access, only the best three NoSQL graph databases, i.e., Neo4j, OrientDB and ArangoDB are compared taking examples from disease-symptom graphs. Comparisons are done with some benchmark operations, such as searching or traversing and querying the database. The main contributions of this paper are two fold. At first, as per popularity trend, eight NoSQL graph databases are studied in depth and their properties are compared. Secondly, based on this study, three most efficient NoSQL graph databases are evaluated to compare their performances. The ranking of the database management systems are found in [18], which shows that among graph databases, the ranks of Neo4j, ArangoDB and OrientDB are 1, 3 and 4, respectively, (as in August 2020). The three graph databases are open source, multimodal, distributed with document store and key-value database. Figure 1 shows the popularity of the graph databases.

Table 1 Categories of *NoSQL* databases

NoSQL database	Examples
Column store/column family	Hadoop/Hbase [3], Cassandra [4]
Document store	MongoDB [5], CouchDB [6]
Key value store	DynamoDB [7], Riak [8], Azure Table [9]
Graph database	Neo4j [10], InfiniteGraph [11], ArongoDB [12], OrentDB [13], Titan [14], AllegroGraph [15], HypergraphDB [16], DEX [17]

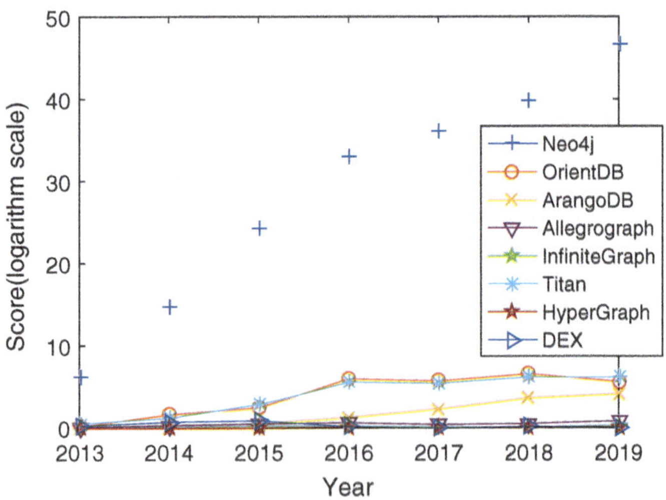

Fig. 1 Popularity trend to use three NoSQL graph databases [18]

The remainder of the paper is organized as follows. In Sect. 2, taxonomy of NoSQL graph databases is given with examples of each category. A survey of related works is given elaborately in Sect. 4 where different NoSQL graph databases are compared. Based on the review, three NoSQL graph databases are chosen and compared, and the results are shown in Sect. 4.3. Finally, we conclude the paper in Sect. 5.

2 Taxonomy of NoSQL Graph Databases

NoSQL graph databases are used for storing complex, unstructured and heterogeneous data using graph data model. Many NoSQL graph databases exist with different characteristics. They can be open source or close source. Document based, key-value store, wide-column, SQL graph based, distributed and multimodal based are the categories of open-source graph databases. Again, multimodal based, document based and key-value store belong to the close source graph databases. A graph database inherits all the properties of graph theory. By definition, it uses a graph data structure with vertices and edges and is distinguished by the graph data model. The graph data models can be property graph, RDF graph and hypergraph [20] as shown in Fig. 2. Examples of different NoSQL graph databases are shown in Table 2, and different properties are shown in Tables 3 and 4.

3 Study on NoSQL Graph Databases

Many research works [21–27] have been conducted for evaluation of performance of different graph databases. The main concerns of these works are storage and access. Different performance measures have been taken from the literature survey,

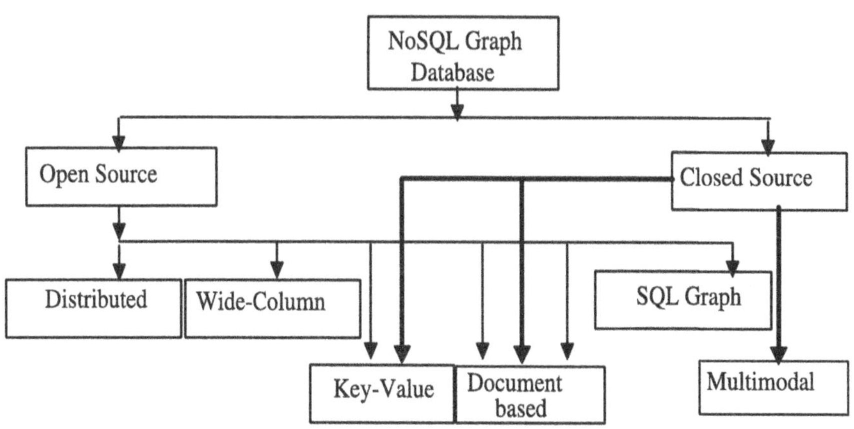

Fig. 2 Taxonomy of NoSQL graph databases [19]

Table 2 Examples of each category of NoSQL databases

NoSQL database	Distributed	Wide column	Key-value	Document store	SQL graph	Multi-modal	Close source	Open source
Neo4j	✓	X	X	X	X	X	X	✓
OrientDB	✓	X	✓	✓	✓	X	X	✓
ArangoDB	✓	X	✓	✓	✓	✓	X	✓
DEX or Sparkshre	✓	X	X	X	X	X	✓	X
InfiniteGraph	X	X	X	X	X	X	✓	X
Titan	✓	X	✓	X	X	✓	X	✓
AllegroGraph	✓	X	X	✓	X	✓	✓	X
HypergraphDB	✓	X	✓	X	X	X	X	✓

and it can be seen that *Loading*, *Query Searching* both are used for storage and access. The *Loading* term used to mean to load the whole graph database in NoSQL graph databases. *Query searching* is the non-traversal queries in which results are varied depending upon the performances of *NoSQL* graph databases. *Max Storage* of any graph database is the maximum capability to store the data locally if it is not supported by cloud, and it can be used only for the storage of the graph database. The traversal techniques, includes *BFS*, *DFS* and *Shortest Path*, *Read*, *Write*, etc., are used to measure the performances as access of the graph database.

There are lot of benchmarks evaluated to compare the efficiency of the graph databases. The study of the research works for the storage and access are described concisely in Table 3.

In [21], in order to find efficient NoSQL graph databases, the authors have made in comparison with six NoSQL graph databases; these are Tinkerpro Stack, DEX, InfiniteGraph, Neo4j, OrientDB and Titan after proposing *bluebench benchmark*. The authors used amazon's co-purchasing network and synthetic data for measuring the performances. In terms of storage efficiency of NoSQL graph databases, the higher to lower order are as DEX< Neo4j< InfiniteGraph < OrientDB < Titan. In terms of access time, *Shortest Path* algorithm, non-traversing query as well as delete operation, Neo4j is the most efficient compared to five other databases.

In [22], the authors have evaluated *Graph-DB benchmark* and shown in their work that Neo4j is faster compared to Tinker graph and OrientDB.
In another work [23], the authors have shown that *DEX* is the best performer among four databases like Neo4j, Jena and OrientDB.

In [24], the authors have compared DEX, OrientDB, Native Sail and SGDB graph databases. While loading a database, performance is better in case of Sail and SGDB than Neo4j, and when *BFS* traversal is considered, SGDB performs better than Neo4j.

Similarly, in [25], the authors have compared four NoSQL graph databases like Neo4j, OrientDB, Titan and DEX. In case of *load workload*, DEX required minimum time and Titan also required short time. Neo4j has taken less time than OrientDB. In traversal methods like *Neighbor search*, *BFS* and *Shortest Path*, Neo4j needed

Table 3 Different properties of graph database

Graph database	Interface	Storing	Query	Schema	Cloud supported	Concurrency	Written language	API
Neo4j	HTTP	Disk	Cypher	Schema free	Cloud enabled	MVCC	Java	Native Java API
ArangoDB	HTTP or REST	Memory	AQL	Schema free	Cloud supported	MVCC	C, C++, Java	Java RESTful HTTP API
OrientDB	HTTP or REST	Memory	SQL lite	Flexible schema	Cloud supported	MVCC	Java	Java API
InfiniteGraph	HTTP	Disk	Predicate language and graph navigation API	Schema-full	Cloud enabled	Update locking and concurrent and non-blocking ingest	C++, Java	Java API
DEX	REST	Disk	SPARQL	Schema supported	Cloud supported	Yes	C++, Java	DEX API
TITAN	HTTP or REST	Disk	GREMLIN	Yes	None	Cloud not supported	ACID tunable C	Titan API
HypergraphDB	REST	Disk	JSON	Yes	None	Cloud not Supported	STM	Java
AllegroGraph	REST	Disk	SPARQL and prolog	Yes	Cloud supported	Lock	Common Lisp	Java API

Table 4 Comparasion of different research work for different parametric measures

References	Access time								Storage		Database
	Loading	BFS	N. Search	S. Path	Update	Delete	Read	Queries	Loading	Query	
[21]	Neo4j	Neo4j		Neo4j	Neo4j			Neo4j	DEX		Amazon's N/W
[22]	Neo4j	Neo4j		Neo4j		Neo4j		Neo4j			Amazon's N/W
[23]		Sail & SGDB	SGDB & Neo4j								Synthetic dataset
[24]	DEX	Neo4j	Neo4j	Neo4j							Mico & Yeast
[25]						Neo4j	Neo4j				Social N/W
[26]	ArangoDB	Neo4j	Neo4j	Neo4j	Neo4j						Mico & Yeast
[27]					Neo4j						Social N/W

less time than other four graph databases. In case of *intensive workload*, DEX and OrientDB outperform Neo4j and Titan.

In another research paper [26], the authors compared the efficiency of the graph databases as well as document store or key-value store databases to observe storage and access performance. It is tested on social networking sites like Facebook and LinkedIn. The authors only compared against insertion and deletions of node. Neo4j has taken less time for insertion, and also it has taken less time for computation during edge insertion, compared to others like AllegroGraph, OrientDB, and FlockDB. Neo4j and FlockDB have taken less time for read operation.

In [27], the authors have focused on storage and querying tasks where the priority is on high-throughput and transactional operations. The authors compared six NoSQL graph databases, namely Neo4j, OrientDB, Sparksee or DEX, Titan, ArangoDB and BlazerGraph. The experiments are done with *MiCo* and *Yeast* real-life datasets. In terms of *loading* time, ArangoDB was the fastest, and Neo4j was almost equally fast. OrientDB and Sparksee achieved the least space consumption. In *micro-benchmark* result, in query completion rate, Neo4j is the best and OrientDB is the second best. Also for insertion, update and deletion operations, Sparkshree, Neo4j and ArangoDB took same time. Also the traversal time in almost all *micro-benchmark* conditions, the most efficient is Neo4j, followed by OrientDB and ArangoDB in this order.

The authors in [27] have compared FlockDB, OrientDB, Neo4j, InfoGrid and AllegroGraph. They have shown in the result that Neo4j took less time for computation and more time to do edge insertion compared to other graph databases in social network databases. Neo4j and FlockDB took less time for read operation.

Now, by observing the related research works, popularity of graph databases and different advantages of different graph database, it is clear that the three NoSQL graph databases, namely Neo4j, OrientDB and ArangoDB are most efficient than any other databases. In this work, these three NoSQL databases are compared for storage and accesses on disease-symptom graph data.

4 Experimental Evaluation

In this section, the specified NoSQL graph databases are compared to show the performances in terms of storage and access of health data. The data is represented as a *disease-symptom* [28] graph database, and analysis is done by some existing parametric measures on it. All the experiments have been done with Intel Core i3-5010U CPU 2.10GHZ, 4GB RAM and 64 bit windows machine.

4.1 Dataset

The dataset is a *disease-symptom* graph data. The *disease-symptom* graph database is defined as a directed acyclic graph in which *symptom* nodes are directed to *sign of symptom* and *sign of symptom node*s are directed to *disease* nodes. Now, it consists of around 200 nodes with approximately 15 *diseases*, 60 *symptoms* and 100 *sign of the symptoms* that are present in this growing dataset.

4.2 Metrics for Performance Analysis

We consider some parametric measures for the storage and access to compare the three efficient NoSQL graph databases. The parametric measures are related for the storage and access which are the memory requirements and time needed for executing all the operations. The operations are depicted as follows.

1. To execute set of queries to these NoSQL graph databases.
2. To load the full *Disease-Symptom* graph database into NoSQL graph databases to measure the *time* and *memory* needed.
3. *BFS* traversal technique is done for *disease-symptom* graph database in specified NoSQL graph databases.
4. Some operations like shortest path from source to destination, the access time and memory are considered. Further, deletion of a particular node, only memory is considered. Comparisons for the execution time for three NoSQL graph databases are done as well.

The following queries are tested on *disease-symptom* graph databases.
Q_1: To get the full disease-symptom graph database.
Q_2: To get the *symptoms* and *sign-of-the symptoms* for the disease "Diabetes."
Q_3: To get all the *symptoms* of the disease "Dengue."
Q_4: To remove the *sign-of-the-Symptom* "Left Lower" from the database.
Q_5: To get all the diseases for the symptom of "Fever."

We used to load the databases by .CSV file for all three NoSQL databases. However, different procedures are also available like JSON. In ArangoDB, *disease-symptom* graph dataset can be loaded by using the JSON or .CSV file as well as OrientDB. However, it can extract from JSON and an RDBMS directly through JDBC drivers. So, the loading time of any graph dataset into specified NoSQL graph databases is same. The *disease-symptom* graph database is loaded as .CSV file into three NoSQL graph databases and compared the memory storage and execution time.

Also *BFS*-based traversal technique can be used for access time for the Neo4j, ArangoDB and OrientDB. The *BFS* traversal technique is done for only with *Fever* as single source *Symptom* input.

4.3 Results and Discussion

The queries from Q_1 to Q_5 are tested on *disease-symptom* graph database for Neo4j, OrientDB and ArangoDB, NoSQL graph databases.

It is observed that for the Neo4j graph database, *query time* is less than any other two NoSQL graph databases. But *memory* requirement is greater than the two databases. Hence, it is concluded that for the small queries, *Neo4j* is the best for *query time* as well as *memory* requirement. ArangoDB and OrientDB have the highest access *time* as well as *memory* requirement. The *query time* and *memory* are shown in Figs. 3 and 4, respectively.

For the *Loading*, *BFS*, *Shortest Path*, *Deletion* of a node and *Neighbor Search* operations required less access time for Neo4j than OrientDB and ArangoDB, and it is shown in Figs. 5 and 6.

Hence, it is concluded from the experimental results that Neo4j is the best NoSQL graph database.

Fig. 3 Query time for the *disease-symptom* graph

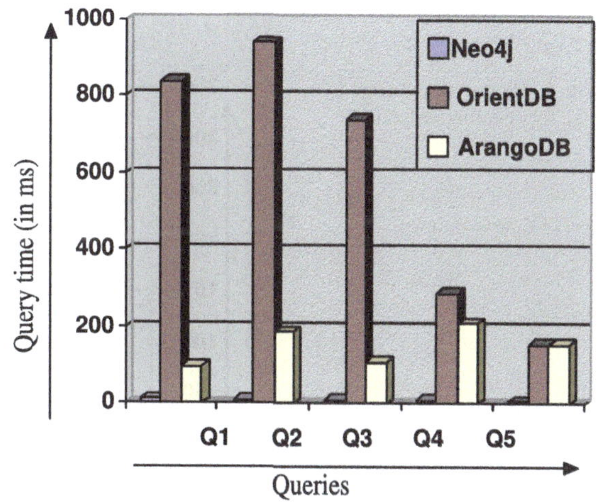

Fig. 4 Memory for the *disease-symptom* graph

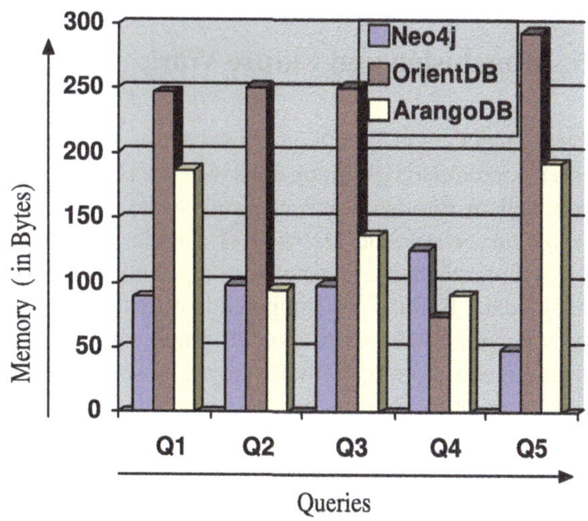

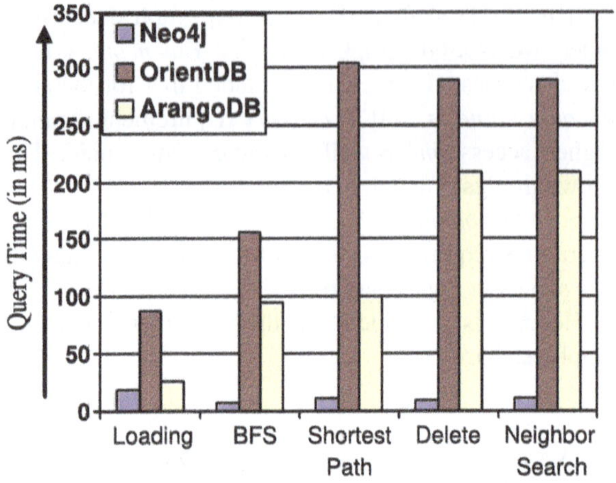

Fig. 5 Access time measure for specified operations

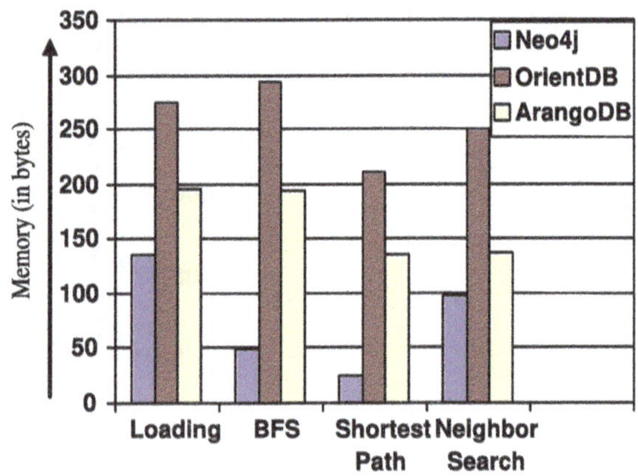

Fig. 6 Required memory size for specified operations

5 Conclusion and Future Work

In this paper, eight NoSQL graph databases are studied thoroughly, and among them, three popular NoSQL graph databases are chosen to compare the storage and access mechanism through some operations like *Loading*, *queries*, *BFS*, *Shortest Path*, *Neighbor Search* and *Deletion* of a node. It is shown that Neo4j is the efficient NoSQL graph database in terms storage and access mechanism for *disease-symptom* health data. But it is not compared in distributed environment and also not included in comparison with more NoSQL graph databases. Our future work will be introduced much more operations among more NoSQL graph databases in distributed environment for *disease-symptom* graph database.

References

1. Bartholomew, D.: Sql vs. nosql. Linux J. **2010**, 195 (2010)
2. Bernardino, J., Furtado, P., Abramova, V.: A performance overview. open j. databases. Open J. Databases, 1
3. White, T.: Hadoop: The Definitive Guide. O'Reilly Media, Inc. (2012)
4. Lakshman, A., Malik, P.: Cassandra: a decentralized structured storage system. SIGOPS Oper. Syst. Rev. **44**(2), 35–40 (2010)
5. Kristina, C., Michael, D.: MongoDB: The Definitive Guide, 1st edn. O'Reilly Media, Inc. (2010)
6. Chris Anderson, J., Lehnardt, J., Slater, N.: CouchDB: The Definitive Guide Time to Relax, 1st edn. O'Reilly Media, Inc. (2010)
7. Sivasubramanian, S.: Amazon dynamodb: a seamlessly scalable non-relational database service. In: Proceedings of the 2012 ACM SIGMOD International Conference on Management of Data, SIGMOD '12, pp. 729–730, New York, ACM (2012)
8. In http://basho.com/products/riak-kv/
9. Oh, G., Seo, C., Mayuram, R., Kee, Y.-S., Lee, S.-W.: Share interface in flash storage for relational and nosql databases. In: Proceedings of the 2016 International Conference on Management of Data, SIGMOD '16, pp. 343–354. ACM, New York (2016)
10. Webber, J.: A programmatic introduction to neo4j. In: Proceedings of the 3rd Annual Conference on Systems, Programming, and Applications: Software for Humanity, SPLASH '12, pp. 217–218. ACM, New York (2012)
11. Koloniari, G., Pitoura, E.: Transaction management for cloud-based graph databases. In: Revised Selected Papers of the First International Workshop on Algorithmic Aspects of Cloud Computing, vol. 9511, ALGOCLOUD 2015, pp. 99–113, Springer, Berlin, Heidelberg (2016)
12. In https://www.arangodb.com/
13. In https://orientdb.com/
14. Wu, J., Chen, L., Xie, Y., Zheng, Z.: Titan: a system for effective web service discovery. In: Proceedings of the 21st International Conference on World Wide Web, WWW '12 Companion, pages 441–444. ACM, New York (2012)
15. In https://franz.com/agraph/allegrograph/
16. In http://hypergraphdb.org/
17. Martínez-Bazan, N., Gómez-Villamor, S., Escalé-Claveras, F.: Dex: a high-performance graph database management system. In: 2011 IEEE 27th International Conference on Data Engineering Workshops, pp. 124–127 (2011)
18. In https://db-engines.com/en/
19. In http://nosql-database.org/
20. Angles, R.: The property graph database model. In: Proceedings of the 12th Alberto Mendelzon International Workshop on Foundations of Data Management, Cali, Colombia, May 21–25, 2018 (2018)
21. Kolomičenko, V., Svoboda, M., Mlýnková, I.H.: Experimental comparison of graph databases. In: Proceedings of International Conference on Information Integration and Web-based Applications & #38; Services, IIWAS '13, pp. 115, 115–115:124. ACM, New York (2013)
22. In http://code.google.com/p/graphdb-bench/
23. Dominguez-Sal, D., Urbón-Bayes, P., Giménez-Vañó, A., Gómez-Villamor, S., Martínez-Bazán, N., Larriba-Pey, J.L.: Survey of graph database performance on the hpc scalable graph analysis benchmark. In: Shen, H.T., Pei, J., Tamer Özsu, M., Zou, L., Lu, J., Ling, T.-W., Yu, G., Zhuang, Y., Shao, J., (eds) Web-Age Information Management, pp. 37–48, Berlin, Heidelberg (2010)
24. Ciglan, M., Averbuch, A., Hluchy, L.: Benchmarking traversal operations over graph databases. In: 2012 IEEE 28th International Conference on Data Engineering Workshops, pp. 186–189 (2012)
25. Jouili, S., Vansteenberghe, V.: An empirical comparison of graph databases. In: 2013 International Conference on Social Computing, pp. 708–715 (2013)

26. Lissandrini, M., Brugnara, M., Velegrakis, Y.: Beyond macrobenchmarks: Microbenchmark-based graph database evaluation. Proc. VLDB Endow. **12**(4), 390–403 (2018)
27. Mathew, A.B., Madhu Kumar, S.D.: Analysis of data management and query handling in social networks using nosql databases. In: 2015 International Conference on Advances in Computing, Communications and Informatics (ICACCI), pp. 800–806 (2015)
28. Mondal, S., Mukherjee, N.: Mobile-assisted remote healthcare delivery. In: 2016 Fourth International Conference on Parallel, Distributed and Grid Computing (PDGC), pp. 630–635 (2016)

Finding MST by Considering Increasing Cost Order of the Edges Incident on Vertices

Rayudu Srinivas, R. V. S. Lalitha, T. Rama Reddy, and B. Durga Anuja

Abstract Finding a minimum cost spanning tree (MST) plays a key role in many applications. For a given graph $G(V, E)$, there may be more than one spanning tree (ST) that can be constructed, but out of all these STs, the ST which has least cost edges is called MST. Finding MST in linear time is important. In this paper, a novel approach is proposed to find MST. The approach used in finding the MST is based on the selection of vertices and least cost edge incident on that vertices. The approach proposed is simple and easy to implement. The time complexity of this algorithm is $O(n\log n)$.

Keywords Cycle · Graph · Minimum cost spanning tree · Spanning tree · Tree

1 Introduction

Spanning tree of a graph is a tree which includes all the vertices of a graph and minimum number of edges that connects all these vertices. A minimum spanning tree of a weighted graph is a spanning tree in which the sum of the weight of all its edges is a minimum of all such possible spanning tree of the graph [1–3].

Given an undirected graph $G(V, E)$, where V, E denote set of vertices and edges, and a real number is associated with each edge called weight. The weight of an edge e is represented as $w(e)$ where $e \in E$. Minimum spanning tree is defined $T^*(V, E')$ on G such that $w(E') = \min T \sum w(e)$, the minimum taken over all possible spanning trees of G. Similarly, maximum spanning tree can also be defined.

R. Srinivas (✉) · T. R. Reddy
Aditya Engineering College, Surampalem, Andhra Pradesh, India
e-mail: srinivas.rayudu@aec.edu.in

R. V. S. Lalitha
Aditya College of Engineering and Technology, Surampalem, India
e-mail: rvslalitha@acet.ac.in

B. D. Anuja
Govt. Degree College for Womens Srikalahasti, SriKalahasti, Andhra Pradesh, India

V. Bhateja et al. (eds.), *Computer Communication, Networking and IoT*, Lecture Notes in Networks and Systems 197, https://doi.org/10.1007/978-981-16-0980-0_15

In literature, a number of methods were proposed. These methods can be classified as classical methods and modern method. MST has long history and first algorithm is proposed in 1926 [4]. The popular classical methods are Boruvka's, Prim's, and Kruskal's algorithms [5–9]. The Boruvka's algorithm is suitable for parallel implementation. The time complexities of classical algorithms are $O(m\log n)$ [5]. Where m is the number of edges and n is the number of vertices. The maximum number of edges in an undirected graph without a self-loop is $n(n - 1)/2$ [10]. Heap can be used in Prim's and Kruskal's algorithms to further minimize the complexity.

In recent years, researchers focused on developing new algorithms by using advanced data structures like F-heap, H-heap, etc., Boruvka's algorithm performance is improved by using heap. These concepts were developed by Yao and Cheriton and Tarjan [11–13].

The objective of this paper is to propose a novel approach to find the spanning tree (ST) of a graph. Second contribution to this paper is to find the cycles while considering edges to find the ST.

2 Literature Survey

To build an efficient and effective network, constructing MST is essential. To fulfill the need based on the application, maximum cost spanning tree or minimum cost spanning tree may be required to build. While considering network efficiency, maximum cost spanning tree is considered, whereas minimum cost spanning tree is considered if the cost/distance between two nodes is used as parameter. Not only in network problem, it can also be used in the design of computers, power distribution, transportation, communication, pipeline network, links road network, electrical circuits, islands connection, etc. [14].

Construction of ST is not a new problem. This problem is solved by number of researchers. Still there is great scope to work on this problem to develop a method which will take linear time to implement and simple to implement.

In literature, a number of methods were proposed out of these methods, and three methods are popularly used. These are Boruvka's, Prim's, and Kruskal's. The time complexity of these algorithms is not linear. Various problems on tree and MST are addressed by Pierce et al. [15–18].

In literature, a number of algorithms are proposed which are easy to implement. Algorithms and computational experiment reports found in [19–21].

In the last few years, researchers are focused on developing faster algorithm by using the power of new tools and advanced data structures. A novel method was proposed by Karger et al. [22]. This is restricted random-access computation model and has linear expected time.

3 Methodology

A novel algorithm is proposed to find MST by considering one least cost edge for one node and continuing the same approach until ST is formed. The proposed algorithm is explained by considering the following graph as shown in Fig. 1. This graph has eight vertices and ten edges. The adjacency list is given in Table 1.

3.1 Proposed Algorithm

Step 1: For each vertex, find the least cost edge incident on that vertex.
Step 2: Construct components using the selected edges. If a cycle is formed, then find the costly edge in the cycle and remove that edge.
Step 3: Repeat step until single component is formed.

i. Using adjacency list, find the least cost edge that connects the components and constructs single component.

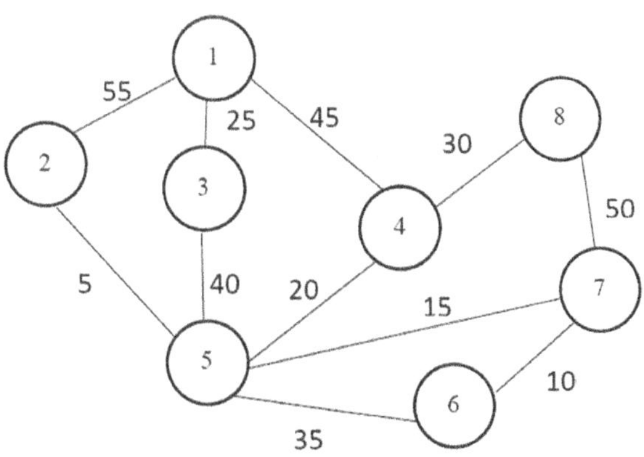

Fig. 1 Example graph

Table 1 Vertex in degree and adjacency list

Vertex	In degree	Adjacency vertices
1	3	2, 3, 4
2	2	1, 5
3	2	1, 5
4	3	1, 5, 8
5	5	2, 3, 4, 6, 7
6	2	5, 7
7	3	5, 6, 8
8	2	4, 7

To illustrate how this algorithm will work, consider the graph as given in Fig. 1. Apply step 1 on this graph.

For vertices 3 and 5, least cost edges are already selected while considering vertices 1 and 2 (Fig. 2).

Apply step 2 on the result of step 1. The components constructed are given in Fig. 3a, b.

Prepare adjacency list for each component and cost of the edge in ascending for connecting these edges. Now apply step 3 for obtain MST (Table 2).

After applying step 2, three components are formed these are {1, 3}, {2, 4, 5, 8} and {6, 7}. Find the least cost edges that connect these components. To connect first and second components, three edges with the cost 40(3, 5), 45(1, 4), and 55(1,2) are there. To connect second and third components, three edges are there. These are (6, 5), (7, 8), and (7, 5) and cost of these edges are 35, 50, and 15, respectively. But there

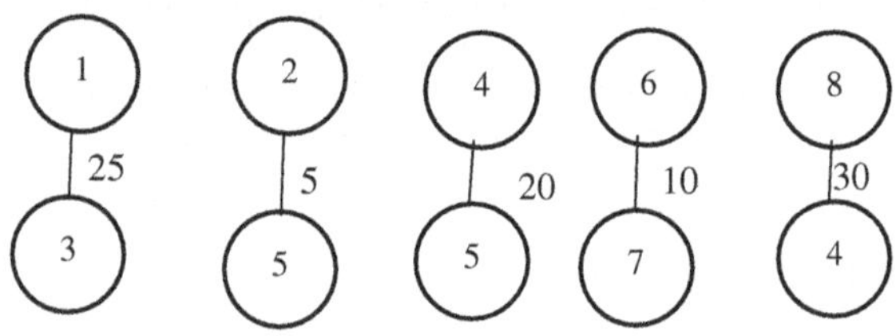

Fig. 2 Components formed by step 1

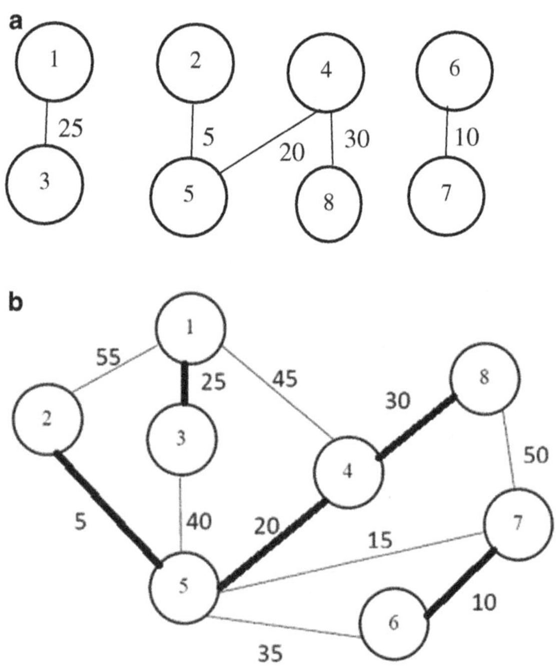

Fig. 3 **a** Individual components formed by step 2. **b** Components formed by step 2 in graph

Table 2 Components in adjacency list

Component	Adjacency vertices	Description
1, 3	2, 4, 5	Second component nodes are adjacent
2, 4, 5, 8	1, 3, 6, 7	Both first and third component nodes are adjacent
6, 7	5, 8	Only second component nodes are adjacent

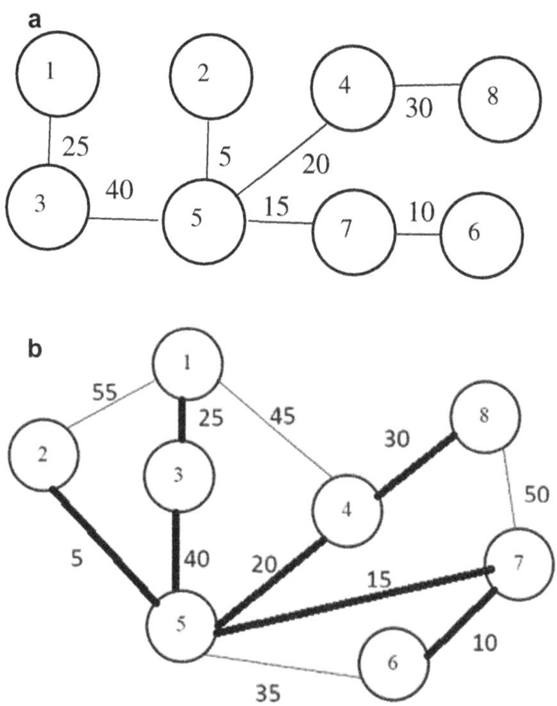

Fig. 4 **a** Minimum cost spanning tree of Fig. 1 by considering components. **b** Minimum cost spanning tree of Fig. 1

are no edges to connect component one and three. Consider least cost edges (3, 5), (7, 5) to connect these components. Final result is shown in Fig. 4a, b is MST.

3.2 *Time Complexity of the Proposed Algorithm*

If the edges incident on vertex are in sorted order, then the selection of this least cost edge for once vertex will take $O(1)$ time. For N vertices, it will be $O(N)$. It is time complexity for step 1 of this algorithm. After this step, maximum $N/2$ components will be formed, and for combining these components will take time $O(N/2)$.

This procedure is repeated until one component is formed. To form one component, maximum $\log(N)$ steps are required. So the overall time complexity is $O(N) + O(N/2) + \cdots O(1) = O(N \log N)$.

4 Conclusion

In this paper, a novel algorithm is proposed to construct MST by processing edge of the graph incident on the vertices based on increasing order of the cost of the edges. Further, the performance of this algorithm can be improved by using MinHeap. The proposed algorithm time complexity is good compared to existing algorithm. The proposed algorithm is easy to understand and requires less time to generate code and execute. The proposed method can also be used to construct minimum cost spanning tree and maximum cost spanning tree.

References

1. Cormen, T.H., Leiserson, C.E., Rivest, R.L., Stein, C.: Introduction to Algorithms, 3rd edn. PHI Learning Private Ltd. pp. 624–642 (2009)
2. Motwani, R., Raghavan, P.: Randomized algorithms. Chapman & Hall/CRC (2010)
3. Wu, B.Y., Chao, K-M.: Spanning Trees and Optimization Problems. CRC Press (2004)
4. Boruvka, O., Jistem, O.: Problem minimalnim. Praca Moravske Prirodovedecke. Spolecnosti **3**, 37–59 (1926)
5. Pettie, S., Ramachandran, V.: An optimal minimum spanning tree algorithm. In: Automata, Languages and Programming, pp. 49–60. Springer, Berlin (2000)
6. Tarjan, R.E.: Data structures and network algorithms. In: CBMS-NFS Regional Conference Series in Applied Mathematics. Society for Industrial and Applied Mathematics, Philadelphia, p. 44 (1983)
7. Prim, R.C.: Shortest connection networks and some generalizations. Bell Syst. Techn. J. 1389–401 (1957)
8. Dagar, S.: Modified Prim's algorithm. IJCIT **03**(02). ISSN 2218-5224 (2012)
9. Kruskal, J.B.: On the shortest spanning subtree of a graph and the traveling salesman problem. Proc Am Math Soc **7**, 48–50 (1956)
10. Mandal, A., Dutta, J., Pal, S.C.: A new efficient technique to construct a minimum spanning tree. Int. J. **2**(10) (2012)
11. Yao, A.: An O(|E|log log |V|) algorithm for finding minimum spanning trees. Inf. Process. Lett. **4**, 21–23 (1975)
12. Fredman, M.L., Tarjan, R.E.: Fibonacci heaps and their uses in improved network optimisation algorithms. J. Assoc. Comput. Mach. **34**(3), 596–615 (1987)
13. Cheriton, D., Tarjan, R.E.: Finding minimum spanning trees. SIAM J. Comput. **5**, 724–742 (1976)
14. Patel, N., Patel, K.M.: A survey on: enhancement of minimum spanning tree. Int. J. Eng. Res. Appl. **5**(1), 06–10. ISSN: 2248-9622 (2015)
15. Pierce, A.R.: Bibliography on algorithms for shortest path, shortest spanning tree and related circuit routing problems (1956–1974). Networks **5**, 129–149 (1975)
16. Graham, R.L., Hell, P.: On the history of the minimum spanning tree problem. Ann. History Comput. **7**, 43–57 (1985)

17. Maffioli, F.: Complexity of optimum undirected tree problems: a survey of recent results. In: Ausiello, G., Lucertini, M. (eds) Analysis and design of algorithms in combinatorial optimization. International Center for Mechanical Sciences. CISM Courses and Lectures, vol. 266, pp. 107–28. Springer, New York (1981)
18. Bazlamacci, C.F., Hindi, K.S.: Minimum weight spanning tree algorithms a survey and empirical study. Comput. Oper. Res. **28**, 767–785 (2001)
19. Haymond, R.E., Jarvis, J.P., Shier, D.R.: Computational methods for minimum spanning tree algorithms. SIAM J. Sci. Stat. Comput. **5**(1), 157–174 (1984)
20. Moret, B.M.E., Shapiro, D.: How to find a minimum spanning tree in practice. In: Maurer, H. (ed.) New results and new trends in computer science: proceedings, Graz, Austria, June 1991. Lecture notes in computer science, vol. 555, pp. 192–203. Springer, Berlin (1991)
21. Glover, F., Klingman, D., Krishnan, R., Padman, A.: An in-depth empirical investigation of non-greedy approaches for the minimum spanning tree problem. Eur. J. Oper. Res. **56**, 343–356 (1992)
22. Karger, D.R., Klein, P.N., Tarjan, R.E.: A randomized linear-time algorithm to find minimum spanning trees. J. Assoc. Comput. Mach. **42**(2), 321–328 (1995)

Analysis of Sybil Attacks in Online Social Networks Using SyPy

Christina Roseline, Blessy Antony, Bijitha Balakrishnan, Chaithra C. V., and P. V. Bhavyasree

Abstract Sybil attacks in online social networks have crucial security implications. The problem of pernicious activities in these networks, such as Sybil attacks and use of bogus identities, can severely affect the social activities in which users interact. Hence, these types of spiteful accounts must be observed and deactivated. In this paper, we simulate an online social network with an honest region and a Sybil region. This paper mainly focuses on the various parameters in the online social network and scrutinizing the impact of these parameters on Sybil attack. It is equally important to detect the most vulnerable nodes in the network to avert them from malicious users. This paper requires extensive knowledge about graphs and its properties are probability triads, clustering coefficient, degree distribution, etc.

Keywords Sybil attacks · Graph based · Vulnerable nodes · Degree distribution

1 Introduction

Widening of online social networks (OSNs), for example, Twitter, Facebook, and YouTube, has changed the manners by which individuals associate, think, and lead business. Recently, people have encapsulated tremendous action via Web-based media destinations, bringing about an extraordinary measure of client produced content at a persistent movement. The tremendous amount of users on OSNs has made them the greatest wellspring of solid character administrations for the Internet. As the ubiquity and utilization of OSNs have expanded, the incitement to assault these frameworks have likewise developed. A malevolent client makes different counterfeit characters to get admittance to significant private data to perform different sorts of cybercrimes like trading off information honesty, "savaging" that is making purposely provocative or analysis of feelings or hostile online postings, misleading, fixing notoriety, and violating trust in online affiliations. These attacks are types of Sybil attacks [1]. Sybil attack derives from the book Sybil that emphasizes a case study concerning a woman called Sybil Dorset who was treated with multiple identity disorder [1].

C. Roseline (✉) · B. Antony · B. Balakrishnan · Chaithra C. V. · P. V. Bhavyasree
SCMS Engineering and Technology, Ernakulam, Kerala, India

V. Bhateja et al. (eds.), *Computer Communication, Networking and IoT*, Lecture Notes in Networks and Systems 197, https://doi.org/10.1007/978-981-16-0980-0_16

OSNs have become an essential aspect of our everyday life, with numerous individuals totally depending on them in the domain of work, social communications, data sharing, and different parts of day-by-day living [2]. Any depressing impact on these zones because of the assaults harms the client experience and furthermore the respectability and security of the given organization. They are made as open stages. To horrendous clients, the most captivating some portion of OSNs is that they can be promptly associated with numerous clients at an attainable expense in contrast with other Web channels. Another major OSN proviso is that online clients will in general confide in their social associations and aimlessly esteem them. In this way, they may fall into a snare set by counterfeit social associations submitted by digital aggressors.

Different explores have indicated that nearly every individual who has invested some energy in Twitter and Facebook has interacted with bogus profiles. In actuality, 1 out of 20 user accounts on Twitter is believed to be Sybil, while on Facebook, somewhat more than 1 of every 100 dynamic users are bogus personalities [3]. The Sybil accounts are typically run by low-paid people or even via computerized bots with the point of swelling devotees, pushing spam or malware and in any event, impacting public talk. The fascinating piece is that Sybil accounts are very simple to get on the Web. The utilization of these records to make social help and to impact genuine clients is raising. A genuine model can be found in a recent report on 12 million clients of Weibo—a Chinese informal community like Twitter. This examination found that up to million clients partook in crusades that hoped to impact clients on items and administrations. The greater part of them engendered messages and referenced items and administrations from certain persuasive records—prone to be the brand behind these records. Sybil assault will undoubtedly get a lot greater and more extensive in the coming years, as informal communities become much more firmly combined with individual spending. In other words, users might need to be cautious previously tolerating any new companion demands, as the underground market in purchasing and selling counterfeit records will undoubtedly get considerably greater. Hence, it is very much essential to detect and eradicate all the Sybil accounts from online social networks.

In general, the Sybil detection [4] schemes are divided into four main categories:

- Graph-based schemes
- Machine learning-based schemes
- Manual verification
- Prevention approaches

In this project, we make use of graph-based methods [4]. Graph-based methods with the employment of social network information, represent interdependencies between objects using edges or links, within the case of Sybil. These techniques emphatically rely on social chart properties to distinguish Sybil clients from real clients [5]. Inside the social graph, Sybil and honest region exist, and furthermore the associations between them are free, as appeared in Fig. 1.

In our work, using graph-based methods, we initially simulate an online social network and then analyze the various parameters in this network which can have a

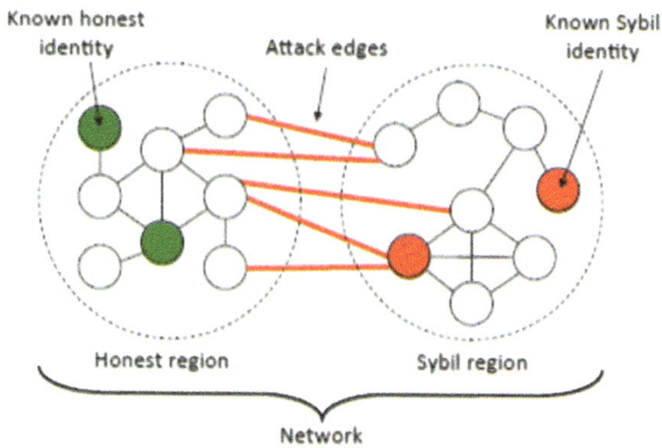

Fig. 1 Honest and Sybil region

great impact on Sybil attacks. This paper also makes use of SyPy module which is an easy-to-use Python package for Sybil node detection in social and information networks. Using this simulated network and with the utilization of SyPy Python package, we could efficiently determine how Sybil attacks can accelerate and decelerate with modification in these parameters. Once the impact of different parameters on Sybil attack is identified, the users vulnerable to these attacks can be detected effortlessly in real life.

This paper is sorted out as follows: Sect. 2 provides the related work of this topic; Sect. 3 provides the overview of the system architecture; Sect. 4 discusses the experiments and results; and Sect. 5 concludes the paper.

2 Related Works

Graph-based Sybil detection [4] has gained greatest attention in today's world. In graph-based Sybil detection, Y. Boshmaf in 2013 proposed a unified framework for evaluation and analysis of machine learning algorithms. The changes made over graphs are analyzed continuously and local communities are identified by making use of graph-based Sybil detection (GSD) [6]. Here, different models are stipulated to analyze, configure, design, implement, and assess the existing graph-based Sybil detection algorithms. Many of the real-world algorithms are evaluated and in such a way Facebook is analyzed to take up the necessities of designing and analysis of algorithms. This model basically focuses on open online social networks. Facebook, Twitter, Dropbox, etc., are the examples of such online social networks (OSNs). In OSNs, the user identification and their liaisons are prototyped as graphs. A graph $G = (V, E)$, where every node $u \in V$ represents a user identity and every edge $e = (u, v) \in E$ represents a relationship between two identities. GSD algorithms are handled by finding communities around known honest nodes [3]. There are basically two communities: global or local community detection. The global community detection is done by using Girvan and Newman algorithm [7], and the local community detection is done by Andersen, Chung, and Lang algorithms [8]. This paper enables the analysis and design of the existing and the new GSD algorithms that provides security and privacy that can be easily evaluated empirically.

The use of trust relationships, such as social links, to counter undesirable communication is one of the major threats pointed by fake accounts. Ostra [9], proposed by Mislove et al. in 2008, delves into this concern. Ostra limits the total amount of undesirable connection a user can create that is based on the number of social links the user has. It then relies on the matter that it is challenging for a malicious user to create arbitrarily many such trust relationships. Here, a method is explained that can exploit the already existing trust relationships between users to impose a cost on the senders of unwanted communication in a way that avoids the drawbacks of existing solutions. This system depends on extending trust relationships to link senders and receivers through chains of pairwise trust relationships. It then uses a pairwise, link-based credit scheme that imposes a cost on originators of undesirable communications without demanding any sender authentication or global identities. The system also relies on criticisms and ideas from receivers to distinguish unwanted communication. Any user who still wishes to continue sending such communication risks isolation and eventually loses the ability to communicate. This system can make use of such existing social links because achieving and maintaining a relationship often demands some effort. With respect to Ostra, this property of a social network ensures that an attacker cannot achieve and maintain randomly many relationships or replace the lost relationships effortlessly. Ostra is widely applicable. It is suitable for both content-sharing systems like YouTube as well as messaging systems such as e-mail.

Integro [10] proposed by Boshmaf in 2008 is a scalable defense system that helps online social networks (OSNs) to detect automated fake accounts using a robust user ranking scheme. Integro is mainly designed for the social networks whose social relationships are bidirectional and whose ranking processes are fully transparent to the users. It starts with estimation of the accounts of the victim's user-level activity like gender, time, number of friends, which are cheap to extract from the user, since the last update. These are extracted from the user in order to train the classifier to predict the unknown labels in an online social network. After the activities of a user is detected, Integro combines these predictions into graphs as their weights so that the edges incident to predicted victims have much lower weights than the others. At last, it lists user accounts based on a modified random walk [11] that starts from the real account. Integro also guarantees that most of the real accounts ranks higher [2] than fake accounts so that the OSN operators can take actions against the low-ranking fake accounts. In Integro, as victims are directly connected to the fake accounts, they form a border line [12] separating real accounts from the fake accounts in an online social network. Based on the landing probability of the modified random walk that begins from a real known account, it also ranks the user accounts. Integro achieves linear scalability with the number of nodes in a graph. Integro computationally practical even for large OSNs such as Facebook [13] and Tuenti as it thwarts fakes in the wild with at least 10 times more precision.

GANG [14], proposed by Wang et al. in 2017, is a guilt-by-association technique that is utilized in coordinated charts, to perceive misrepresentation clients in OSNs. In GANG, a parallel irregular variable is related with every client and a mark is allocated, and afterward a novel pairwise Markov random field (pMRF) [11] is designed to model the probability distribution of all the random variables based on the directed social graph. This pMRF assimilates unique characteristics of the fraud users detection problem. The pMRF produces a larger joint probability, if two users in a graph have the same label and are linked by bidirectional edges. But if u and v are linked by a unidirectional edge (u, v), and if u is fraudulent or v is normal or vice versa, then whether the unidirectional edge (u, v), does not influence the joint probability of pMRF and this can be because of the fact that a malicious user can follow any other users without being followed back, whereas a honest user can be followed by any other users without following them back. In GANG, to measure the posterior probability distribution for each binary random variable, loopy belief propagation (LBP) is used [15] and for the corresponding user, it is used to predict the label. But there were many problems with the basic version [15]. Therefore, the later versions of GANG are optimized in such a way that it includes removing message maintenance and estimating GANG as a concise matrix form. The later versions of GANG are also evaluated and compared with various existing guilt-by-association methods using a large-scale Twitter dataset and a large-scale Sina Weibo dataset. Both datasets have labeled fraudulent and normal nodes and have directed edges. Results additionally exhibit that GANG outperforms existing guilt-by-association methods.

3 Architecture

The proposed architecture is shown in Fig. 2. The social network is defined by a graph $G = (V, E)$ consisting of a set of vertices V, denoting the user accounts on the social network and E, a set of edges, that corresponds to the social relationships between users. A bidirectional relationship exists between edges and is incarnated with an undirected edge denoting that two nodes trust each other. The subgraph of G consists of two regions—honest region and the Sybil region. The honest region contains honest nodes and the Sybil region contains the Sybil nodes in the network.

The network consists of two kinds of nodes—the honest nodes and the Sybil nodes. The honest nodes are the benign users in the network, some of the honest nodes are considered to be known nodes/labels. The honest users that accept the friend requests from the malicious users are said to be the victim nodes. The Sybil nodes are the bogus identities in the network that perform various adversarial activities in the network like stealing the user data, spreading the misinformation and even distributing the malware.

There are also three different kinds of edges—honest edges, Sybil edges, and the attack edges. The honest edges are the edges that connect the honest nodes with each other. The Sybil edges are the edges that connect the Sybil nodes with each other. The edge that connects the Sybil node in the Sybil region and the honest node in the

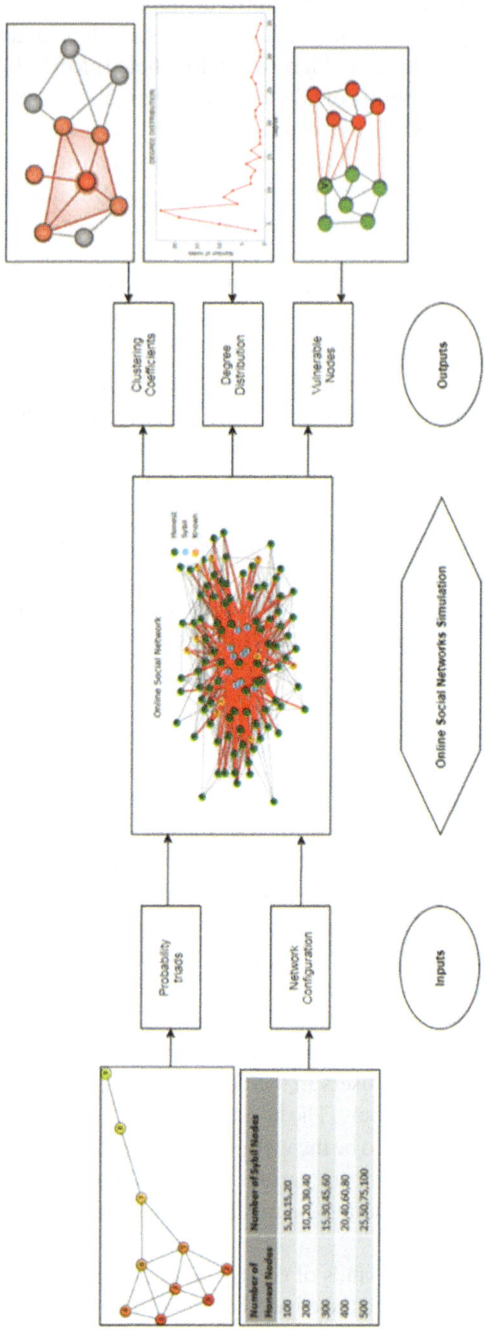

Fig. 2 Proposed architecture

honest region is the attack edge, that is the attack edges create a connection between these two regions.

Our mechanism uses a Python package SyPy for the simulation and the analysis of the network. It is built around the three assumptions:

1. The honest region is fast mixing, that is it is well connected.
2. The defender knows at least one honest node.
3. The network structure consists of two regions—Sybil region and the honest region.

Our model works in two steps. The process starts with a network simulation. In the first step we create a network consisting of honest nodes, Sybil nodes, some known labels and randomly stitches the attack edges between them. Both the honest and Sybil regions are connected by attack edges between them. In the second step, a performance analysis is done on the above simulated network. This can be performed by varying the various parameters in the simulated network which incorporates the network configuration and also the probability of forming triads to determine the vulnerable nodes and the clustering coefficient, respectively.

3.1 Network Simulation

The network simulation deals with the creation of both the Sybil and also the honest regions. The regions are created using power law graph functions included within the network.

Power Law Graphs

In the power law graph, most nodes have a relatively low level of connectivity, but a few nodes will have a high degree of connectivity. The nodes with the large degree form the hubs in the network. The existence of hubs will give the degree distribution a long tail (Fig. 3).

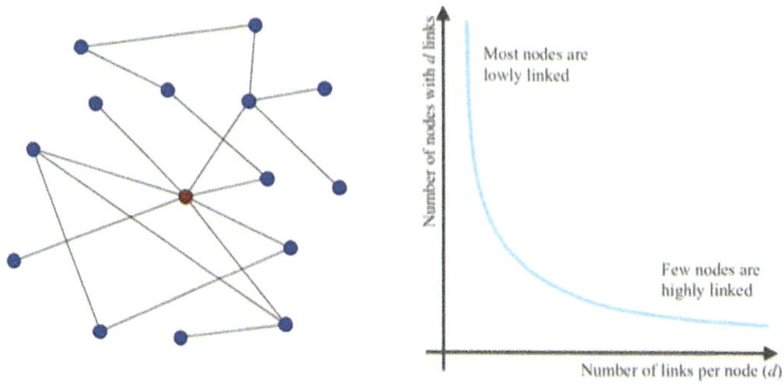

Fig. 3 Power law graph

Table 1 Different network configurations

Number of honest nodes	Number of Sybil nodes
100	5, 10, 15, 20
200	10, 20, 30, 40
300	15, 30, 45, 60
400	20, 40, 60, 80
500	25, 50, 75, 100

After the regions are created, we randomly stitch the attack edges between these two regions. It is done by randomly picking one node from each of the two regions and stitching edges between these two regions.

3.2 Vulnerable Nodes Detection

The most vulnerable nodes in the network are found out. Vulnerable nodes are more likely to be the victims than others. They are the honest nodes that have more number of connections with the Sybils and are more sensitive to Sybil attacks. We then make variations in the network configuration.

Configurations used for detecting the vulnerable nodes in various iterations are as follows:

From Table 1, it is clearly visible that for the given number of honest nodes, its 5, 10, 15, and 20% of Sybil nodes will be used in the various iterations.

3.3 Determining Clustering Coefficient

A clustering coefficient is the degree to which nodes in a graph tend to form a cluster. In simple words, it is said to be the extent to which one's friends are also friends of each other.

Clustering coefficient is given by the equation:

$$\text{Clustering coefficient} = \frac{\text{Number of triads formed}}{\text{Total no. of triads possible}} = \frac{2T(u)}{\deg(u)\deg(u-1)} \quad (1)$$

Clustering coefficient for node u can also be defined as the fraction of possible triangles that exist for the node u. In Eq. (1), $T(u)$ is the triangles through u and $\deg(u)$ is the degree of the node u. Clustering coefficient is determined by varying the probability triads.

Probability Triads

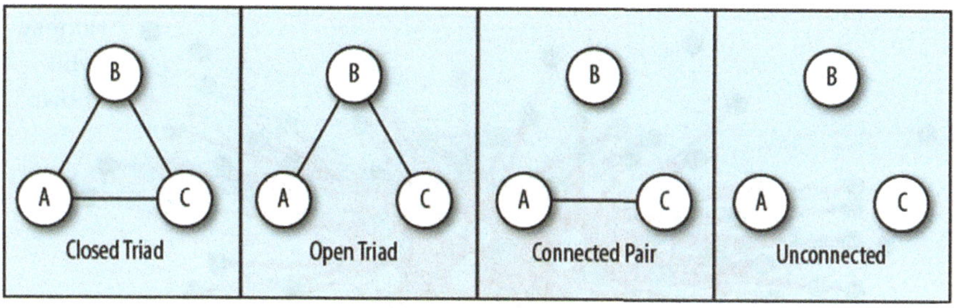

Fig. 4 Four possible triadic states in a graph

Probability triad is the probability of adding a triangle between the node connections. It can be defined as the property among the three nodes *A*, *B* and *C* such that if a strong relation exists between *A* and *B* and also between *B* and *C*, then there is a probability of a strong relation between *A* and *C* (Fig. 4).

3.4 Degree Distribution

The degree of a node in a network is the number of connections it has to other nodes and the degree distribution is the fraction of vertices with a particular degree. It describes a relationship between the degree of a node and its frequency of occurrence. Most of the real-world networks exhibit power law degree distribution, that is there are few nodes with higher degree and many nodes with very less degree.

4 Experiments and Results

This is the visualization of our simulated graph. We simulated our graph with the base configuration being 100 honest nodes represented by green color, 10 Sybil nodes represented by blue and we randomly selected 10 known honest nodes represented in yellow. The attack edges are in red color and honest connections are in black color (Fig. 5).

4.1 Vulnerable Nodes

Vulnerable nodes in the various iteration are determined in the graphs given in Fig. 6.

We have inferred from the above graphs, when the number of Sybils are increased, simultaneously the vulnerable nodes are also increased.

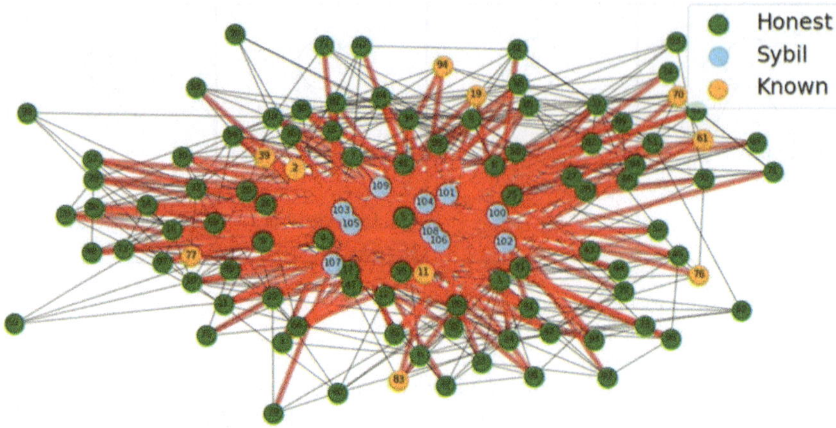

Fig. 5 A simulated online social network

Exception: In Fig. 6a, we have observed that when the number of honest nodes is 100 and the number of Sybil nodes is 5, there is a huge growth in the number of vulnerable nodes. In this configuration, there are 200 attack edges (double the number of honest nodes) and even though there are only 5 Sybil nodes, they will try to make 200 connections, thus there is an exceptional increase in the vulnerable nodes (blue curve).

4.2 Clustering Coefficient

Clustering coefficient by varying the probability triad is determined in the graph is given in Fig. 7.

From the above graphs, we have observed that when the probability triad is varied in each iteration, gradually there is an increase in the clustering coefficient since this parameter is dependent on the value of triads. As a result, it can accelerate the rate of Sybil attack. If the maximum value of the clustering coefficient for a node is 1 then that node will have maximum degree (Fig. 7c).

4.3 Degree Distribution

We plotted a graph for degree distribution as shown in Fig. 8. Since we use power graphs for the simulation of the two regions—honest and the benign regions—the degree distribution follows a power law distribution. Hence, we can observe that there is a long tail in the plotted graph.

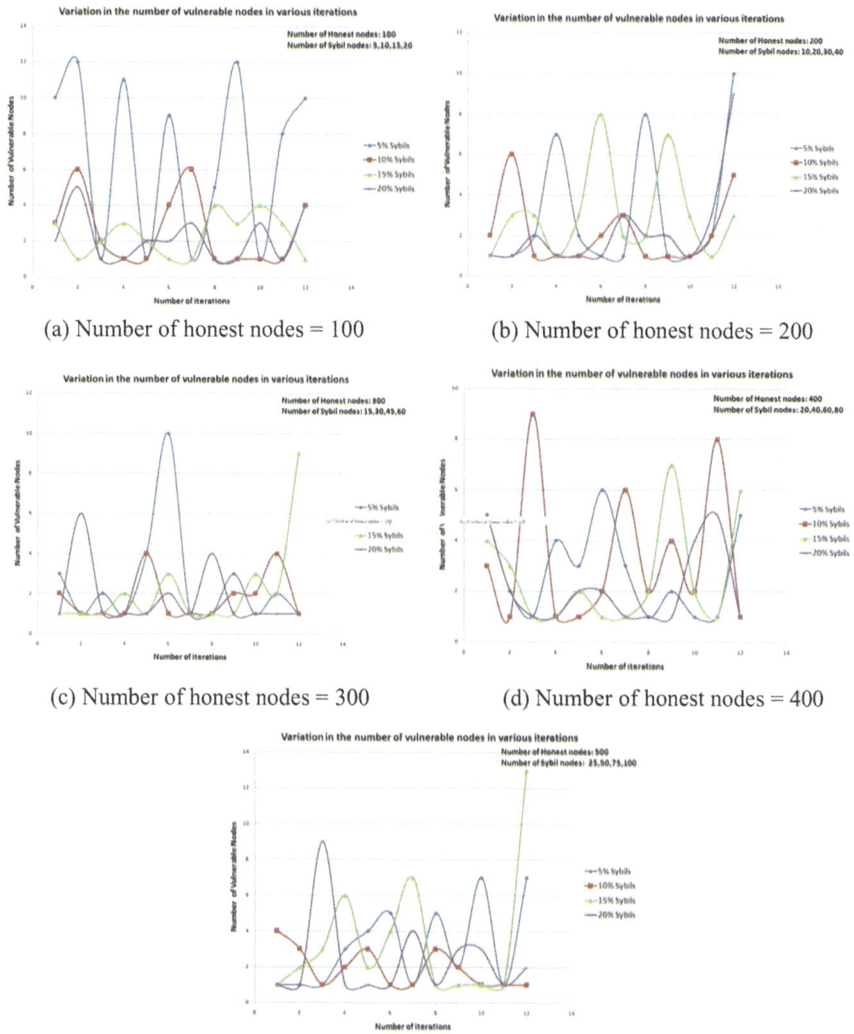

(a) Number of honest nodes = 100

(b) Number of honest nodes = 200

(c) Number of honest nodes = 300

(d) Number of honest nodes = 400

(e) Number of honest nodes = 500

Fig. 6 Variation in the number of vulnerable nodes in various iterations

5 Conclusion

Sybil attack is one of the major threatening problems in this era of social networks such that it questions the security and integrity of both the personal and public life of online users. In our work, we examined Sybil attacks in online social networks by applying SyPy Python package and thus we identified and analyzed the impacts of different parameters in the online social network that influences the Sybil attack.

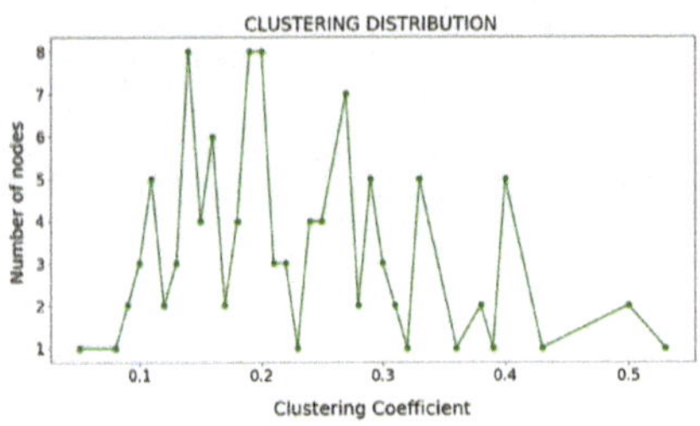

(a) Probability triad = 0.25

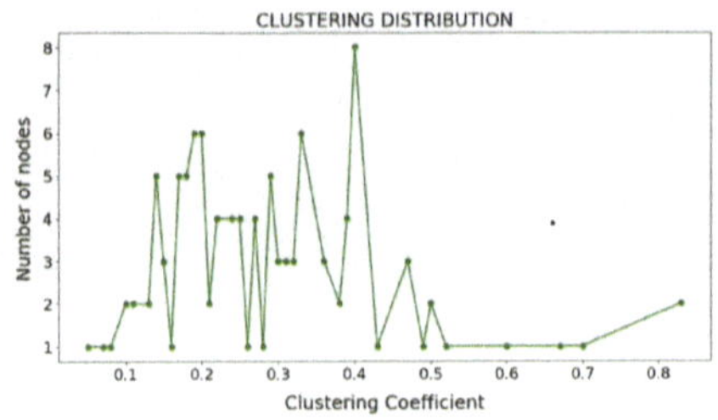

(b) Probability triad = 0.5

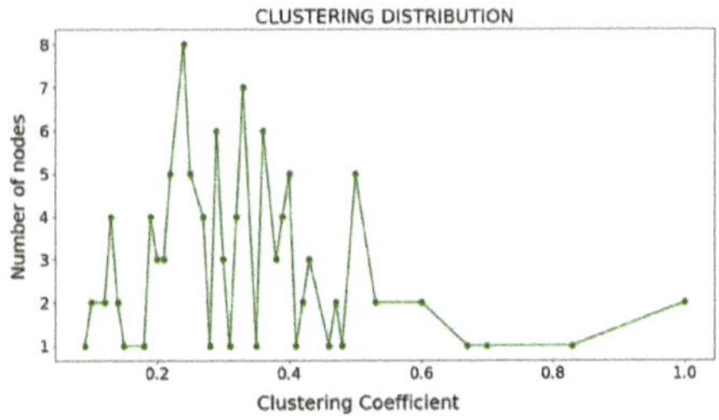

(c) Probability triad = 0.75

Fig. 7 Clustering coefficient when probability triads is varied

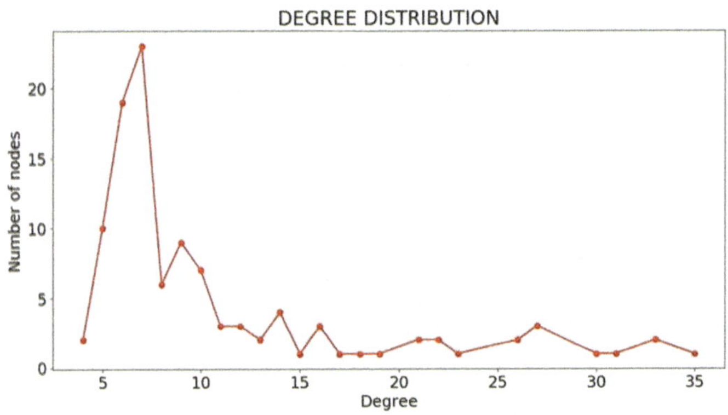

Fig. 8 Plot against the degree of the node and its frequency of occurrence

Our model can be applied to the real-world networks so as to detect users that are vulnerable to Sybil attack and avert them from these attacks.

References

1. Douceur, J.R., et al.: The Sybil attack. In: Druschel, P. (ed.) Peer-to-peer systems. Springer, Heidelberg, pp. 251–260 (2002)

2. Yu, H.: Sybil defenses via social networks: a tutorial and survey. ACM SIGACT News **42**(3), 80–101 (2011)
3. Viswanath, B., Post, A., Gummadi, K.P., Mislove, A.: An analysis of social network-based sybil defenses. In: Proceedings of the ACM SIGCOMM 2010 Conference on SIGCOMM, ser. SIGCOMM '10, pp. 363–374. ACM, New York (2010)
4. Boshmaf, Y., et al.; Graph-based sybil detection in social and information systems. In: Proceedings of ASONAM'13. IEEE (2013)
5. Boshmaf, Y.: A Quick Survey of Social Network-Based Sybil Defenses. University of British Columbia, Vancouver (to appear)
6. Viswanath, B., Mondal, M., Clement, A., Druschel, P., Gummadi, K.P., Mislove, A., Post, A.: Exploring the design space of social network based sybil defenses. In: 2012 Fourth International Conference on Communication Systems and Networks (COMSNETS), pp 1–8, Jan 2012
7. Girvan, M., Newman, M.: Community structure in social and biological networks. Proc. Natl. Acad. Sci. **99**(12), 7821–7826 (2002)
8. Andersen, R., Chung, F., Lang, K.: Local graph partitioning using pagerank vectors. In: 47th Annual IEEE Symposium on Foundations of Computer Science, 2006. FOCS'06, pp. 475–486. IEEE (2006)
9. Mislove, A., Post, A., Druschel, P., Gummadi, K.P.: Ostra: leveraging trust to thwart unwanted communication. In: NSDI (2008)
10. Boshmaf, Y., et al.: Integro: leveraging victim prediction for robust fake account detection in OSNs. In: Proceedings of NDSS, pp. 8–11 (2015)
11. Jia, J., Wang, B., Gong, N.Z.: Random walk based fake account detection in online social networks. In: IEEE DSN (2017)
12. Alvisi, L., Clement, A., Epasto, A., Sapienza, U., Lattanzi, S., Panconesi, A.: SoK: the evolution of sybil defense via social networks. In: Proceedings of the IEEE Symposium on Security and Privacy (2013)
13. Boshmaf, Y., Muslukhov, I., Beznosov, K., Ripeanu, M.: Design and analysis of a social botnet. Comput. Netw. **57**(2), 556–578 (2013)
14. Wang, B., Gong, N.Z., Fu, H.: GANG: detecting fraudulent users in online social networks via guilt-by-association on directed graphs. In: IEEE ICDM (2017)
15. Pearl, J.: Probabilistic reasoning in intelligent systems: networks of plausible inference (1988)

A Fusion Method for Automated Segmentation of the Lungs from the Chest X-Rays

Prashant A. Athavale and P. S. Puttaswamy

Abstract Lungs remain to be one of the most imaged parts of a human body owing to its connection with several diseases and its ability to readily equip the diagnostician with the useful insight about the patient. The inherent advantages of computer-aided diagnostics can be brought to its full use in the case of lungs, if the lungs' segmentation, which is usually the first stage in computer-aided interpretation, be automated. Statistical models and rule-based approaches are by far the traditional methods, and this paper presents a fusion of the two methods for segmenting lungs. In this fusion method, the results of active shape models' detection of the lungs are further refined using morphological operations. The segmentation results show an accuracy of around 96% on the images from a publicly available dataset. This approach has the potential to be extended to other organs or object of interest in non-medical image segmentation problem.

Keywords Computer-aided diagnosis · Active shape models · Lungs segmentation · Morphological operations · Jaccard index

1 Introduction

Chest X-rays reveal most of the disorders of the lungs including an outline of the heart and some of the major blood vessels. The prognosis of lungs is vital as the lungs are associated with some of the common diseases like asthma, chronic bronchitis, emphysema, chronic obstructive pulmonary diseases (COPD), pulmonary fibrosis, sarcoidosis, etc., which are a result of infection and pollution. The September 2018 report published by Indian Council of Medical Research (ICMR) and others [1] puts India with 32% of the world's total COPDs. The respiratory diseases-related

P. A. Athavale (✉)
Electrical and Electronics Engineering Department, BMS Institute of Technology and Management, Bengaluru, India
e-mail: prashanth@bmsit.in

P. S. Puttaswamy
Electrical and Electronics Engineering Department, GSSS Institute of Engineering and Technology for Women, Mysore, India

V. Bhateja et al. (eds.), *Computer Communication, Networking and IoT*, Lecture Notes in Networks and Systems 197, https://doi.org/10.1007/978-981-16-0980-0_17

deaths are also increasing substantially. The top three contributors for these deaths are ambient air pollution, smoking, and household air pollution. Some research [2] had predicted rise in the deaths due to smoking, especially in the middle age adults. Respiratory diseases-related death will be around 6.5% in India [3].

An unknown genetic change happening in the basic building block of the human body, i.e., the cell, leads to a change in the normal life cycle of it. An uncontrollable growth of these cells forms a mass or a nodule commonly known as tumor. Such tumors are classified as benign or malignant; the latter is cancerous and can develop in any part of the human body. The medical field classifies the cancer based on the origin of such tumor. Carcinomas are the cancer prevalent in organs like lungs, breast, prostate, etc., and form solid masses, and can spread through metastasis. A diagnostician can conduct several tests to diagnose the condition of the patient with medical imaging is one of the non-invasive methods.

Of the 200 varieties of cancer, lung cancer affects millions of people around the world. A timely diagnosis and detection together with proper treatment can cure the lung cancer. But the detection of the cancer in the lungs could be accidental in many cases. A low cost non-invasive method of the diagnosis of the lung cancer as against the biopsy is the analysis of digital chest X-rays (CXR). The computed tomography (CT) is the expensive imaging alternative, seldom found in rural areas. The location, size, and other details of the nodules are visible on a CXR, and the diagnostician is trained in interpreting the images and classifying them. But during routine check-up, the nodules could be missed by a human observer, and there is interobserver variability in the interpretation, Qin et al. [4].

This gives rise to a requirement of a computer-aided detection of such nodules on the CXR. The computer can be trained to detect the nodules and classify the same. Early research work aimed at replacing the doctor with the computers, which could mostly be attributed to the early hype created by the artificial intelligence. But the later works have been focusing on making the computer generated information available to the doctor and assist him in his decision making. This could reduce the actual nodules being missed by the doctor. This use of computers in the diagnosis of the medical condition is known as computer-aided diagnosis (CAD).

There are multiple surveys on the performance and efficacy of CAD [5–9]. Few of the commercially available CAD systems are: iCAD SecondLook™ (for mammography), RapidScreen™ (for CXR), CADstream™ (by Confirma for breast MRI), ImageChecker™ (for mammography by R2 Technology), MeVis, Medicsight, Fujita, Zhang et al. [10].

The segmentation of the lungs from a CXR is still an unsolved problem. Hence, this paper presents a fusion method for segmenting the lungs from a given CXR in which the results given by active shape model (ASM) are refined with morphological operations to yield a better result, in a shorter time, overcoming the disadvantages of the ASM. In the further sections, a brief description of the dataset is given and the ASM is described. The next section is about morphological operations. Finally the segmentation results are presented in terms of standard performance indicators and then conclusions.

2 Material and Method

In computer-aided diagnostic systems for lungs, the computer analyzes the given CXR and returns its findings to the diagnostician. If there are any candidate nodules or lesions, the output of the CAD is further analyzed by the diagnostician. This reduces the burden on the human being and can enhance the productivity, thereby improving the patient care. An accurate determination of even the subtle nodules on the CXR at an early stage could be lifesaving.

A typical CAD system has the following sequence of operations for detecting the nodules on the lung images: Preprocessing, segmentation of the lungs, removal of the rib shadows, detection of the nodules, characterization and classification of the nodules as shown in Fig. 1.

Each stage of the above-mentioned crucial steps has a plethora of possible approaches [11, 12]. The choice and the implementation could be guided by the problem at hand. For the study purpose, the CXRs from the Japanese Society of Radiological Technology (JSRT) have been used [13]. The database has 247 images of which a number of nodule and non-nodule are 154 and 93, respectively.

Each image is of dimension 2048 × 2048 pixels, and the pixels are of 0.175 mm size. These images, which have also have patient specific information, have been used for training and testing. Table 1 summarizes the data distribution of JSRT database.

There are a multitude algorithms and different methodologies to segment the lungs from the CXR. Few of them are mentioned here.

The interdependency of pulmonary abnormalities and lung region symmetry has been explored for screening the diseased population in PA radiographs by Santosh and Antani [14]. It is based on the hypothesis that asymmetry in lung region could indicate abnormality. An accuracy of 91% in abnormality detection has been reported.

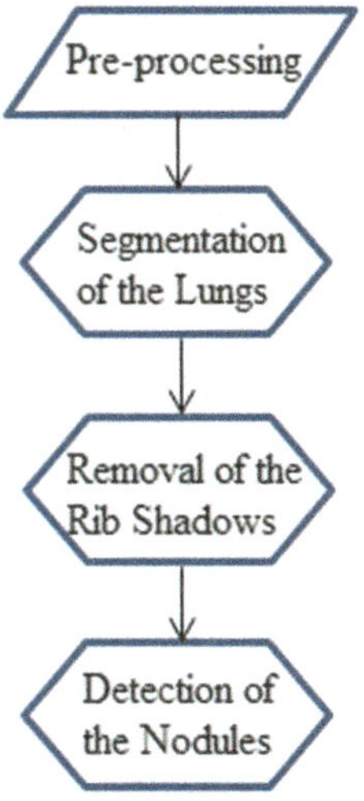

Fig. 1 A flowchart of the typical steps in a CAD system. Preprocessing is to prepare the image for the next stages and enhances the signal content in it

Table 1 Metadata of JSRT database

S. no.	Parameters	Cases	No. of CXRs	%
1	Degree of subtlety of the nodules	5 (most obvious)	12	7.79
		4	38	24.67
		3	50	32.46
		2	29	18.83
		1(extremely subtle)	25	16.23
2	Malignancy of the nodules	Malignant	100	64.93
		Benign	54	35.06
3	Nodule size	~10 mm dia	83	53.90
		~20 mm dia	50	32.47
		~30 mm dia	19	12.34
		~40 mm dia	1	0.65
		~50 mm dia	1	0.65
4	Nodule location	Left upper lobe	28	18.18
		Right upper lobe	52	33.77
		Left lower lobe	22	14.29
		Right lower lobe	27	17.53
		Right middle	13	8.44
		Left	9	5.84
		Right	3	1.95
5	Final diagnosis	Lung cancer	99	64.29
		Others (Tuberculoma, metastasis, inflammatory mass, granuloma, hematoma, pneumonia, abscess, cryptococcosis, unknown)	55	35.71

The method compares the left and right lungs by segmenting the two regions using graph cut algorithm as an atlas.

The saliency of the lung region has been used as the basis for its detection by Li et al. [15] on JSRT images. The work is based on pixel intensities and used a saliency map to classify the region of lungs. Active contour model is used to estimate edges and have reported 0.91 overlap score.

A deep learning-based method is used by Saidy and Lee [16] for lungs' segmentation from the CXR suing the encoder–decoder convolutional network. The algorithm has been tested and results on five JSRT images have been reported.

Usage of convolutional neural networks to detect the nodules and an effort to reduce false positives has been shown by Li [17]. Apart from the shape model, the local gray-level distribution is also modeled in the statistical models in the work reported by Seghers [18]. The appearance model-based segmentation of the lungs is reported by Dieter et al. Another work reported is of Nagata et al. [19], which uses rule-based algorithm. The marking of the landmark points is dealt with considerable detail in the work of Starcevic et al. [20]. Yan et al. [21] also present a segmentation algorithm for the two-dimensional CXR.

The CXRs from the database pose various challenges to the segmentation algorithms, and hence in the following discussion, the same is highlighted. Each of the CXRs is having unique multimodal distribution of the gray-level intensities. Hence, a simple thresholding operation cannot produce a satisfactory result.

An example image is shown in Fig. 2a and its histogram in Fig. 2b. Figure 2c–h in the next rows shows the binary images obtained based on various threshold values, 'T'. The operation is to classify the pixels as '1' if the value of it in input image I_m is above the T, else group it under '0', shown mathematically as,

$$I_o(i, j) = \begin{cases} 0, & I_m(i, j) < T \\ 1, & I_m(i, j) \geq T \end{cases} \tag{1}$$

A satisfactory result is not given by any of those chosen values. This is because of the distribution of the gray-level values across the image, with great variability. The

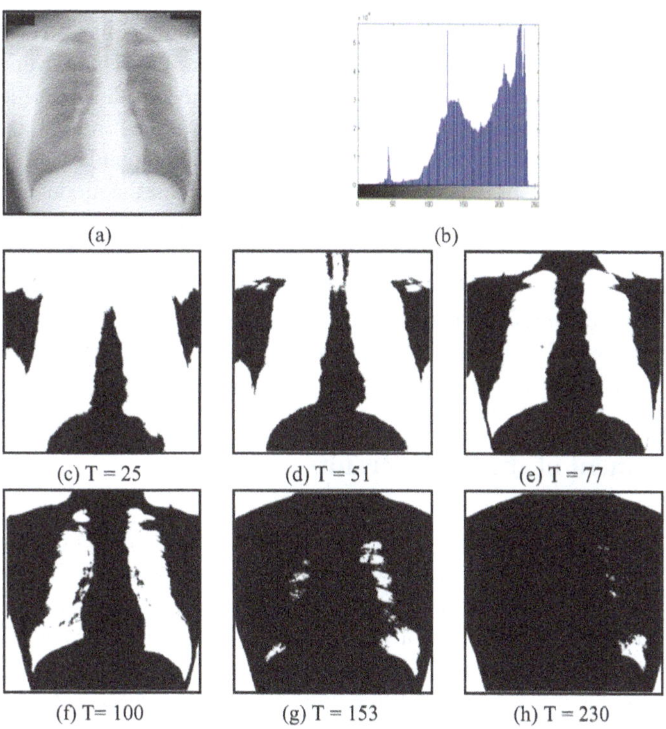

Fig. 2 **a** A sample CXR from the database, **b** its histogram. **c–h** are the binary images obtained by thresholding the input image at various thresholds gray level 'T'

pixel value in the image depends on attenuation of the X-rays by various muscles internally and also depends on the position of the patient with respect to the imaging device. Hands and neck portion are visible in some images of the database, where as it not seen in the remaining. A generalized segmentation algorithm may fail to work in this case.

Variability in Shapes:

The 3D thorax is superimposed on a 2D image, and hence there is an overlap of the rib cage, lungs, heart and other organs like vertebrate, windpipes, etc. In spite of this, the lungs to be segmented are having a characteristic shape which is easily recognizable by a human observer. The diagnosticians are able to observe lungs alone by mentally subtracting the overlapping structures because of the training and experience. The same operation can be done by a computer but at some compromise in quality.

Hence, only one approach cannot solve the problem of segmentation and requires multiple operations either in series or parallel, to obtain a satisfactory result.

The ASM has been used for the segmentation of the lungs in CXRs and has given good results. But the limitation of these statistical models is its vulnerability to local intensity features. All the landmark points on the statistical model are annotated by an expert, and an average of all such shapes from each training image is the shape model. The shape model so built is initially placed on the target image. The model has to stretch itself to fit to an instance of the lung in it, if present. The movement of the shape, in or out, is guided by local gray-level profile. The gray-level profile is a one-dimensional sample of the image at a landmark point averaged from all the training images. This sample is searched along the normal to the boundary of the object, before settling at a best-fit location. At each of the iterations, the model adjusts itself to the available instance of the lung and an energy minimization function decides the number of further iterations. In case if the point deviates from the actual boundary, then it will never return to the correct edge of the object as there is no reset option in ASM as described in [22]. The results after first iteration of the ASM on the CXRs are depicted in Fig. 3. The image number has been retained as per the database.

3 Refinement of the ASM Results

The use of the ASM is restricted to only a coarse shape of the lungs. Further refinement of the results to obtain an accurate segmentation of the lungs which delineates its salient features like diaphragm, costophrenic (CP) angle, and the apex of the two lobes is done by the application of morphological operations. This is interspersed with rule-based approaches also.

This fusion method approach of fine tuning the segmentation outcomes has the following advantages: (i) The statistical model can be built with only few annotated points, and hence the complexities of computations are overcome. The lateral sides of the lungs are almost linear and hence few points are sufficient to describe them. The major curvature is at the CP angle, and a couple of landmark points are used

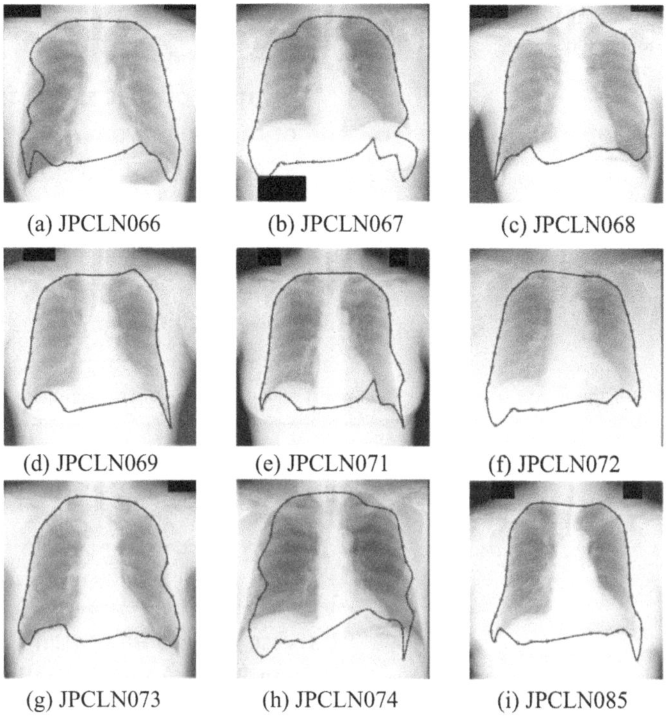

Fig. 3 Lungs boundary detection by ASM in one iteration, on CXRs

in modeling it. (ii) The problem of the shape model being swayed away by a non-edge location is overcome. (iii) The method is faster. (iv) The operations are guided by the local gray-level distributions only. For processing a portion of the image, a suitable region is carved out and it makes the process independent of the gray-level distribution elsewhere in the image. (v) The two lungs are not modeled separately in the shape model, but are considered as a bell-shaped structure. This will remove the two lungs from the rest of the image in first stage and then separate them more easily. (vi) The accuracy of the method is higher. A flowchart of the proposed method is shown in Fig. 4.

The first iteration of the ASM gives an approximate location of the lungs on the target CXR. 30 landmark points have been used in building the ASM and these points stretch and modify according to the variability in the target shape as shown in Fig. 3. As indicated in the flowchart, a suitable rectangular region around these set of points is carved out for locating the diaphragm, shown in red color box in Fig. 5. Once the region is carved out, a simple thresholding operation suffices to segment the rectangular image content in to lungs and background, with diaphragm as the dividing line. Morphological operations like opening and closing help in removal of noise and obtaining an accurate detection of diaphragm.

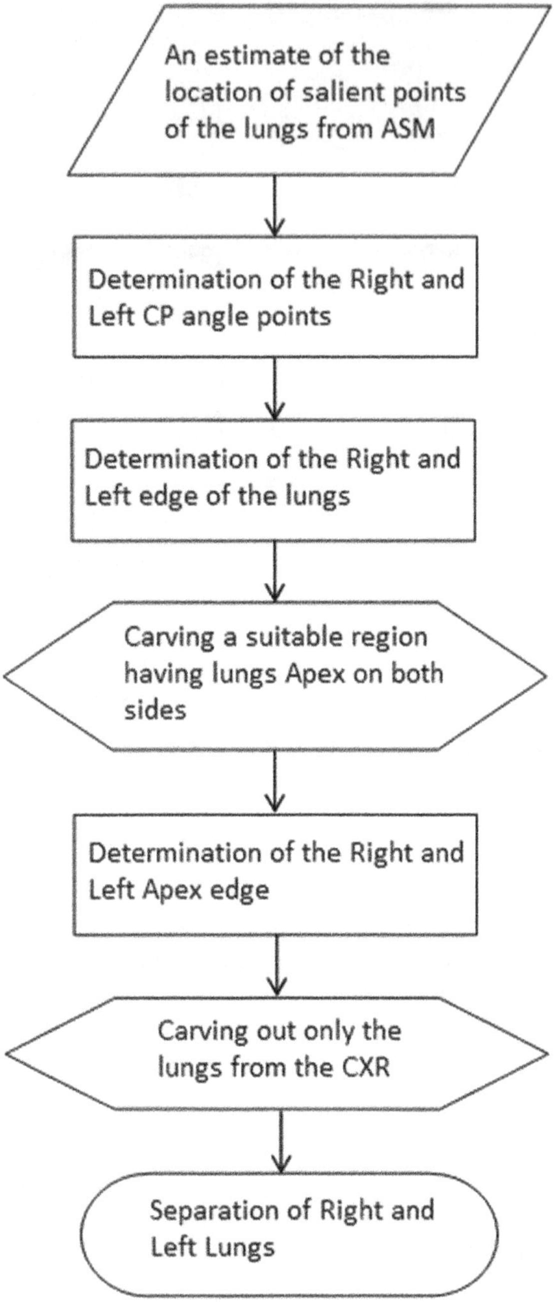

Fig. 4 Flowchart of the proposed fusion method

Fig. 5 A schematic representation of the prominent components

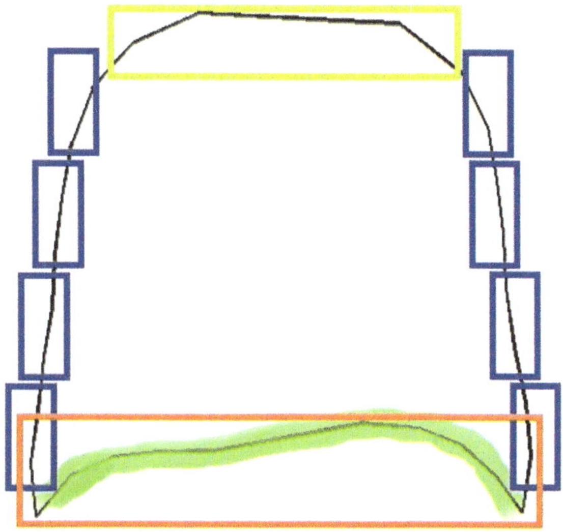

A dynamic program then has to divide the original image in to smaller portions and then process them individually to get the side edges of the lungs. This is shown with the blue color boxes. The first box closer to the diaphragm represents the region which will be processed for detecting the edge of the lungs. As can be observed, the area that is carved out is a small region and contains a portion of the lungs which is distinct from the background. Hence, here too, a thresholding operation based on Otsu's method is sufficient in highlighting the edges. The number of such vertical boxes for detecting the side edges of the lungs depends on the height of the lung, which in turn is directly obtained from the first iteration of the ASM. The first iteration gives an approximate coverage of the lungs in the target CXR (Fig. 3). Once the lateral edges are determined, a suitable rectangular region is dynamically identified by the program, which would possibly house the lobes of the lungs, depicted by yellow color box in Fig. 5. The upper lobes are not simple lines unlike diaphragm or lateral edges, but very complex structure as it contains prominent impressions of cervical bones, shadows of the collar bones and wind pipes. The cluttered nature of this region makes it imperative to use a mix of thresholding, morphological operations and a dynamic program to accurately decide the edges of the upper lobes.

It should be observed that till this step the lungs outline resembles an inverted convex hull, and has been determined. The right and left lungs are still fused with each other and have the heart silhouette in between. The gray-level distribution of the lungs is different from that of the heart, and hence can be separated by binarizing the image. Through this entire set of steps, the whole CXR is divided in smaller segments and processed separately, to look for a specific feature, i.e., edge.

The proposed fusion method has been tested on a select 55 CXRs from the same database, and these were not part of the set of training images. One of the authors has segmented the right and left lung regions manually for each of the test images, under the guidance of an expert. This has been considered as the gold standard for comparison of the automated segmentation [23]. The quantification of the results has been done based on the standard evaluation metrics: Sensitivity, Specificity and Accuracy. These are defined in Eqs. 2–4, in terms of four basic outcomes: True Positive, True Negative, False Negative, False Positive.

These cardinals are defined as follows: True Positive is the set of pixels rightly classified as belonging to the lungs and True Negative are the set of pixels rightly classified as not belonging to the lungs. On the same lines, False Positive is the total number of pixels wrongly classified as belonging to the lungs and False Negative is the total number of pixels wrongly marked as non-lung pixels (Fig. 6).

$$\text{Sensitivity} = \frac{\text{TP}}{\text{TP} + \text{FN}} \tag{2}$$

$$\text{Specificity} = \frac{\text{TN}}{\text{TN} + \text{FP}} \tag{3}$$

$$\text{Accuracy} = \frac{\text{TP} + \text{TN}}{\text{TP} + \text{TN} + \text{FP} + \text{FN}} \tag{4}$$

These parameters and the corresponding values have been tabulated in Table 2. The proposed algorithm was tested on three image sizes as depicted. These images were obtained by down-sampling the original image. Correspondingly the time requirement is also less for the segmentation for lower size image.

Yellow color pixels indicate the TPs, and blue color pixels are part of the lungs but omitted by the machine classifier during segmentation as not belonging to the lungs. Hence, these are the FN pixels. On the same lines, the red color pixels are the background pixels but incorrectly classified as lung pixels, and hence FPs. The white pixels in the same image are the background pixels and classified accordingly. These are the TNs. Random colors have been chosen for an effective representation and expected to give a good contrast on print or electronic medium.

All the above-mentioned performance indices have been shown in Table 3 for 2048 × 2048 images.

4 Conclusion

The problem of segmentation of the lungs can be approached in multiple ways, but each approach has its own limitations, and a hybrid approach is required to refine the results, to be of any clinical importance. All the objects of interest in a medical image are bound by a characteristic shape, but vary within a permissible variability. Modeling the shape and including in it the variability is a challenging task, as the model depends on the training dataset. In the proposed method, the model is dependent on the training dataset, but not its accuracy as it is a hybrid method. The

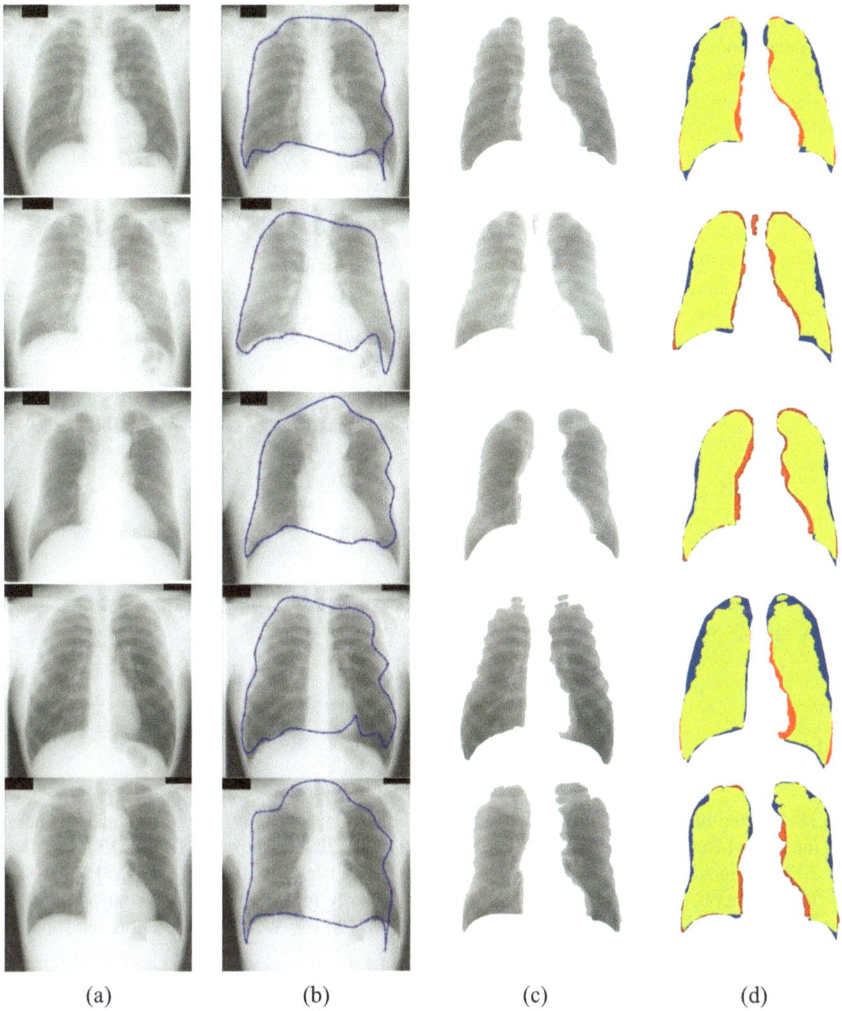

Fig. 6 The results of application of the fusion method on select CXRs. **a** The input images from the database. **b** The corresponding outermost edges detected by single iteration of ASM. **c** Separation of the right and left lungs using the proposed. **d** Color coded overlap region between the machine segmentation and the gold standard segmentation

Table 2 Performance of the proposed method on nodule CXRs of three different sizes

Parameter	512 × 512	1024 × 1024	2048 × 2048
Sensitivity	0.911	0.92	0.935
Specificity	0.89	0.93	0.977
Accuracy	0.921	0.916	0.964

Table 3 Gist of the performance indices on the CXRs from JSRT database

Performance indices	For non-nodule CXRs	For nodule CXRs
	Mean ± std. deviation	Mean ± std. deviation
Sensitivity	93.9 ± 2.67	93.51 ± 2.71
Specificity	96.7 ± 1.47	97.75 ± 1.05
Accuracy	95.8 ± 1.1	96.41 ± 0.68
Jaccard index	0.87 ± 0.02	0.89 ± 0.021
Dice coefficient	0.93 ± 0.013	0.945 ± 0.01
Matthews correlation coefficients	0.90 ± 0.02	0.916 ± 0.01
Active shape model (in s)	3.77 ± 0.24	4.17 ± 0.6
Proposed fusion method (in s)	7.78 ± 0.6	

simple framework for detection of shapes as explained in this paper can be extended to any shape on medical or other images.

References

1. The burden of chronic respiratory diseases and their heterogeneity across the states of India: the Global Burden of Disease Study 1990–2016, vol. 6, issue 12, pp. PE1363–E1374, 01 Dec 2018. Open Access Published: 11 Sept 2018. https://doi.org/10.1016/S2214-109X(18)30409-1
2. Jha, P., Jacob, B., Gajalakshmi, V., et al.: A nationally representative case-control study of smoking and death in India. New England J. Med. (2008). WHO-India Report on Tobacco Free Initiative, GATS India (TFI) 2009–2010
3. WHO-India Report on Tobacco Free Initiative, GATS India (TFI), 2009–2010
4. Qin, C., Yao, D., Shi, Y., Song, Z.: Computer aided detection in chest radiography based on artificial intelligence: a survey. BioMed. Eng. On line (2018)
5. Al Mohammad, B., Brennan, P.C., Mello-Thoms, C.: A review of lung cancer screening and the role of computer-aided detection. Clin. Radiol. **72**(6):433–442 (2017)
6. van Ginneken, B., ter Haar Romney, B.M., Viergever, M.A.: Computer-aided diagnosis in chest radiography: a survey. IEEE Trans. Med. Imaging **20**(12) (2001)
7. El-Baz, A., Gimelfarb, G., Falk, R., El-Ghar, M.A.: A new CAD system for early diagnosis of detected lung nodules, 2007. In: ICIP 2007. IEEE International Conference on Image Processing, vol. 2, pp. II-461–II-464 (2007)
8. Doi, K.: Current status and future potential of computer-aided diagnosis in medical imaging. Rev. Article Br. J. Radiol. **78**, S3–S19 (2005)
9. Lee, N., Laine, A.F., Márquez, G., Levsky, J.M., Gohagan, J.K.: Potential of computer-aided diagnosis to improve CT lung cancer screening. IEEE Rev. Biomed. Eng. **2** (2009)
10. Fujita, H., Zhang, X., Kido, S., Hara, T., Zhou, X., Hatanaka, Y., Xu, R: An introduction and survey of computer-aided detection/diagnosis (CAD). In: 2010 International Conference on Future Computer, Control and Communication (2010)

11. Story, A., Aldridge, R.W., Abubakar, I., Stagg, H.R., Lipman, M., et al.: Active case finding for pulmonary tuberculosis using mobile digital chest radiography: an observational study. Int. J. Tuberculosis Lung Diseases **16**, 1461–1467 (2012)
12. Shi, Y., Qi, F., Xue, Z., Ito, K., Matsuo, H., Shen, D.: Segmenting lung fields in serial chest radiographs using both population and patient-specific shape statistics. In: International Conference on Medical Image Computing and Computer-Assisted Intervention, pp. 83–91. Springer, Berlin (2006)
13. Shiraishi, J., Katsuragawa, S., Ikezoe, J., Matsumoto, T., Kobayashi, T., Komatsu, K., Matsui, M., Fujita, H., Kodera, Y., Doi, K.: Development of a digital image database for chest radiographs with and without a lung nodule: receiver operating characteristic analysis of radiologists' detection of pulmonary nodules. Am. J. Roentgenol. **174**, 71–74 (2000)
14. Santosh, K.C., Antani, S: Automated chest X-ray screening: can lung region symmetry help detect pulmonary abnormalities? IEEE Trans. Med. Imaging **37**(5), 1168–1177 (2018)
15. Li, X., Chen, L., Chen, J.: A visual saliency-based method for automatic lung regions extraction in chest radiographs. In: 2017 14th International Computer Conference on Wavelet Active Media Technology and Information Processing (ICCWAMTIP), Chengdu, pp. 162–165 (2017)
16. Saidy, L., Lee, C.: Chest X-ray image segmentation using encoder-decoder convolutional network. In: 2018 IEEE International Conference on Consumer Electronics-Taiwan (ICCE-TW), Taichung, pp. 1–2 (2018)
17. Li, C., Zhu, G., Wu, X., Wang, Y: False-positive reduction on lung nodules detection in chest radiographs by ensemble of convolutional neural networks. IEEE Access 6 (2018)
18. Seghers, D.: Image segmentation using local shape and gray-level appearance models. In: Medical Imaging 2006: Image Processing, vol. 6144. International Society for Optics and Photonics (2006)
19. Nagata, R., Kawaguchi, T., Miyake, H: A rule-based algorithm for detection of ribcage boundary in chest radio-graphs. In: ITC-CSCC: International Technical Conference on Circuits Systems, Computers and Communications, pp. 1001–1004 (2009)
20. Starcevic D.S., Ostojic, V.S., Petrovic, V.S.: Open-source digital diagnostic radiography image annotation software. In: 24th Telecommunications Forum (TELFOR), 2016, pp. 1–4. IEEE (2016)
21. Yan, Z., Zhang, J., Zhang, S., Metaxas, D.N.: Automatic rapid segmentation of human lung from 2D chest X-ray images. In: MICCAI Workshop on Sparsity Techniques in Medical Imaging (2012)
22. Athavale, P.A., Puttaswamy, P.S.: Median values of gray levels for detection of lung contours. In: International Conference on Emerging Research in Electronics, Computer Science and Technology (ICERECT-2015) PESCE. ISBN: 978-1-4673-9563-2
23. Muyoyeta, M., Maduskar, P., Moyo, M., Kasese, N., Milimo, D., Spooner, R: The sensitivity and specificity of using a computer aided diagnosis program for automatically scoring chest X-rays of presumptive TB patients compared with Xpert MTB/RIF in Lusaka Zambia. PLoS ONE **9**(4), e93757 (2014)

Metamaterial-Inspired Circularly Polarized Microstrip Patch Antenna

Chirag Arora

Abstract This article presents a metamaterial-based single-feed circularly polarized microstrip antenna with improved gain and enhanced bandwidth. To obtain circular polarization, the two diagonally opposite corners of a square-shaped traditional microstrip patch antenna are truncated, and then the ground plane of this truncated square microstrip patch antenna is loaded with an array of proposed splitted circular metamaterial rings. The traditional patch antenna, with mitered corners, resonates at 2.4 GHz presenting the gain and bandwidth of 4.8 dBic and 500 MHz, respectively. However, on loading splitted circular metamaterial rings on its ground plane, an improvement in gain by 0.4 dBic and enhancement in bandwidth by 15 MHz is observed. The three dB axial ratio bandwidth of this composite structure is 610 MHz. The proposed metamaterial-inspired microstrip patch antenna possesses the overall size of 37.53 mm × 37.53 mm × 1.6 mm.

Keywords Microstrip patch · Circular polarization · Metamaterial · High-frequency structure simulator (HFSS)

1 Introduction

Attention of antenna researchers toward the circularly polarized antennas is increasing day by day as they are less susceptible to abnormal weather conditions and other hindrances. Moreover, they are more reliable as compared to traditional linearly polarized patch antennas. The performance parameters of these circularly polarized patch antennas can further be increased by incorporating them with various metasurfaces like reactive impedance surface comprising of metallic patches arranged in a periodic manner [1, 2], artificial magnetic surface [3, 4], uniplanar compact photonic band gap surface [5], single/double negative metamaterials [1–4], etc. Out of all these techniques, the performance enhancement by using metamaterials is generally preferred as most of the remaining techniques make the antenna bulky or produce difficulty in fabrication. Therefore, since last few years, metamaterials have

C. Arora (✉)
KIET Group of Institutions, Delhi-NCR, Ghaziabad, UP, India

V. Bhateja et al. (eds.), *Computer Communication, Networking and IoT*, Lecture Notes in Networks and Systems 197, https://doi.org/10.1007/978-981-16-0980-0_18

been widely utilized in improving the performance parameters of circularly polarized patch antennas and their polarization conversion from linear to circular.

Metamaterial-based circularly polarized patch antennas with single feed have been designed to obtain an overall size of about 0.3λo × 0.3λo × 0.03λo. However, their three dB axial ratio bandwidths are much smaller than 5% [6–9]. A patch antenna with square shape, possessing a diagonal slot reactive impedance surface with 45° orientation, has attained an axial ratio bandwidth of about 15% with overall size of 0.77λo × 0.77λo × 0.06λo [10]. A square patch with truncated corner and artificial magnetic conductor ground structure was proposed for the overall dimension of 0.78λo × 0.80λo × 0.096λo and achieved 20.4% AR bandwidth [11]. In [10, 11], a metasurface has been used to convert linearly polarized antenna to a circularly polarized. This is achieved by placing the square-shaped patch, having truncated corners, between the ground plane and lattice of square metallic plates. However, the fabrication of this structure is quite tedious.

This article presents a circularly polarized microstrip patch antenna based on metamaterial. The proposed patch antenna is composed of a truncated square-shaped radiator, the ground plane of which is loaded with an array of proposed circular metamaterial rings with splits. Microstrip patch is truncated at two diagonally opposite ends to generate proper perturbation for circular polarization. The metamaterial substrate is a periodic structure with nine elements arranged in 3 × 3 fashion. Introduction of the paper has been given in the first section. The second section discusses the geometry of conventional and proposed metamaterial-inspired patch antenna. The second last section presents various simulated results. The conclusion of the whole paper has been discussed in the final section.

2 Proposed Antenna Design

This segment describes the design of (i) a traditional microstrip patch antenna with truncated corners, (ii) the unit cell of splitted circular metamaterial ring, (iii) metamaterial substrate and (iv) the proposed composite structure. A traditional circularly polarized patch antenna has been designed using a square patch with two diagonally opposite edges being cut out. The traditional patch antenna resonates at 2.4 GHz, and the dimensions (L) of the square patch required to produce this resonant frequency are determined to be 28.65 mm × 28.65 mm, with truncated length of $x = y = 5.1$ mm. Overall size of the ground plane is computed to be 37.53 mm × 37.53 mm. Transmission line model is used to calculate these dimensions [12]. Geometric structure of this perturbed microstrip square patch is shown in Fig. 1.

Mitering of the diagonally opposite corners of the square-shaped patch leads to the phase shift between two orthogonal modes of the square-shaped patch, i.e., TM_{10} and TM_{01}. Feeding of patch is done with a microstrip line of dimensions $a = 8.25$ mm and $b = 3.1$ mm. Moreover, this technique of using single feeding port and proper perturbation can also be used to generate dual bands which are not too much apart from each other.

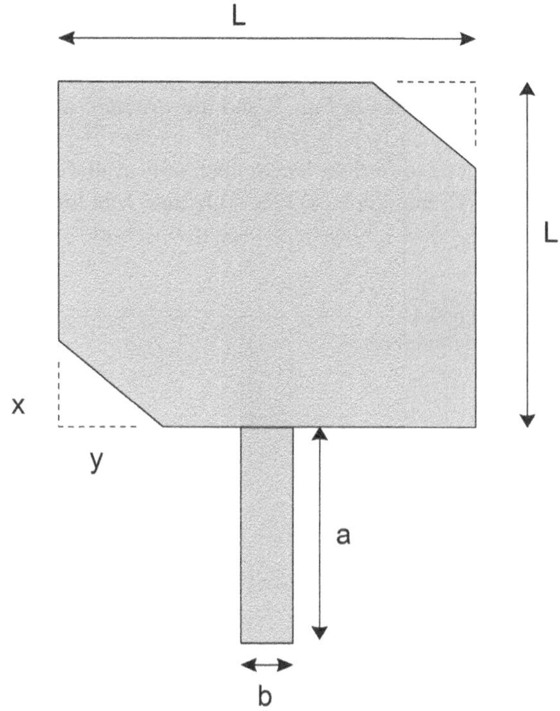

Fig. 1 Geometric sketch of perturbed square patch

Figure 2 presents the unit cell of the splitted circular metamaterial ring. The proposed splitted circular metamaterial unit cell has thickness of 0.3 mm and split

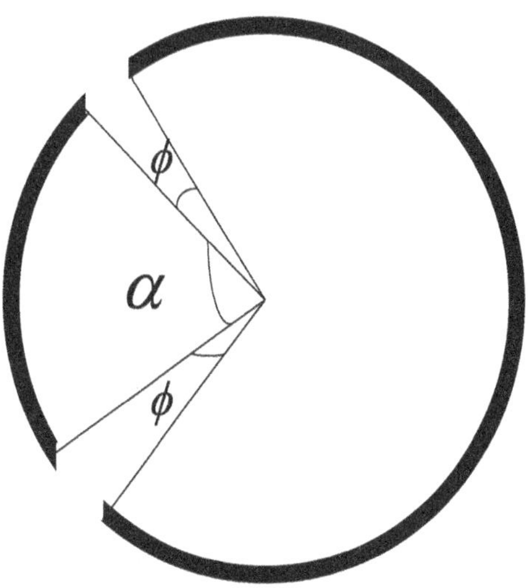

Fig. 2 Top view of proposed splitted circular metamaterial unit cell

edges form the angle $\alpha = 24°$ and $\beta = 5°$. The inner diameter of each circular ring is 6 mm. A metamaterial unit cell array of 3 × 3 elements is etched on the ground of mitered patch and it acts as the metamaterial substrate. Structure of this metamaterial substrate is shown in Fig. 3, and the structure of proposed metamaterial substrate-inspired circularly polarized patch antenna is presented in Fig. 4. Simulations of these structures are done by commercially available electromagnetic software, Ansoft HFSS. 1.48 mm thick (h) FR-4 substrate with loss tangent (δ) = 0.01 and dielectric constant (ε_r) = 4.3 has been used in this work.

Fig. 3 Proposed metamaterial-inspired substrate

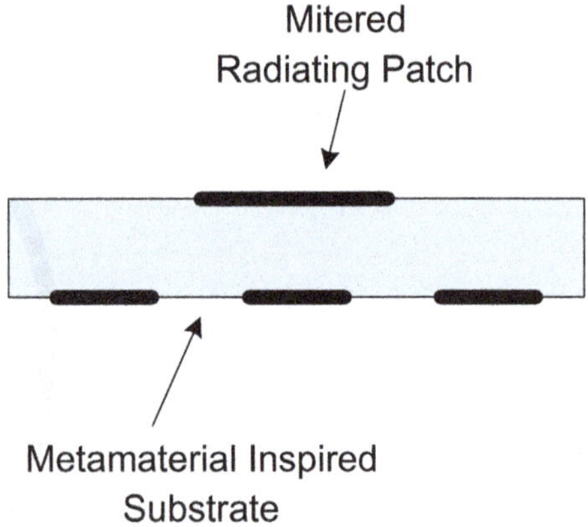

Fig. 4 Proposed metamaterial substrate-inspired circularly polarized patch antenna

3 Simulation Results and Discussion

Figures 5 and 6 present the return loss characteristics (simulated) of the traditional circularly polarized microstrip patch antenna and the designed metamaterial substrate loaded patch antenna, respectively. These curves depict that unloaded circularly polarized square patch antenna possesses impedance bandwidth of 500 MHz at resonant frequency of 2.4 GHz while the proposed metamaterial loaded circularly

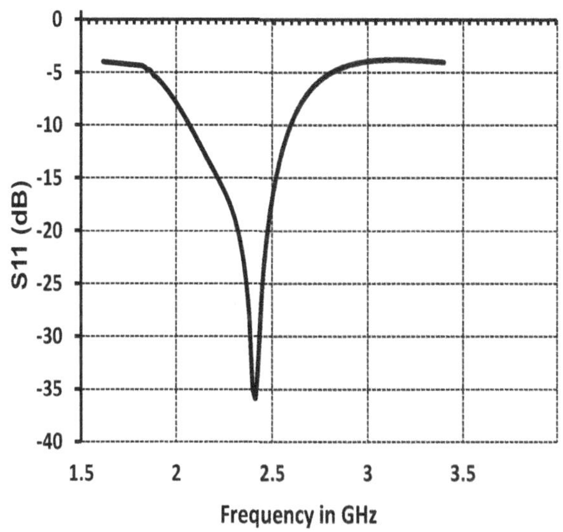

Fig. 5 Return loss characteristics of unloaded circularly polarized patch antenna

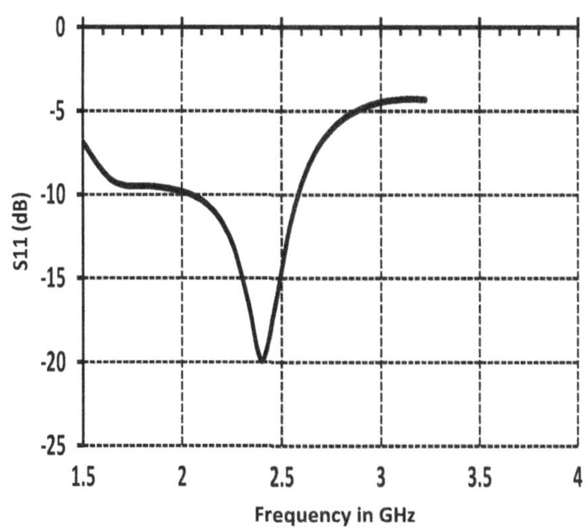

Fig. 6 Return loss curve of proposed metamaterial substrate loaded circularly polarized patch antenna

polarized patch antenna possesses the impedance bandwidth of 515 MHz at same resonant frequency.

Similarly, Figs. 7 and 8 present the E-plane radiation characteristics of both these microstrip patch antennas. From these figures, it is concluded that unloaded patch antenna radiates with overall gain of 4.8 dBic. But the gain of 5.2 dBic is obtained, on loading this traditional circularly polarized microstrip patch antenna with the designed metamaterial substrate.

Axial ratio of loaded patch antenna is plotted to determine its polarization and is represented in Fig. 9. As depicted from Fig. 9, it is obvious that the proposed

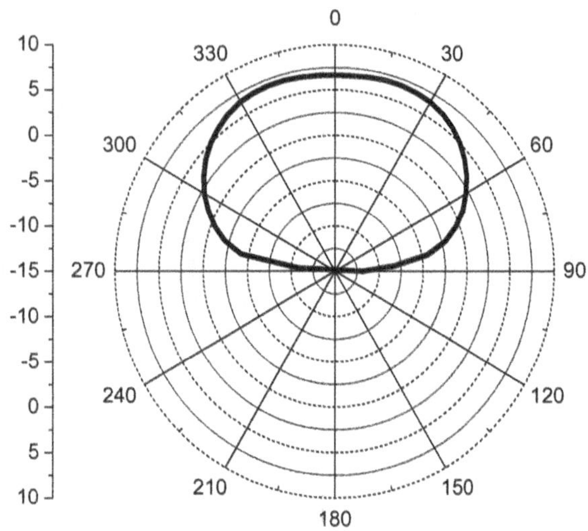

Fig. 7 E-plane radiation patterns of conventional circularly polarized patch antenna

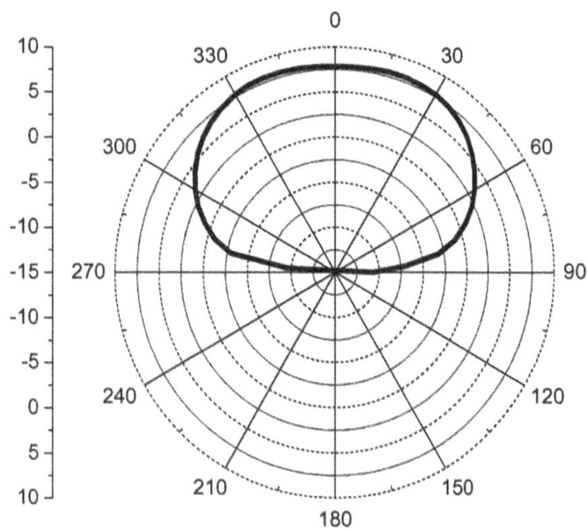

Fig. 8 E-plane radiation patterns of proposed metamaterial-inspired circularly polarized patch antenna

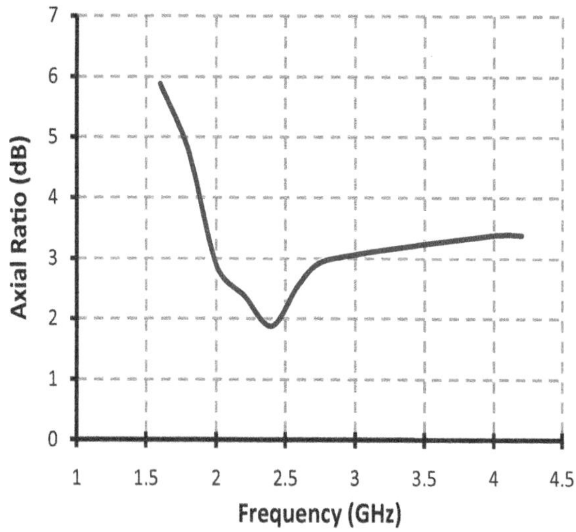

Fig. 9 Axial ratio curve of metamaterial substrate loaded circularly polarized patch antenna

metamaterial-inspired square patch antenna possesses the axial ratio of 1.9 dB at 2.4 GHz, which is below 3 dB, illustrating that the loaded patch antenna is circularly polarized in nature. The three dB axial ratio (AR) bandwidth of loaded patch antenna is 610 MHz.

The novel features of the metamaterials are responsible for the gain improvement and bandwidth enhancement of the designed metamaterial substrate-based microstrip patch antenna with circular polarization. Whenever a metamaterial geometry is etched on ground plane of a traditional patch antenna possessing circular polarization, negative permittivity is produced at the frequency of resonance. It happens due to the interaction of this metamaterial with electric field. Since majority of the current is concentrated near the metamaterial substrate, disturbance in the original current distribution on the square microstrip patch occurs. Therefore, by choosing the appropriate split gap between two asymmetrical splits and proper dimensions of the proposed substrate, sub-wavelength resonance of the microstrip patch antenna is obtained, providing improvement in bandwidth and gain.

4 Conclusions

This article presents a novel design technique for improving the performance of a traditional circularly polarized patch antenna. For this purpose, the edge mitered circularly polarized square microstrip patch antenna has been loaded with the metamaterial substrate, which has been designed by 3×3 array of splitted circular rings. The uniqueness of this proposed composite structure is present in the fact that gain and bandwidth of the conventional circularly polarized microstrip patch antenna have been increased without any extra hardware cost and profile. The proposed circularly polarized antenna will be a good candidate for various S band applications. Equivalent circuit and measured results of this proposed patch antenna will be published later in an international journal of repute.

References

1. Al-Zoubi, A., Yang, F., Kishk, A.: A low-profile dual band surface wave antenna with a monopole-like pattern. IEEE Trans. Antennas Propag. **55**(12), 3404–3412 (2007)
2. Bernard, L., Chetier, G., Sauleau, R.: Wideband circularly polarized patch antennas on reactive impedance substrates. IEEE Antennas Wireless Propag. Lett. **10**, 1015–1018 (2011)
3. Yang, F., Rahmat-Samii, Y.: Reflection phase characterizations of the EBG ground plane for low profile wire antenna applications. IEEE Trans. Antennas Propag. **51**(10), 2691–2703 (2003)
4. Nakamura, R., Fukusako, T.: Broadband design of circularly polarized microstrip patch antenna using artificial ground structure with rectangular unit cells. IEEE Trans. Antennas Propag. **59**(6), 2103–2110 (2011)
5. Lamminen, A., Vimpari, A.R., Saily, J.: UC-EBG on LTCC For 60-Ghz frequency band antenna applications. IEEE Trans. Microw. Theory Tech. **57**(10), 2904–2912 (2009)
6. Dong, Y., Toyao, H., Itoh, T.: Compact circularly-polarized patch antenna loaded with metamaterial structures. IEEE Trans. Antennas Propag. **59**(11), 4329–4333 (2011)
7. Dong, Y., Toyao, H., Itoh, T.: Design and characterization of miniaturized patch antennas loaded with complementary split-ring resonators. IEEE Trans. Antennas Propag. **60**(2), 772–785 (2012)
8. Agarwal, K., Nasimuddin, Alphones, A.: RIS based compact circularly polarized microstrip antennas. IEEE Trans. Antennas Propag. **61**(2), 547–554 (2013)
9. Xu, H., Wang, G., Liang, J., Qi, M., Gao, Z.: Compact circularly polarized combining metasurfaces and strong space-filling meta-resonators. IEEE Trans. Antennas Propag. **61**(7), 3442–3450 (2013)
10. Zhu, H., Cheung, S., Chung, K., Yuk, T.: Linear-to-circular polarization conversion using metasurface. IEEE Trans. Antennas Propag. **61**(9), 4615–4623 (2013)
11. Jing, Y., Li, Y., Zhang, J., et al.: Achieving circular-to-linear polarization conversion and beam deflection simultaneously using anisotropic coding metasurfaces. Sci. Rep. **9**, 12264 (2019). https://doi.org/10.1038/s41598-019-48812-y
12. Garg, R., Bhartia, P., Bhal, I., Ittipiboon, A.: Microstrip Antenna Design Handbook. Artech House, Boston (2001).

A Geometry Based Algorithm for Comparison of Tetrahedral Metal Binding Sites

Swati Adhikari and Parthajit Roy

Abstract In this study, a geometry-based novel algorithm has been proposed to compute structural deviation of two tetrahedral metal-binding sites on protein's structure that compares one user-provided protein structure with standard one. It is proposed to change the orientation of both structures by applying some mathematical transformation in such a way so that minimum mean square error (MSE) will be produced due to their deviation. Total 998 tetrahedral zinc-binding sites on protein's structures have been tested to evaluate the performance of the proposed method. Deviation statistics for individual sites as well as a summary report for all sites have also been presented.

Keywords Bioinformatic algorithm · Protein-metal interaction · Metal-binding site · Shape of metal-binding site

1 Introduction

Features of Metalloproteins are structurally and functionally very complex in nature. As the metals are highly versatile, challenges lie with the computational prediction and analysis of their coordination sites on protein's structure. Various features of different types of metalloproteins are discussed in detail in [1]. Generally, these sites form varying geometric shapes that include tetrahedral, octahedral, planar, etc. which have great influence on the proper functioning of metalloproteins.

Detection and exploration of such coordination sites are very critical. A number of software exists for the purpose. Among them, CavityPlus [2], Fpocket [3] and CASTp [4] work irrespective of the ligand type whereas CheckMyMetal (CMM) [5], MIB [6], MetalDetector [7] etc. are specifically meant for metal ions.

S. Adhikari · P. Roy (✉)
Department of Computer Science, The University of Burdwan, Purba Bardhaman, West Bengal 713104, India

V. Bhateja et al. (eds.), *Computer Communication, Networking and IoT*, Lecture Notes in Networks and Systems 197, https://doi.org/10.1007/978-981-16-0980-0_19

Many algorithms have also been developed in parallel to the development of such software like CHED and SeqCHED [8]. CHED is a structure-based predictor whereas SeqCHED is a sequence-based predictor.

When structure of a new coordination site has come into existence, it is required to predict type or functioning of the unknown site. The new site can be compared with known one. On the other hand, it may be required to know the structural properties of some coordination sites that belong to proteins with some specific diseases. In this case, the concerned sites can be compared with sites that are known to be for normal human beings. If coordination geometries of both sites are compatible, some other property like how much does the unknown site (comparable) deviates from the known one (standard) can also be measured. In other words, when both sites deviate minimally from each other, the prediction of functional information of the unknown site will become easier.

Although the above-mentioned softwares are capable of detecting binding sites for some of the metal ions with respective merits and demerits, none of them is competent for comparing two structures at a time.

In the present paper, an algorithm has been proposed that compares one tetrahedral coordination site with that of a standard one. It is proposed to apply mathematical transformations on both sites in order to bring them into same orientation so that their deviation produces minimum MSE.

By considering the varied biological function of zinc (Zn^{2+}) [9], its binding sites in different protein's structures are chosen for evaluation of the proposed method. The structures are taken from the database of RCSB PDB [10].

2 Materials and Methods

The proposed algorithm for measuring minimum MSE for structural deviations of two tetrahedral metal-binding sites employs the concept as stated below:

To measure the distance deviation of two lines, the lines can be aligned in multiple ways, two ways of which are as shown in Fig. 1a, b. In Fig. 1a, one of the end point of two lines are aligned and in Fig. 1b, lines are aligned by leaving equal amount of spaces on both sides. It is found that when they are aligned as shown in Fig. 1b, the MSE of their distance deviation will be at minimum, which can be proved by using Differential Calculus.

This concept can be extended for structures with number of atoms more than two. In case of tetrahedron, all vertices will be at same distances from the circumcentre (CC) of the circumsphere of the tetrahedron. So, if it is possible to align circumcentres of two tetrahedra at the same point and to make one of their faces parallel to one of the coordinate plane as well as to make respective sides of that faces, in order of their length, parallel to each other, then all six sides of the two tetrahedra will be of the form of Fig. 1b and MSE for distance and angular deviations will be at minimum.

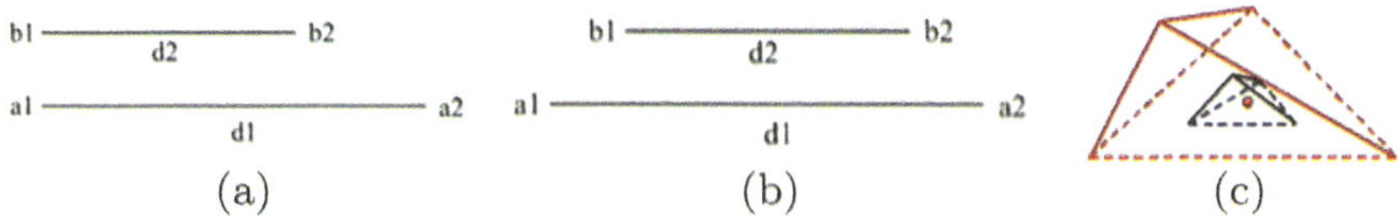

Fig. 1 Possible alignments: **a** Alignment of one of the end points, **b** aligning by leaving equal amount of spaces on both sides, **c** aligning circumcentres of two tetrahedral object

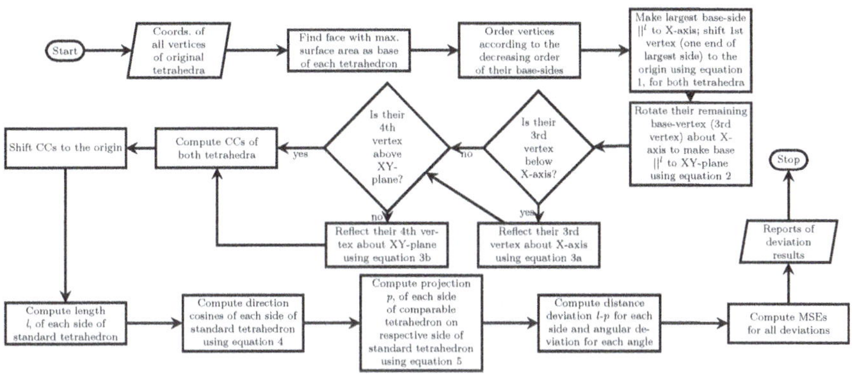

Fig. 2 Flowchart of the proposed algorithm

Figure 1c depicts such arrangements of vertices for two tetrahedra. In Fig. 1c, the position of red circle is the position of the circumcentres of circumspheres of the two tetrahedra.

Flowchart of the Proposed Algorithm
The flowchart of Fig. 2 is used to transform standard and comparable tetrahedra in the form of Fig. 1c and to find deviation statistics of the comparable tetrahedron as proposed.

Illustration of the Proposed Arrangements of Coordinating Atoms
Figure 3 shows Wedge-Dash representation of proposed arrangements of coordinating atoms for tetrahedral Zn-sites [Comparable site—PDB/CIF Accession Number 3dgd; PDB Sr. No.: 4279, Chain Id: C; Residue Name: ZN; Residue No.: 134; Standard site—PDB/CIF Accession Number 6ela; PDB Sr. No.: 5244, Chain Id: C; Residue Name: ZN; Residue No.: 302]. This figure is generated by using the software CHEMFIG [11].

Here, only the positions of binding atoms are shown. The atoms of standard and comparable sites are shown in red and in blue colour respectively. At the centre of the diagram, the CC, which resides at the origin, is shown. The base of the tetrahedron that is parallel to XY-plane is shown as dashed representation and the atom that is situated at the top of base is represented as wedge representation. In this representation, the bond length is set equal to the circumradius of the transformed tetrahedral site. The bond angles are set equal to the angles made by any two atoms with CC.

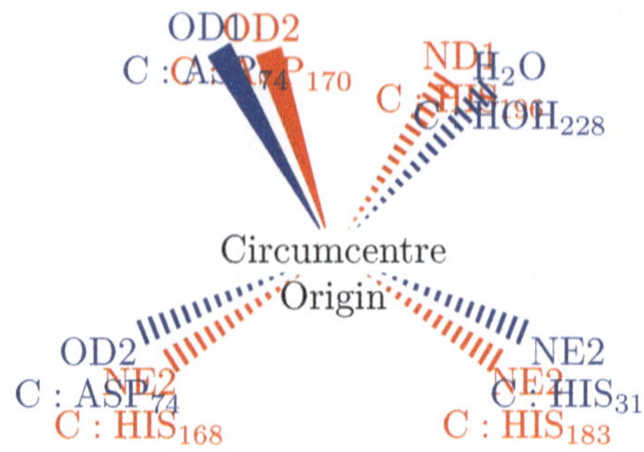

Fig. 3 Wedge-Dash representation of proposed concept [Structure 3dgd]

Some Formulas of Three Dimensional (3D) Coordinate Geometry

Following 3D transformations are used to transform the tetrahedron:

Let (*x, y, z*) and (x', y', z') be the initial and new coordinates of any vertex v, respectively.

Shift: To shift v to the new position, if (t_x, t_y, t_z) be the shift vector, then

$$x' = x + t_x; \qquad y' = y + t_y; \qquad z' = z + t_z \tag{1}$$

Rotation: If θ be the angle of rotation, then to rotate v with respect to the origin about the X-axis, equation 2 is used. Equations for rotation about Y-axis and Z-axis follow accordingly.

$$x' = x; \qquad y' = y\cos\theta - z\sin\theta \qquad z' = y\sin\theta + z\cos\theta \tag{2}$$

Reflection: If v is reflected about the X-axis and XY-plane, then equations 3a and 3b are used respectively.

$$x' = x, \quad y' = -y, \quad z' = -z \tag{3a}$$

$$x' = x, \quad y' = y, \quad z' = -z \tag{3b}$$

To compute deviation of distances between two transformed tetrahedra, following two measures are used:

Direction Cosines (D.C.) of a Line: If (x_1, y_1, z_1) and (x_2, y_2, z_2) be the two end points of a line and *l*, *m* and *n* be the D.C. of the line, then

$$r = \sqrt{(x_2 - x_1)^2 + (y_2 - y_1)^2 + (z_2 - z_1)^2} \tag{4a}$$

$$l = (x_2 - x_1)/r; \quad m = (y_2 - y_1)/r; \quad n = (z_2 - z_1)/r \tag{4b}$$

Projection of a line on another line: If (x_1, y_1, z_1) and (x_2, y_2, z_2) be the two end points of a line AB and *l*, *m* and *n* be the D.C. of a line A′B′, then projection PQ of AB on A′B′ is given by

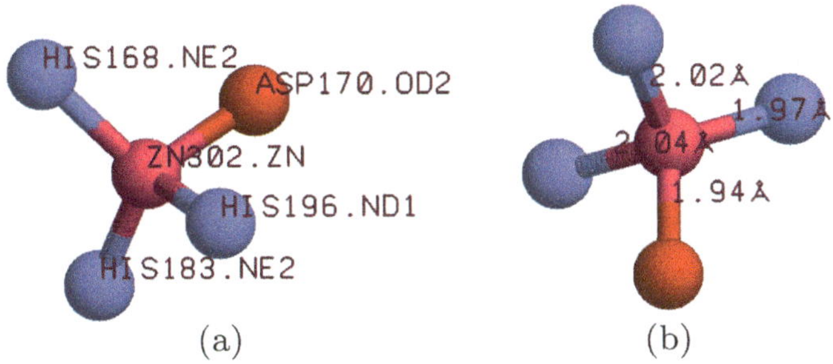

Fig. 4 Zn-site of the structure 6ela: **a** binding atoms, **b** Zn-atom distances

$$PQ = l \times (x_2 - x_1) + m \times (y_2 - y_1) + n \times (z_2 - z_1) \qquad (5)$$

Selection of Dataset

As the proposed model is for metal binding sites with coordination number 4, any metal ion for which tetrahedral coordination is possible can be chosen for testing purpose. In the present paper, the tetrahedral coordinating sites of Zn^{2+} ions are considered for evaluation purpose. In total, 798 different protein structures are taken from RCSB PDB [10] that contain 998 tetrahedral Zn-coordinating sites. These structures are downloaded either in PDB or mmCIF format.

Tetrahedral Zn-site

In tetrahedral coordination geometry of Zn-site, most common ligands are Aspartic Acid (Asp), Histidine (His), Cysteine (Cys), and Glutamic Acid (Glu) residues and water molecules [1]. Any combination of these ligands is observed in several Zn-sites.

Figure 4a shows the tetrahedral Zn-coordination sites of the structure with PDB/CIF accession no. 6ela, which is considered as standard site for evaluation of the proposed model. Figure 4b shows the distance of coordinating atoms from Zn^{2+} ion in this structure. These figures are created in Rasmol [12].

Although distances of Zn^{2+} ion from any coordinating atom are varied with type of ligands, the mean of such distances in standard site is noted as 1.99 Å and the same for 998 test Zn-sites are observed as 2.15 Å. In standard site, the distance between Zn^{2+} and CC is noted as 0.08 Å.

3 Results and Discussion

Deviation Statistics for Two Tetrahedral Sites

To measure deviation statistics, CC to atom and inter-atom distances, angles made by any two atoms with CC and with any other atom are considered. Root MSEs (RMSEs) produced due to two types of distance and angular deviations are taken into account for analysis purpose. CC to Zn^{2+} ion distances are also analyzed. The results of investigation are presented in Tables 1, 2 and 3.

Table 1 Statistics for distances of Zn^{2+} ions from the CC of Zn-sites

Mean distance	Standard deviation (σ)	No. of sites in distance from mean RMSE		
		1σ	2σ	$>2\sigma$
0.14	0.10	363	304	331

Table 2 Statistics for RMSEs for distance and angular deviations

RMSE type	Mean RMSE	Standard deviation (σ) of RMSE	No. of sites in dist. from mean RMSE		
			1σ	2σ	$>2\sigma$
RMSE[a]	0.17	0.13	495	84	419
RMSE[b]	1.67	0.61	0	224	774
RMSE[c]	0.08	0.04	141	304	553
RMSE[d]	0.05	0.03	145	384	469

RMSE for [a]dist. dev. for atom-CC dist., [b]dist. dev. for atom-atom dist. [c]Angular dev. for angle atom-CC-atom, [d]angular dev. for angle atom-atom-atom

Table 3 Sizes of all Zn-sites compared to standard site

Total sites	Larger	Smaller	Same
998	908	89	1

Statistics related to distances between Zn^{2+} ion and the CC and that of 4 types of RMSEs are recorded in Tables 1 and 2 respectively whereas Table 3 stores comparative statistics regarding overall sizes of all Zn-sites.

Generally, metal ions reside at the centre of the coordinating site [13]. But in test protein structures, it has been noticed that sometimes reverse happens. Table 1 shows the fact. In Table 1, such distances are divided into 3 groups—distances those are within σ, 2σ and beyond 2σ distances from mean, σ is the standard deviation. Table 1 helps in grasping idea about how many Zn^{2+} ions are nearer to or at distant from CC.

Table 1 also shows that the mean Zn^{2+} to CC distance for 998 Zn-sites is 0.14, which is quite larger than the standard one. So, this distance can be set as new standard for position of Zn^{2+} ion in tetrahedral coordinating sphere, although it seeks data from more structures to reach any concluding decision.

To determine the position of coordinating atoms with respect to standard site, the 4 types of deviation results are analyzed on the basis of RMSEs produced due to respective deviations. In Table 2, data for each of the 4 types of RMSEs are also divided into 3 groups—RMSEs those are within σ, 2σ and beyond 2σ distances from mean RMSE, σ is the standard deviation.

In general, sites for which RMSEs are closer to zero, are treated as most compatible. Table 2 shows number of Zn-sites that are almost structurally compatible with the standard site. For example, from Table 2, it is seen that the mean RMSE for inter-atom distances are quite larger than other RMSEs and the sides of 224 Zn-sites are quite deviating from the standard site, as their RMSEs are within 2σ distance from mean RMSE. The sides of rest of the sites are deviating more, as their RMSEs are beyond 2σ distance from mean RMSE.

Table 3 compares the overall sizes of test Zn-sites with the standard site. Tables 2 and 3 combinedly helps in identifying the sites which are most compatible with the standard one. It also helps in taking decision about setting new standard, if it is found that most of the sites are deviating from the standard site but they have some kind of structural similarity. For that, more structures are needed to be compared.

4 Comparative Study of the Proposed Method

To show that the proposed technique generates minimum MSE, another result is considered which is obtained by computing RMSEs for deviations with original coordinating atoms.

The results of computation of RMSEs for inter-atom distances and angular deviations for angles made by any two atoms with another atom, for original as well as transformed coordinating atoms by applying proposed technique are given in Table 4. The results are obtained for six selected Zn-sites.

To visualize two types of deviation results, as given in Table 4, Fig. 5a, b are used. Table 4 and Fig. 5 clearly show that the deviations made by the proposed algorithm produce minimum RMSEs for both type of deviations.

In the field of metallobiomolecules, different aspects of metal sites are studied by many researchers. Two such works are as presented in [14, 15]. Antarip et al. [14] have investigated the role of Mg^{2+}-ion's coordination with RNA and have provided

Table 4 Results of RMSEs for original Zn-sites versus proposed method

Zn-site[a]	Atom-atom distance dev.		Atom-atom-atom angular dev.	
	Original	Proposed	Original	Proposed
Site 1: 1i76:A-998	4.37	1.55	0.03	0.02
Site 2: 2oxu:A-265	2.43	1.62	0.07	0.03
Site 3: 3gz0:A-262	3.77	2.34	0.14	0.07
Site 4: 4qiy:C-301	2.12	1.51	0.09	0.05
Site 5: 5pkh:A-401	3.95	0.21	0.04	0.02
Site 6: 6f3b:B-301	4.58	1.88	0.07	0.04

[a]Site name: PDB/CIF Accn.: chain name-residue number

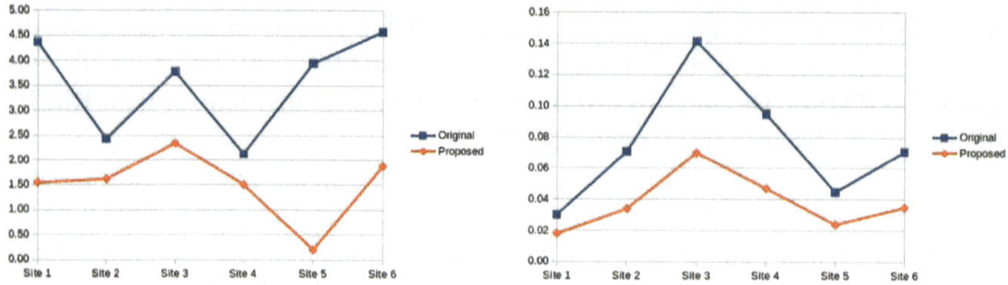

(a) (b)

Fig. 5 Comparative study: RMSEs for deviation of **a** distances **b** angles

results from different angles in which Mg^{2+}-ion regulates the stability of the system of RNA folding. Their findings help in enhancing the existing algorithms to predict RNA structure.

Alan et al. [15] have developed a website, BioMe, that is capable of generating statistical reports about the searched information for the specified metal ions and its ligands in biomolecule's structure. These reports can be used as a valuable research tool relevant to metallobiomolecules. The present study is comparable with such type of workings.

5 Conclusion

After analyzing all tables and figures it can be concluded that the proposed algorithm is competent for measuring structural deviations between two sites with minimum MSE. The applicability of this algorithm is not limited only to tetrahedral shape of binding sites, there is a scope of further improvement. The power of this algorithm can also be utilized for other shapes.

Acknowledgements Authors are thankful to RCSB PDB for their online standard dataset and Department of Computer Science, The University of Burdwan.

References

1. Kretsinger, R.H., Uversky, V.N., Permyakov, E.A. (eds.): Encyclopedia of Metalloproteins. Springer, New York Heidelberg, Dordrecht, London (2013)
2. Youjun, X., Wang, S., Qiwan, H., Gao, S., Ma, X., Zhang, W., Shen, Y., Chen, F., Lai, L., Pei, J.: CavityPlus: A web server for protein cavity detection with pharmacophore modelling, allosteric site identification and covalent ligand binding ability prediction. Nucl. Acids Res. **46**(W1), W374–W379 (2018)
3. Le Guilloux, V., Schmidtke, P., Tuffery, Pi.: Fpocket: An open source platform for ligand pocket detection. BMC Bioinform. **168**(10) (2009)

4. Tian, W., Chen, C., Lei, X., Zhao, J., Liang, J.: CASTp 3.0: Computed atlas of surface topography of proteins. Nucl. Acids Res. **46**(W1), W363–W367 (2018)
5. Zheng, H., Cooper, D.R., Porebski, P.J., Shabalin, I.G., Handing, K.B., Minor, W.: CheckMyMetal: a macromolecular metal-binding validation tool. Acta Crystallogr. D. Struct. Biol. **73**(3), 223–233 (2017)
6. Lin, Y.-F., Cheng, C.-W., Shih, C.-S., Hwang, J.-K., Chin-Sheng, Y., Chih-Hao, L.: MIB: Metal ion-binding site prediction and docking server. Journal of Chemical Information and Modeling **56**(12), 2287–2291 (2016)
7. Lippi, M., Passerini, A., Punta, M., Rost, B., Frasconi, P.: MetalDetector: A web server for predicting metal-binding sites and disulfide bridges in proteins from sequence. Bioinformatics **24**(18), 2094–2095 (2008)
8. Sobolev, V., Edelman, M.: Web tools for predicting metal binding sites in proteins. Israel J. Chem. **53**(3–4), 166–172 (2013)
9. Vallee, B.L., Auld, D.S.: Zinc coordination, function, and structure of zinc enzymes and other proteins. Biochemistry **29**(24), 5647–59 (1990)
10. Berman, H.M., Westbrook, J., Feng, Z., Gilliland, G., Bhat, T.N., Weissig, H., Shindyalov, I.N., Bourne, P.E.: The Protein data bank. Nucl. Acids Res. **28**(1), 235-242. www.rcsb.org (2000)
11. Brefo-Mensah, E.K., Palmer, M.: mol2chemfig, a tool for rendering chemical structures from molfile or SMILES format to LATE X code. J. Cheminform. **24**(4) (2012) https://doi.org/10.1186/1758-2946-4-24
12. Roger Sayle, E., Milner-White, J.: RasMol: Biomolecular graphics for all. Trends Biochem Sci (TIBS) **20**(9), 374 (1995)
13. Mason, M., Yamashita, L.W., Eisenman, G., Eisenberg, D.: Where metal ions bind in proteins. Proc. Nadl. Acad. Sci. USA **87**(15), 5648–5652 (1990)
14. Halder, A., Roy, R., Bhattacharyya, D., Mitra, A.: Consequences of Mg^{2+} binding on the geometry and stability of RNA base pairs. Phys. Chem. Chem. Phys. **20**, 21934–21948 (2018)
15. Tus, A., Rakipovic, A., Peretin, G., Tomic, S., Šikić, M.: BioMe: biologically relevant metals. Nucl. Acids Res. **40**(W1), W352–W357 (2012)

MICR:Mobile-Assisted Immunization of Child in Remote Healthcare

Safikureshi Mondal and Nandini Mukherjee

Abstract Immunization is an effective way to prevent or resist in infectious diseases in children. Proper tracking of vaccination of children in remote areas and remembering vaccination time schedules by new mothers or caregivers are two real challenges for immunization system in India. To alleviate the challenges in immunization system, a mobile based child immunization system is implemented. An android-based *mobile app* is developed to send *Push Message* alerts to the registered mobile numbers of new mothers or caregivers for reminder of vaccination of each child within a given time schedule.

Keywords Immunization · Remote healthcare · Mobile app · GCM · *Push-Alert* message

1 Introduction

"Prevention is better than cure" is the general truth which is introduced by a Romanian doctor [1]. Immunization is one of the most successful achievements in medical science. Vaccine provides an antigenic material that causes to develop the person's immune system against diseases. Vaccines keep children healthy and protect from life-threatening diseases. Due to vaccination, every year 3 million children are saved [2]. The four life-threatening diseases like measles, diphtheria, pertussis (Whooping cough), and tetanus have been the causes of 218,400 deaths worldwide in 2013 [3]. DTP and MCV are used to protect a child from the aforementioned diseases. The government of India and many other countries have taken up immunization programs. The *Global Vaccine Action Plan* (*GVAP*), developed by key stakeholders including *UNICEF* and *WHO*, sets out two key measures of success for immuniza-

S. Mondal (✉)
Department of Computer Science and Engineering, Narula Institute of Technology, Kolkata, India

N. Mukherjee
Department of Computer Science and Engineering, Jadavpur University, Kolkata, India
e-mail: nmukherjee@cse.jdvu.ac.in

V. Bhateja et al. (eds.), *Computer Communication, Networking and IoT*, Lecture Notes in Networks and Systems 197, https://doi.org/10.1007/978-981-16-0980-0_20

tion by 2020. First one is "At least 90% coverage must be achieved for all vaccines nationally". The second one is "Achieving at least 80% coverage for all district". Unfortunately, there are many reasons which create gaps and disparities in immunization system. The first and foremost reason is the awareness of people, including mothers. Another reason is proper tracking of vaccination process which remains a big challenge worldwide due to reach in remote areas. The third reason is lack of cold chain process in the society. Finally, the forgetfulness of parents, or caregivers to vaccinate a child is another cause of Immunization target failure. In a recent research paper [4], the authors have conducted a study in a rural area of India. It is shown that two major reasons for missed vaccination are—prior reminder not given (32.9%) and mothers' forgetfulness (26.6%). The author in [5] also pointed out that vaccinations are missed mostly due to mothers' forgetfulness.

The above mentioned challenges can be overcome by combination of cloud computing and mobile devices. Mobile internet users will increase from 390.9 in 2018 to 500.9 million by 2023 [6]. Moreover, number of mobile users in rural India is increasing day by day including Internet users. A recent trend in research focuses on the remote health care using ICT-enabled devices and cloud computing technologies [7–9]. The objective of this paper is to demonstrate the use of an android based Child-Care mobile application for immunization system in cloud environment. The mobile app reminds the users who register in the application with the Push Notification about every new vaccination, news feed and provide a mother with the best way to take care of her child with auto update databases.

The remainder of the paper is organized as follows. Related work is given along with the state of the art of health mobile ap in Sect. 2. Section 3 discusses the immunization program for children to understand its necessity for child health. Next an immunization system is proposed with brief description of components for the implementation of immunization system and flow of messages by different components as described in Sect. 4. Implementation is described in Sect. 5 with some snapshot results of this android app. The paper is concluded in Sect. 6.

2 Related Work

State of the art of mobile applications in healthcare domain is shown in Fig. 1. Among the different healthcare applications, immunization apps are much popular, like *Ready Vax* [10], *CDC Vaccine* [11] and *Canimmunize* [12]. However, these apps do not have features like reminding the vaccine in time or sending *Push Message* alerts for reminder of vaccination to the mother or caregiver.

In [13], the authors implemented and evaluated text message reminders for the second HPV2 and third HPV3 vaccine doses. Abbas et al. [14] proposed a model for vaccination reminder via SMS alert to the mobile devices. This prototype was developed using Visual Basic .NET 2008 and *SQL* Server environment. The *CVRS-A-SMS-A System* is a manual system. However, as cloud is not used in this paper, the use of this service is limited.

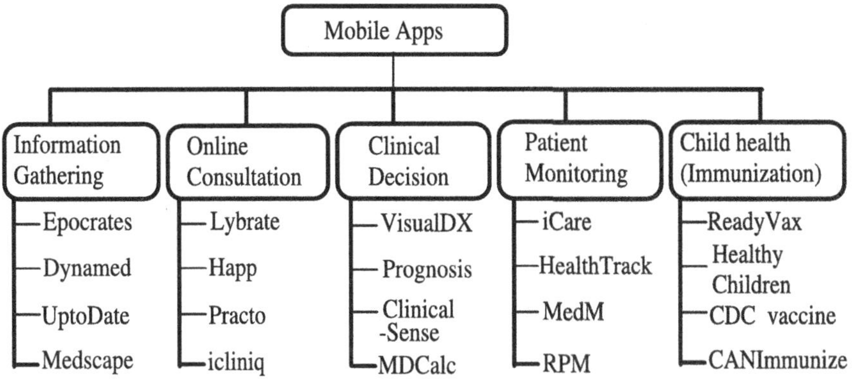

Fig. 1 The state of the art of healthcare mobile apps

In [15], the author proposed e-health elderly care *app* by using Google Cloud Messaging (GCM). E-health services, like health information, prescription, medication instruction, medicine taking time, reminding schedule back, etc. are provided to the elderly patient via *Push Message Notification* through smartphone automatically. Through the push notification, elderly patient can get their own prescription, personal health information after successful authentication of personal information by hospital.

Nevertheless, the existing literature does not provide a complete solution for a immunization system.

3 Immunization of Children

As per census 2011, 26 millions of children are born every year in India and children of 0–5 years of age are 13% of the total population of India. The grave situation can be understood from the fact that 12.7 lakh children under 5 years are dying every year. The World Health Organization (*WHO*) and *UNICEF* estimated that 100 million children receive at least basic immunization and 2.5 million lives are saved every year. A proper immunization system can help to keep away any neurological complications associated with diseases like measles, mumps, rubella, and even chickenpox. In the absence of large-scale immunization program, highly infectious diseases such as measles, mumps, rubella, and chickenpox affect nearly all children. Approximately 132 million babies need to be fully immunized each year and in order to meet this, immunization systems must have adequate resources.

4 Immunization System

In this section, an immunization system is described. The objective of this paper is to develop and design an android mobile application based on immunization system for child care using cloud computing. The app also helps to send the *Push Messages* for reminder of vaccination schedules to the caregiver or mother for a newborn baby. The implementation architecture for immunization of children with *push alert* messages is shown in Fig. 2. The architecture is composed of three main components, mobile app, Cloud, and Google Cloud Messaging(GCM). The components or requirements of this immunization process are described below.

4.1 Mobile Apps for Child Care

The app for immunization is intended for use by a caregiver or mother. The immunization app starts with the combination of sign-in and registration module. A caregiver or mother can sign up providing the child's demographic information and the mobile number of the mother. After signing up she can sign-in and can view all the immunization information like immunization chart, babies healthcare tips and also height/weight chart. Mobile app is developed using android studio.

4.2 Cloud Computing

Cloud is another important component of this architecture. It provides services over the internet or dedicated network, with delivery on demand, and payment based on usage. Cloud computing services range from full applications and development platforms to servers, storage, and virtual desktop.

Firebase is a cloud service provider. There are lots of services provided by the firebase service provider. Such as *Firebase Analytics*, *Firebase Auth*, *Real-time Database*, *Firebase Storage*, *Firebase Cloud Messaging* etc. for Android.

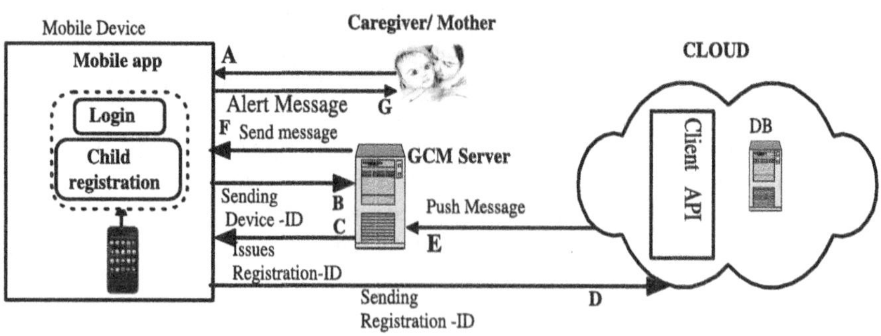

Fig. 2 Architecture of immunization system

4.3 Google Cloud Messaging (GCM and Push Notification)

Google Cloud Messaging for android (GCM) is a service that helps developer to send data from server to their android applications on android devices. There are three components of this service. Android Mobile device, Application server, and GCM server. First the android mobile device sends sender id and application id to the GCM server for registration. This sender id is taken from google after applying API key in google. After registration GCM server issues a registration id to the android device. Therefore, the mobile device sends this id to the server. The server stores registration id in the database for later usage. *Firebase* is used for the cloud storage in this application.

Push notification is the ability of a mobile device to receive messages that are pushed from a server. The most common form of notification is SMS (*Short Message Service*). *Push notification*s can be sent in several forms. For example,

1. *Alert* which is a pop-up text message.
2. *Badge*, *Tile* which is a graphical representation that includes short text or image, Banner.
3. *Toast* which is a pop-up text message at the top of the device display that disappears after it is read.

If any Android device wants to get *push notification*, it sends the sender id to the GCM server along with registration id. This registration id is stored previously in database. GCM server will deliver the messages to respective mobile devices. The communication between mobile, cloud, and GCM server of immunization process are shown in Fig. 2.

5 Implementation

Mobile based child care immunization system is implemented using Google Cloud in remote health care. The architecture of this immunization system is shown in Fig. 1 and it is seen from this immunization system that there are basic four important processes that must be implemented. The four functions are

1. Authentication followed by sign-in /sign-up: In our system *Firebase* makes it easy for authentication. In *Firebase*, apps have built-in support for logging in with email and password. The Authentication procedure is the combination of some sub-procedures like setup on *Autoxidation Provider*, *Enabling Provider*, *Monitoring Authentication*, *Storing User data*, *Authenticating user* with email and password etc
2. Using a line of code *ref.createUser (email, password, new Firebase.ResultHandler())*; a user will be created at the reference when *createUser()*; is called. Here *ref* is the reference of the *Firebase* database.

3. Connecting firebase with android mobile: There are few steps to connect the *Firebase* with immunization app. Such as *Add Firebase* , *add Android permission*, *setup Firebase* in android etc.
4. Sending / retrieving data to/ from Firebase : *Firebase* data is retrieved by attaching an *asynchronous listener* to a *Firebase* reference. The *listener* will be triggered once for the initial state of the data and again anytime the data changes. The *listener* receives a *DataSnapshot*, which is a picture of the data at a particular location in a *Firebase* database in any instance. Calling *getValue()* on a *snapshot* returns the Java object representation of the data. We used the *addValueEventListene()*. With *Firebase* queries, it can selectively retrieved data based on various factors. To construct a query, we start by specifying how we want our data to be ordered using one of the ordering functions *orderByChild()*, *orderByKey()*, *orderByValue()*, or *orderByPriority()*. We can then combine these with five other methods to conduct complex queries *limitToFirst()*, *limitToLast()*, *startAt()*, *endAt()*, and *equalTo()*. At first we are going to retrieving the name and Age of the child. Age is calculated by the Date of Birth given by the user at the time of Registration and it will be shown in the main page of the application after login authentication.
5. Sending *PUSH message* notifications: On Android devices, when a device receives a push notification, an icon and a message appear in the status bar. *Batch* is a push notification platform built around predictive analytic and smart segmentation algorithms. With *Batch*, we can efficiently send transactional (one to one) and marketing notifications (one to many) while leveraging our other modules to run a successful retention strategy. The *Batch* dashboard needs to have our GCM Server API Key to be able to push users of our application and we will need our GCM Sender ID to place within our app code. First, it needs to go to the google services wizard to enable google services and GCM. After configuring batch push, it needs to add permissions. GCM uses the *Google Play Services* library. If we haven't included it yet, we must integrate it to continue. *Batch* uses GCM via *Google Play Services* which is not compatible with the old *C2DM API*. If mobile app is using *C2DM API*, we must migrate to GCM via *Google Play Services* before using *Batch* Push. *GCM* needs *4* permissions to setup push. These will handle notifications and display them for us without any additional code. Activate *Batch* push by setting the *GCM Sender Id* that we obtained earlier, before setting the *Batch API Key*.

6 Result and Discussion

In this paper, mobile-based immunization system with push notification is implemented using google cloud. There are two activities. One is login activity and another is register activity. If one is not registered then clicking on the "New Here First Sign up" text a registering page is open to register the parents information with baby's demographic information to access the app.

The snapshots of the registration page, and when an account is successfully created are shown in Figs. 3, 4 and 5.

There are some fields that need to be filled in the registration activity, like email, password, child name, child date of birth, *Mobile no* etc. This Immunization app also provides vaccination chart along with wieght/height chart of baby according to age which is shown in Figs. 6 and 7.

Fig. 3 Login page in mobile

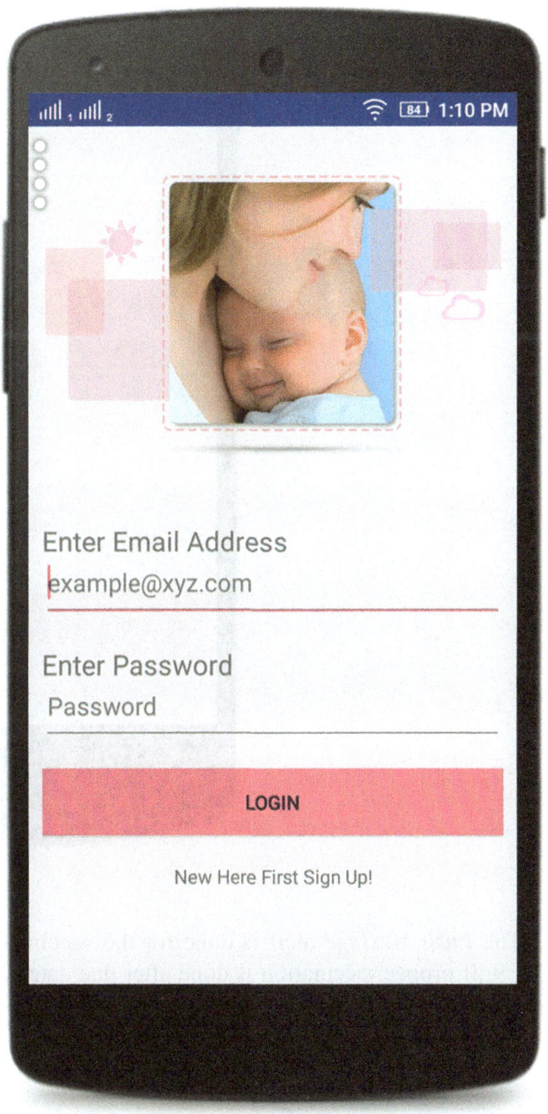

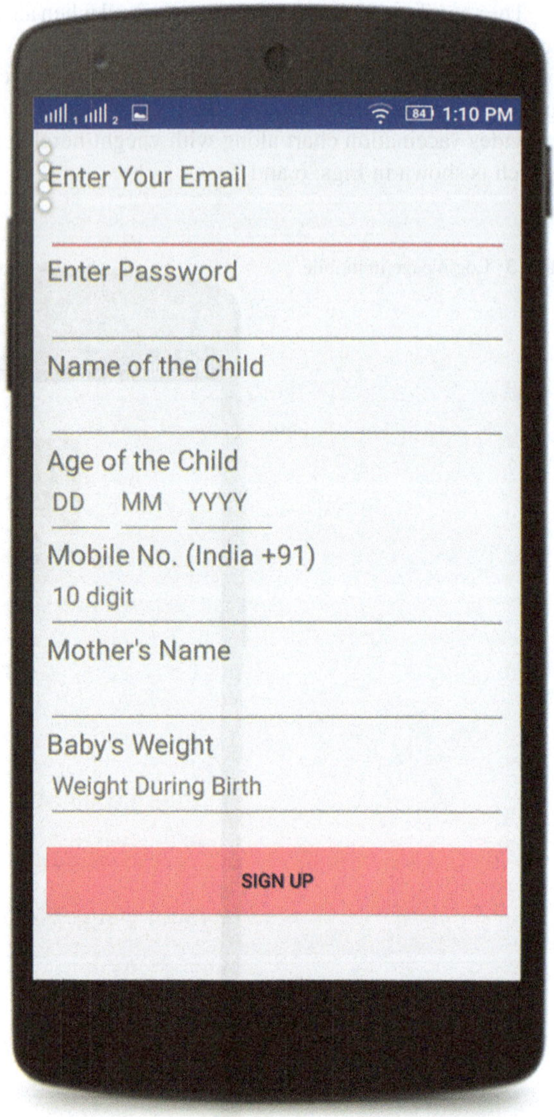

Fig. 4 Registering to input demography information

The *Push Message* alert is done for the vaccinations that are due as shown in Fig. 9. If proper vaccination is done after due date, then parents can stop the *Push Message* alert by clicking on *Stop Notification* as shown in Fig. 8.

Fig. 5 Classification services for immunization

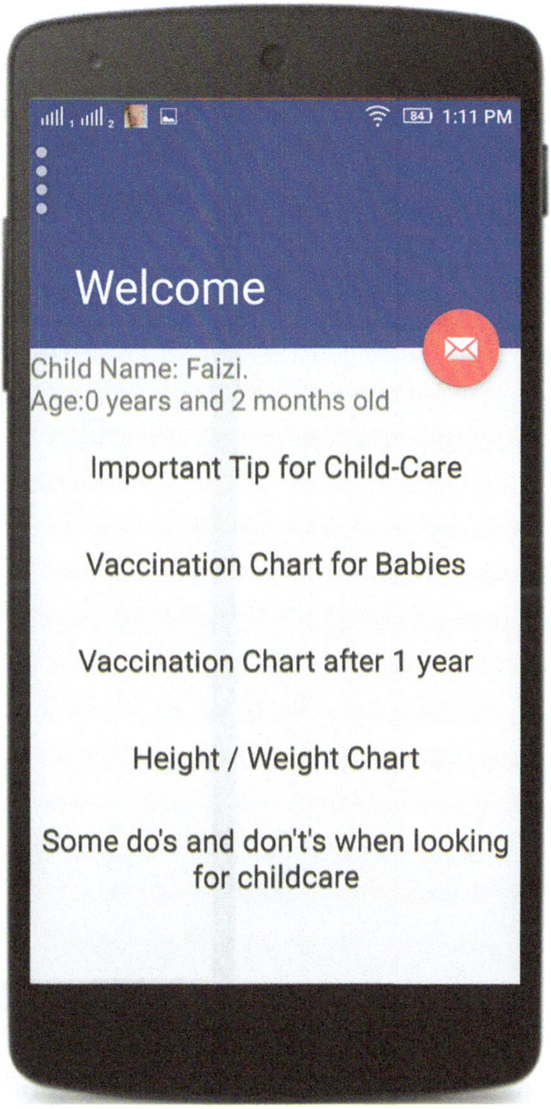

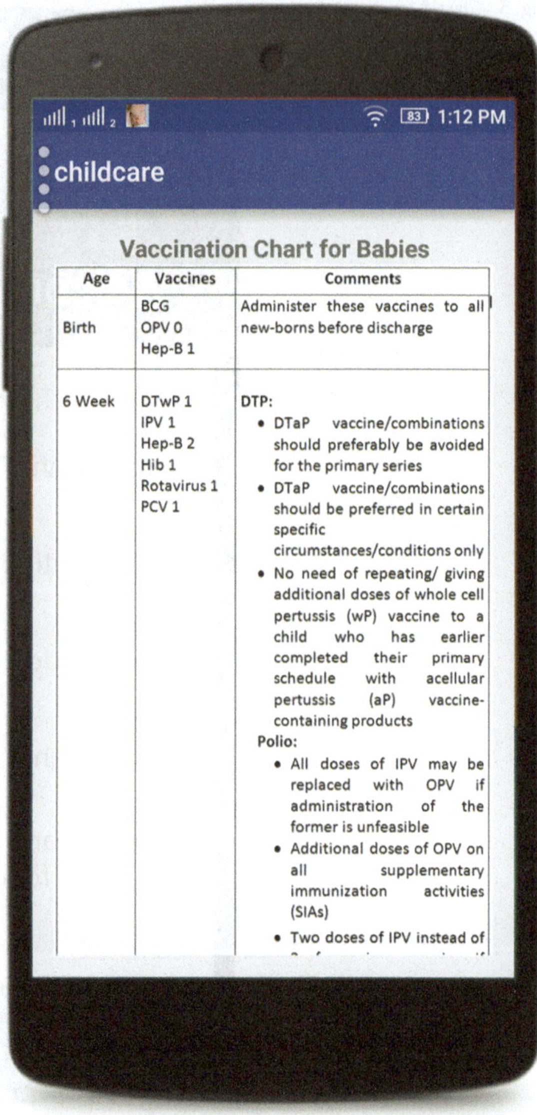

Fig. 6 Vaccination chart for babies

Fig. 7 Weight/height chart

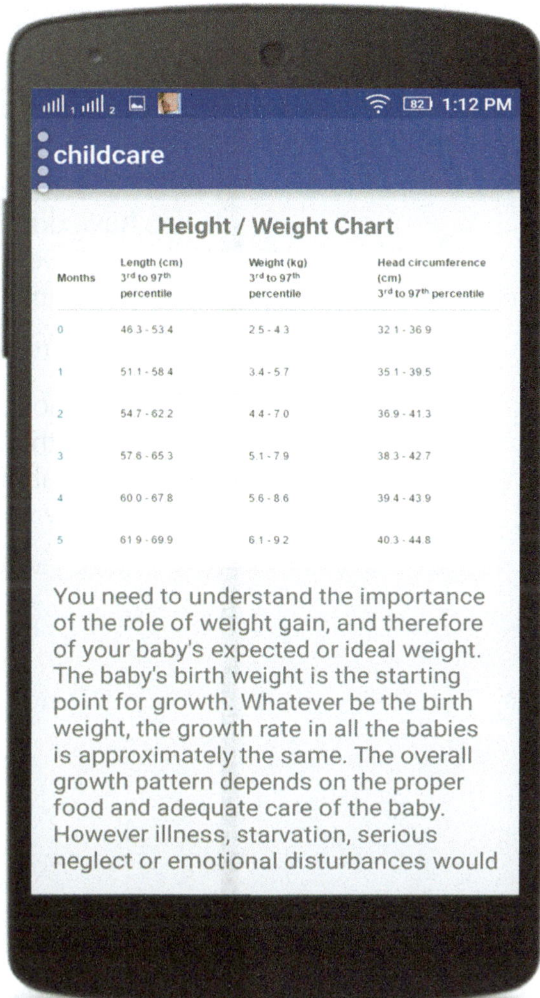

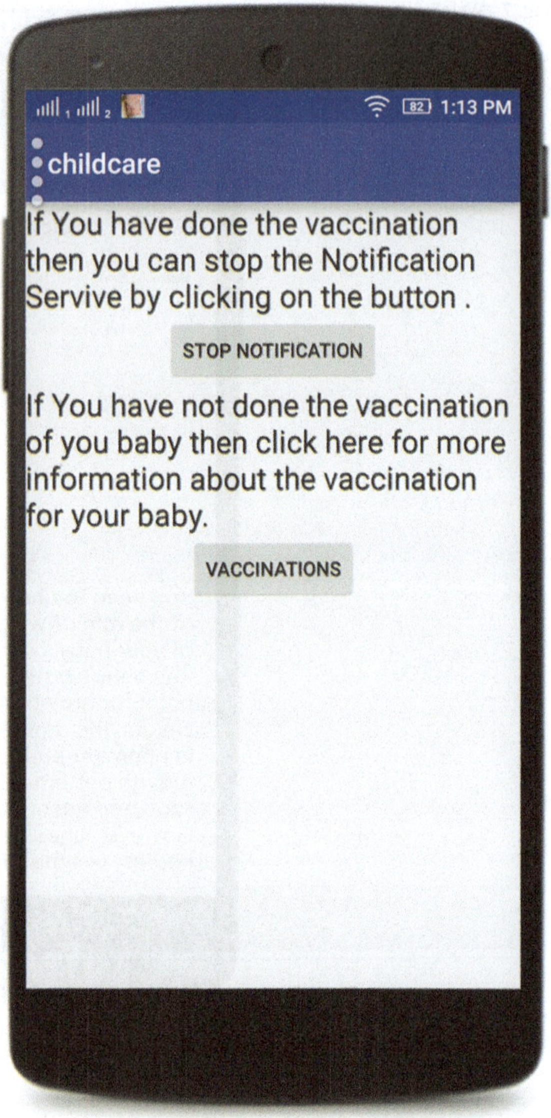

Fig. 8 Setting vaccination page

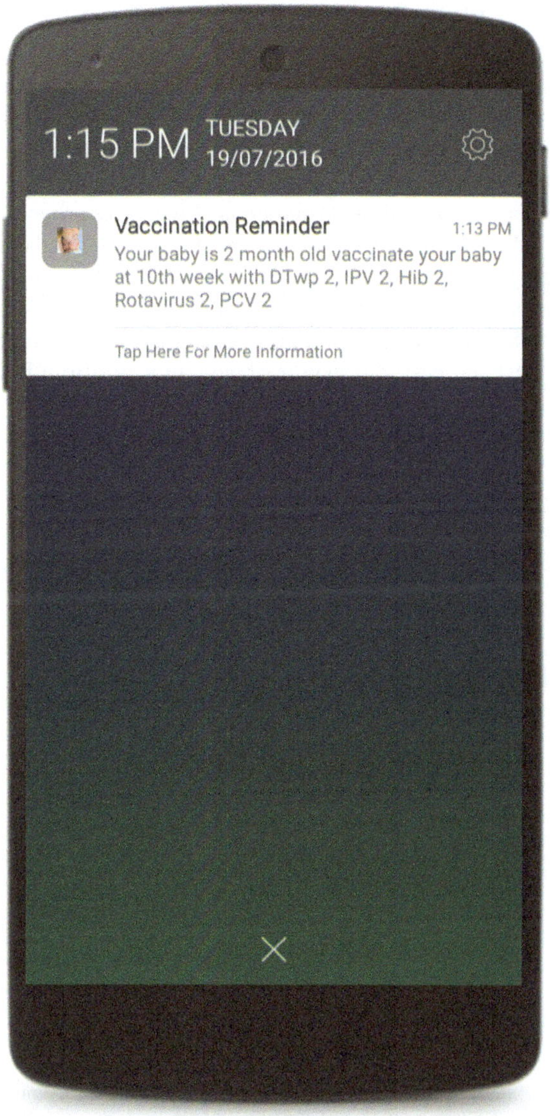

Fig. 9 Push messages generation

7 Conclusion

A mobile-assisted immunization system is implemented using google cloud. It is a mobile-based healthcare application for mothers that provides child immunization notifications, child-care tips, and latest news about child-care starting from the birth of the child to 12 years of the age. It helps the mothers to know the detailed information about vaccinations and automatically notifies proper vaccination schedule according to the child's age to the users who have registered their children with this application. Providing proper guidance to the mothers for taking care of the children for better health is important. It is also important to remind the mothers, who are busy in their daily life, to provide vaccinations to their children. Our system provides the latest and best information about child-care from Indian Academy of Pediatrics. It also reminds the mothers by sending push notifications about the vaccinations as issued by Indian Academy of Pediatrics. There are some limitations which need to be overcome later. There is no online user support for better care of children. Also Feeding System for the babies and medicare system may be included in future. The parents who have more than one children need to do multiple registrations for each child. These limitations may be taken care of in future.

References

1. In: Bulletin of the World Health Organization, vol. 89, pp. 248–249 (2011)
2. In: http://www.unicef.org/immunization/
3. Murray, C.J.L., Ortblad, K.F., Guinovart, C., Lim, S.S., Wolock, T.M., Roberts, D.A.: Global, regional, and national incidence and mortality for HIV, tuberculosis, and malaria during 1990–2013: a systematic analysis for the global burden of disease study 2013. Lancet **384**(9947), 1005–1070 (2014)
4. Patel, T., Pandit, N.: Why infants miss vaccination during routine immunization sessions? Study in a rural area of Anand District, Gujarat. vol. 55, pp. 321–323 (2011)
5. Pareek, R.P., Manjunath2, U.: Maternal knowledge and perceptions about the routine immunization programme—A study in a semi urban area in Rajasthan. Ind. J. Med. Sci. **57**, 158–163 (2003)
6. In: https://www.statista.com/statistics/558610/number-of-mobile-internet-user-in-india/
7. Das, R., Mondal, S., Mukherjee, N.: More-care: mobile-assisted remote healthcare service delivery. In: 2018 10th International Conference on Communication Systems Networks (COMSNETS), pp. 677–681 (2018)
8. Mondal, S., Mukherjee, N.: Mobile-assisted remote healthcare delivery. In: 2016 Fourth International Conference on Parallel, Distributed and Grid Computing (PDGC), pp. 630–635 (2016)
9. Mondal, S., Mukherjee, N.: A BFS based pruning algorithm for disease-symptom knowledge graph database. In: 2018 The 3rd International Conference on Information and Communication Technology for Intelligent Systems(ICTIS 2018), 06–07 Apr 2018
10. Salmon, D.A., Whitney, E., Omer, S.B., Bednarczyk, R.A., Frew, P.M.: Readyvax: a new mobile vaccine information app. Hum. Vacc. Immunotherap. **15**(3), 1149–1154
11. In: https://www.cdc.gov/globalhealth/immunization/default.htm/
12. Atkinson Ph.D K., Michelle B.sc., Paradis BA, Houle BSP, S.K.D., Wilson Msc, FRCS MD, K.: Canimmunize: a digital tool to help patients manage their immunizations. Can. Pharm. J. (Ott), **15**(4), 236–238 (2017)

13. Kharbanda, E.O., Stockwell, M.S., Fox, H.W., Andres, R., Lara, M., Rickert, V.I.: Text message reminders to promote human papillomavirus vaccination. Vaccine **29**(14), 2537–2541 (2011)
14. Abbas, A.H., Yusof, Y.: Children vaccination reminder via SMS alert. In: 2011 International Conference on Research and Innovation in Information Systems, pp. 1–5 (2011)
15. Yang, C.N,, Wu, F.H., Tsai, S.Y., Kuo, W.C.: E-health services for elderly care based on google cloud messaging. In: 2015 IEEE International Conference on Smart City/SocialCom/SustainCom (SmartCity), pp. 9–12 (2015)

IOT-Based Traffic Signal Violation and Over Speed Recognition

A. R. Aswatha and J. Shwetha

Abstract Internet of Things (IoT) uses Unique Identifiers (UIDs) assigned on every device to communicate between objects equipped with sensors and digital devices around the world through the Internet. Faster Internet speeds and wider bandwidths allow more opportunities for inventions and marketability opportunities. The world is quickly urbanizing. This has brought about a complex increase in the number of vehicles on city streets causing engendering traffic violations to turn out to be progressively basic these days. This causes extreme devastation of property and more accidents that may endanger the lives of the individuals. To take care of the disturbing issue and prevent such incomprehensible results, criminal traffic offense identification frameworks are required. It must be realized in real time as the authorities track the streets constantly. Subsequently, traffic authorities will not just be quiet in actualizing safe streets precisely, but also efficiently; as the traffic violations will be detected faster than humans. This work proposes a useful approach for monitoring and detection of traffic rules violations like traffic signal jumping (red signal) and over speeding of the vehicles.

Keywords Street light · LED · Raspberry Pi · Node MCU

1 Introduction

To define in simple terms, IoT is a system which is of interrelated computing devices and machines, and it is provided with a unique identifier. It will also have ability to exchange the data over a network which will not involve interaction between human to computer or even between humans. As the technology improved, people developed nature of focusing on to improvement of the different commercial areas like lighting systems, traffic signals, hospitals, and industries. Also, this thought made everyone to make use of more advanced and efficient smart systems which has convenient web applications in it. Using such web applications, users can control

A. R. Aswatha · J. Shwetha (✉)
Dayananda Sagar College of Engineering, Bangalore, Affiliated to VTU, Belgaum, Karnataka, India

V. Bhateja et al. (eds.), *Computer Communication, Networking and IoT*, Lecture Notes in Networks and Systems 197, https://doi.org/10.1007/978-981-16-0980-0_21

or monitor the different equipment's which will in turn helps to save time and early detection in case of any errors. This system will be connected to the IoT cloud server in which the required data can be stored. Nowadays, smart cities are more interested in developing user-friendly and efficient systems like, for example, optimal energy-based traffic signals and light control systems. Major problem the world is facing nowadays is increase in population ration due to which several problems are caused like scarcity of food, clothing and shelter, illiteracy problem, and traffic problems.

It is impossible to maintain the traffic with a single or couple of traffic police officers. It is due to negligence or hard to monitor, each vehicle or even government cannot afford too many traffic police. So, one method to avoid manpower in this case is to design a machine which can automatically monitor the traffic signals day and night. Here, a similar kind of approach is being proposed, where, with the help of RFID readers which are placed at every traffic signals, it will read the RFID tag placed in every vehicle. So that it can detect any traffic rules violations like jumping the signals or over speeding. So that using this approach manpower can also be reduced, and this system scans each vehicle which is passing the traffic signals. Therefore, monitoring is done 24 * 7.

2 Applications of the System Proposed

IoT has many applications in different fields, for instance.

2.1 Monitoring

Since the application uses RFID technology user who is handling the device can determine the exact location of the objects to which the RFID tag is implanted and to be noted, this will be applicable only when the tagged asset is in the specified range. And therefore, it will monitor and makes note of all the vehicles which is being violated the traffic principles. It can even detect high traffic areas.

2.2 Less Manpower

As the number of humans involved is less, unwanted errors due to negligence can be reduced. This machine does not need anyone to monitor now and then, no need to monitor the data because once the traffic rules are violated, the challan or fine will be issued to the person via email or text message.

2.3 Operability

24 * 7 working in the absence of any internal machinery errors.

2.4 No Communication Loss

Since RFIDs are placed near the traffic signals within that defined range of RFID, there will be no communication loss until there is an Internet loss, as here Wi-Fi modules are used.

2.5 Secured

As password locked database is used the information will be secured.

3 Literature Survey

The IoT devices as to be provided with high security and only the authenticated users should be able to get the access to it. This is because, nowadays as all know almost all the applications will be implemented using IoT which are used in private and even public environment. There may be threats like leakage of sensible information which leads to hacking the application itself [1]. Hence, security factors like access only to authenticate user, confidentiality of information, privacy, and authorization must be ensured.

If specific application can be design based on wide areas or small area regions. Then one can make use of two methods based on available area. For confined premises, wireless technology can be used as all appliances will be associated with a common network. In the other method where there is wide area make use of wired approach [2]. As wired configuration avoids range issues.

RFID is a fast and most accurate device. It is used for various application like in industries, educational institutes, and hospitals. Other than this, the main application of RFID is its role in avoiding traffic problems like it can detect the high traffic regions, traffic violations. The main advantage of this is authentication between RFID scanner/reader and the RFID tag. The information about tag like identifier number will be highly secured to avoid the tracking by unauthorized users [3].

A cost-effective system as to be developed to monitor traffic congested junctions. As a single person cannot look after all the activities in traffic signals, there should be a system which monitors any traffic rules violation or any vehicle failures and accidents which will result in road blockage, and in worse case, there may be loss of many lives also [4].

RFID technology has wide range of applications in the traffic management systems. An efficient system is required to clear the traffic in case like, when the ambulance, fire engine or any emergency vehicle is stuck in the traffic which must save people in difficulty [5]. When there is no such system, then the vehicles will be stuck in traffic for hours in worst cases like peak hours of traffic, and this must not happen. So, RFID tags can be installed at every traffic signals and when an emergency vehicle arrives, RFID detects and makes path of it to pass saving lives.

When the distance parameter between the RFID reader and the RFID tag varies constantly, the performance of the RFID system will be affected [6]. Initially, one must know how it effects on the performance of the system based on that application must be designed. So that in the preliminary stage only drawbacks can be avoided.

Old technologies use radar for measuring vehicular speed but the major drawback here is radar needs LOS (line of sight). But one cannot assure that the vehicles will always stay in one direction. As the area increases, the performance of radar eventually decreases. When there are many vehicles in the radar beam, then the velocity measurement will fail. This is the major disadvantage of radar technology [7]. So usage of RFID has wide advantages.

Most of the application includes use of LAMP (Linux, Apache, MySQL, PHP) architecture as it has advantage that they are all commercial-grade software with perfect stability which if proven and it free to use [8]. Layer 7 attacks can be eliminated more quickly using LAMP, and it also reduces the malicious messages which are sent to the victims [9]. In high traffic congestion areas, priority-based traffic management system can be used where based on the priority (like emergency vehicles), vehicles can pass [10]. It is always the best way to develop and use a traffic monitoring system based on web technology [11].

4 Introduction About Components Used

RFID stands for radio frequency identification. RFID consists of a reader and a tag, where a reader/scanner reads the content of the tag whenever it is bought near the reader. RFID does not require LOS. It can detect the tag from predefined distance. But for barcode reader, it requires LOS, so in our application vehicles will not stay in LOS always so we make use of radio frequency identification system and not barcode technology.

RFID tags consists of a single antenna and an integrated circuit. The antenna is used to transmit the data from the tag to the RFID reader. After this transmission, the reader will convert the radio waves into data which is more understandable and usable form. This data in turn transmitted to the host computer via stored in database (Fig. 1).

RFID tag consists of a single antenna and chip which contains the tag number and other information is shown in Fig. 2. The RFID reader is shown in Fig. 3, whenever

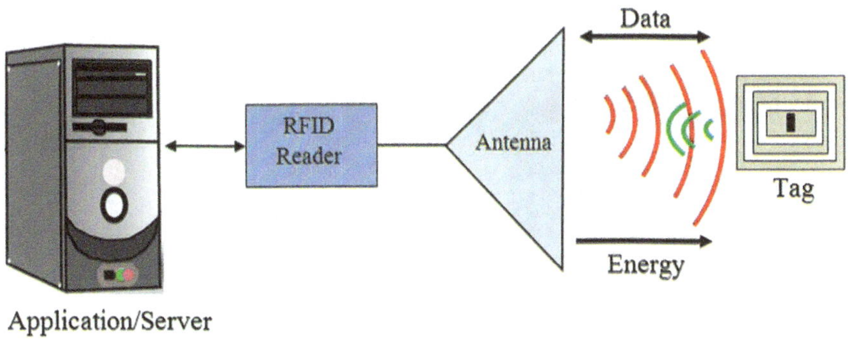

Fig. 1 Block diagram showing RFID system

Fig. 2 RFID tag

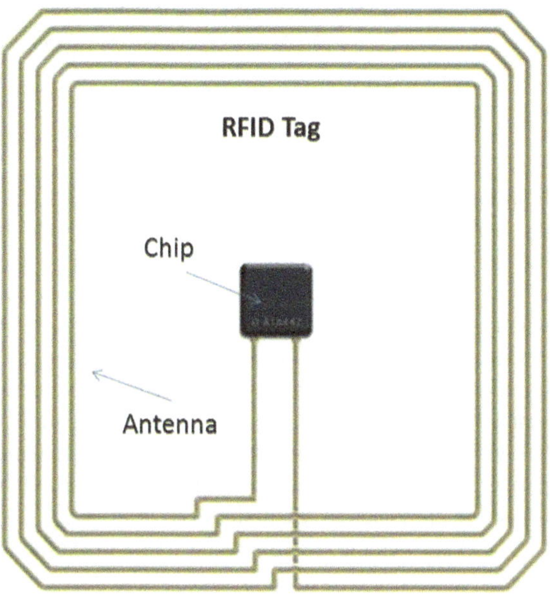

the RFID tag is brought near the reader, it reads the tag number and other required information and stores it in the database.

In this work, EM-18 reader module is being used, and it is shown in Fig. 3.

5 System Overview

Now concentrating on how to identify whether a vehicle as violated traffic rule or not. Generally, the traffic rule violation is done if he/she jumps a signal (when traffic

Fig. 3 RFID reader

light is red), this is the key here. Keep the RFID readers in the junction, and when the vehicle arrives, the RFID reader will scan the RFID tag which is being placed in the vehicle. When the traffic light is green disable RFID because vehicles should be moving when the signal is green. But when it becomes red activate the RFID, and when vehicle arrives RFID reader placed at one side of junction will scan the tag and when the vehicle crosses junction the RFID scanner placed at the other side of the junction will also scan the tag attached to the vehicle, when both the RFID readers scans the tag simultaneously during red signal, than conclusion is made that he/she has violated the traffic rules. And the message with fine will be sent to person's phone. This data consists of person's name that will be sent to the database. This is fed to the cloud across the node MCU (Fig. 4).

For every vehicle, there will be RFID tag number and when this tag is bought near the reader, it will read the data available in the tag. So initially store the name, contact details, vehicle number, and every required data in the tag when the vehicle is bought, so that when the RFID scanner reads the tag, it can get the required details (Fig. 5).

In the same way do it for over speed warning, but here two RFID scanners are placed at some predefined distance. And with known formula:

$$\text{Speed} = \frac{\text{distance}}{\text{time}} \tag{1}$$

As the distance between two RFIDs is known. Because they are placed at fixed positions and none of them are moving components. When a vehicle approaches first RFID scanner, start the timer flag and when it approaches second RFID scanner, stop

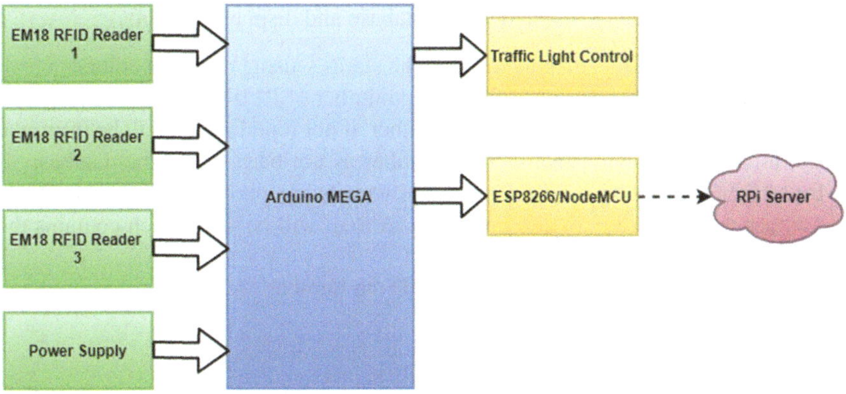

Fig. 4 Block diagram to monitor traffic rules violation

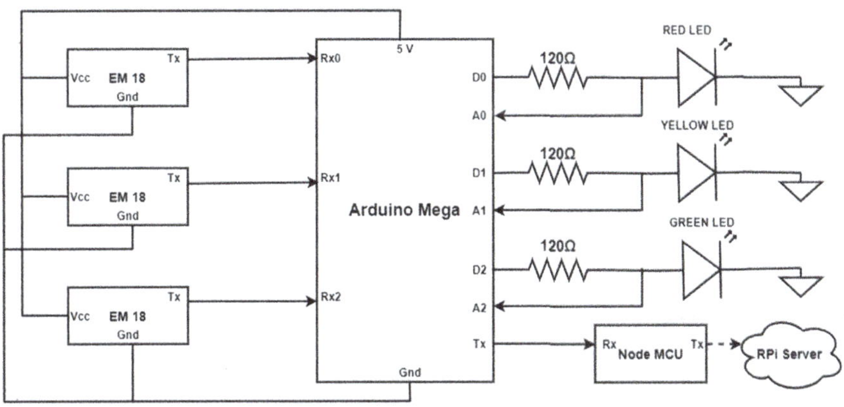

Fig. 5 Schematic representation for traffic rules violation

the timer flag. Using this time taken to travel the fixed distance can be obtained. So, time and distance are available so calculate the speed using above equation. And if this calculated speed exceeds the limit (threshold speed fixed for the area), then conclusion is done that the over speeding is done and send a notification message to the user.

Here RFID is used and not radar because radar can detect only one object at a time, and it also requires line of sight (LOS), whereas RFID scans multiple objects at a time. And it does not require LOS.

Now cloud will have three data's:

- Status of LED's.
- Signal Violation (RFID tag number).
- Over Speeding (RFID tag number).

Algorithm followed to search data in database and display the data:

- If it is status of LED—display directly the status—using different colors.
- Get data for signal violation—get the tag number of RFID.
- Search in database for respective tag number, if not found then it will be informed to traffic police, saying that vehicle number as not been fed to the database. If found then the corresponding details like owner name, phone number, and vehicle number will be loaded into the program. A mail will be sent which contains the fine details.
- Get data for over speeding—get the RFID tag number.
- Step 5 is same as the step 3.

6 Flowchart of the Application

See Fig. 6.

7 Results Obtained

See Fig. 7.

8 Conclusion

Traffic rules violation is detected; that is, if the vehicle has crossed the signal when it is red is identified, respective vehicles ID number will be noted. And over speed detection is also done, where the vehicles which are crossing the speed limit is being identified with respective vehicle ID number.

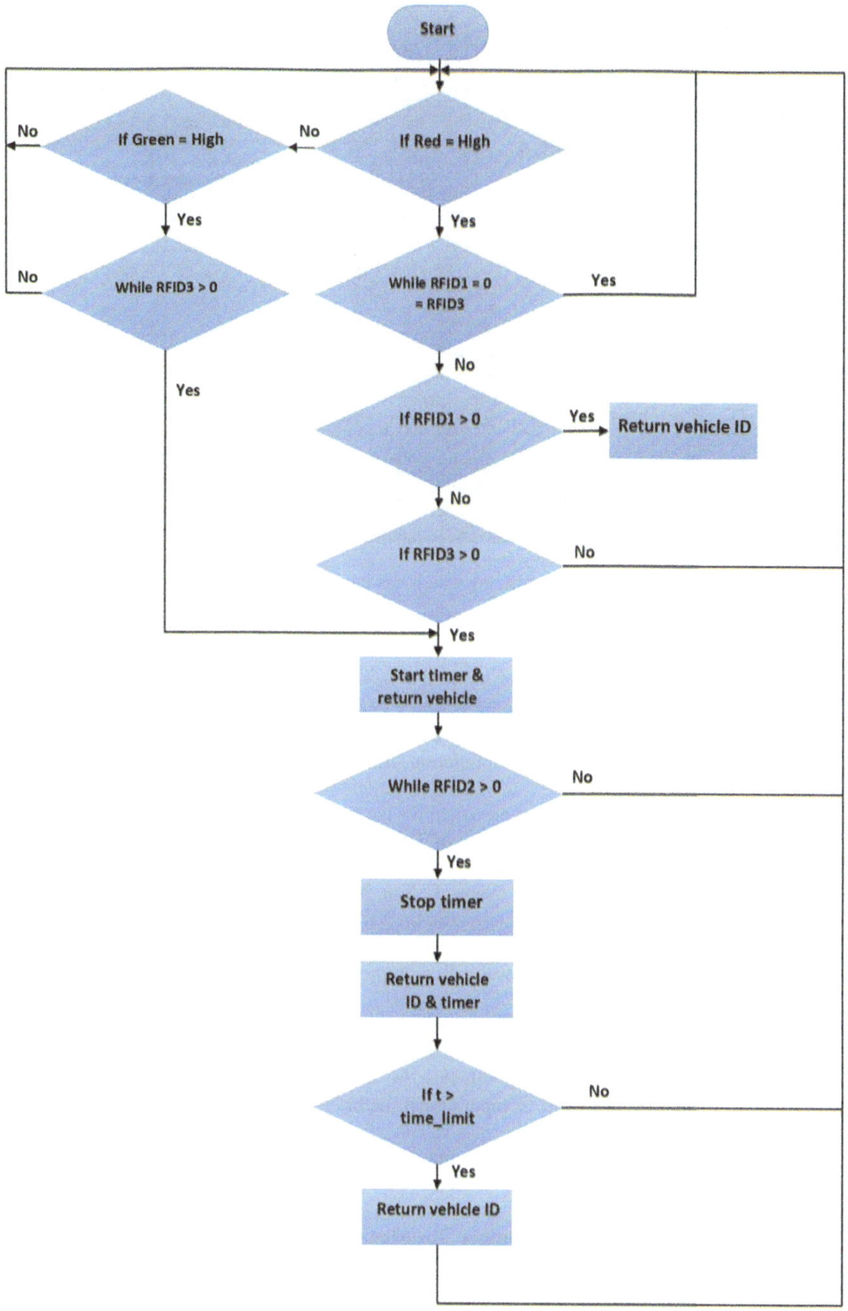

Fig. 6 Flowchart of the algorithm

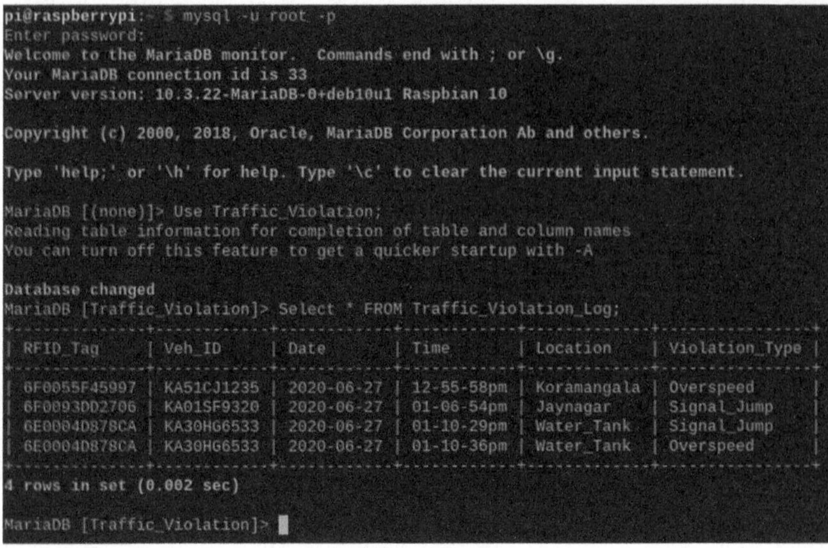

```
pi@raspberrypi:~ $ mysql -u root -p
Enter password:
Welcome to the MariaDB monitor.  Commands end with ; or \g.
Your MariaDB connection id is 33
Server version: 10.3.22-MariaDB-0+deb10u1 Raspbian 10

Copyright (c) 2000, 2018, Oracle, MariaDB Corporation Ab and others.

Type 'help;' or '\h' for help. Type '\c' to clear the current input statement.

MariaDB [(none)]> Use Traffic_Violation;
Reading table information for completion of table and column names
You can turn off this feature to get a quicker startup with -A

Database changed
MariaDB [Traffic_Violation]> Select * FROM Traffic_Violation_Log;
+--------------+------------+------------+------------+-------------+----------------+
| RFID_Tag     | Veh_ID     | Date       | Time       | Location    | Violation_Type |
+--------------+------------+------------+------------+-------------+----------------+
| 6F0055F45997 | KA51CJ1235 | 2020-06-27 | 12-55-58pm | Koramangala | Overspeed      |
| 6F0093DD2706 | KA01SF9320 | 2020-06-27 | 01-06-54pm | Jaynagar    | Signal_Jump    |
| 6E0004D878CA | KA30HG6533 | 2020-06-27 | 01-10-29pm | Water_Tank  | Signal_Jump    |
| 6E0004D878CA | KA30HG6533 | 2020-06-27 | 01-10-36pm | Water_Tank  | Overspeed      |
+--------------+------------+------------+------------+-------------+----------------+
4 rows in set (0.002 sec)

MariaDB [Traffic_Violation]>
```

Fig. 7 Database table showing the vehicles which has been violated the traffic rules

References

1. Kassa, W.-E., Billabert, A.-L., Faci, S., Algani, C.: Electrical modelling of semiconductor laser diode for heterodyne ROF system simulation. IEEE J. Quantum Electron. **49**(10), 894–900 (2013)
2. Gupta, A.K., Johari, R.: Iot based electrical device surveillance and control system. In: 2019 4th International Conference on Internet of Things: Smart Innovation and Usages (IoT-SIU). IEEE, pp. 1–5 (2019)
3. Li, Q., Sun, Z., Huang, J., Liu, S., Wang, J., Yan, N., Wang, L., Min, H.: Secure UHF-RFID tag for vehicular traffic management system. In: 2017 IEEE International Conference on RFID (RFID). IEEE, pp. 26–29 (2017)
4. Meghana, B., Kumari, S., Pushphavathi, T.: Comprehensive traffic management system: Real-time traffic data analysis using RFID. In: 2017 International Conference of Electronics, Communication and Aerospace Technology (ICECA), vol. 2. IEEE, pp. 168–171 (2017)
5. Naik, T., Roopalakshmi, R., Ravi, N.D., Jain, P., Sowmya, B. et al.: RFID-based smart traffic control framework for emergency vehicles. In: 2018 Second International Conference on Inventive Communication and Computational Technologies (ICICCT). IEEE, pp. 398–401 (2018)
6. Zhiyuan, Z., Jie, T., Hongsheng, Z., Qiang, G., Na, L.: A dynamic RFID performance test system. In: 2010 IEEE International Conference on RFID-Technology and Applications. IEEE, pp. 31–36 (2010)
7. Felguera-Martín, D., Gonzalez-Partida, P., Almorox-Gonzalez, J.-T., Burgos-García, M.: Vehicular traffic surveillance and road lane detection using radar interferometry: IEEE Trans. Veh. Technol. **61**(3), 959–970 (2012)
8. Tsai, S.-J., Luo, C.: Synchronized power-quality measurement network with lamp. IEEE Trans. Power Delivery **24**(1), 484–485 (2008)
9. Grigoryan, G., Liu, Y.: Lamp: Prompt layer 7 attack mitigation with programmable data planes. In: 2018 IEEE 17th International Symposium on Network Computing and Applications (NCA). IEEE, pp. 1–4 (2018)

10. Nagmode, V.S., Rajbhoj, S.: An IoT platform for vehicle traffic monitoring system and controlling system based on priority. In: 2017 International Conference on Computing, Communication, Control and Automation (ICCUBEA). IEEE, 2017, pp. 1–5.
11. Li, Y., Li, Z.: Design and development of the monitoring system for traffic signal controller based on web technology. In: 2011 International Conference on Multimedia Technology. IEEE, pp. 3920–3924 (2011)

Analysis of DEEC Deviations in Heterogeneous WSNs: A Survey

R. K. Yadav and Rashmi Mishra

Abstract In the auspicious years, the most famous research area is wireless sensor environments (WSNs). WSNs sensor knobs are positioned randomly in the zone. The key feature of these knobs is to intellect the information from the environment for which they are meant for. Sensor knobs are either mobile in nature or in static mode. Sensor knobs have following components: storage capacity, processing power, transmitter unit, and power unit. Depending on the following components, the environment lifetime will be reduced or prolong. So, for prolong the environment lifetime, sensor knobs with the embedded battery should have high storage capacity, high transmission power, processing power, and high-power unit because once the sensor knobs are arranged in the pitch are difficult to replace the knobs. There are usually two parts of the sensing units such as sensor and analog-to-digital converters. The indicators generated by sensor knobs are analog signals which are converted into numerical signal by analog-to-digital conversion and forwarded to the processing unit. Processing unit is the smallest unit which accomplish the job in collaboration with the other sensor knobs in the environment. To connect the knobs in the environment, the transceiver unit is used. Power sifting unit is supported by the primary component of the sensor knobs. For securing the energy efficiency of the network, energy-efficient clustering etiquettes are designed for the HWSN. In this paper, author projected the variants of DEEC such as DDEEC, EDEEC, TDEEC, IDEEC, EDDEEC, and SDEEC in terms of cluster head selection which is shown in Fig. 1.

Keywords WSN · DEEC · DDEEC · EDEEC · TDEEC · EDDEEC · IDEEC · SDEEC · Distributed · Heterogeneous

R. K. Yadav · R. Mishra (✉)
Delhi Technological University, Delhi, India

R. K. Yadav
e-mail: rkyadav@dtu.ac.in

V. Bhateja et al. (eds.), *Computer Communication, Networking and IoT*, Lecture Notes in Networks and Systems 197, https://doi.org/10.1007/978-981-16-0980-0_22

1 Introduction

Most of the routing algorithms and the identification task, accruing the correct evidence of the position and the environment. Sensor knobs, which are far away from the BS, consume extra influence to communicate information to the BS. Therefore, to resolve the problem of more vigor consumption, routing protocols are intended for the WSNs such as hierarchal gathering routing, weighted gathering routing, and dynamic clustering routing. As per the Akkaya, to move the sensor knobs to complete the assigned task, mobilizer is used. Cluster head collects all the information with the help of the sensor knobs in its region, remove duplicate values, process them, and send it to the base station in the centralized system, but information which is conventional by the BS is period in nature. As far as the performance of the WSNs is concern, it is measured by considering many constraints such as vigor expended by the sensor knobs, distance from the base station to the CH and from CH to the sensor knobs, transmission range, collection creation, assortment of the cluster head, load balancing, synchronization, fault tolerance, and numerals of numerals of cluster formed. Among the following constraints, vigor saving of the sensor knobs is one of the main happenstances confronted by the researchers [1]. Further, the clustering algorithm has played a significant character to reduce the vigor consumption by the knobs in the literature. It can also grow the scalability and period of the system. With the increase of the technology in micro-electro mechanical sensors, the ideas of the different researchers, the sensor knobs are divided into smaller clusters, dense sensor knobs, and the lower cost sensors are used. The WSNs are widely speared not in the field of engineering and science, but also in the area of military, traffic, temperature, agricultural, heath sector, pressure, industries, child care, surveillance, disaster area, and in many more files. Clustering stretches the top result in footings of balancing the period of the environment. Aggregated information is sent to the cluster head, and cluster head sends the information to the BS. Long-distance sensors knobs only have to send the information to the base station. This mechanism increases the period of the system. Clustering can be applied in two types of the system, that is, homogeneous and heterogeneous environments. The environment having similar types of vigor near of the sensor knobs is recognized as homogeneous environment, and knobs having different level of vigor level are known as heterogeneous environments. Some of the protocols is published by the researchers for the homogeneous environment such as low-vigor adaptive clustering hierarchy (LEACH), an emergent algorithm for highly uniform cluster formation (ACE), hybrid vigor-efficient distributed clustering (HEED), power efficient gathering in sensor information systems (PEGASIS), threshold-sensitive vigor-efficient sensor environment protocol (TEEN), adaptive threshold-sensitive vigor-efficient environment (APTEEN), etc. [2, 3] (Fig. 1).

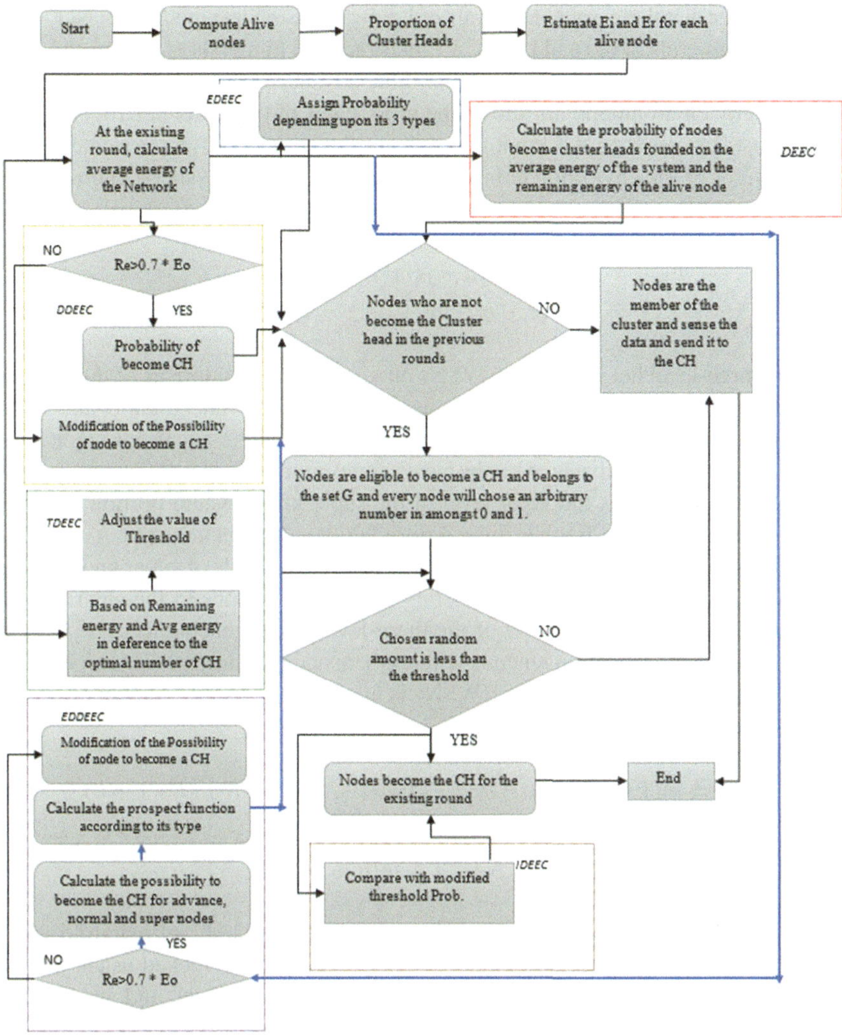

Fig. 1 Flowchart of all the protocols [2]

1.1 Heterogeneous WSN Model (HWSNs)

HWSNs environment is made up of sensor knobs having different vigor level, and each knob lies in the three categories such as in two level, three level, and multi-level. In the two-level heterogeneous WSN environment, there are alteration two types of knobs present, normal knob and advanced knob. In the two level of heterogeneity, E_o is the normal knobs vigor level, $E_o\ (1 + a)$ is the advanced vigor level where a shows that the advanced knobs have a times more vigor level than the normal knobs.

If we consider that there are N numerals of knobs in the environment, then N(1 − m) is the total numerals of normal knobs, $N * m$ is the total numerals of advanced knobs in the environment, and m is the segment of the advanced knobs. In the two level of heterogeneity environment, am shows the more vigor of two-level heterogeneous.

WSN is associated to the homogeneous and heterogeneous environement. The primary vigor of the environment is sum of the vigor of the advanced knob and the normal knobs, that is Eq. (1),

$$\begin{aligned} E_{\text{total}} &= N(1-m)E_o + Nm(1+a)E_o \\ &= NE_o(1-m+m+am) = NE_o(1+am) \end{aligned} \tag{1}$$

In the three-level heterogeneous WSNs environment, it comprises of three types of sensor knobs in the environment such as normal knob, advanced knob, and super knobs. *Eo* is the normal knobs vigor level, E_o $(1 + a)$ is the advanced vigor level where a shows that the advanced knobs have a times more vigor level than the normal knobs, and E_o $(1 + b)$ is the vigor level of the super knobs where portion of mo is partaking an influence of b times more vigor than the normal knobs [4]. As N is the entire numerals of knobs in the environment, the total numerals of super knobs in the environment are Nmm_o, and total numerals of advanced knobs in the environment are Nm $(1 - m_o)$. The vigor level of the three-level heterogeneous is $(a + m_o b)$ is more as associate to the homogeneous wireless sensor environments. Therefore, total primary vigor of the three level of HWSNs is measured by using following formula (2), (3):

$$E_{\text{total}} = N(1-m)E_o + Nm(1-m_0)(1+a)E_o + Nm_oE_o(1+b) \tag{2}$$

$$E_{\text{total}} = NE_o(1 + m(a + m_ob)) \tag{3}$$

The multi-level heterogeneous environment comprises knobs of manifold vigor level. The primary vigor of all the knobs is dispersed into the locked usual, that is, $[E_o, E_o(1 + a_{\text{max}})]$, Eo shows the lower bound of the vigor level, and amax is the maximum level of the vigor. The knobs having vigor level of $E_o(1 + a_i)$ which is ai times additional vigor than the lower level of the vigor level [5, 6]. As the cluster head consumes more vigor in the information aggregation, processing than after some iterations, the vigor level of the knobs gets changed as associate to the other knobs in the environment. So, heterogeneity of the knobs is introduced in the homogeneous environment. The primary vigor of the multi-level heterogeneous environment is considered as (4):

$$E_{\text{total}} = \sum_{i=1}^{N} E_o(1 + a_i) = E_o\left(N + \sum_{i=1}^{N} a_i\right) \tag{4}$$

Nearby, there are three categories of heterogeneity of the sensor knobs such as computational, link, and vigor heterogeneity. The sensor knobs having higher computational level means that knobs have powerful microprocessor and extra storage, and it is different from the normal knobs in terms of information processing and in terms of storage. In the link heterogeneity, the knobs have more bandwidth and travel long distance as associate to the other knobs in the environment. Link heterogeneity is used where reliable information transmission is required [7]. Further, in the vigor heterogeneity, the knobs in this environment have line powered or the battery of these knobs would be replaced. Among these resources of heterogeneity, the vigor heterogeneity is most essential among three because it consumes less vigor as associate to the computational and link heterogeneity.

1.2 Active Heterogeneous Wireless Sensor Environment Model

The model shown in the describing that the numerals of bits transmitted over the distance d, estimations of environment lifetime, numerals of information packets send to the base stations, and how to calculate the clusters present in the environment.According to the radio vigor dissipation model, l-bits are transmitted over the environment through the distance *d*. The vigor dissipation is given by (5):

$$E_{T_x}(l,d) = \begin{cases} lE_{\text{elec}} + l\varepsilon_{fs}d^2 & \text{if } d < d_0 \\ lE_{\text{elec}} + l_{\varepsilon_{mp}}d^4 & \text{if } d \geq d_0 \end{cases} \tag{5}$$

where

- E_{elec} is the energy dissipation to route the transmitter.
- $\varepsilon fs\ and \varepsilon mp$ is the announcement energy influence using the free space model, and multi-path model to contingent upon the power amplifier model? The factor named as free space model, and multi-path model is introduced by Heinzelman et al.
- $d_0 = \sqrt{\varepsilon_{fs}/\varepsilon_{mp}}$ is the distance threshold and for receiving the information, vigor dissipation for the received signal is $E_{Rx}(l) = l.E_{\text{elec}}$.

2 Overview of Distributed Heterogeneous Protocols

2.1 Distributed Energy-Efficient Clustering Protocol (DEEC)

Qing et al., in 2006 [8] have proposed an algorithm named as DEEC which is well suited in the heterogeneous environment. Heterogeneous environment contains of sensor knobs with dissimilar vigor level and divided into multiple clusters having cluster head (CH). CHs are selected on the base of likelihood function (*Pf*), i.e., ratio of remaining vigor and the middling vigor of the environment [9, 10]. The *Pf* is calculated for each sensor knobs in the environment. The knob having the higher *Pf* will have advanced possibility to become the CH. The algorithm for the selection of the cluster head is applied periodically. The sensor knobs will sense the information and send it to the CH, and CH sends the information to the base station. DEEC uses remaining vigor and the primary vigor of the knob for selecting the CH. The knob si will become the CH in numerals of iterations *ni,* and during each iteration, numerals of CHs are *poptN*. The knob having probability to CH is *pi*, that is, knob having higher amount of vigor has larger value of pi as associate to the popt. Each knob in the environment having same vigor level in each iteration, then the average probability of the environment is equal to the optimal probability of the knobs, i.e., *popt*. If knobs having dissimilar amount of vigor, then *pi* > *popt*, i.e., knob having higher pi as associate to the knobs having lesser amount of vigor. For calculating the average vigor of the environment, each knob should have knowledge of all the other knobs in the environment. Each knob calculates the average vigor of the environment by using following Eq. (6):

$$\bar{E}(r) = \frac{1}{N}\sum_{i=1}^{N} E_i(r) \tag{6}$$

Each knob in the environment calculates its election probability P_i by seeing P_{opt} primary and remaining vigor of the environment (7).

$$P_i = P_{\text{opt}}\left[1 - \frac{\bar{E}(r) - E_i(r)}{\bar{E}(r)}\right] = P_{\text{opt}}\frac{E_i(r)}{\bar{E}(r)} \tag{7}$$

The average vigor of the CHs per iteration is calculated by using following Eq. (8):

$$\sum_{i=1}^{N} P_i = \sum_{i=1}^{N} P_{\text{opt}}\frac{E_i(r)}{\bar{E}(r)} = P_{\text{opt}}\sum_{i=1}^{N}\frac{E_i(r)}{\bar{E}(r)} = P_{\text{opt}}N \tag{8}$$

During each iteration, if a knob s is selected for the CH, then it belongs to the set *G,* and in each iteration, every knob has to select a random numeral in between 0 and 1. If the chosen numerals are less than the predefined threshold, then the knobs are eligible to select as a CH otherwise not. Threshold would be calculated by using the following Eq. (9):

$$T(s_i) = \begin{cases} \frac{p_i}{1 - p_i\left(r \bmod \frac{1}{p_i}\right)} & \text{if } s_i \in G \\ 0 & \text{otherwise} \end{cases} \tag{9}$$

If homogeneous environment is considered, then the value of p_{opt} is same for all the knobs. But it is different in heterogeneous environment and calculated by using the following formula for the two-level heterogeneous environment (10):

$$p_{\mathrm{adv}} = \frac{p_{\mathrm{opt}}}{1 + am}, \; p_{\mathrm{nrm}} = \frac{p_{\mathrm{opt}}(1 + a)}{(1 + am)} \tag{10}$$

where p_{nrm} and p_{adv} are normal and advance knob in the HWSNs and we can calculate the probability of normal and advance knob by using the following formula (11):

$$p_i = \begin{cases} \frac{p_{\mathrm{opt}} E_i(r)}{(1+am)\overline{E}(r)} & \text{if } s_i \text{ is normal node} \\ \frac{p_{\mathrm{opt}}(1+a) E_i(r)}{(1+am)\overline{E}(r)} & \text{if } s_i \text{ is the advance node} \end{cases} \tag{11}$$

For finding the average vigor of the environment with respect to the iteration, r is calculated as follows (12):

$$\bar{E}\,(r) = \frac{1}{N} E_{\mathrm{total}}\left(1 - \frac{r}{R}\right) \tag{12}$$

where R is total round in network lifetime and $R = \frac{E_{\mathrm{total}}}{E_{\mathrm{round}}}$.
Where $\frac{E_{\mathrm{total}}}{E_{\mathrm{round}}}$ is total vigor of the environment/vigor spending in apiece iteration.

2.2 *Developed Distributed Energy-Efficient Clustering Protocol (DDEEC)*

DDEEC [11] is same as DEEC in terms of calculation of average of the environment and consider remaining vigor for assortment of the cluster head. The only transformation among both the protocol is only possibility expression for the normal knob and the advance knob. In DEEC, the knob having higher vigor level is more likely to develop the CH as associate to the knob having less amount of vigor. If the advance knobs having vigor level equal to the advance knob vigor level. In this case, DEEC only considers advance knob for selecting the CH, so the advance knob will die rapidly as associate to the normal knob. Furthermore, the DDEEC changes in p_i of DEEC. The threshold value of the DEEC is measured using following Eq. (13):

$$\mathrm{Th}_{\mathrm{REV}} = E_0\left(1 + \frac{a E_{\mathrm{disNN}}}{E_{\mathrm{disNN}} - E_{\mathrm{disAN}}}\right) \tag{13}$$

If the threshold value of both the types of knobs is equal, then both the knobs will become the CH. Therefore, for balancing the cluster head selection process, DDEEC has introduced $\mathrm{Th}_{\mathrm{REV}} \cong \left(\frac{7}{10}\right) E_o$, and the probability p_i for selecting the

CH is calculated by using the following formula (14):

$$p_i = \begin{cases} \frac{p_{\text{opt}} E_i(r)}{(1+am)\overline{E}_i(r)} \text{ for Nml nodes, } E_i(r) > \text{Th}_{\text{REV}} \\ \frac{(1+a) p_{\text{opt}} E_i(r)}{(1+am)\overline{E}(r)} \text{ for Adv nodes, } E_i(r) > \text{Th}_{\text{REV}} \\ c\frac{(1+a) p_{\text{opt}} E_i(r)}{(1+am)\overline{E}(r)} \text{ for Adv, Nml nodes, } E_i(r) \leq \text{Th}_{\text{REV}} \end{cases} \tag{14}$$

2.3 *Enhanced Distributed Energy-Efficient Clustering Protocol (EDEEC)*

The three-level heterogeneous environment model is used by the EDEEC [12]. Threshold for selecting the CH is measured by using the following formula (15):

$$T(s_i) = \begin{cases} \frac{p_i}{1-p_i(r \bmod \frac{1}{P_i})} & \text{if } p_i \in G' \\ \frac{p_i}{1-p_i(r \bmod \frac{1}{P_i})} & \text{if } p_i \in G'' \\ \frac{p_i}{1-p_i\left(r \bmod \frac{1}{P_i}\right)} & \text{if } p_i \in G''' \\ 0 & \text{Otherwise} \end{cases} \tag{15}$$

In EDEEC, there are three categories of knobs, first is normal knob, advanced knob, and the super knob. Here, p_{opt} is reference for the probability. Different values of p_{opt} are used for all the types of knobs. Probability for selecting the CH is measured by using the following formula (16):

$$p_i = \begin{cases} \frac{p_{\text{opt}} E_i(r)}{(1+m(a+m_o b))\overline{E}(r)} & \text{if } s_i \text{ is the normal node} \\ \frac{p_{\text{opt}}(1+a) E_i(r)}{(1+m(a+m_o b))\overline{E}(r)} & \text{if } s_i \text{ is the advanced node} \\ \frac{p_{\text{opt}}(1+b) E_i(r)}{(1+m(a+m_o b))\overline{E}(r)} & \text{if } s_i \text{ is the super node} \end{cases} \tag{16}$$

2.4 *Threshold Distributed Energy-Efficient Clustering Protocol (TDEEC)*

TDEEC [13] working is similar to the DEEC for calculating the average vigor of the environment at the iteration r is: $\overline{E}(r) = \frac{1}{N} E_{\text{total}}\left(1 - \frac{r}{R}\right)$, where R is the total iterations, and it is estimated by $\frac{E_{\text{total}}}{E_{\text{round}}}$. For the selection of CH, each knob chose the numerals among 0 and 1. If the chosen numerals are fewer than the predefined numerals, then that knob will develop the cluster head in that iteration, and the threshold is calculated as follows (17):

$$T(s) = \begin{cases} \frac{p}{1-p.(r \bmod \frac{1}{p})} * \frac{\text{Residual Energy of a node}*k_{\text{opt}}}{\text{Average energy of network}} & \text{if } s \in G \\ 0 & \text{Otherwise} \end{cases} \quad (17)$$

TDEEC is work for the two level, three level, and for multi-level. For the two-level possibility for normal knob and for advanced knob is estimated by using the following formula (18):

$$p_i = \begin{cases} \frac{p_{\text{opt}} E_i(r)}{(1+am)\bar{E}(r)} & \text{if } s_i \text{ is normal node} \\ \frac{p_{\text{opt}}(1+a)E_i(r)}{(1+am)\bar{E}(r)} & \text{if } s_i \text{ is the advance node} \end{cases} \quad (18)$$

For the two-level possibility for normal knob, advanced knob and for super knob is estimated by using the following formula (19):

$$p_i = \begin{cases} \frac{p_{\text{opt}} E_i(r)}{(1+m.(a+m_o.b))\bar{E}(r)} & \text{if } s_i \text{ is the normal node} \\ \frac{p_{\text{opt}}(1+a)E_i(r)}{(1+m.(a+m_o.b))\bar{E}(r)} & \text{if } s_i \text{ is the advanced node} \\ \frac{p_{\text{opt}}(1+b)E_i(r)}{(1+m.(a+m_o))\bar{E}(r)} & \text{if } s_i \text{ is the super node} \end{cases} \quad (19)$$

And for the multi-level heterogeneity, the probability is estimated by using the following formula (20):

$$p_i = \frac{p_{\text{opt}}.N.(1+a_i).E_i(r)}{\left(N + \sum_{i=1}^{N} a_i\right) \bar{E}\ (r)} \quad (20)$$

2.5 *Enhanced Developed Distributed Energy-Efficient Clustering (EDDEEC)*

EDEEC [14] uses the perception of three-level heterogeneity that is normal, advanced, and super knobs. The solitary transformation is estimating the threshold value. In the starting of each iteration, every knob will calculate the threshold value for CH election process, and it is calculated by using following formula (21):

$$T(s_i) = \begin{cases} \frac{p_i}{1-p_i\left(r \bmod \frac{1}{p_i}\right)} & \text{if } s_i \in G \\ 0 & \text{Otherwise} \end{cases} \quad (21)$$

G is the set of knobs who are eligible for the CH. For the selection of the three type of knob, following equation is used (22):

$$p_i = \begin{cases} \frac{p_{\text{opt}} E_i(r)}{(1+m(a+m_o.b))\bar{E}(r)} & \text{if } s_i \text{ is the normal node} \\ \frac{p_{\text{opt}}(1+a) E_i(r)}{(1+m(a+m_o.b))\bar{E}(r)} & \text{if} s_i \text{ is the advanced node} \\ \frac{p_{\text{opt}}(1+b) E_i(r)}{(1+m(a+m_o.b))\bar{E}(r)} & \text{if } s_i \text{ is the super node} \end{cases} \tag{22}$$

To resolve the uncertainty and instability in the three-level heterogeneity of the DDEEC, EDDEEC has come up with the solution. Author changes the function used by the EDEEC for estimating the probabilities of all the three types of knobs. The changes are based on the T_{absolute} because under this value, the vigor level of all the kinds of knobs is equal. Therefore, the anticipated probabilities for choosing the cluster head, EDDEEC used the following formula (23):

$$p_i = \begin{cases} \frac{p_{\text{opt}} E_i(r)}{(1+m(a+m_o b))\bar{E}(r)} & \text{for } N_{ml} \text{ nodes if } E_i(r) > T_{\text{absolute}} \\ \frac{p_{opt}(1+a) E_i(r)}{(1+m(a+m_o b))\bar{E}(r)} & \text{for Adv nodes if } E_i(r) > T_{\text{absolute}} \\ \frac{p_{opt}(1+b) E_i(r)}{(1+m(a+m_o b))\bar{E}(r)} & \text{for Sup nodes if } E_i(r) > T_{\text{absolute}} \\ c\frac{p_{opt}(1+b) E_i(r)}{(1+m(a+m_o b))\bar{E}(r)} & \text{for } N_{\text{ml}}, \text{ Adv, Sup nodes if } E_i(r) \le T_{\text{absolute}} \end{cases} \tag{23}$$

The value $T_{\text{absolute}} = zE_0$, if the value of z is 0, then the protocol is behaved like EDEEC.

2.6 Improved Distributed Energy-Efficient Clustering Protocol (IDEEC)

IDEEC is similar to DEEC; only difference is scaling factor which means the simplification power $\in_{fs}$ is compact by factor 10. The scaling factor could be measured using following equation:

$$\text{Scaling factor} = \left\{ \text{rand}() \times \frac{\text{Area of Network Field}}{\text{area of the cluster} \times \text{no. of nodes in a cluster}} \right\} \tag{24}$$

And the possibility of knob should develop the cluster head using equation:

$$P_i = \frac{\text{Energy of the } i\text{th node}}{\sum_{i=1}^{n} \text{Total energy of all the nodes within a cluster}} \tag{25}$$

Threshold value will decide whether the particular knob will become the cluster head or not.

$$T(s_i) = \begin{cases} \frac{P_i}{1-P_i(r \bmod \frac{P_i}{\sum_{i=1}^{n} P_i}} & \text{if } S_i \in G \\ 0 & \text{otherwise} \end{cases} \quad (26)$$

2.7 *Stochastic Distributed Vigor-Efficient Clustering (SDEEC)*

SDEEC [15] working is same as the DEEC scheme. It uses the primary and remaining vigor of the knob for the cluster head election process and for estimating the average environment lifetime. In SDEEC, the environment topology is fixed, and the BS is situated at the center of the environment. The average vigor of the environment at rth iteration is estimated as:

$$\bar{E}(t) = \frac{1}{N} E_{\text{total}} \left(1 - \frac{r}{R}\right) \text{ where } R = \frac{E_{\text{total}}}{E_{\text{Round}}} \quad (27)$$

where E_{Round} is the total vigor degenerate and calculated using Eq. (28):

$$E_{\text{round}} = L(2N E_{\text{elec}} + N E_{\text{DA}} + k_{\text{Emp}} d^4_{toBS} \quad (28)$$

where k is the numerals of clusters (Table 1).

As knobs are dispersed consistently, therefore, we get $d_{toCH} = \frac{M}{\sqrt{2k\prod}}$, $d_{toBS} = 0.765\frac{M}{2}$ and numerals of clusters are $k_{\text{opt}} = \frac{M}{d^2_{toBS}} \frac{\sqrt{N}}{\sqrt{2\prod}} \frac{\sqrt{E_{fs}}}{\sqrt{E_{mp}}}$.

The probability of the knob become the cluster head is calculated by using the following formula

$$p_i = \begin{cases} \frac{p_{\text{opt}} E_i(r)}{(1+am)\bar{E}(r)} & \text{for Nrml node} \\ \frac{(1+a)p_{\text{opt}} E_i(r)}{(1+am)\bar{E}(r)} & \text{for Adv node} \end{cases} \quad (29)$$

3 Conclusion

Author explains the distributed heterogeneous wireless sensor environments. For the operative spreading of vigor among the sensor knobs, an adaptive vigor aware protocol is used for the vigorously change the possibility of knob to develop a cluster head. The existing work can be prolonged comparing the optimization technique and contrast with the existing work. The comparison parameters would be environment lifetime, stability periods, packets direct to the cluster heads, numerals of knobs

Table 1 Evaluation chart of heterogeneous WSNs protocol [17–20]

Protocol	Year	Mobile	Hopping	Clustering	Heterogeneity	Selection of CH	Stability	Location awareness	Heterogeneity level	Primary vigour	Remaining vigour	Avg vigour
DDLE	2019	Static	Multi-hop	Inter cluster	Three-level	Deterministic	Medium	No	Vigour	No	No	No
DEEC	2006	Static	Single hop	Inter cluster	Two and Three-level	Probabilistic	Medium	No	Vigour	No	Yes	Yes
DDEEC	2010	Static	Single hop	Inter cluster	Two-level	Probabilistic	High	No	Vigour	Yes	Yes	No
SDEEC	2009	Static	Single hop	Inter cluster	Two-level	Probabilistic	Medium	No	Vigour	Yes	Yes	No
EDEEC	2010	Static	Single hop	Inter cluster	Three-level	Probabilistic	Medium	No	Vigour	Yes	Yes	No
TDEEC	2010	Static	Single hop	Inter cluster	Two, Three-level	Probabilistic	Medium	No	Vigour	No	Yes	No
EEHC	2009	Static	Single hop	Inter cluster	Three-level	Probabilistic	High	No	Vigour	No	Yes	No

alive, and information packets send to the base station. In the literature, EDEEC and TDEEC are well worked in all the situation, and DEEC and DDEEC will best work in the situation where vigor dissimilarity between the normal, advanced, and super knobs are very high [16]. Stability period of TDEEC and EDDEEC is best among all the protocol. EDDEEC is more reliable as associate to the DEEC, DDEEC, and EDEEC.

References

1. Heinzelman, W.B., Chandrakasan, A.P., Balakrishnan, H.: An application-specific protocol architecture for wireless microsensor environments. IEEE Trans. Wireless Commun. **1**(4), 660–670 (2002)
2. Qureshi, T.N., Javaid, N., Malik, M., Qasim, U., Khan, Z.A.: On performance evaluation of variants of DEEC in WSNs. In: 2012 Seventh International Conference on Broadband, Wireless Computing, Communication and Applications (2012). https://doi.org/10.1109/bwcca.2012.35
3. Fanian, F., Rafsanjani, M.K.: Cluster-based routing protocols in wireless sensor environments: a survey based on methodology . J. Environ. Comput. Appl. **142**(15), 111–142 (2019). https://doi.org/10.1016/j.jnca.2019.04.021
4. Gherbi, C., Aliouat, Z., Benmohammed. M.: An adaptive clustering approach to dynamic load balancing and vigour efficiency in wireless sensor environments. Vigour **114**, 647–662 (2016). https://doi.org/10.1016/j.vigour.2016.08.012
5. Akkaya, K., Younis, M.: A survey of routing protocols in wireless sensor environments. Ad Hoc Netw. **3**(3), 325–349 (2005)
6. Gambhir, A.P., Arya, R.: Performance analysis of artificial bee colony optimization-based clustering protocol in various scenarios of WSN. Procedia Comput. Sci. **132**, 183–188 (2018)
7. Kumar, D., Aseri, T.C., Patel, R.B.:EEHC: Vigour efficient heterogeneous clustered scheme for WSNs. Comput. Commun. **32**, 662–667 (2009)
8. Qing, L., Zhu, Q., Wang, M.: Design of a distributed vigour-efficient clustering algorithm for heterogeneous wireless sensor environment. Comput. Commun. **29**, 2230–2237 (2006)
9. Jafari, H., Nazari, M., Shamshirband, S.: Optimization of vigour consumption in wireless sensor environments using density-based clustering algorithm. Int. J. Comput. Appl. (2018). https://doi.org/10.1080/1206212X.2018.1497117
10. Amudhavel, J., Rajaguru, D., Sampath Kumar, S., Lakhani, S.H., Vengattaraman, T., Prem Kumar, K.: A Chaotic Krill Herd optimization approach in VANET for congestion free effective multi hop communication. In: Proceedings of the 2015 International Conference on Advanced Research in Computer Science Engineering & Technology (ICARCSET 2015)—ICARCSET '15, pp. 1–5, Article no. 27, March 2015. https://doi.org/10.1145/2743065.2743092
11. Elbhiri, B., Saadane, R., El Fkihi, S., Aboutajdine, D.: Developed distributed vigour-efficient clustering (DDEEC) for heterogeneous wireless sensor environments. In: 5th International Symposium on I/V Communications and Mobile Environment (ISVC) (2010)
12. Saini, P., Sharma, A.K.: E-DEEC-enhanced distributed vigour efficient clustering scheme for heterogeneous WSN. In: 2010 1st International Conference on Parallel, Distributed and Grid Computing (PDGC—2010)
13. Saini, P., Sharma, A.K.: Vigour efficient scheme for clustering protocol prolonging the lifetime of heterogeneous wireless sensor environments. Int. J. Comput. Appl. **6**(2), 30–36 (2010)
14. Javaid, N., Qureshi, T.N., Khan, A.H., Iqbal, A., Akhtar, E., Ishfaq, M.: EDDEEC: enhanced developed distributed vigour-efficient clustering for heterogeneous wireless sensor environments. Procedia Comput. Sci. **19**, 914–919 (2013). https://doi.org/10.1016/j.procs.2013.06.125

15. Brahim, E., Rachid, S., Zamora, A.-P., Aboutajdine, D.: Stochastic distributed vigour-efficient clustering (SDEEC) for heterogeneous wireless sensor environments. ICGST-CNIR J. **9**(2) (2009)
16. Engmann, F., Katsriku, F.A., Abdulai, J., Adu, K.S., Banaseka, F.K.: Prolonging the lifetime of wireless sensor networks: a review of current techniques. Wirel. Commun. Mob. Comput. **2018**, 1–23 (2018)
17. Javaid, N., Rasheed, M.B., Imran, M., Guizani, M., Khan, Z.A., Alghamdi, T.A., Ilahi, M.: A vigour-efficient distributed clustering algorithm for heterogeneous WSNs. EURASIP J. Wirel. Commun. Environ. **2015**(1). https://doi.org/10.1186/s13638-015-0376-4
18. McDonald, B., Znati, T.F.: Design and performance of a distributed dynamic clustering algorithm for ad-hoc networks. In: Proceedings of 34th Annual Simulation Symposium, Seattle, WA, USA, 2001, pp. 27–35, https://doi.org/10.1109/SIMSYM.2001.922111
19. McDonald, B., Znati, T.F.: A mobility-based framework for adaptive clustering in wireless ad hoc networks. IEEE J. Sel. Areas Commun. **17**(8), 1466–1487 (1999). https://doi.org/10.1109/49.780353
20. Das, S.R., Castaneda, R., Yan, J., Sengupta, R.: Comparative performance evaluation of routing protocols for mobile, ad hoc networks. In: Proceedings 7th International Conference on Computer Communications and Networks (Cat. No.98EX226), Lafayette, LA, USA, 1998, pp. 153–161. https://doi.org/10.1109/ICCCN.1998.998772
21. Khan, M.K., Shiraz, M., Ghafoor, K.Z., Khan, S., Sadiq, A.S., Ahmed, G.: EE-MRP: Energy-efficient multistage routing protocol for wireless sensor networks. Wirel. Commun. Mob. Comput. (2018)

Phase-to-Phase Fault Classification in Transmission Line

A. Naresh Kumar, P. Shiva Kumar, S. Srikanth, and K. Devender Reddy

Abstract The performance of conventional relay especially for phase-to-phase faults on transmission lines is adversely affected. This paper presents a method for classification of phase-to-phase faults in parallel transmission line PTL using k-nearest neighbor (k-NN) approach. A 68 km length, 138 kV PT line has been simulated using a Simulink® and SimPowerSystems® toolbox of the MATLAB software. The impact of various fault distance location and fault inception angle has been validated on the performance of k-NN protection approach. The results of k-NN-based fault classifier show that it correctly classified all types of phase-to-phase faults within half of cycle. It verifies the accuracy of k-NN approach.

Keywords Faults · PT line · K-NN

1 Introduction

Nowadays, in electrical power system, increasing the reliability of power transmission line is of importance. Nowadays, we have witness growing electrical energy systems' demands. It is critical to develop the power capacity of accessible line to facilitate boost demand of the electrical energy. PTL is the potential selection along with the multi-phase transmission line, for transfer extra power energy with the presented configurations [1–9] of a PTL. Preliminary analysis on the implementation and designing has explained the feasibility of PTL.

The complexities of protective system for a PTL are supplementary as comparing to single three-phase lines. The different faults normally may effect on PTL that been proposed in various paper works [10–12]. The different classification modules have been discussed in Refs. [13, 14] using the voltage phasors and current phasors. Some protection algorisms apply the faulty current phasors [15, 16] only. The shunt fault classifications as an input of current phasors in PTL has been analyzed in Ref. [17] widely. As published in [18–21] a lot of fault classification modules are discussed for PTL mainly. A few of the approaches did not give the classifications of shunt fault

A. Naresh Kumar (✉) · P. Shiva Kumar · S. Srikanth · K. Devender Reddy
Institute of Aeronautical Engineering, Hyderabad 500043, Telangana, India

V. Bhateja et al. (eds.), *Computer Communication, Networking and IoT*, Lecture Notes in Networks and Systems 197, https://doi.org/10.1007/978-981-16-0980-0_23

in PTL. Fault classification using k-NN in three-phase circuit lines are explained in [22]. In fact, the proposal of this paper work is to boost classifications accuracy. Therefore, a k-NN module for fault classifications in PTL is planned in this paper work.

The rest of our paper as follows. Section 2 gives the PTL configurations, model design, and phase-to-phase fault simulations. Section 3 gives the k-NN module for classifying the phase-to-phase fault group. Then, Sect. 4 gives the simulation result study for classifying the phase-to-phase fault group. Next our conclusions are given in Sect. 5.

2 Simulation of PTL

A diagram of PTL is given in Fig. 1. The PTL consists of 138 kV, 50 Hz, length 68 km. The PTL is supplied from two source sides at both received side and source side with two loads of 120 MVAR and 260 MW at received side. The reactance/resistance ratio has short-circuit capabilities of source side at the two ends are 9 and 1260 MVA, respectively. Basic descriptions and configurations data of PTL are seen in Table 1. The PTL is executed using Simulinks, SimPowerSystem toolboxes with MATLAB. The current signal waveform of PTL phases during phase-to-phase b and a fault in phases b and a at 50 km from source position are given in Fig. 2.

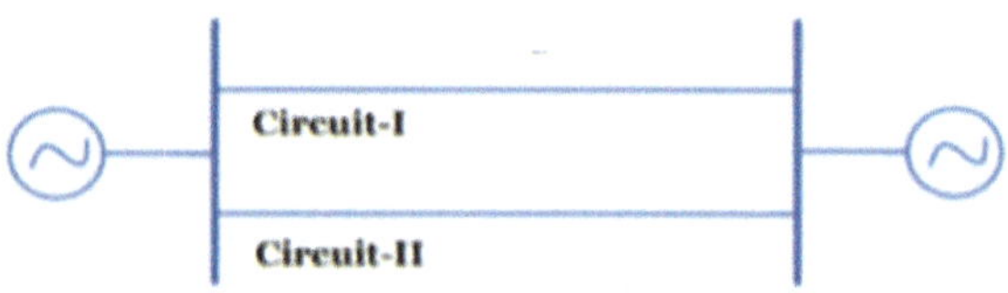

Fig. 1 Power system network considered

Table 1 Model report records

Parameters	Nominal values
Voltage	138 kV
Power	110,000 KVA
Frequency	50 Hz
Earth (ground) resistivity	110 Ω-m
Line length	68 km
Short circuits capability	1100
Reactance/resistance ratio	9
Bus load	120,000 W
	120,000 VAR

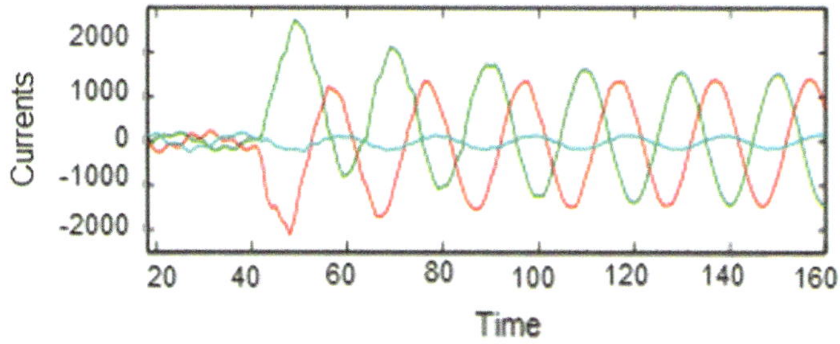

Fig. 2 PTL current (phase-to-phase faults)

3 Designing of k-NN Module

The k-NN module fault classifications are calculated for classify a variety of types of phase-to-phase faults. The K-NN is having two dissimilar k-NN systems: k-NN-A and k-NN-B. The k-NN-A is allocated to categorize the fault phases, whereas k-NN-B is allocated whether the phase faults involving earth or not. The training pattern used for modeling k-NN-A and k-NN-B system is current phasors. The respective output of target for faulty phase is allocated as no faults-0 and faulty phases-1. The peak training accuracies for k-NN-A and k-NN-B systems is achieved with $k = 3$, and thus, $k = 3$ is elected for k-NN-A and k-NN-B as their k-nearest neighbor.

4 Results

The test is vital to authenticate the k-NN module-based fault classification performance. The k-NN fault classifications is verified using tests choosing all types with transform in a faults distance locations from 2 to 66 km and FIA between 1 and 359°. Total 56 faults are tested. The training module is authenticated by altering k value for both k-NN-A and k-NN-B systems as scheduled in Tables 2 and 3, respectively. Subsequently, to checking the k-NN phase-to-phase faults classified, it has been authenticated that it can able to classify all possible faults (phase-to-phase faults) correctly, i.e., accuracy of 99.832%. The accuracy is considered using Eq. (1) in fault classifications.

$$\text{Accuracy} = \frac{\text{Number of testing samples correctly classified}}{\text{Number of testing samples}} \times 100 \qquad (1)$$

Table 2 k-NN classification for FIA

Fault type	FIA	k-NN fault classification					
		A I	B I	C I	A II	B II	C II
A II C II	31°	0	0	0	01	0	01.1
A I B I C I	51°	0.98	01	01	0	0	0
A II B II	121°	0	0	0	01	01	0
A I B I C I	161°	0.97	01	01	0.03	0.01	0
A II C II	331°	0.04	0.02	0.04	0.001	0.95	01
A II C II	21°	0	0	0	0.98	0	0.97
A II B II C II	81°	0	0	0	01	01	0.988
A II B II C II	151°	0	0.03	0	01	0.99	01
A I C I	231°	0.96	0	0.969	0	0	0
A II B II C II	321°	0.021	0.011	0	0.989	1.11	01

Table 3 k-NN classification for fault location

Fault type	L	k-NN fault classification					
		A I	B I	C I	A II	B II	C II
A II C II	3	0	0	0	1.11	0	1.12
A I B I C I	7	0.98	0.959	01	0	0	0
A II B II	4	0	0	0	01	01	0
A I B I C I	11	0.94	01	1	0.31	0	0
A II C II	27	0.05	0.02	0	0.03	9.91	1
A II C II	43	0	0	0	0.96	0	0.96
A II B II C II	46	0	0	0	0.96	1	0.96
A II B II C II	57	0	0.03	0	01	0.96	1
A I C I	66	0.94	0	0.95	0	0	0
A II B II C II	3	0.02	0.03	0	0.96	1.11	01
A I B I	9	0.95	0.95	0	0	0	0

5 Conclusions

This paper presented an accurate unit for phase-to-phase fault classification in PTL based on k-NN. The frequent shunt faults under shifting all faults like locations of faults from sources (2–66 km) and FIA (1–359°) have been inspected. The current measured at source position is given as input to k-NN system. The output of k-NN module has been authenticated by the test results. The offline tests authenticate the planned k-NN suitability.

References

1. Kang, N., Liao, Y.: Double-circuit transmission-line fault location utilizing synchronized current phasors. IEEE Trans. Power Deliv. **28**(2), 1040–1047 (2013)
2. Naresh Kumar, A., Sridhar, P., Anil Kumar, T., Ravi Babu T., Chandra Jagan Mohan, V.: Adaptive neuro-fuzzy inference system based evolving fault locator for double circuit transmission lines. IAES Int. J. Artif. Intell. **9**(3), 448–455 (2020)
3. Apostolopoulos, C., Korres, G.: A novel fault-location algorithm for double-circuit transmission lines without utilizing line parameters. IEEE Trans. Power Deliv. **26**(3), 1467–1478 (2011)
4. Naresh, A., Sanjay, Ch., Chakravarthy, M.: Fuzzy inference system-based solution to locate the cross-country faults in parallel transmission line. Int. J. Electr. Eng. Educ. (2019). https://doi.org/10.1177/0020720919830905
5. Barnes, H.C., Barthold, L.O.: High phase order transmission. Electra **24**, 139–153 (1973)
6. Naresh Kumar, A.: Series (open conductor) fault distance location in three phase transmission line using artificial neural network. Int. J. Sci. Res. Eng. Technol. **3**(7), 1067–1071 (2014)
7. Stewart, J.R., Grant, I.S.: High phase order—ready for application. IEEE Trans. Power Appar. Syst. **101**(6), 1757–1764 (1982)
8. Tiaras, N., Singh, L.P.: Mathematical modelling and analysis of multi phase systems. IEEE Trans. Power Appar. Syst. **101**(6), 1784–1793 (1982)
9. Stewart, J.R., Kallaur, E., Grant, I.S.: Economics of EHV high phase order transmission. IEEE Trans. Power Appar. Syst. **103**(6), 3386–3392 (1984)
10. Tiwari, S.N., Singh, G.K.: Multi-phase transmission research—a survey. Electr. Power Syst. Res. **24**, 207–215 (1992)
11. Bhatt, N.B., Venkata, S.S., Guyker, W.C., Booth, W.H.: Six-phase (multi-phase) power transmission system: fault analysis. IEEE Trans. Power Appar. Syst. **96**(3), 758–767 (1977)
12. Venkata, S.S., Guyker, W.C., Kondragunta, J., Saini, N.K., Stanek, E.K.: 138-kV, Six Phase Transmission System: Fault Analysis. IEEE Trans. Power Appar. Syst. **101**(5), 1203–1216 (1982)
13. Onogi, Y., Isaka, K., Chiba, A., Okumoto, Y.: A Method of Suppressing Fault Currents and Improving the Ground Level Electric Field in a Novel Six-Phase Transmission System. IEEE Trans. Power Appar. Syst. **102**(4), 870–879 (1983)
14. Ekici, S.: Support vector machines for classification and locating faults on transmission lines. Appl. Soft Comput. **12**(6), 1650–1658 (2012)
15. Jana, S., Nath, S., Dasgupta, A.: Transmission line fault classification based on wavelet entropy and neural network. Int. J. Electr. Eng. Technol. **3**(2) (2012)
16. Naresh Kumar, A., Sanjay, Ch., Chakravarthy, M.: Mamdani fuzzy expert system based directional relaying approach for six phase transmission line. Int. J. Interact. Multimed. Artif. Intell. **6**(1), 41–50 (2020)
17. Naresh Kumar, A., Sanjay, Ch., Chakravarthy, M.: A single end directional relaying scheme for double circuit transmission line using fuzzy expert system. Complex Intell. Syst. **6**(2), 335–346 (2020)
18. Jain, A., Thoke, A.S, Patel, R.N., Modi, P.K.: Classification and location of single line to ground faults in double circuit transmission lines using artificial neural networks. Int. J. Power Energy Convers. **2**(2), 109–125 (2010)
19. Naresh Kumar, A., Sanjay, Ch., Chakravarthy, M.: Fuzzy expert system based protection for double circuit incomplete journey transmission lines. Int. J. Comput. Intell. IoT **2**(1) (2019)
20. Naresh Kumar, A., Chakravarthy, M.: Simultaneous fault classification and localization scheme in six phase transmission line using artificial neural networks. J. Adv. Res. Dyn. Control Syst. **10**(3), 342–349 (2018)
21. Koley, E., Jain, A., Thoke, A.S., Jain, A.: Detection and classification of faults on six phase transmission line using ANN. In: International Conference on Computer & Communication Technology, Allahabad, India, pp. 100–103 (2011)

22. Naresh Kumar, A., Chakravarthy, M.: Fuzzy inference system based distance estimation approach for multi location and transforming phase to ground faults in six phase transmission line. Int. J. Comput. Intell. Syst. 11(1), 757–769 (2018)

Two-Dimensional Analytical Expression of Threshold Voltage for Un-doped Double Gate MOSFET

Vikas Maheshwari, Somashekhar Malipatil, Narendra K. Garg, and Rajib Kar

Abstract In this work, new two-dimensional (2-D) analytical solution of threshold voltage (V_{th}) for un-doped Double Gate MOSFETs is proposed. In this research work, Green's function based techniques are applied to solve the Poisson equations in 2-dimensional and we derived the threshold voltage (V_{th}) expression by using the concept of surface potential. This design is supposed uniform doping profile in Si area. The design is compared with existing experimental results and we got better solutions compared to existing works.

Keywords Minimum surface potential · Green's function · Symmetric DG-MOSFET · Threshold voltage · 2D Poisson equations and Un-doped

1 Introduction

In last two decades, VLSI (very large-scale integration) Technology CMOS has been used as a fundamental building block for low power system designs. Double-Gate (DG) MOSFET is emerging as a latest research domain in VLSI because the scaling of these devices is possible to the minimum channel length L possible for provided process technology parameters [1]. The scaling of MOSFET at deep submicron level has shown new and serious challenges for the designing and fabrication of future integrated circuits. When the MOSFET dimensions are scaled down, both the supply voltage and the gate t_{ox} be scaled down. The short channel effects [2] are controlled by scaling down both the gate-oxide thickness and channel length. The

V. Maheshwari
Department of ECE, Guru Nanak Institutions Technical Campus, Hyderabad, India

S. Malipatil (✉)
Department of ECE, Malla Reddy Engineering College & Management Sciences, Medchal, Hyderabad, India

N. K. Garg
Department of ECE, Amity School of Engineering & Technology, Gwalior, MP, India

R. Kar
Department of ECE, NIT Durgapur, Durgapur, West-Bengal, India

V. Bhateja et al. (eds.), *Computer Communication, Networking and IoT*, Lecture Notes in Networks and Systems 197, https://doi.org/10.1007/978-981-16-0980-0_24

scaling of gate-oxide thickness is limited to about 1 nm because of high gate tunneling current [3, 4]. Scaling down the device dimension in nanometer regime, leakage currents plays an important role which needs more attention from the designer. Scalability of DG-MOSFETs is limited by sub-threshold swing which is an important design parameter. Formation of two channels in the symmetric DG MOSFET [5] provides steep sub-threshold swing, high drive current, and trans-conductance. The sub-threshold swing and other short channel effects can be optimized by using appropriate gate dielectric [6]. Therefore new MOSFET structures such as Dual Gate and Tri-Gate MOS transistors [7] are proposed to replace conventional planer MOSFET. It is also important that these new structure of MOSFETs are compatible with latest designing and fabrication techniques [8] of silicon integrated circuits. Dual gate MOSFETs also referred to as inversion charge transistors have many advantages over conventional MOSFETs. First, there is a considerable reduction in the device area. Second, DG MOSFET's miller capacitance and output conductance can be further considerably reduced which makes it useful device for the designing of analog integrated circuits. Third, the breakdown voltage of the device can be made very high by using proper design methodology. Fourth, short channel effects in scaled devices are drastically minimized. Comparable to conventional MOSFET, in case of Dual Gate MOSFET, both gates control the properties of channel. Thus more scaling of gate length can be performed in DG MOSFETs. DG MOSFETs is better alternative of conventional MOSFETs because DG MOSFETs have improved on short channel effects. Also DG MOSFET has higher current density. Therefore performance of the devices can be maintained low leakage and higher current density by employing DG MOSFETs. In DG MOSFETs better manage short channel effects and V_{th} is achieved ideally without altering the concentration of channel dopants. This technique diminished the statistical dopant fluctuations [9] an also reduces the phenomena of impurity scattering. Many structures and techniques were used SOI devices with less short channel effects [10, 11]. Using DG MOSFETs can minimize short channel effects. The conventional MOSFET cannot be scaled down below 10 nm process technology. Therefore DG MOSFETs are widely employed in the analog electronic circuits where reduced short-channel effects, electronic gain control capability, high breakdown voltages are required. DG MOSFETs are fabricated in below 20 nm fabrication process technology by using metal gate technology for both the gates that diminished the poly depletion effects. The effects are degradation [12] and mobility of dopant atoms are eliminated of DG MOSFETs. Due to these advantages, an analytic and simple threshold-voltage model is highly desirable for un-doped DG MOSFETs as one of the key point parameters for the design of such nano-scale devices. More advantages of Dual Gate MOSFETs are improved trans-conductance, early voltage, and carrier transport efficiency. Accurate and physics-based formulations are required for implementation of high-speed VLSI circuits comprising of compact MOSFET model. Due to these requirements for the device, the un-doped symmetric DG MOSFETs are best-suited device structure because of the thin and un-doped nature of the channel. In the situation of surface inversion near oxide-silicon interface, the energy band bends up to $2\Phi_{B}$ at the silicon-oxide interface. The same phenomena can be utilized for the analysis of the uniformly doped carrier concentration in the substrate. For the analytical modeling of the 2-D characteristics of DG MOSFETs, the two-dimensional Poisson's equations are analyzed by using suitable initial and final boundary conditions. Concept of Green's function techniques are used to get the solution for the two-dimensional Poisson's equation considering the uniform doping profile. The modeling of DG MOSFET's reported by authors [13–19]. For the optimization of the performance of the devices in nanometer regime for low power applications, V_{th} of the DG MOSFETs should be properly modeled. The analytical expression for 2-Dimensional V_{th} in the substrate is derived explicitly and verified by previous existing 2-Dimensional analytical models.

2 Structure of DG MOSFET

Figure 1 discusses the general structure of DG MOSFETs. In this structure of DG MOSFET 2 gates are used. Coupling is enhanced between gate to channel and hence short channel effects are considerably suppressed. DG MOSFETs are categorized as either symmetric DG MOSFET or asymmetric depends upon the voltages are applied to both gates. These two metal gates may or may not have same work functions. In this research work, analysis is done for symmetric DG MOSFETs. The symmetric DG MOSFETs means work function is similar for both gates. The thickness of the gate oxide is equal with same bias voltage is applied. After application of V_{ds}, V_{gs}, and with constant Quasi-Fermi level is shown in Fig. 1.

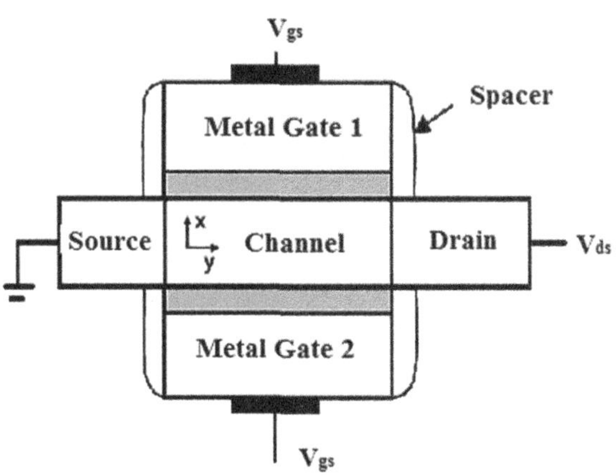

Fig. 1 Structure of general DG-MOSFET

3 Modeling of DG-MOSFET

3.1 he Basic Analysis

The cross-sectional view of a thin-film Dual Gate MOS device for two-Dimensional analytical model is shown in Fig. 2. The simplified domains for the analytical solution for the 2D Poisson's equations [20–23] along with the suitable boundary conditions are shown in Eq. (1). The below equation shows the 2D Poisson's equation with respect to rectangular coordinate system.

$$\Delta^2\varnothing(x,y) = -\frac{\rho(x,y)}{\varepsilon_{si}} = \frac{qN_A f(y)}{\varepsilon_{si}}, 0 \le y \le t_{si}, 0 \le x \le L$$

where N_A is substrate acceptor and $f(y)$ is doping profile in Si area. The 2-D Poisson's equations are converted into 2-D Laplace equations in the back and front oxide regions.

$$\left.\begin{array}{ll} \varnothing(0,y) = V_{bi}(y) & 0 < y < t_{si} \\ \varnothing(L,y) = V_{bi}(y) + V_{ds} & 0 < y < t_{si} \\ D_{sf}(x,0) = \epsilon_{si}\ E_y(x,0) & 0 < x < L \\ D_{sb}(x,t_{si}) = \epsilon_{si}\ E_y(x,t_{si}) & 0 < x < L \end{array}\right\} \tag{1}$$

$$\begin{aligned} E_y(x,0) &= \frac{\partial\phi(x,y)}{\partial y} = \frac{C_{ox}}{\varepsilon_{si}}\left[V_{gs} - V_{fb} - \phi_s - \frac{Q_0}{C_{ox}}\right], \\ E_y(x,t_{si}) &= \frac{\partial\phi(x,y)}{\partial y} = -\frac{C_{ox}}{\varepsilon_{si}}\left[V_{gs} - V_{fb} - \phi_s - \frac{Q_0}{C_{ox}}\right] \end{aligned} \tag{2}$$

The concept of Green's function technique is utilized for solution of the 2-Dimensional Poisson's potential distribution. Simplifying the Green's function

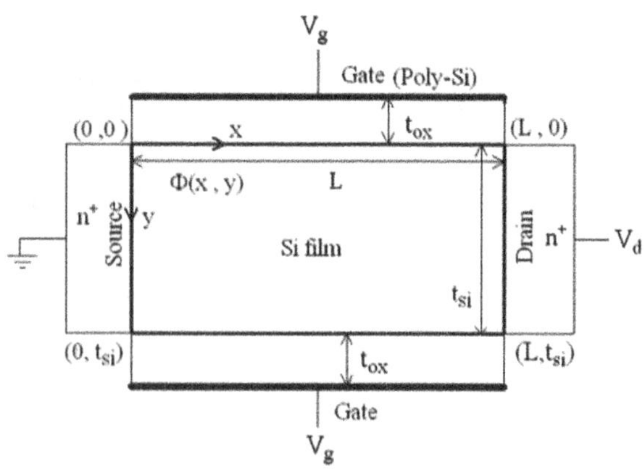

Fig. 2 Cross section view of DG-MOSFET

solution into Green's theorem [16], this is shown below

$$\emptyset(x,y) = \iint \frac{\rho(x',y')}{\varepsilon} G(x,y;x',y')\mathrm{d}x'\mathrm{d}y' + \int G(x,y;x',y')\frac{\partial\phi}{\partial n'}\mathrm{d}s' - \int \Phi(x',y')\frac{\partial G}{\partial n'}\mathrm{d}s' \quad (3)$$

where $G(x, y; x', y')$ is the Green's function satisfying $\Delta^2 G = -\delta(x-x')\delta(y-y')$, n' is the outward direction.

$$G_x(x,y;x',y') = \frac{2}{L}\sum_{n=1}^{\infty} \frac{\sin(K_n x')\sin(K_n x)\sinh(K_n y)\sinh K_n(t_{si}-y')}{K_n \sinh K_n t_{si}}, y < y' \quad (4)$$

$$G_x(x,y;x',y') = \frac{2}{L}\sum_{n=1}^{\infty} \frac{\sin(K_n x')\sin(K_n x)\sinh(K_n y')\sinh K_n(t_{si}-y)}{K_n \sinh K_n t_{si}}, y > y' \quad (5)$$

$$G_y(x,y;x',y') = \frac{2}{t_{si}}\sum_{m=1}^{\infty} \frac{\sin(K_m y')\sin(K_m y)\sinh(K_m x)\sinh K_m(L-x')}{K_m \sinh K_m L}, x < x' \quad (6)$$

$$G_y(x,y;x',y') = \frac{2}{t_{si}}\sum_{m=1}^{\infty} \frac{\sin(K_m y')\sin(K_m y)\sinh(K_m x')\sinh K_m(L-x)}{K_m \sinh K_m L}, x > x' \quad (7)$$

Equation (3) is expression for electrostatic potential distribution and $\rho(x', y') = -qN_A f(y')$ defined as the charge density of Si region. Assuming uniform doped ($f(y') = 1$). 2D potential equation is shown below.

$$\emptyset(x,y) = \frac{(-qN_A)}{(2\varepsilon_{si})}x(L-x) + V_{bi} + V_{ds}\frac{x}{L} + \sum_{n=1}^{\infty} \frac{(\sin K_n x)}{\varepsilon_{si}k_n \sinh k_n t_s i}\left[D_{sf}^{m}\cosh k_n(t_{si}-y) - D_{sb}^{m}\cosh k_n y\right] \quad (8)$$

where D_{sf}^{m} and D_{sb}^{m} is electric displacement of front & back gate. The $\varnothing(x, y)$ have to satisfy the boundary conditions,

$$\left.\frac{\partial\phi(x,y)}{\partial x}\right|_{y=0^+} - \left.\frac{\partial\phi(x,y)}{\partial x}\right|_{y=0^-} = 0 \quad (9)$$

$$-\varepsilon_{si}\frac{\partial\phi(x,y)}{\partial y}\bigg|_{y=0^+} + \varepsilon_{ox}\frac{\partial\phi(x,y)}{\partial y}\bigg|_{y=0^-} = 0 \tag{10}$$

After solving Eqs. (9) and (10) and we can obtain

$$D_{sf}^{m} = \frac{\varepsilon_{ox}[B.D\varepsilon_{si}k_n \sinh k_n t_{si} - C.\sin K_n x]}{[B.\varepsilon_{si}\sinh k_n t_{si} + B.\varepsilon_{ox}\cosh k_n t_{si} - A.\varepsilon_{ox}].\sin K_n x} \tag{11}$$

where coefficients

$$A = \frac{\sin K_n x}{K_n}\left\{\frac{1}{\varepsilon_{ox}} + \frac{1}{\varepsilon_{si}\tanh k_n t_{si}} - \frac{1}{\varepsilon_{si}\sinh k_n t_{si}}\right\},$$
$$B = \frac{\sin K_n x}{K_n}\left\{\frac{1}{\varepsilon_{si}\sinh k_n t_{si}} - \frac{1}{\varepsilon_{si}\tanh k_n t_{si}} - \frac{\tanh k_n t_{ox}}{\varepsilon_{ox}}\right\} \tag{12}$$

$$C = \frac{2\sin K_m t_{ox}}{t_{ox}K_m \sinh k_m L}\{V_{bi}\sinh K_m(L-x) + (V_{bi}+V_{ds})\sinh K_m x\}$$
$$D = \frac{4(V_{gs}-V_{fb})\sinh K_n x}{n\pi \cosh k_n t_{ox}} + \frac{qN_A}{2\varepsilon_{si}}x(L-x) - V_{bi} - V_{ds}\frac{x}{L} \tag{13}$$

3.2 Threshold Voltage Model

The location of X_{min} lies on Si surface

$$\frac{\partial\Phi(x,y)}{\partial x}\bigg|_{x=x_{\min},\,y=0,t_{si}} = 0 \tag{14}$$

From Eq. (9), the position of the minimum surface potential x_{min} is

$$\frac{\varepsilon_{ox}\cosh k_n t_{si}}{E}\left\{\frac{qN_A}{\varepsilon_{si}}\frac{x_{\min}(L-x_{\min})K_n}{\tan K_n x_{\min}} - \frac{qN_A}{2\varepsilon_{si}}x_{\min}(L-2x_{\min}) - \frac{V_{ds}}{L} - \frac{2V_{ds}}{L}\frac{x_{\min}K_n}{\tan K_n x_{\min}} - \frac{2V_{bi}K_n}{\tan K_n x_{\min}} + \frac{4(V_{gs}-V_{fb})}{L.\cosh k_n t_{ox}\tan K_n x_{\min}}\right\}$$
$$-\frac{qN_A}{2\varepsilon_{si}}x_{\min}(L-2x_{\min}) + \frac{V_{ds}}{L} = 0 \tag{15}$$

the expression for minimum surface potential $\phi_{sf,\min}$ is simplified as from above Eq. (15)

$$\Phi_{\min}(x_{\min},0) = \frac{-qN_A}{2\varepsilon_{si}}x_{\min}(L-x_{\min}) + V_{bi} + V_{ds}\frac{x_{\min}}{L}$$

$$+\sum_{n=1}^{\infty}\frac{\sin K_n x_{\min}}{\varepsilon_{si}k_n\sinh k_n t_{si}}\left[D_{sf}^m\cosh k_n t_{si}-D_{sb}^m\right]=\Phi_{s,\min} \tag{16}$$

The V_{th} for the DG-MOSFET is derived from [24]. The V_{th} in terms of surface potential is expressed by

$$V_{\text{th}}=V_{fb}+\left\{2\Phi_f+\frac{-qN_A}{2\varepsilon_{si}}x_{\min}(L-x_{\min})-V_{bi}-V_{ds}\frac{x_{\min}}{L}\right\}$$
$$\frac{\{\sin K_n x_{\min}\}^{-1}}{2G_f}-\frac{P}{2G_f} \tag{17}$$

where

$$G_f=\left[1-(-1)^n\right]\frac{R}{d_0}\left[\frac{2}{m\pi\cosh k_n t_{ox}}+\sum_{m=1}^{\infty}\frac{t}{(m-.5)\pi}\left[(-1)^m-\frac{1}{(m-.5)\pi}\right]\right],$$
$$R=-\frac{\varepsilon_{si}}{\varepsilon_{ox}}\frac{\tanh k_n t_{ox}}{\sinh k_n t_{si}},\ t=\frac{4}{n\pi}\left[1+\frac{L^2(m-.5)^2}{t_{ox}^2n^2}\right] \tag{18}$$

$$d_0=\frac{1}{(\sinh k_n t_{si})^2}-\left\{\frac{\varepsilon_{si}\tanh k_n t_{ox}}{\varepsilon_{ox}}+\frac{1}{\tanh k_n t_{si}}\right\}^2,$$
$$P=\frac{1}{d_0}\left\{\left[1-(-1)^n\right]\frac{qN_AL^2}{2\varepsilon_{si}}\frac{8R}{(n.\pi)^3}+T\left[V_{bi}\left(1-(-1)^n\right)+V_{ds}(-1)^{n+1}\right]\right\} \tag{19}$$

$$T=\sum_{m=1}^{\infty}2Rt((m-.5)\pi)^{-2}-\frac{4R}{n\pi} \tag{20}$$

$$V_{th,\text{long}}=V_{fb}-\frac{P}{2G_f} \tag{21}$$

The threshold voltage roll-off ΔV_{th} expression is the difference between the values of threshold voltage for short channel and long-channel devices given as

$$\Delta V_{\text{th}}=\left\{2\Phi_f+\frac{-qN_A}{2\varepsilon_{si}}x_{\min}(L-x_{\min})-V_{bi}-V_{ds}\frac{x_{\min}}{L}\right\}\frac{\{\sin K_n x_{\min}\}^{-1}}{2G_f} \tag{22}$$

4 Results and Discussion

Figures 3 and 4 shows the various levels of threshold voltage roll-off ΔV_{th} in the substrate the length of the channel length for various values of *Si* film t_{si}= (1.5, 5,

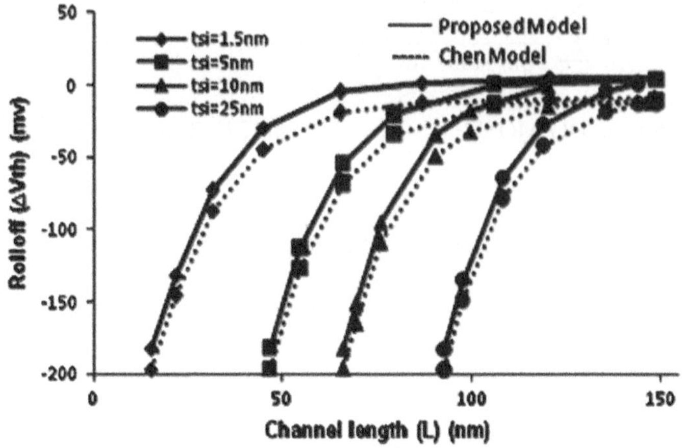

Fig. 3 Proposed design compared with Chen [24], $t_{ox} = 1$ nm

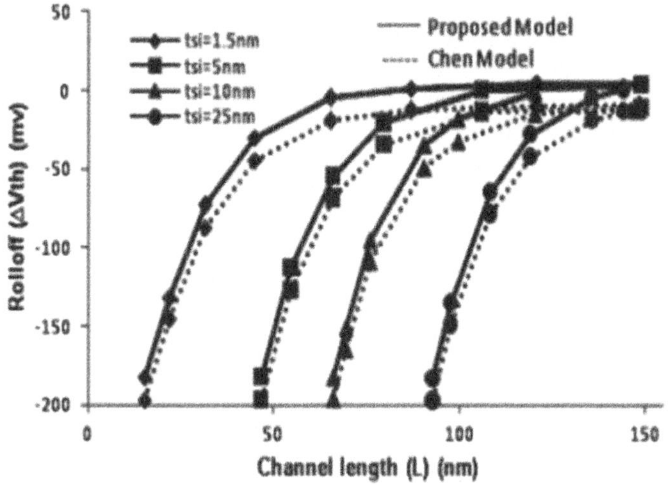

Fig. 4 Proposed design compared with Chen [24], $t_{ox} = 1.5$ nm

10, and 25 nm) at Gate $t_{ox} = 1$ nm and $t_{ox} = 1.5$ nm at same $V_{ds} = 0.05$ V and $N_a = 10^{17}$ cm^{-3} for proposed analytical model and Chen [24]. The V_{th} roll-off increases with increase in t_{si}. With the increase in silicon film thickness, the curves are shifted towards right side for constant $V_{ds} = 0.05$ V and constant gate $t_{ox} = 1$ nm and $t_{ox} = 1.5$ nm. The proposed model shows Threshold Voltage roll-off ΔV_{th} is 5–7% greater than Chen design.

Figure 5 shows the parametric analysis for threshold voltage roll-off for various drain bias voltages V_{ds}= (0.005, 1.0 V) and the parameters are $t_{ox} = 1.5$ nm and silicon film thickness $t_{si} = 10$ nm. Figures 3, 4, and 5 shows the threshold voltage

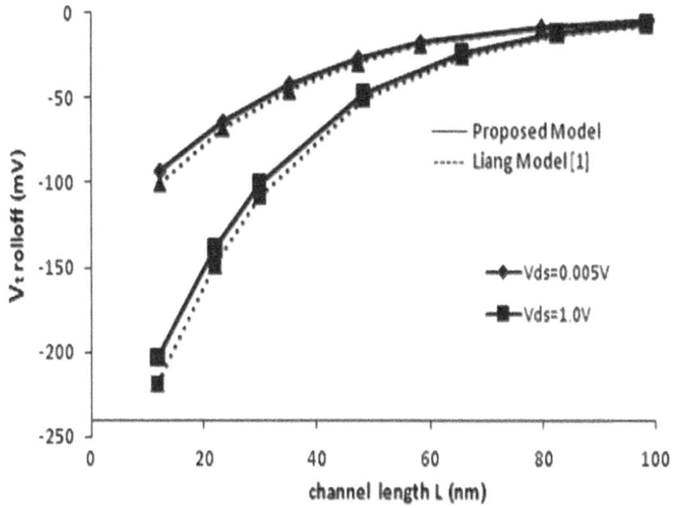

Fig. 5 Proposed design compared with Liang [1], $t_{si} = 10$ nm and $t_{ox} = 1.5$ nm

roll-off ΔV_{th}. The proposed model accurately predicts the device behavior compared with the existing works Liang [1] and Chen [24], with 3–5% more accurate.

5 Conclusion

The two-dimensional Poisson's equations have been analytically derived using the concept of Green's function incorporating suitable initial and final boundary conditions in Silicon area. The accuracy of the proposed model in the substrate (Si-film) is analyzed and verified by existing models. This design is valid for uniform doping profile in Si area. Due to the symmetry of DG-MOSFETS, the distributions of analytic potential for both sides of the surfaces in the substrate are same. It is obvious from simulation results so as to iterative method can only be used to find the minimum surface potential. It can be easily analyzed that better results are produced between the proposed and Chen [24] design analysis for parametric analysis of device structure for various biasing. The proposed analytic voltage model can be applied as a basic concept for high-speed physical analysis of the short-channel effects. It will define the deep-sub micrometer optimized scaling rule for thin-film Silicon on Insulator based MOS transistor in ULSI Technology.

References

1. Jiang, C., Liang, R., Wang, J., Xu, J.: A two-dimensional analytical model for short channel junctionless double-gate MOSFETs. AIP Adv **5**(5), 0571223-13 (2015)
2. Gupta, K.A., Anvekar, D.K., Venkateswarlu, V.: Modeling of short channel MOSFET devices and analysis of design aspects for power optimization. Int. J. Model. Optim. **3**(3), 266–271 (2013)
3. Avci, U.E., Morris, D.H., Young, I.A.: Tunnel field-effect transistors: prospects and challenges. J. Electr. Dev. Soc. **3**(3), 88–95 (2015)
4. Narendiran, A., Akhila, K., Bindu, B.: A physics-based model of double-gate tunnel FET for circuit simulation. IETE J. Res. **62**(3) (2016)
5. Hosseini, R.: Analysis and simulation of a junctionless double gate MOSFET for high-speed applications. **67**(9), 1615–1618 (2015)
6. Salmani-Jelodar, M., Ilatikhameneh, H., Kim, S., Ng, K., Klimeck, G.: Optimum high-k oxide for the best performance of ultra-scaled double-gate MOSFETs. IEEE Trans. Nanotechnol. 1–5 (2015)
7. Ture, E., Brückner, P., Godejohann, B.-J., Aidam, R., Alsharef, M., Granzner, R., Schwierz, F., Quay, R., Ambacher, O.: High-current submicrometer tri-gate GaN high-electron mobility transistors with binary and quaternary barriers. J. Electr. Dev. Soc. **4**(1), 1–6 (2016)
8. Bauman, S.J., Novak, E.C., Debu, D.T., Natelson, D., Herzog, J.B.: Fabrication of sub-lithography-limited structures via nanomasking technique for plasmonic enhancement applications. IEEE Trans. Nanotechnol. **14**(5), 790–793 (2015)
9. Saha, S.K.: Modeling statistical dopant fluctuations effect on threshold voltage of scaled JFET devices. IEEE Access J. **4**, 507–513 (2016)
10. Xie, Q., Lee, C.-J., Xu, J., Wann, C., Sun, J.Y.-C., Taur, Y.: Comprehensive analysis of short-channel effects in ultrathin SOI MOSFETs. IEEE Trans. Electr. Dev. **60**(6), 1814–1819 (2013)
11. Colinge, J.-P.: Multiple-gate SOI MOSFETs. Solid-State Electron. **48**(6), 897–905 (2004)
12. Cros, A., Romanjek, K., Fleury, D., Harrison, S., Cerutti, R., Coronel, P., Dumont, B., Pouydebasque, A., Wacquez, R., Duriez, B., Gwoziecki, R., Boeuf, F., Brut, H., Ghibaudo, G., Skotnicki, T.: Unexpected mobility degradation for very short devices : a new challenge for CMOS scaling. IEEE Int. Electr. Dev. Meet. 663–666 (2006)
13. Rajabil, Z., Shahhoseini, A., Faez, R.: The non-equilibrium green's function (NEGF) simulation of nanoscale lightly doped drain and source double gate MOSFETs. In: IEEE International Conference on Devices, Circuits and Systems (ICDCS), pp. 25–28 (2012)
14. Karatsori, T.A., Tsormpatzoglou, A., Theodorou, C.G., Ioannidis, E.G., Haendler, S., Planes, N.G.: Development of analytical compact drain current model for 28nm FDSOI MOSFETs. In: 4th International Conference on Modern Circuits and Systems Technologies, Thessaloniki Greece, pp 1–4,14–15 (May 2015)
15. Tripathi, S.L., Kumar, M., Mishra, R.A.: 3-D channel potential model for doped symmetrical ultra-thin quadruple gate-all-around MOSFET. J. Electr. Dev. **21**, 1874–1880 (2015)
16. Ávila-Herrera, F., Cerdeira, A., Paz, B.C., Estrada, M., Íñiguez, B., Pavanello, M.A.: Compact model for short-channel symmetric double-gate junction-less transistors. Solid-State Electron. **111**, 196–203 (2015)
17. Zhang, G.-M., Su, Y.-K., Hsin, H.-Y., Tsai, Y.-T.: Double gate junctionless MOSFET simulation and comparison with analytical model. In: IEEE Regional Symposium on Micro and Nanoelectronics (RSM-2013), Langkawi, Malaysia, pp. 410–413, 25–27 (2013)
18. Yadav, V.K.S., Baruah, R.K.: An analytic potential and threshold voltage model for short-channel symmetric double-gate MOSFET. In: International Symposium on VLSI Design and Test, Coimbatore, India, 16–18 July, 2013
19. Bhartia, M., Chatterjee, A.K.: Modeling the drain current and its equation parameters for lightly doped symmetrical double-gate MOSFETs. J. Semicond. **36**(4), 1–7 (April 2015)
20. Liang, X., Taur, Y.: A 2-D analytical solution for SCEs in DG MOSFETs. IEEE Trans. Electr. Dev. **51**(9), 1385–1391 (2004)

21. Lo, S.-C., Li, Y., Shao-Ming, Yu.: Analytical solution of nonlinear poisson equation for symmetric double-gate MOSFET. Math. Comput. Model. **46**(1–2), 180–188 (2007)
22. Chiang, T.K.: A novel scaling-parameter-dependent sub threshold swing model for double-gate (DG) SOI MOSFETs: including effective conducting path effect (ECPE). Semicond. Sci. Technol. **19**(12), 1386–1390 (2004)
23. Huaxin, Lu., Taur, Y.: An analytic potential model for symmetric and asymmetric DG-MOSFETs. IEEE Trans. Electr. Dev. **53**(5), 1161–1168 (2006)
24. Chen, Q., Harrell, E.M., Meindl, J.D.: A physical short-channel threshold voltage model for undoped symmetric double-gate MOSFETs. IEEE Trans. Electron Dev. 50(7) (2003)
25. Maheshwari, V., Malipatil, S., Gupta, N., Kar, R.: Modified WKB approximation for Fowler-Nordheim tunneling phenomenon in nano-structure based semiconductors. In: 2020 International Conference on Emerging Trends in Information Technology and Engineering (ic-ETITE), Vellore, India, 2020, pp. 1–5. https://doi.org/10.1109/ic-ETITE47903.2020.460

Sentiment Analysis through Word Vectors: A Study on Movie Reviews from IMDb

Abhijit Roy and Parthajit Roy

Abstract The present study emphasizes the challenges of labeling the correct semantic orientation of movie review texts due to the high degree of ambiguity and varied narratives of the reviewers. The study uses a vector-based model of sentiment analysis on IMDb movie review data set. The model computes the attributes of text reviews based on singular words and the longest common subsequence of words. We present a comparative assessment of unsupervised vector-based model (K-means clustering) and supervised models (support vector machine with linear kernel and naïve Bayesian classifier) for sentiment-based classification of movie review data set. The study reports that supervised models are outperforming unsupervised model in terms of accuracy. Despite the high degree of ambiguity of movie reviews, the results are encouraging in the context of natural language processing.

Keywords Sentiment analysis · Text analytics · Word sentiments · Semantic orientation · Semantic analysis

1 Introduction

Sentiment analysis and processing of emotions are attracting many fields of studies in business to reveal new insights [1]. Vector-based word representations using lexicological identification of words are critical components in deciphering the semantics in the context of natural language processing [2]. A vector-based approach classifies words based on distance or angle in a multidimensional space. The present paper addresses the rating inference problem over and above the classification of data. The similarities of the mood of the words are further classified in this paper with the degree of similarity. Over and above the binary categorization of the opinion-based words in 'good' or 'bad' [3, 4], the study captures more information by extracting

A. Roy
Dr. Bhupendra Nath Dutta Smriti Mahavidyalaya, Burdwan, West Bengal 713407, India

P. Roy (✉)
Department of Computer Science, The University of Burdwan,
Burdwan, West Bengal 713104, India

V. Bhateja et al. (eds.), *Computer Communication, Networking and IoT*, Lecture Notes in Networks and Systems 197, https://doi.org/10.1007/978-981-16-0980-0_25

the degrees of similarities between words [5]. The objective of our model is to learn semantic vectors with sentiment information. In our model, using a movie review data set, we have shown how it can be classified primarily in binary classification of positive and negative reviews and then we further analyze the degree of semantic similarities between positive reviews and negative reviews.

In the following sections, we first present the related works in the field of text-based sentiment analysis. In the next section, we specify our model and mention the data set that we use for our experiments followed by results analysis and conclusions.

2 Review of Related Work

The models we present in this paper draw inspiration from both unsupervised and supervised vector space models for extracting the mood of a textual review. One area of research in text analytics tries to identify the text genre, that is, whether a document is editorial of a newspaper, a personal blog or a product review [6–8]. Another important development in this area is the identification of the subjective language in a text, like emotions or opinions [9–11]. Subjectivity analysis can identify the presence of opinions or emotions in a text but they do not identify the mood of such opinions or emotions. So at this point, the studies on sentiment analysis of texts are natural extension of research on identification of language subjectivity. Majority of the previous researches on sentiment classification of texts identify semantic orientation of words based on linguistic heuristics or some preselected mood-based lexicons [3]. A feature-based heuristics algorithm assigns aspect-based sentiment labels to each document and measures the aggregate scores of all the aspects to understand the overall sentiment [12]. Prior research also used sentiment classification models based on the principles of cognitive linguistics or manually constructed word lexicon [13].

In a supervised vector spaced model, term frequency (tf) and delta-inverse document frequencies (didf) are used as weights to transform word vectors to values for better performance [2, 14]. The proposed model uses unsupervised vector spaced latent semantic analysis where the documents are decomposed into word vectors, and thereafter analysis of the word vectors is done with a predefined set of linguistic heuristics. Section 3 describes the model in detail.

3 Model Specification

Let us now discuss the proposed model for sentiment analysis. The proposed model is a vector-based sentiment analysis model also known as "bag of word" model. That means every word in the documents has been converted to its representative numeric vectors, and a specific mood is tagged with the word based on a mood-based lexicon. Once the vectors are prepared, both unsupervised and supervised classifications have been applied on them to separate them into adequate number of classes. In this study, we have considered two types of vectors to represent a single text. These are word frequency and longest common subsequence (LCS) of words. The word frequency has been computed on the whole document, whereas LCS has been computed on sentences and finally been accumulated for the whole document.

3.1 Learning Word Sentiments

For word-based sentiment vector, we have considered a lexicon with positive and negative tagging of each word [15, 16]. The positive list contains 2006 words, whereas the negative list contains 4783 words. Given a document, we have first tokenized the entire document into words. We have considered "-" separated word as a single word. For example, the word "brand-new" has been considered as a single word rather than two words "brand" and "new". We have also considered ",.:()[]" as well as space, tab and newline as word separators. The character ' (apostrophe) has been considered as the part of the word. That is why, the words like "don't" have been left as it is, rather than converting it to "dont". After the tokenization, we have matched each and every word of the document with the list of positive and negative words. The total count of positive and that of negative words have been considered as two components of the document vector.

We have also introduced a novel attribute computation process based on longest common subsequence (LCS). In the context of cognitive understanding, it is not the words in isolation but the sequence of words that gives us the meaning. In a word-based model, the sequence does not matter, whereas in LCS we match the sequence of words and try to recognize the mood of a document. Given two sequences of characters, ABDAB and ADBAB, the LCS is ABAB. This is because the sequence is present in both the strings and that is the longest possible sequence.

We have used LCS for words in sentences. The idea behind this is, in some cases, we can bring the sophisticated sentence semantics into known sentence semantics, as an example, if we have two sentences, "The English movie is very good" and "The Movie is really very good". In the second sentence, the word "really" has been used for fine tuning, but the LCS of the two sentences is "The Movie is very good". Semantically, both the sentences are positive, and we have a long LCS value. One problem may arise, if the second word is "The Movie is not very good" and the LCS will be "The Movie is Very good". In that case, the meaning of will be reversed. This problem is addressed by checking the same sentence with the pool of negative sentences. If a negative sentence is there like, "The English Movie is not Very good", then the LCS will be "The Movie is not Very good" and the LCS match will be five words instead of four. For each sentence, we are taking both the positive LCS and negative LCS as attributes. So, positive sentence will have positive LCS value more than that of negative LCS value and vice versa. The LCS validation rule used is known as winner takes all, i.e., check with positive and negative and assign the sentence to the larger LCS match.

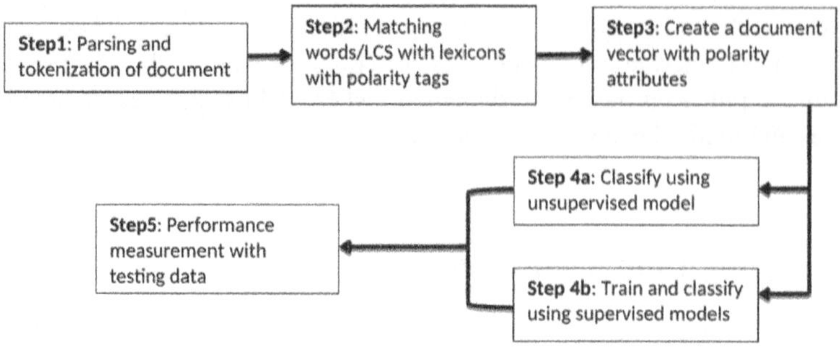

Fig. 1 Flowchart of the proposed model

For LCS computation purpose, we have taken a set of standard positive and negative sentences [5]. In this data set, there are 5331 positive sentences and 5331 negative sentences. For each and every sentence of the movie review, we have computed its LCS value for all the 5331 positive and negative standard sentences. As a cutoff, we have taken minimum LCS match as three. So, if a sentence has an LCS value greater than or equal to three for a standard sentence, we have increased total match by 1. In this way, we have computed the total count for the positive sentences and the total count for the negative sentences. The final counts are the two attributes of the review. A flowchart of the proposed model has been given in Fig. 1.

Finally, we have considered four attributes as document vector for each and every document. These are number of positive words (pw), number of negative words (nw), count of positive LCS greater than equal to three (pl) and count of negative LCS greater than equal to three (nl), i.e., our document vector is a four-dimensional numeric vector (pw, nw, pl, nl).

3.2 Data and Measurement Indexes Used for the Study

In our study, we have considered a standard benchmark data set of movie review by Pang and Lee [17]. Pang and Lee's work [17] is considered to be standard in sentiment analysis of movie reviews [18, 19], and the data set has been popular with other researchers [20]. This data set has a total of 1000 positive review documents and 1000 negative review documents, each kept in a separate ASCII text files. The average number of lines per document for a positive review is 32.93, and average number of lines per document for a negative review is 31.78. Average words per positive review are around 787, where average words per negative review are around 705 (punctuation marks included). The total size of the whole data set is 7.9 MB.

The second important thing is the performance measurement index. In our research, we have taken two similarity indexes to measure the performances. Both of them are external similarity indexes. We have considered Czekanowski-Dice (Cz-Dice) and Fowlkes-Mallows (F-M) index [21]. The formulas for the two indexes are as follows: $\text{CzDice} = \frac{2\times \text{TP}}{(2*\text{TP}+\text{TN}+\text{FP})}$ and $\text{FM} = \frac{\text{TP}}{\sqrt{(\text{TP}+\text{TN})*(\text{TP}+\text{TP})}}$, where TP is true positive or truly selected, FP is false positive of wrongly selected, and TN is true negative of truly rejected.

Once the document vectors have been computed and the indexes have been chosen, the next task is to go for classification. The present uses both unsupervised and supervised clustering to compare the performances. We have used K-means clustering as unsupervised model. Here we have selected two arbitrary cluster centers, and then for each of the 2000 document vectors (1000 for positive and 1000 for negative) we have computed the vector distances from both the centers. The smaller distance decides the assignment of the document to the cluster. Once all the vectors have been assigned to the clusters, the cluster centers have been updated to the cluster mean. This process has iteratively been executed until the cluster centers become stationary. Then the final cluster sets have been considered as positive and negative review results as per the proposed model. The results produced by the proposed model have been tested with the known results to measure performance. Then we have compared the results of unsupervised model with a set of supervised model for clustering. In Bayes classification, we use a set of training data to determine prior probabilities and then those prior probabilities are used to classify documents [22]. If we have D different documents to segregate in C categories, on a mood-based lexicon with T distinct words, then d_i is the no. of documents with $i = 1, 2, \ldots D$, t_j is words in the lexicon with $j = 1, 2, \ldots T$ and c_k are categories of moods where $k = 1, 2, \ldots C$. The probabilities of each document to fall in a particular mood category can be expressed using Eq. 1. This formula is used to judge the posterior distribution based on the prior and the sample.

$$p(c|d) = \frac{p(d|c) \times p(c)}{\sum_{k=1}^{C} p(d|c_i)p(c_i)}, \forall k = 1, 2, \ldots C \tag{1}$$

Support vector machines (SVMs) are very popular classification method that uses partition space to separate classes. If there exist inputs $D = d_1, d_2, \ldots, d_n$ with n dimensions, $D \in R^n$ and these inputs are to be categorized in $C = c_1, c_2, \ldots c_k$ with k dimensions, then SVM classifies n dimensional input D, using hyperplane that maximizes the distance between classes C. In the present study, we use support vector machine with linear kernel as the classifier.

As K-means is an unsupervised method, the full data set of 1000 positive and 1000 negative reviews has been passed to the model for their outcomes. For other three models, we have used 60% documents as training and 40% as testing. For supervised models, we have also used most frequent words as attributes. For the computer simulation, we have used Python and C language.

4 Experiments and Result Analysis

Table 1 shows the classification accuracy of the movie review data set using unsupervised K-means clustering. The results suggest that the movie review data set is pretty hard to classify in unsupervised model. So, the obvious question is: What makes a movie review unique and different from a product and service review? This is because of the high degree of subjectivity in movie review text. A movie may have a villain who has made an unpleasant scene very realistic which the reviewer has expressed in her review but in machine-based learning algorithm the semantic orientation of the text may turn out to be negative. In the performance classification, we see that the word-based classification reports more accuracy than the LCS-based classification. This is simply because a sentence explaining a perfectly crafted violent or negative scene may wrongly be counted as negative sentence in LCS mechanism. This conforms to the work of Turney [4] where the author has presented a confusion matrix for movie review data set. Turney [4] reports that out of 40.83% sentences with positive in its semantic orientation 12.50% are false positive. Remaining 59.17% sentences are negative in semantic orientation out of which 21.67% are false negative. So, collectively there is an ambiguity of 34.17%. We must read the performance of our unsupervised model in light of this ambiguity. Figures 2, 3 and 4 report the clustering results. A "circle" represents negative review, and "triangle" represents positive review. A red circle represents correct negative, and blue circle represents false negative classification. Similarly, a blue triangle represents correct positive but a red triangle reports a false positive classification.

Next we have classified the movie review data set supervised models. We have used both support vector machine with linear kernel and Naïve Bayesian classifier. Tables 2, 3 and 4 report the comparative performance using the Fowlkes–Mallows index. It is evident from the model that the classification performance has improved considerably in supervised models in comparison with unsupervised model. The results are consistent in all three categories of tests, that is, positive and negative words (Pw-nw), longest common subsequence of sentences (LCS) and both with Pw-nw as well as LCS. Unsupervised K-means has the poorest performances among all, and the Naïve Bayesian classifier has the highest accuracy. We use soft models, so the output varies with every run as we shuffle the document. We have taken the best of the first five runs.

Table 1 Performance measurement of the the proposed unsupervised K-means-based model in various measurement scales

Performance measures	$pw - nw$	$pl - nl$	$pw - nw + pl - nl$
Czekanowski dice	0.5464	0.5253	0.5320
Fowlkes mallows	0.5484	0.5258	0.5329
Precision	0.5032	0.5031	0.5025
Rand	0.5041	0.5036	0.5030
Recall	0.5977	0.5496	0.5652

Fig. 2 K-means model on positive and negative words

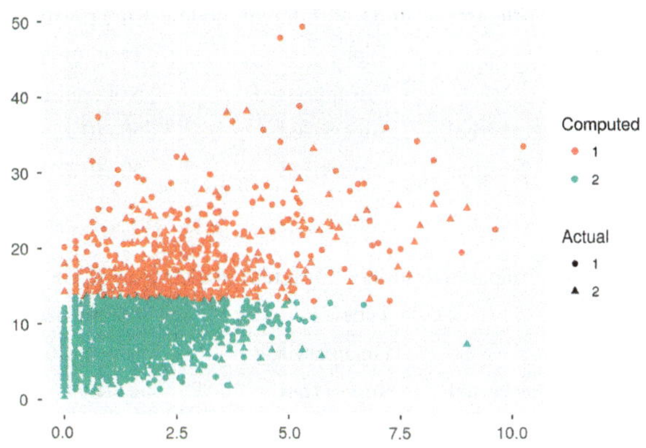

Fig. 3 K-means model on positive and negative LCS

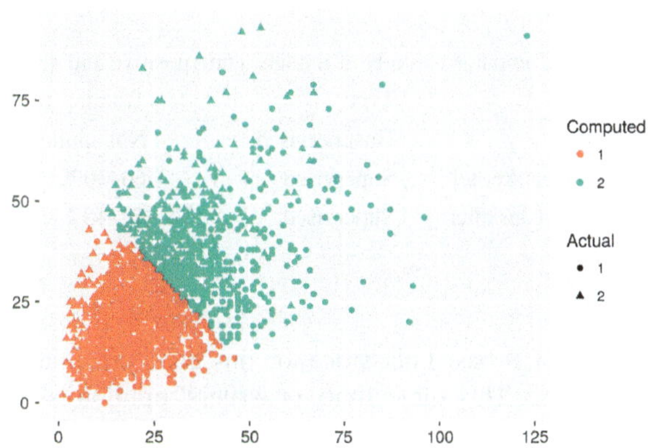

Fig. 4 K-means model on positive and negative words and LCS

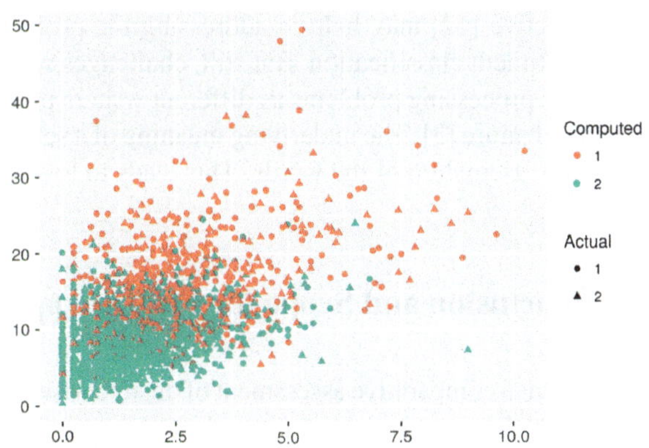

Table 2 Comparative study of different models using positive and negative words

Model	Type	Training-testing	Accuracy (%)
K-Means	Unsupervised	Not applicable	54.84
SVM-linear kernel	Supervised	60–40%	76.5
Bayesian classifier	Supervised	60–40%	84

Table 3 Comparative study of different models using LCS counts

Model	Type	Training-testing	Accuracy (%)
K-Means	Unsupervised	Not applicable	52.58
SVM-linear kernel	Supervised	60–40%	75.25
Bayesian classifier	Supervised	60–40%	84.75

Table 4 Comparative study of models using positive and negative words and LCS counts

Model	Type	Training-testing	Accuracy (%)
K-Means	Unsupervised	Not applicable	53.29
SVM-linear kernel	Supervised	60–40%	81.25
Bayesian Classifier	Supervised	60–40%	85.0

The LCS-based classification has good potential in understanding the mood of a sentence. This subjectivity or writers' opinion is a very basic feature of a review that makes the machine learning difficult to identify the correct semantic orientation of a sentence. Combination of adverb and verb or adjective and noun or adverb and adjective [12] may help in addressing the problem of ambiguity in document level sentiment classification. But still, challenges will remain in this type of natural language processing problems as different writers use different narratives to explain a certain feature [3]. The underlying meaning of a sentence is different than the literal semantic orientation of the words. This leads to low precision and recall.

5 Conclusion and Scope for Future Applications

We present a comparative assessment of unsupervised vector-based model and supervised models to capture semantic and sentiment information through words and the longest common subsequent sentences. Supervised classification models are recommended over unsupervised models because of better accuracy in classification considering the high degree of ambiguity in the movie review texts. The algorithm for the automated classification of subjective review data has shown a satisfactory result with significant accuracy.

Though the results are significant, future research should be focused on improving the accuracy and precision of the classification. Constructing lexicons for niche areas of text analytics is required for advancing the research with best-fit models for document classification. The ability of a machine-based classification has the potential for the inferential statistics also. For example, ordinal regression such as the ordinal "logit" model considers polarity-based classification. This model can extensively be used in economic studies.

References

1. Khoo, C.S.G., Nourbakhsh, A., Na, J.C.: Sentiment analysis of online news text: a case study of appraisal theory. Online Inf. Rev. (2012)
2. Maas, A., Daly, R.E., Pham, P.T., Huang, D., Ng, A.Y., Potts, C.: In: Proceedings of the 49th Annual Meeting of the Association for Computational Linguistics: Human Language Technologies, pp. 142–150 (2011)
3. Pang, B., Lee, L., Vaithyanathan, S.: Thumbs up? Sentiment classification using machine learning techniques. arXiv preprint cs/0205070 (2002)
4. Turney, P.D.: Thumbs up or thumbs down? Semantic orientation applied to unsupervised classification of reviews. arXiv preprint cs/0212032 (2002)
5. Pang, B., Lee, L.: Seeing stars: exploiting class relationships for sentiment categorization with respect to rating scales. arXiv preprint cs/0506075 (2005)
6. Lieungnapar, A., Todd, R.W., Trakulkasemsuk, W.: Genre induction from a linguistic approach. Indonesian J. Appl. Linguist. **6**(2), 319 (2017)
7. Pritsos, D., Rocha, A., Stamatatos, E.: In: European Conference on Information Retrieval, pp. 3–11. Springer (2019)
8. Worsham, J.M.: Towards literary genre identification: Applied neural networks for large text classification. Ph.D. thesis, University of Colorado Colorado Springs. Kraemer Family Library (2018)
9. Chaturvedi, I., Ragusa, E., Gastaldo, P., Zunino, R., Cambria, E.: Bayesian network based extreme learning machine for subjectivity detection. J. Frank. Inst. **355**(4), 1780 (2018)
10. Chaturvedi, I., Cambria, E., Welsch, R.E., Herrera, F.: Distinguishing between facts and opinions for sentiment analysis: survey and challenges. Infor. Fusion **44**, 65 (2018)
11. Kamil, A.Z., Rustamov, S., Clements, M.A., Mustafayev, E.: Recent Developments and the New Direction in Soft-Computing Foundations and Applications, pp. 63–70. Springer (2018)
12. Singh, V.K., Piryani, R., Uddin, A., Waila, P.: In: 2013 International Mutli-Conference on Automation, Computing, Communication, Control and Compressed Sensing (iMac4s), pp. 712-717. IEEE (2013)
13. Das, S., Chen, M.: In: Proceedings of the Asia Pacific Finance Association Annual Conference (APFA), vol. 35, p. 43. Bangkok, Thailand (2001)
14. Martineau, J.C., Finin, T.: In: Third International AAAI Conference on Weblogs and Social Media (2009)
15. Hu, M., Liu, B.: In: Proceedings of the Tenth ACM SIGKDD International Conference on Knowledge Discovery and Data Mining, pp. 168–177 (2004)
16. Liu, B., Hu, M., Cheng, J.: In: Proceedings of the 14th International Conference on World Wide Web pp. 342–351 (2005)
17. Pang, B., Lee, L.: A sentimental education: Sentiment analysis using subjectivity summarization based on minimum cuts. arXiv preprint cs/0409058 (2004)
18. Sahu, T.P., Ahuja, S.: In: 2016 International Conference on Microelectronics, Computing and Communications (MicroCom), pp. 1–6. IEEE (2016)

19. Zhuang, L., Jing, F., Zhu, X.Y.: In: Proceedings of the 15th ACM International Conference on Information and Knowledge Management, pp. 43–50 (2006)
20. Wang, S.I., Manning, C.D.: In: Proceedings of the 50th Annual Meeting of the Association for Computational Linguistics (Volume 2: Short Papers), pp. 90–94 (2012)
21. Desgraupes, B.: Univ. Paris Ouest-Lab Modal **1**, 34 (2013)
22. Das, S.R., et al.: Found. Trends® in Finance **8**(3), 145 (2014)

A Design Approach for Real-Time Data Level Multi-target Trajectory Simulator for Bi/Multistatic Radar

Jailaxmi Shankar, B. V. Srividya, and Paramananda Jena

Abstract Modeling and simulation play a vital role in any radar design and development. It enhances the confidence level of any subsystem or software before it is actually integrated to the main system. Radar system performs multi-level data processing or funneling to abstract accurate position of the target by means of various software algorithms. It becomes important to test these algorithms before deployment into the system. This paper presents an approach for design and development of multi-target trajectory simulator for bi/multistatic radar system. The results shown are various trajectories that can be generated to test various radar parameters.

Keywords Bistatic · Multistatic · Plot · Track · Detections · Radar controller · Plot extractor · Radar data processor

1 Introduction

The radar operates in an environment which consists of targets of interest, clutter which affects the probability detection and line of site between the radar and target [1]. It is important to simulate the attributes of each one of them in the laboratory environment to test the real-time performance of radar software modules. It also enhances the speed of development and testing of radar software modules before the deployment of the actual radar in the field. This multi-target simulator for bi/multistatic radar simulates complete radar-operating scenario in real time by simulating multiple target trajectories, radar operation, and clutter distribution with respect to common grid reference [2].

1.1 Role of Simulation

This simulator generates the real-time detection reports in order.

J. Shankar (✉) · B. V. Srividya · P. Jena
Department of Telecommunication, Dayananda Sagar Collage of Engineering, Bangalore, India

V. Bhateja et al. (eds.), *Computer Communication, Networking and IoT*, Lecture Notes in Networks and Systems 197, https://doi.org/10.1007/978-981-16-0980-0_26

(1) To evaluate the peak load condition.
(2) To evaluate the various real-time scenarios in the laboratory environment.
(3) To evaluate several tracker parameters like extreme target maneuver and speed conditions, the testing under simulated environment is the only option as it will be very expensive and in some cases impractical to create a real-life environment.
(4) To evaluate the resolving capability of multiple targets by the radar.

The bistatic radar will have a transmitter and a receiver placed at considerable distance. The target range will be the range sum of transmitter to target range and receiver to target range. The target azimuth will be receiver angle to target. Multiple pair of transmitter and receiver constitutes the multistatic radar. The whole signal processing is carried out at receiver station.

The returned signal received at the receiver which crosses the threshold strength is considered to be a valid detection by the signal processor. The bistatic receiver constitutes following of various software modules for processing the received data.

1.2 Radar Software Modules

Figure 1 shows the block diagram of software modules present at the receiver.

(a) Radar controller:

The radar controller acts as a communicating agent among the software modules of the various subsystems of the radar.

(b) Plot Extractor:

The returned signal from the target spreads across the multiple range, azimuth cells and in multiple receives beams. The spread is due to RCS of the target and range

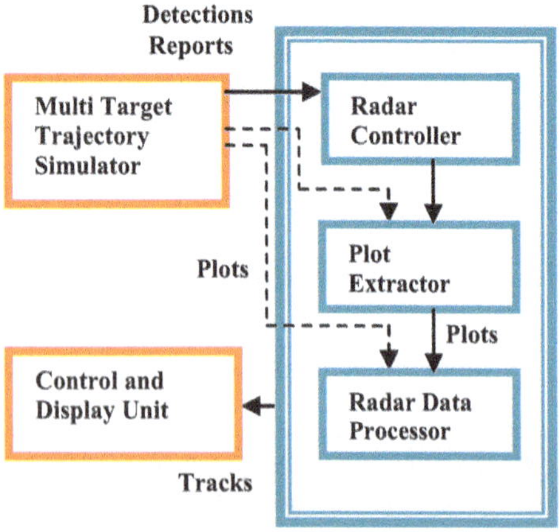

Fig. 1 Block diagram of radar data processing

straddling and multiple overlapped beams. The plot extractor cluster related detections pertaining to single target and find the true position of the target. Formed plot is sent to data processor to form the track.

(c) Radar Data Processor:

The radar data processor initiates, maintains, and terminates the tracks based on the continuous updation from scan to scan [3].

2 Existing Simulator Developed

One of the classifications for radar is based on the location of transmitter and receiver. Monostatic radar where transmitter and receiver are co-located on the same platform provides target position with respect to radar location. A number of target trajectory simulators are developed for monostatic radar to test radar data processing algorithms. The monostatic radar found to be decreased resilience to electronic counter measures and reduced RCS of the target due to geometrical effects. The bi/multistatic radar is considered to optimal solution to overcome these issues.

Hence, it becomes important to develop real-time target trajectory simulator to test the performance of bi/multistatic radar data processing algorithms.

3 Proposed Simulator for Bi/Multistatic Radar

The method proposed here is to design real-time target trajectory simulator to simulate bi/multistatic radar-operating scenario.

Bistatic Radar

The bistatic radar consists of a transmitter and a receiver kept at considerable distance. Receiver being passive in nature, the jammer will not be able to locate receiver and find difficult to jam receiver, and it provides resilience to electronic counter measures. The simulator defines the geometry of a bistatic radar system as shown in Fig. 2. The range sum is obtained by adding the distance between transmitter-target (R_{T}) and transmitter-receiver (R_{R}) i.e. $\rho = R_{\mathrm{T}} + R_{\mathrm{R}} = 2a$. This sum defines an ellipsoid of constant range with receive angle of the target θ_{R}.

Multistatic Radar

A multistatic radar system is composed of multiple numbers of transmitters and receivers distributed across defined geographical area. The RCS of the target will be enhanced as it is detected by multiple sensors. As the radar system forms network of sensors, detections from all the receivers are fused and will provide larger picture of coverage.

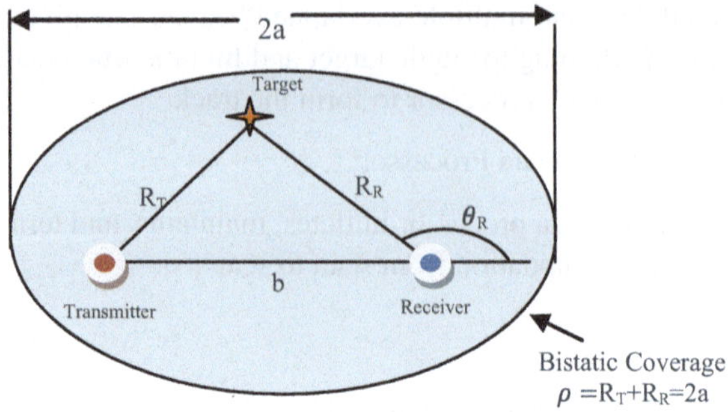

Fig. 2 Bistatic surveillance radar coverage

The simulator generates the target detections for radar controller to test the full chain of all software modules. The detection consists of information about the target include position, velocity, and heading [4].

Detections Generated for Radar Controller and Plot extractor:

The radar coverage is discretized in terms of multiple cells in three dimension for processing. The moving target will be sampled at every scan period. This point target is spread in to multiple adjacent cells as shown in Fig. 3. When the beam hits the target, the target position is acquired and converted from Cartesian coordinate to elliptical coordinate. Then, the range sum and receive angle are required to send as a part of detection report along with other parameters.

These detections can be sent to PEX directly or through RCN with different identifier to test performance of PEX algorithms.

Detections Generated for Radar Data Processor

The detections corresponding to the true target position is sent to radar data processor for estimation of true position of the target in presence of clutter [5].

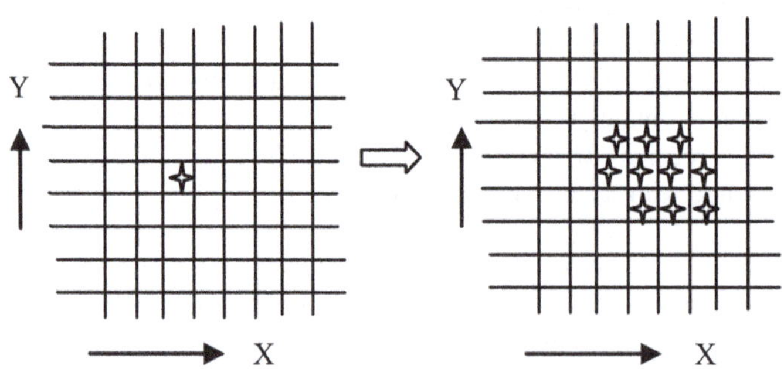

Fig. 3 Spreading of point target in to multiple detections

4 Methodology Used for Implementation

The simulator is designed and developed in c# .net on windows platform using six distinct models as shown in Fig. 4, which allows the user to define the target trajectories, radar parameters, clutter distribution, and interacts while generating the reports.

4.1 Area Definition Module

This module allows the user to create an operational area. Radar, multiple targets and clutter zones are created with respect to this area (Fig. 5).

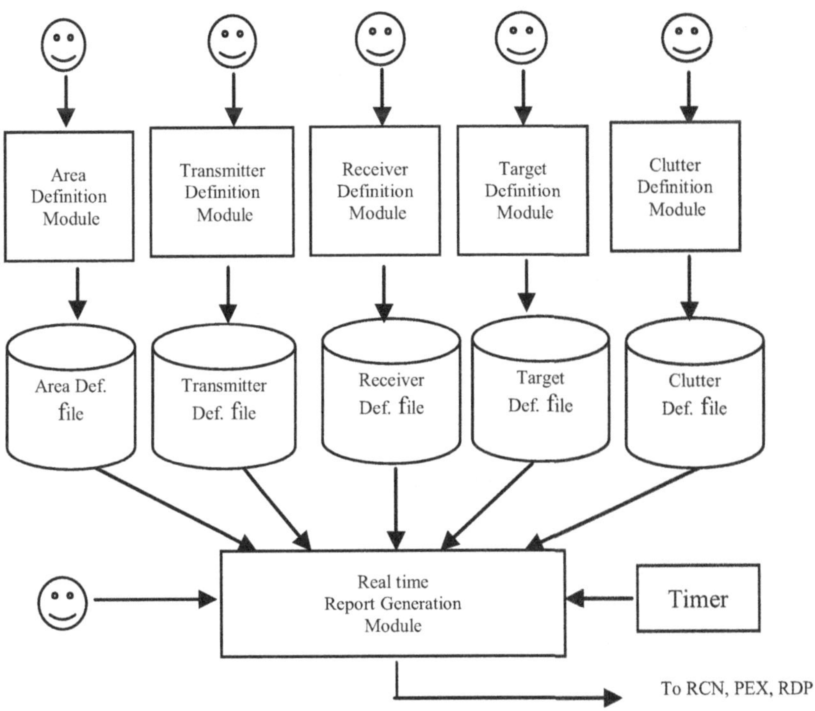

Fig. 4 Design of multi-target trajectory simulator

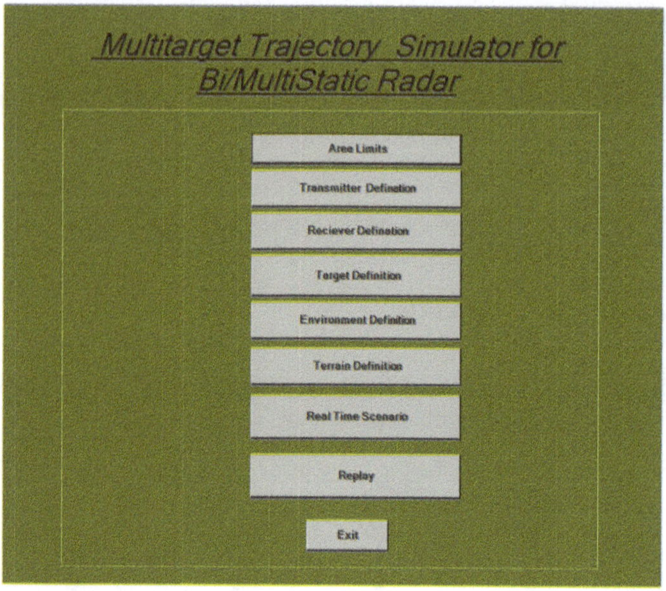

Fig. 5 GUI of multi-target trajectory simulator

4.2 *Target Definition Module*

The entire target profile is divided in to multiple segments based on the type of the maneuver [6]. This module allows the user to define the target trajectories such as type, location, speed, and heading and type of maneuver (constant velocity, acceleration, climb, and turn) with respect to the common grid reference as shown in Figs. 6 and 7 [7].

Modeling of Target Trajectory Segments

(a) Constant velocity segment:

Seg_end_pos = Seg_start_pos + $(v * (t))$
where $v \rightarrow$ Velocity of the target
$t \rightarrow$ Sampling period.

(b) Constant acceleration segment:

Seg_end_pos = Seg_start_pos + $v * t + \frac{1}{2} * a * t^2$ where $a \rightarrow$ Acceleration rate.

(c) Turn segment:

$X_N = X_{N-1} + F1 * \sin(H_{N-1} + F2)$
$Y_N = X_{N-1} + F1 * \cos(H_{N-1} + F2)$ where $F1 = 2 * v * |\sin(\mathrm{H} * \Delta\mathrm{t}/2)/\mathrm{H}|$
$F2 = \mathrm{H} * \Delta\mathrm{T}/2$
$\mathrm{H} = \mathrm{g} * (\sqrt{G^2} - 1)/v$ where G is the turn rate, $g = 9.81$ m/s^2 and N is the scan number.

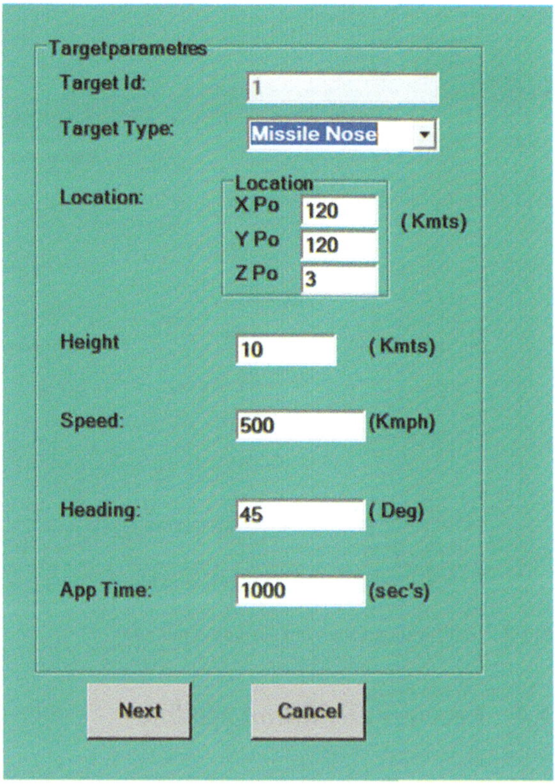

Fig. 6 GUI for target definition module

(d) Climb segment:

$V_Z = V * \sin(\alpha)$ where α will be the climb angle.

The turn rate and climb rates are selected by the user as parameters to segment generation, and it is based on the system requirements [8, 9].

4.3 Transmitter Definition Module

Transmitter definition module allows the user to define the radar transmitter characteristics to respect to the common grid reference. It allows the user to enter the location, maximum and minimum range, pulse repetition frequency, scan rate, etc.

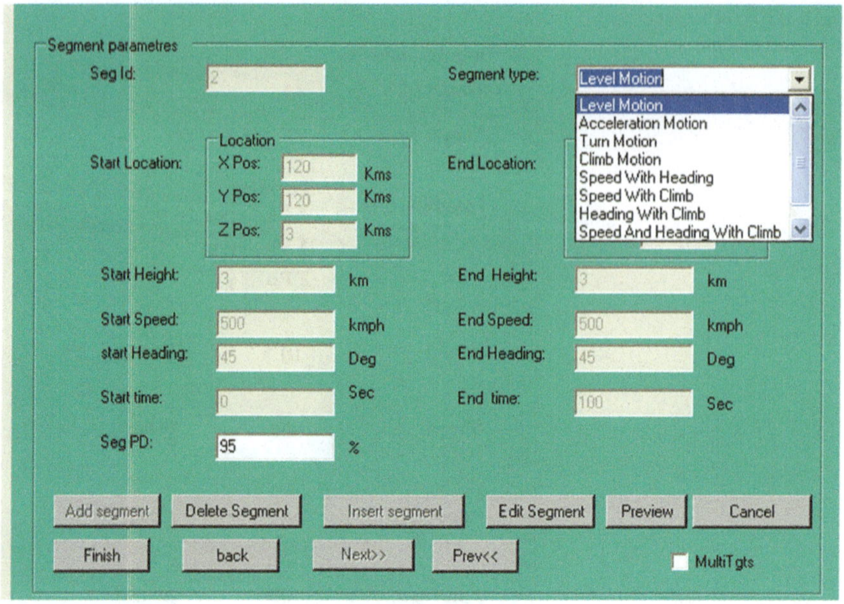

Fig. 7 GUI for target segment definition module

4.4 Receiver Definition Module

Receiver definition module allows the user to define the receiver characteristics to respect to the common grid reference. It allows the user to enter the location, maximum and minimum range, maximum number of beams, and corresponding beam widths.

4.5 Clutter Definition Module

Environmental definition module allows the user to define the clutter information with respect to the given operational area. It allows the user to enter number of clutter points to be distributed in the clutter zone, probability of detection, and type of distribution, i.e., normal or chi-square.

These offline-generated scenarios are stored in the files and accessed independently, while generating real-time reports.

4.6 Real-Time Report Generation Module

The real-time report generation module updates the target positions, transmitter antenna position in real-time independently. Whenever transmit and receive beam overlaps and finds the detection or clutter point in its cone of surveillance, it generates the plot which consists of 3D information range sum, receive angle, strength, and receive beam number, which is used to test data processor algorithms. The simulator spreads the detection across discrete cells depending on the type and height of target, which is used to test plot extractor algorithms [10].

4.7 Replay Module

The replay module replays the data which is logged by the radar in real time. This facility enables the debugging for the errors which encountered in real-time scenario.

5 Results

The target paths to be tested by the software modules are decided by the user based on which parameter of the radar is of interest. One of the scenario is demonstrated below.

Scenario:

Objective: Generation of detections to test target turn handling capacity by the bi/multistatic tracker

Target position: (72 km, 72 km, 3 km)

Number of segments: 3.

Segment 1:

Type of segment: Level

Speed: 250 kmph

Heading: 45°.

Segment 2:

Type of segment: Turn

Turn Rate: 1

Turn angle: 360°.

Segment 1:

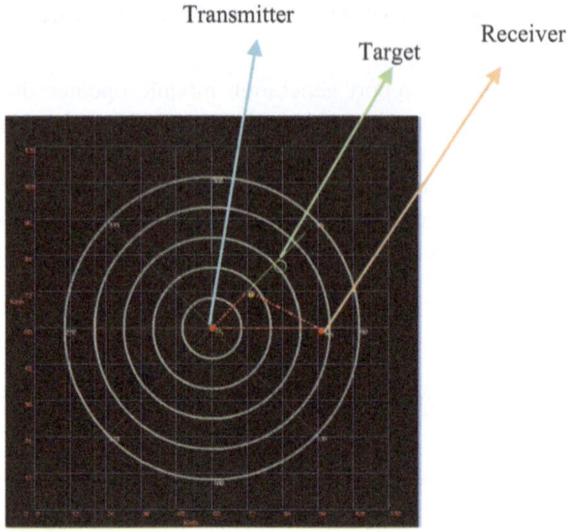

Fig. 8 Bistatic surveillance radar coverage

Type of segment: Level

Heading: 45°.

Figure 8 shows GUI of bi/multistatic radar-operating scenario. The real-time module-generated detection reports are plotted with respect to transmitter and receiver. Figure 9 shows range versus azimuth plot with respect to transmitter, and Fig. 10 shows range versus azimuth plot with respect to receiver. Figure 11 shows bi/multistatic range sum and receive angle. These detections are sent to test the performance of radar software modules.

6 Conclusion

Radar technology is used in measuring position of the moving objects by radiating electromagnetic energy from an antenna to propagate in space. The multi-target trajectory simulator plays a vital role in generating the radar detections which is used extensively in testing the performances of radar software modules. It simulates the complex target manures, radar transmitter antenna sweep, and receive beam pattern at data level which enables the feasibility to test the critical performance evaluation of the internal radar software modules as well as multi-sensor data fusion software which provides the enhanced coverage of surveillance.

The simulator reduces the time delay in order to wait for the readiness of actual hardware/firmware and able to test the scenarios which are practically expensive to test.

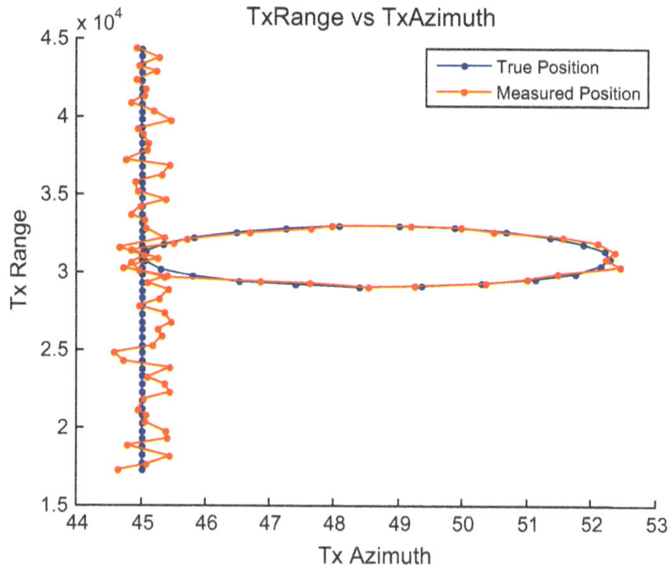

Fig. 9 Detections of simulated target profile (true and measured) with respect to transmitter

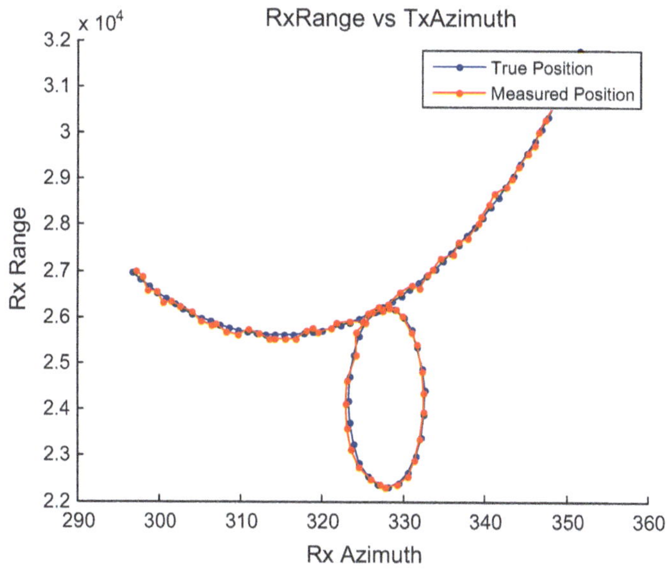

Fig. 10 Detections of simulated target profile (true and measured) with respect to receiver

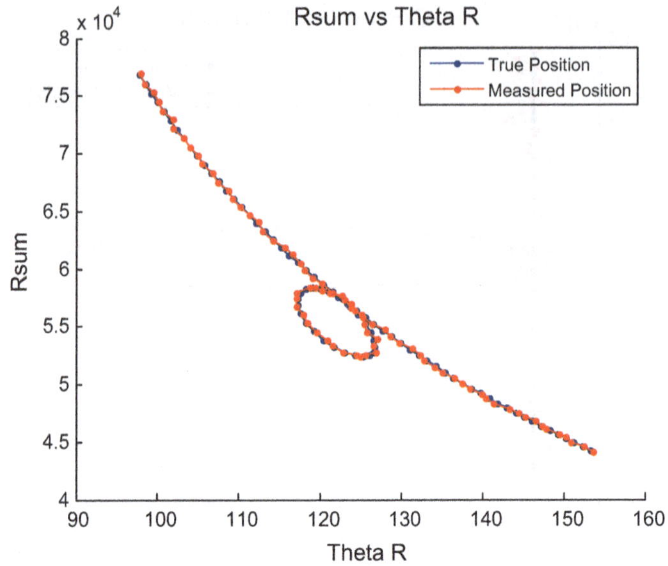

Fig. 11 Bistatic detections (Rsum versus receive angle)

References

1. Skolnik, M.: Introduction to Radar Systems, 2nd edn., McGraw Hill, New York
2. Willis, N.J.: Bistatic Radar, vol. 2. SciTech Publishing (2005)
3. Farina, A.: A Tracking Function in Bistatic and Multistatic Radar Systems. IEEE (1986)
4. Malanowski, M.: Algorithm for target tracking using passive radar. Int J Electron Telecommun (2012)
5. Bar-Shalom, Y., Li, X.R., Kirubarajan, T.: Estimation with Applications to Tracking and Navigation: Theory, Algorithms, and Software. Wiley, New York (2001)
6. Li, XR., Jilkov, V.P.: A survey of maneuvering target tracking. Part I: dynamic models. IEEE Trans. Aerosp. Electron. Syst. **39**, 1333–1364 (2003)
7. Singer, R.A.: Estimating optimal tracking filter performance for manned maneuvering targets. IEEE Trans. Aerosp. Electron. Syst. **AES-6**, 473–483 (1970)
8. Wang, T.C., Varshney, P.K.: A tracking algorithm for maneuvering targets. IEEE Trans. Aerosp. Electron. Syst. **29**, 910–924 (1993)
9. Zhou, H., Kumar, K.S.P.: A current statistical model and adaptive algorithm for estimating maneuvering targets. AIAA J. Guidance **7**, 596–602 (1984)
10. Mehrotra, K., Mahapatra, P.R.: A jerk model for tracking highly maneuvering targets. IEEE Trans. Aerosp. Electron. Syst. **33** (1997)
11. Alaee, M.B., Alaee, M., Amiri, R., Shokooh, R.K., Mahdlo, A.: Modeling and simulation of trajectory in RADAR systems. In: Fourth Asia International Conference on Mathematical/Analytical Modelling and Computer Simulation (2010)

Proof of Authentication for Secure and Digitalization of Land Registry Using Blockchain Technology

K. Nandini and G. S. Girisha

Abstract The aim is to secure land records and eliminate the fraudster to create duplicate land records and sell to another party. By using blockchain technology, we can solve this problem, the current land registries are still it is paper-based, and performance of document verification process is slow and unreliable in terms of accuracy, security, and occasionally corrupt. In this paper, we propose an architecture to improve the performance of record management. Architecture is separated by application domain and core system so that it improves the performance. The consensus algorithms play an important role in the blockchain for creating new blocks and appending them to the blockchain. The proposed New Block generate algorithm for creating a new block provides an efficient way of storing data in blocks. Before placing the block in the blockchain network, the block is digitally signed by using cryptographic techniques. The second algorithm is proof of authentication consensus protocol; it is used for the block validation process.

Keywords Blockchain · Cryptographic technique · Proof of authentication · New block generate algorithm · Architecture of land record management

1 Introduction

The evolution of blockchain technology, we have a chance to fix many problems such as the financial sector, health care, agriculture, cyber security, supply chain management, forecasting, and more. This revolutionary technology can also transform to land registrations. Blockchain makes things efficient, transparent, decentralized, democratic, and secure.

K. Nandini (✉) · G. S. Girisha
Department of CSE, Dayananda Sagar University, Bengaluru 56068, India
e-mail: nandini-cse@dsu.edu.in

G. S. Girisha
e-mail: girisha-cse@dsu.edu.in

V. Bhateja et al. (eds.), *Computer Communication, Networking and IoT*, Lecture Notes in Networks and Systems 197, https://doi.org/10.1007/978-981-16-0980-0_27

Fig. 1 Structure of blockchain

1.1 Blockchain

A blockchain consists of a set of blocks, each block is containing the data, hash, and hash of the previous block as shown in Fig. 1. Those blocks store auditable data, which is not editable (immutable) access. Its own hash value is generated through a complex algorithm on cryptographic and unique value [6].

1.2 Current Process of Land Record Management System (LRMS)

In many countries, the land registries having fraud issues and transparency are still in existence. Property or land registration process in India involves making stamp duty and paying the requisite registration fee for sale. The process of registration depends upon whether the property is directly purchasing from a developer or secondary sale. There are many departments involved in land registration such as 1. Survey Department. 2. Stamps and registry Department. 3. Revenue Department.

Figure 2 shows in the current land registration process, and the owner has to produce property documents for verification. Stamp and registry department will have to calculate the property value, and then we have to buy stamp paper based on the property value and stamp duty [7]. Once stamp paper is ready, we have to pay for stamp duty and registration charges, and then we have to wait for the sub-registrar's

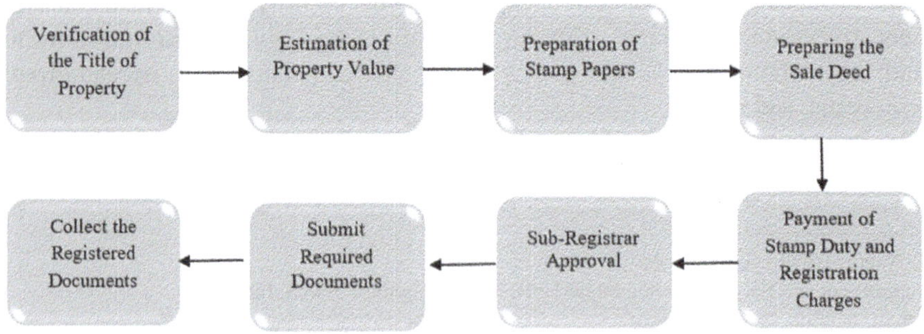

Fig. 2 Land registration process in India

approval. It takes more time to visit the offices, waiting for a civil servant to search records, and the fraudster will create duplicate records of land which is not secure.

2 Related Work

An anonymous developer going by the name of Satoshi Nakamoto released the white paper on "Bitcoin: A peer-to-peer electronic cash system in 2008" [1], and this technology made its public debut. The blockchain has made a lot of improvements in order to overcome some limitations and the most important among the development of permissioned and private blockchain.

2.1 History of Blockchain

The first modern and successful application on blockchain technology is called bitcoin [1]. It is nothing but a form of digital cryptocurrency; for implementing this bitcoin application, they used proof of work consensus protocol to process and validate transactions on the network. After bitcoin, the next version of blockchain is Ethereum. Ethereum is the second largest cryptocurrency platform [3]. Vitalik Buterin was an inventor of Ethereum in late 2013. It is a decentralized open-source blockchain feature that runs smart contract functionality and application. The smart contract, it introduced in Ethereum. The evolution of blockchain 3.0 has provided some of the blockchain platforms such as Hyperledger Fabric (Permissioned Blockchain) [6], Cardano [12], Zilliqa [11], etc.

2.2 Land Registry

See Table 1.

2.3 Consensus Algorithms

Blockchain is a distributed decentralized network that provides privacy, immutability, security, and transparency. There is no central authority to verify and validate the transactions, and every transaction should be considered to be completely verified and secure. This is possible because of the consensus protocol which is present in the core part of blockchain framework (network).

The main aim of a consensus algorithm is to achieve reliability in the blockchain network and also to determine agreement about the entire work. The blockchain

Table 1 Related work on land registry

Author/year	Title	Objective	Pros	Cons
Gunda Abhishek. Dr. Sandeep Shukla (2019)	Property registration and land record management via blockchain [6]	Hyper ledger Open source	Designing. Performance evolution	Many tools have released new and upgraded Not storing the entire data
Khan, Arshee Pandit, Parvati Sampath A. K. (2019)	Land registry using blockchain [7]	Proof of concept	Enhance data security and ensure authenticity	Less performance evolution
Yuan Zhang, Chunxiang Xu, Hongwei Li, Haomiao Yang (2019)	Chronos: secure and accurate time-stamping scheme for digital files via blockchain [8]	Time-stamping	Secure accurate.	Lack in performance
Ruiguo Yu, Jianrong Wang, Tianyi Xu (November 8, 2017)	Authentication with blockchain algorithm and text encryption protocol in calculation of social network [9]	RSA algorithm. MD5 algorithm	User authentication. Text encryption	To improve the security and performance planning to change structure of blockchain and frame work
Qi Tao, Xiaohui Cui, Xiaofang Huang (April 29, 2019)	Food safety supervision system based on hierarchical multi-domain blockchain network [10]	HMDBC network architecture. Two-level verification Mechanism. Improve the overall performance	Security. Smart contract	It can tolerate no more than 33% broken nodes

framework consists of three main parts such as core system, application domain, and smart contracts. The consensus protocol works on the core system side, and it will make sure that is agreed upon by all other nodes whenever a new node is added to the blockchain. There are some of the consensus protocols in Table 2 [17].

2.4 Background of Blockchain Frameworks

The assortment of a framework is complicated and even not everything that is open source, but many open-source frameworks are available to implement a blockchain. Another term for a blockchain is distributed ledger, and it shares the database among untrusted participants. There are many frameworks available Table 3.

Table 2 Comparison of consensus protocol

Consensus protocol	Network settings	Description
Proof of work (PoW)	Public	A first blockchain consensus algorithm, implementing bitcoin used PoW consensus algorithm [1], lot of computational power and PoW is slow
Proof of Stake (PoS)	Public	The most common alternative to PoW. Ethereum has moved from PoW to PoS algorithm. Validators will validate blocks by placing a bet on it then added to the blockchain [6]
Proof of burn (PoB)	Public	It is alternative method of PoW and PoS. It is used in cryptocurrency and very low energy consumptions, and this consensus algorithm no need to invest in powerful hardware. So in this protocol also wastes resources
Proof of capacity (PoC)	Public	In the PoC consensus algorithm validator not inverting expensive hardware, in this protocol hard drive space is providing for validators. It is very similar to PoW [6], PoW consumes more than 500 time of PoC energy-efficient. The storages space is cheap and accessible in PoC
Proof of elapsed time (PoET)	Private	Sawtooth uses PoET, PoET is one of the fairest consensus mechanism algorithm that prevents high energy consumption, keeps the process more efficient and high resource utilization. PoET mainly used in a permission blockchain network, each participating in the network is required and has finished waiting for the given set of time period and then to create a new block
Byzantine fault tolerance (BFT)	Private	Byzantine Fault Tolerance (BFT): This is one consensus algorithm Byzantine fault tolerance to achieve in network even some of the nodes are failed or respond with incorrect information [6]. The failure modes are very less in this algorithms, and it implies no assumptions and no restriction about behavior of a node

Table 3 Comparison of blockchain framework

	Type of blockchain	Smart contract execution	Smart contract language	Data model	Consensus	Application
Eris	Private blockchain	EVM	Solidity, Serpent, LLL	Account-based	Pow	General application
Bitcoin	Public blockchain	Native	Golang, C++	Transaction-based	PoW	Cryptocurrency
NXT	Public blockchain	Built-in	Scripting language	Transaction-based	PoW	Cryptocurrency
Litecoin	Public blockchain	Native	Golang, C-H-	Transaction-based	PoW	Cryptocurrency
Ethereum [3, 14]	Public blockchain	EVM	Solidity, Serpent, LLL	Account-based	Pow	General application
Multichain [4]	Private blockchain	Native	C++	Transaction-based	Trusted validators (RR)	Digital assets
Quorum [16]	Private blockchain	EVM	Golang	Account-based	Raft	General application
Hydra Chain [15]		EVM	Solidity, Serpent, LLL	Account-based	Trusted validators	General application
Corda [12]	Private blockchain	JVM	Kotlin, Java	Transaction-based	Raft	Digital asset
Hyperledger Fabric v0.6.0 [13]	Private blockchain	Dockers	Golang, Java	Key-value	PBFT	General application
Hyperledger V1.0.0 [13]	Private blockchain	Dockers	Golang, Java	Key-value	Ordering service (Kafka)	General application
Hyperledger Sawtooth v1.0.5 [13]	Private blockchain	Native	Python	Key-value	Proof of elapsed time	General application
Hyperledger Besu v2.0 [13]	Consortium or federated blockchain	EVM	Java	Key-value	POW, POA, IBFT, IBFT2.0	Enterprise applications

Hyperledger is a global collaboration, open source, and it is software for developers to build the blockchain platforms and frameworks [13]. Hyperledger is hosted by the Linux foundation, it can be used to build transactional applications, it can establish accountability, trust, and transparency at their core system, and blockchain is a distributed ledger forged by consensus algorithms, combined with smart contracts abstraction in application level [6].

3 Methodology

3.1 Architecture of Land Registry

In the proposed architecture, there is a separation between the application layer and core of the system as shown in Fig. 3, so that we can control data in trust, efficiency, and performance problems [10]. The steps involved for the land registry process are shown below.

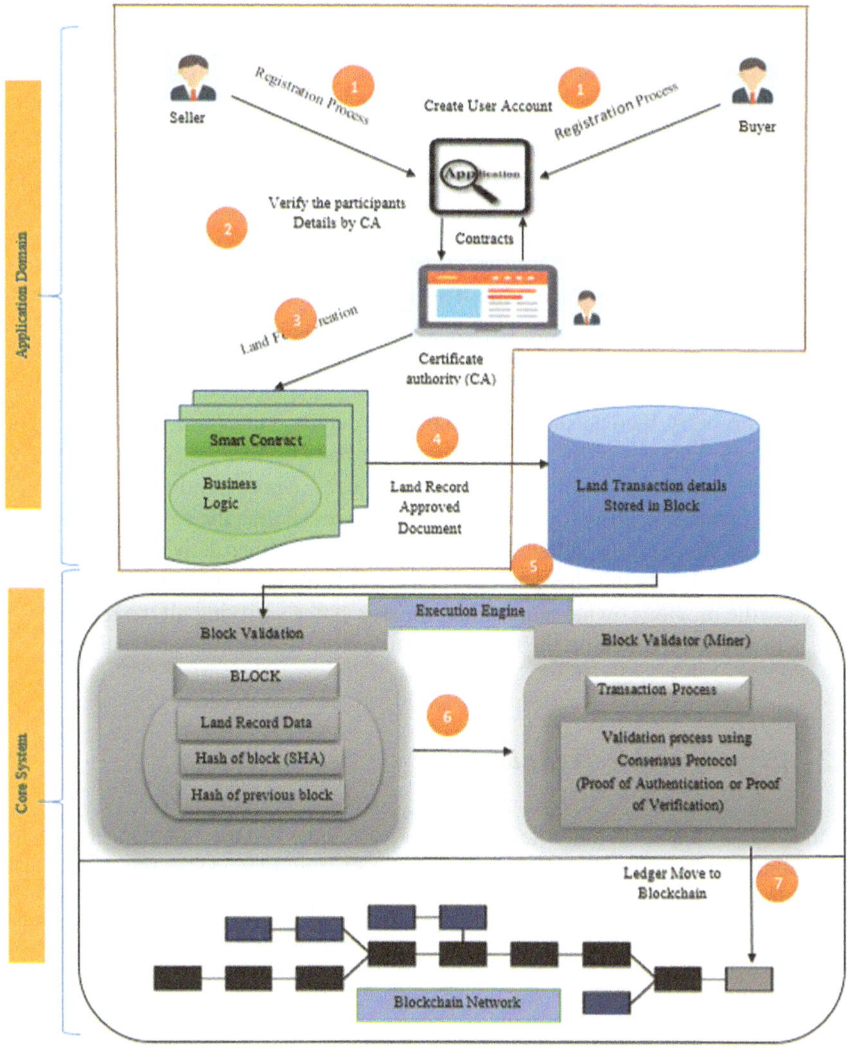

Fig. 3 Architecture of land registry in blockchain network

Step 1: The users who either want to sell or buy properties have to register with the blockchain network with all the documentation. Then the user will get private–public cryptographic keys.

Step 2: Certified authority (CA) will verify all details of the user, the user creates an user transaction; by using these keys, they can do the transaction. CA will verify all the details of seller properties documents, and then it will generate the land form.

Step 3: Transfer documents to the smart contract with business logic, rules, and conditions.

Step 4: After the step 3, land record got approved then it is transferred to block, transaction per block it depends on the size of block and the nature of the network.

Step 5: The new transaction created by the user is added to the smart contract which runs with business logic.

Step 6: The transactions are stored in a block, and all the blocks are added to the blockchain only after validation. Blockchain validation is a process of finding the hash value to the block, and that hash value should be collision-free.

Step 7: The network nodes consensus algorithm validates the new block, containing the transaction details and hash value. This new block is broadcasted and propagated into the other nodes, which are connected in a blockchain (peer-to-peer) network configuration.

Once the seller property's details are uploaded to the blockchain. If a seller wants to sell the property, then he/she has to upload the details. Then it is available to all the users who have signed up as a buyer. The buyer has to select the property from the listed property, then the buyer sends a request to access a property details to the seller. Sellers receive notification for property access requests, then the seller can look into the buyer profile, and the seller will have two options either deny or accept. If the seller accepts, then both seller and buyer can decide the cost of property, even buyer can view the previous ownership of the property. After verifying all the records, the transactions are recorded on the blockchain to ensure traceability and authenticity.

3.2 Cryptographic Techniques

Both symmetric and asymmetric encryption techniques are used in algorithm 1 and algorithm 2 to provide more security to the property record management. Symmetric encryption is 1000 times faster than asymmetric encryption. Asymmetric encryption schemes use different keys for encryption of plaintext and decryption of cipher text. Symmetric is also called private key encryption, using the same key for encryption and decryption [2].

AES Encryption Algorithm: A symmetric key algorithm is used for encryption, i.e., advanced encryption standard (AES) is fast and secure for encryption algorithms, and it is used for encrypting the data for blockchain. This algorithm provides for a variety of block and key sizes. Symmetric key is the K_{sym} given in the algorithm below, using this key we encrypt the plaintext to cipher text.

MD5 (Message Digest algorithm 5) Hashing Algorithm: The hash function takes a large file, has an input, and produces a fixed length of output which is called a message digest. Hash function provide a output of the hash key was high probability and different message digest. This property is useful for verification of digital signatures and message authentication [9].

3.2.1 Block Validation

Each user has a public–private key pair, and it is used for signing/verifying blocks in the network [9]. In below algorithm, we use symmetric key (K_{sym}) to encrypt and decrypt the land data file, asymmetric encryption seller will have one key pair ($S_{pub,}$ S_{priv}), and buyer will have one key pair ($B_{pub,}$ B_{priv}) used for signing and verifying. In a validation step, add $Encrypt_{data}$, H_p to $New_{block.}$ New Block Generate Algorithm 1.

Algorithm 1 New-Block Generate

```
function BLOCK-GENERATE(Data_file)
    if user choose block-chain then
        Generate a Asymmetric public-private key pair such has
        Seller(S_pub, S_priv)  Buyer(B_pub, B_priv)
        Encrypt_data ← Encrypt_sym(Datafile, K_sym)
        H_s ← Hash_fuction(Encrypt_data)
        New_block ← Create a new block then add(Encrypt_data, H_s)
        Signature_block ← Create a digital Signature for New_block using S_priv
        Exchange the Seller public key S_pub to Buyer Using
        Diffie-Hellman Key exchange
        Put Digitally signed block Signature_block in to Blockchain (P2P)
        Network.
    end if
end function
```

Digital Signature: Digital signature is a cryptographic value that is calculated from signer secret key and data [8]. Use of digital signature is for authentication purposes and each user having a public–private key, the key pair used for signing/verifying. In the algorithm, seller will have one key pair (S_{pub}, S_{priv}) and buyer will have one key pair (B_{pub}, B_{priv}). By symmetric key, K_{sym} encrypts the data file, i.e., plaintext, after encryption will get the cipher text is $Encrypt_{data}$. Signer feeds the data file into a hash function, then the hash function will generate the hash value $H_{s.}$ To generate digital signature, we used signer private key S_{priv} is used for a signature key and hash value H_s of plaintext and encrypted data $Encrypt_{data}$ is then fed to the signature algorithm.

3.3 Proof of Authentication Algorithm

During the authentication process, extract the hash value H_s of the seller and $Encrypt_{data}$ from the digital signature block. Generate the hash value H_r of the received data $Encrypt_{data}$ by the same hash algorithm. If the hash value does not match seller and buyer, data has been changed during the transmission process. Proof of Authentication Algorithm 2.

Algorithm 2 Proof of Authentication

```
Input:(Signature_block, S_pub, K_sym)
function AUTHENTICATION(Signature_block, S_pub)
    Using Public key S_pub of seller, extract H_s and Encrypt_data form Signature_block.
    H_r ← calculate hash of the received encrypted data file Encrypt_data
    if H_s == H_r then
        return "Authentication Success return 1"
    else
        return "Authentication failure return 0"
    end if
If hash is match then buyer can decrypts the Encrypt_data using symmetric key K_sym
    data_file ←Decrypt_sym(Encrypt_sym, K_sym)
Buyer will receive all the information about seller then buyer can communicate with seller for decision to buy a land or not.
end function
```

4 Conclusion

This paper provides a new architecture for the blockchain network to improve performance, accuracy, and trust to maintain a record in a network. First algorithm is a block validation, and the structure of the block provides an efficient way of storing the data. Second algorithm is a blockchain-based authentication consensus protocol to keep the user in close proximity and encrypt the relationship with the hash function for keeping property documents in secure. In the algorithm, we used cryptographic techniques such as AES for the encryption scheme. Diffie–Hellman key exchange algorithm used for sharing public keys in blockchain networks.

In future work, we need to select the framework which is suitable for my architecture and algorithms. In this paper, we took an initial look for how we can solve property registration issues using blockchain networks. In the future, we will implement algorithms in an efficient manner to solve the issues of land registration and it provides a real-world security system.

References

1. Nakamoto, S.: Bitcoin: A Peer-to-Peer Electronic Cash System. http://bitcoin.org/bitcoin.pdf (2008)
2. Kosba, A., Miller, A., Shi, E., Wen, Z., Papamanthou, C.: Hawk: the blockchain model of cryptography and privacy-preserving smart contracts. In: 2016 IEEE Symposium on Security and Privacy (SP), San Jose, CA, USA, pp. 839–858 (2016)
3. Khan, R., Ansari, S.J., Sachdeva, S.: Blockchain based land registry system using Ethereum Blockchain
4. Guo, S., Hu, X., Guo, S., Qiu, X., Qi, F.: Blockchain meets edge computing: A distributed and trusted authentication system. IEEE Trans. Industr. Inform. **16**, 1972–1983 (2020)
5. Zhang, S., Lee, J.-H.: Analysis of the main consensus protocols of blockchain. ICT Expr. **6**, 93–97 (2019)
6. Abhishek, G., Shukla, S.: Property Registration and Land Record Management via Blockchain (2019)
7. Khan, A.P., Parvati Sampath, A.K.: Land Registry Using BlockChain
8. Zhang, Y., Xu, C., Li, H., Yang, H.: Chronos: Secure and Accurate Time-Stamping Scheme for Digital Files via Blockchain (2019)
9. Yu, R., Wang, J., Xu, T.: Authentication with Block-Chain Algorithm and Text Encryption Protocol in Calculation of Social Network. (2017)
10. Tao, Q., Cui, X., Huang, X.A.: Food Safety Supervision System Based on Hierarchical Multi-Domain Blockchain Network (2019)
11. The Zilliqa Team: Zilliqa: A Secure, Scalable Blockchain Platform. https://docs.zilliqa.com/positionpaper.pdf (2018)
12. Cardano. https://www.cardano.org/en/home/
13. Kusters, R., Rausch, D., Simon, M.: Accountability in a Permissioned Blockchain: Formal Analysis of Hyperledger Fabric (Full Version). pp. 386–419
14. Ethereum white paper. https://github.com/ethereum/wiki/wiki/white-paper
15. Chain. http://www.chain.com
16. Quorum. http://www.jpmorgan.com/global/Quorum
17. Salimitari, M., Chatterjee, M.: An Overview of Blockchain and Consensus Protocols for IoT Networks. burstwiki.org (2018)

Survey on Cross-Layered Wake-up Receiver-Enabled WSN

B. G. Padmageetha and Mallanagouda Patil

Abstract Energy efficiency in data communication networks is the challenging issue. To execute the coordination between various layers, wake-up radio (WuR)based cross layer design is considered as the most energy-efficient duty-cycled protocol which is available in wireless sensor network (WSNs). It broadcasts the wake-up call (WuC) as wake-up signal (WuS) sensed by designated node that triggers its radio to sleep state on various topology of network. Such wake-up operations eliminate idle listening and overhearing burden for energy consumption in duty-cycled WSNs. For current hardware advancements, WuRs inspire us to assess them as a promising answer for WSNs. This paper compares various WuR-based cross-layered protocols and surveys them on design and attributes.

Keywords WuR · WuC · WuS · MAC · WSN · Cross layer

1 Introduction

WSNs mainly function to sense and monitor the location and directing the collected data to the base station (BS). The topology consists of independent nodes with sensors. The main factor is the attribute that consume excess energy is the bottleneck in WSNs. So, it has become a significant issue in research area. To improve the lifetime of the network, sleep scheduling, on-demand or event-triggered operation of the network are to be considered. This approach is viable with the WuR technology.

The first sensor network was a wired network. The problem of energy wastage was not in existence. But in recent past, WSNs have grown up as one of the leading developments in technology [2]. Each layer's role is important as it should be synchronized or merged and take control on sleep schedule to conserve energy. As per the standard

B. G. Padmageetha (✉)
Department of Computer Science, Surana College, Bengaluru 560004, India
e-mail: padma.cs@suranacollege.edu.in

M. Patil
Department of CSE, Dayananda Sagar University, Bengaluru 560068, India
e-mail: patil-cse@dsu.edu.in

V. Bhateja et al. (eds.), *Computer Communication, Networking and IoT*, Lecture Notes in Networks and Systems 197, https://doi.org/10.1007/978-981-16-0980-0_28

OSI layer model, there is a standard layer that has boundaries. Each layer is having a specific functionality. The main intention is to conserve energy at all layers considering delays. Thus requires Cross layer design across all layers. A single layer may not be efficient for entire network. Goal is to have suitable architecture, designed for better adaptability and optimal value. Several parameters are extracted from two or more layers to achieve optimization.

This survey paper shows the various proposed design in cross layer approach to enhance the performance, address the benefits and the challenges of utilizing the WuR receivers in the WSN applications. The best solution suggested to the problem of OSI model [3] is cross layer design. It is one of the challenging issues in research. So, rigorous study has to be made on the parameters that contribute energy loss. Some of the parameters to improve quality of service (QOS) are energy efficiency, throughput, delay and scalability.

In cross layer design approach, the traditional layer model (OSI) is analyzed, broken down or merged or initiate a new layer. Communication of data between non-contiguous layers results in change of their functionalities as the data are shared. Cross layer design makes the network more adaptive by increasing the energy efficiency in wireless sensor network that is usually wasted by idle listening, overhearing and collision. In this paper, Sect. 2 describes about WuR, protocols, its pros and cons and a comparison study. Section 3 depicts the analysis of J. M. Lebreton with DORA protocol. Section 4 depicts the conclusion and future scope.

2 Related Work

WuR can be implemented in two ways. (1) Hardware—as wake-up circuit and (2) software—with protocols that use WuRs. Various literatures [4] on wake-up receivers are available in market that are applicable to both the categories mentioned above. Also, working principles are discussed in this paper with classification. Wake-up circuit works on battery build internally with WuR monitor with WuS. The medium for the signal can be shared or separate channel. It can adopt frequency division as CSMA/CD. Complexity of sensor node cost increases with a separate channel [2] about 15% of cost of sensor node.

2.1 Duty Cycling Mechanism in WuR

Duty cycling mechanism [4] is used in WSNs. This mechanism consists of turning off the radio component and pausing its microcontroller (MCU). Timer is used to activate the node periodically. A new radio is enclosed with it called 'wake up radio' (WuR) and has wake-up transmitter and receivers named as WuTx and WuRx, respectively. WuS called wake-up signal/ interrupt is sent by the radio, activates the target and its neighboring nodes. WuRx tells micro control unit (MCU) to activate and start data communication. At the end of the communication, the radio is switched off. This is one of the ways to conserve energy. Further, WuR-based MAC protocols can be classified on cross layer approach as duty-cycled and non-duty cycled with transmitter/receiver/bidirectional-initiated protocols.

2.2 Classification of WuR-Based MAC Protocols

The efficient cross layer protocols will also have the described classifications. In this paper, nine protocols belonging to various categories of cross layer are taken and discussed about their proposed method of communication. A comparison study made between these protocols on their attributes that contribute to energy efficiency.

Sparse Topology and Energy Management-Tone (STEM–T) was proposed by Schurgers [5], and it is a range-based duty-cycled protocol.

Proposed Scheme: In this approach, destination address is not available in the busy signal tone sent by the sender. So, all the nodes present in the stipulated area or channel wake-up.

In this case, the target node never sends back an ACK packet to the initiator node compared to Stem-B that sends the acknowledgement (Fig. 1).

Advantages: More energy efficiency compared to Stem-B.

Disadvantages: Unnecessarily wakes all the adjacent nodes and thus results in higher power consumption.

Sparse Topology and Energy Management-Beacon (STEM-B) was proposed by Schurgers [5], and it is an identity-based duty-cycled protocol.

Proposed Scheme: A stream of beacon signal messages is sent by the sender to wake up the desired node, and radio is turned on. Eventually, an ACK is sent back. This packet contains the address of source and destination. Once the transmission is

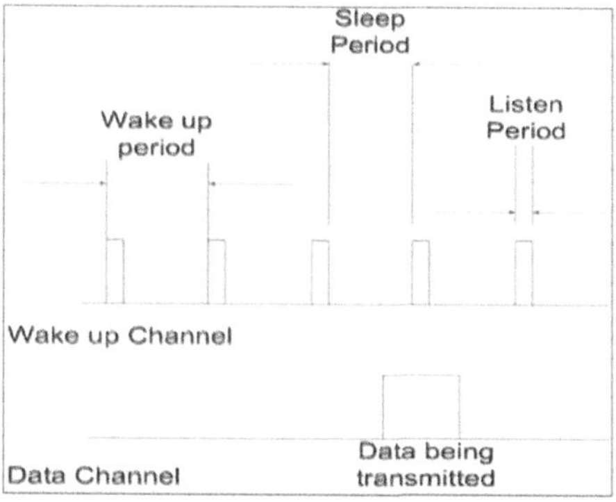

Fig. 1 Stem-T wake-up channel [13]

ended, the node turns off its radio. Later, the initiator node will periodically check for any other node that needs to communicate. The non-targeted nodes, if collision had occurred in the course of time, will come back to its original state as beacon message is not received in each stipulated time.

Advantages: Collision detection can be observed by signal indicator of the radio.

Disadvantages: Suffers from beacon collision.

Dual Channel-based Medium Access Control (DCMAC) was proposed by Zheng [6]. It is duty-cycled, transmitter-initiated and range-based protocol.

Proposed Scheme: In this scheme, the WuR is used to send busy signal transmission and main radio for data transmission. Two radio channels sense the network for a random amount of time with WuR. If the nodes are free, the same signal is used to wake them and become active within its single hop of range suggested. The neighboring nodes within the range turn on their central radio to activate. The single hop selects the next best node depending on the measures such as latency to avoid collision and latency that is calculated during the process.

Advantages: Reduction of the control overhead, through conflict avoid method collision and latency is avoided.

Disadvantages: All the nodes surrounding are awaked unnecessarily in a given range resulting in loss of energy.

Double Radio (DORA) was proposed by Lebreton [7]. It is non-duty-cycled, radio receiver-initiated, range-based and single hop communication.

Proposed Scheme: Initially, all nodes with main radio are in sleep mode. WuM is sent periodically to each node that has pre-mentioned target MAC address. The node wakes up and starts sending data to BS for further processing.

Advantages: Collisions are avoided. Communication between nodes are direct, so wastage of energy is avoided.

Disadvantages: Problem of overhearing persists.

Duty-Cycled Doubled Radio (DC-DORA) was proposed by Lebreton [7]. It is a duty-cycled double radio receiver-initiated, identity-based WuM.

Proposed Scheme: To overcome the problem of overhearing, DC-DORA uses duty-cycled mechanism WuR system. Here, for a certain period, the WuRx functionality is disabled as soon the node sends data to the base station. At this point, node will not receive any WuC as base station (BS) receives the data from node (Fig. 2).

Advantages: Overhearing is eliminated by sending wake-up signal to the destination.

Disadvantages: Since there is a long time gap between the arrival of corresponding WuCs, there are chances of power wastage.

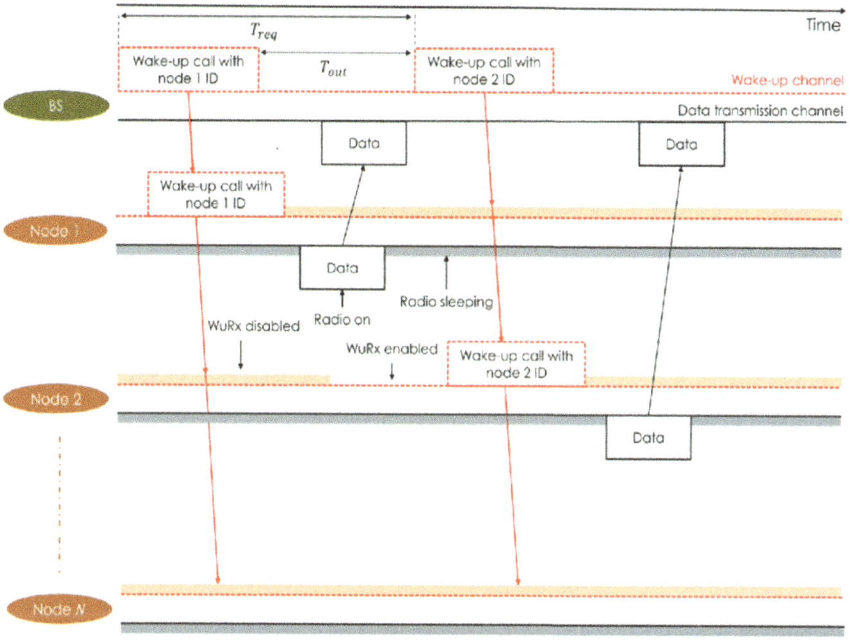

Fig. 2 Data Communication with DC-DORA Protocol [7]

Low Energy Self-organizing Protocol (LESOP) was proposed by Song and Hatzinakos [8]. It is non-duty-cycled, transmitter-initiated and identity-based WuM.

Proposed Scheme: It is used in target tracking in WSNs by estimating the target location. The RF radio and WuR are attached to the nodes. WuR broadcasts a signal to the nodes. As soon as the targeted node senses, starts its data exchange and broadcasts the busy tone from WuR. Now all the adjacent nodes perform sensing. These measurements like linear processing are sent to the new selected target node acting as a front runner. Other nodes get the profile of the target, and so the procedure repeats.

Advantages: Lower complexity.

Disadvantages: Undergoes delay.

Backoff-Enabled Wake-Up Radio (BOWUR) was proposed by Ghose and Li [9]. It is non-duty-cycled, transmitter-based and range-based WuR.

Proposed Scheme: At any point of time, the node can start sending the data. A backoff procedure is adopted before sending the WuCs. It works like CSMA/CA MAC protocol. Before sending the data, clear channel assessment (CCA) is made to scrutinize the network for busy/idle if found idle, and backoff time is calculated from fixed-size contention window (CW). Only the shortest backoff time will activate the main radio. Other nodes wait for the next round. An ACK is sent to the source by the receiver after the exchanges. Retransmission occurs in case ACK not received.

Advantages: High energy and throughput of network by backoff algorithm, Avoid collisions.

Disadvantages: Wastage of energy during transmission.

Correlator-WuR (Cor-WUR) was proposed by Oller [10]. It is non-duty-cycled, bidirectional and range-based WuR. It is operated at transmitter/receiver-initiated approach. It is also known by SCM-WuR.

Proposed Scheme: In the transmitter-initiated mode, a data transmission cycle of the transmitter-initiated WuR lasts from the instant of a WuC initiation to the instant when an acknowledgment (ACK) is received.

In the transmitter-initiated mode, a node which has a packet to transmit sends a WuC to the targeted node using its main radio. After receiving the WuC and decoding the address correctly, the targeted node switches on its main radio from the sleep mode for data communication. Main radio is switched off, goes to sleep mode when data exchanges are completed. However, both nodes keep their WuRx active and continuously listen to the channel. ACK is not sent in Cor-WuR. Data transmission occurs after receiving a WuC correctly. MAC protocol can adopt for instance CSMA/CA, if needed.

Advantages: Flexibility exists as it operates as WuRx or WuTx.

Disadvantages: Collisions occur (Table 1).

3 DORA Performance Analysis

The online version of the volume will be available in LNCS online. Members of institutes subscribing to the Lecture Notes in Computer Science series have access to all the PDFs of all the online publications. Non-subscribers can only read as far as the abstracts. If they try to go beyond this point, they are automatically asked, whether they would like to order the PDF, and are given instructions as to how to do so.

Please note that, if your email address is given in your paper, it will also be included in the meta data of the online version.

$$P(DetS) + P(DetF) = 1 \tag{1}$$

$$P(FAl) + P(CRj) = 1 \tag{2}$$

Table 1 Comparison study of cross-layered WuR-based MAC protocols

Protocol	Attributes for energy conservation	Cross layer support	Topology	Design adopted
Stem-B	Collision	Mac/application	Multi hop	Signal contains MAC address of both sender and receiver. The required node is fixed by the survey of the signal sent in as a stream
Stem-T	Energy conservation	Mac/application	Multi hop	Wake-up message
DC MAC	Collision, latency, control overhead	Mac/routing	Multi hop	Periodic listening/sleeping mechanism and data transmitting data by RTS/CTS/DATA/ACK method
DC-DORA	Collision, delay, overhearing	Mac/physical	Single hop	Uses fixed-size contention window + backoff procedure
LESOP	Lower complexity	Mac/application	Multi hop	Linear processing-target location estimation. Busy tones as a wake-up message
BOWUR	Throughput, energy efficiency	Mac/physical	Single hop	CCA + backoff procedure before transmission
DORA	Energy efficiency, collision	Mac/physical	Single hop	Wake-up message contains destination MAC address
PicoRadio	Energy efficiency	Mac/routing	Multi hop	Coloring algorithm, mobility adaptivity
Cor-WUR	Energy efficiency by swapping as WuRx or WuTx	Mac/physical	Single hop/ multi hop	Communication is on-demand, asynchronous with non-rendezvous

3.1 Energy Repartition of the WuRx Versus WuC Inter Arrival Time Using DORA Protocol and DC-DORA Protocol

As per the analysis done by J. M. Lebreton, simulation with the presence of WuC, with inter-arrival time 50 ms, total energy consumption: with main radio was 91%, sleep mode 1% and decoding 8%. So, DORA consumes energy efficiently. Then,

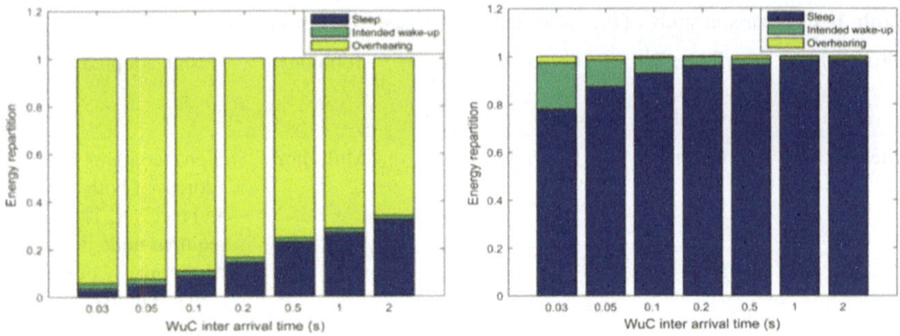

Fig. 3 Energy performance analysis with DORA and DC-DORA [7]

when energy repartition is evaluated with DC, DORA shows that overhearing is all most removed but still remains as the dominating source of energy consumption in DORA protocol (Fig. 3).

4 Conclusion

With the various other WuR methodologies, designing the WuR-based cross-layered MAC protocols is a very challenging and promising task. A great advancement in WSN in energy efficiency is regarded in removal of idle listening. Along, a new problem arises with the progress of wireless circuitry-IoT which is the recent trend. It has been tested that overhearing problem is almost solved in DC-DORA. Similarly, other attributes are to be tested with different protocol combination. In future works, idle listening attribute is implemented as a main parameter and studied along with protocols mentioned in this survey paper. However, reducing the time needed by a node to extract energy is still an open research trend.

References

1. Mendes, L.D.P., Rodrigues, J.J.P.C.: A survey on cross-layer solutions for wireless sensor networks. J. N Comput. Appl. **34**, 523–534 (2010)
2. Saini, R.K.: IQOS: a cross layer design approach for improve quality of service in wireless sensor networks. Int. J. Adv. Sci. Technol. **103**, 1–8 (2017)
3. Fatima, S.G., Fatima, S.K., Sadiq, M.: Wake-up radios for wireless sensor networks. Int. J. Res. Appl. Sci. Eng. Technol. (IJRASET) **7**(IV), 689–698 (2019)
4. Ghribia, M., Meddebb, A.: Survey and taxonomy of MAC, routing and cross layer protocols using wake-up radio. J. Netw. Comput. Appl. **149** (2019)
5. Schurgers, C., Tsiatsis, V., Srivastava, M.B.: STEM: Topology Management for Energy Efficient Sensor Networks. IEEE (2001), pp. 3-1099–3-1108

6. Zheng, G., Fu, J., Tang, S., Li, Y., Dong, Z. (2010). A dual channel-based energy efficient and low latency MAC protocol for WSNs. Second International Conference on Networks Security, Wireless Communications and Trusted Computing. IEEE (2010), pp. 466–469
7. Lebreton, J.M., Kandukuri, S., Murad, N., Lorion, R.: An energy-efficient duty-cycled wake-up radio protocol for avoiding overhearing in wireless sensor networks. Sci Res **8**, 176–190 (2016) (Published Online)
8. Song, L., Hatzinakos, D.: A cross-layer architecture of wireless sensor networks for target tracking. IEEE/ACM Trans. Netw. **15**(1), 145–158 (2007)
9. Ghose, D., Li, F.Y.: Enabling backoff for SCM wake-up radio: Protocol and modeling. Comput. Netw. **153, 22**, 132–144 (2019)
10. Oller, J., Demirkol, I., Casademont, J., Paradells, J., Gamm, G.U., Reindl, L.: Has time come to switch from duty-cycled MAC protocols to wake-up radio for wireless sensor networks. IEEE/ACM Trans. Netw. **24**(2), 674–687 (2016)
11. Saini, R.K., Ritika, S.V.: A survey on cross layer design implementation in wireless sensor network. Int. J. Adv. Inf. Sci. Technol. (IJAIST) **51**(51), 101–107 (2016)
12. Patil, M.: Survey on energy efficiency MAC protocols in wireless sensor network. Test Eng. Manag. 4221–4227, May - June (2020)
13. Bhat, P.: Wireless sensor networks. STEM: Sparse Topology and Energy Management. http://sensors-and-networks.blogspot.com/2011/10/stem-sparse-topology-and-energy.html. Accessed October 5, 2011 (2010)

An Integrated Approach for Building Footprint Extraction from Satellite Imagery

V. R. Jyothish, V. R. Bindu, and M. S. Greeshma

Abstract Satellite imaging has numerous applications in several fields such as urban planning, disaster detection and recovery, agriculture development, 3D GIS, and many more. One of the major challenges of satellite imaging is to extract the building footprints that are vital for disaster planning and recovery. Existing algorithms retain quantitative scores; however, they fail to preserve shape of building footprint. To address this problem, we propose here a building footprint extraction algorithm from satellite images; our objective is to formulate the building pixels in an efficient way, exploiting eccentricity parameter, in order to extract the exact shape of buildings. The proposed hybrid approach comprises of superpixel generation and parameterization such as axis ratio, eccentricity, color, and segment area which have been used to extract the true building footprint. The qualitative and quantitative evaluations demonstrate that the proposed algorithm can extract the building footprint more efficiently over other existing methods, in terms of correctness, completeness, and F-measure. In addition, we also evaluate the Dice similarity coefficients of extracted building footprints, which shows the influence of color parameter for effective extraction providing better results.

Keywords Satellite imaging · Building footprint · Superpixel

1 Introduction

The building footprint extraction from high-resolution satellite images is a challenging task in digital imaging. Due to the heterogeneity in the satellite image, machine visualization and image understanding are complex tasks. Satellite images have minute buildings in small patches of image, thus it is difficult to identify the footprint from images. Building footprint detection is applied for land cover detection, disaster management, urban planning, detecting change in satellite images, etc., with respect to the spectral characteristics of an image [1–4]. Since manual

V. R. Jyothish · V. R. Bindu (✉) · M. S. Greeshma
School of Computer Sciences, Mahatma Gandhi University, Kottayam, Kerala, India
e-mail: binduvr@mgu.ac.in

V. Bhateja et al. (eds.), *Computer Communication, Networking and IoT*, Lecture Notes in Networks and Systems 197, https://doi.org/10.1007/978-981-16-0980-0_29

extraction of building footprint is a time-consuming process and very hard to implement, there is a need for automatic building footprint extraction with better accuracy. Building footprint extraction is a demanding task with respect to size, color, structure, and surrounding objects. The contrast of the roof is high when compared with surrounding objects. Generally different filters are used to extract building footprint from satellite images. The proposed method generally concentrates on low-level image processing.

In this proposed method, satellite image is subjected to building footprint extraction. Here, a combined approach of the colorization technique, eccentricity parameter, and superpixels is used for the extraction. This method focuses on the shape of the superpixel. Eccentricity parameter is used to estimate the shape of superpixel. Usually eccentricity is applied in elliptical structured objects to determine shape of objects. Here, eccentricity factor is considered since superpixels are irregularly shaped areas. The literature survey of building footprint extraction algorithms is described in Sect. 2. The working principle of the proposed approach is deliberated in Sect. 3. Section 4 describes the implementation details and analysis, and the concluding remarks are given in Sect. 5.

2 Traditional Versus State-of-the-Art Approaches

Numerous algorithms have been proposed based on the building footprint detection. Literature discloses the numerous applications in satellite imagery for building footprint extraction. An automated building detection system from single grayscale satellite image is proposed by Saeedi et al. [5]. Initially border lines of satellite image are extracted. Then they are combined to generate quadrilateral rooftop hypothesis. This method mainly concentrates on rooftop gradient values for extracting building footprint.

The method proposed by Dahiya et al. [6] is an automatic building extraction from an object-oriented approach. By applying filters, region of interest is extracted as raster image, and it is transformed into vector image for processing. This technique generates accurate results from high-resolution satellite images, since buildings have greater reflective intensity with respect to other objects. It generates adverse effect when the image has unique intensity for both objects and building. Here, the significance of shape is highlighted. An automatic approach using shadow extraction is proposed by Gao et al. [7]. This technique mainly concentrates on the extraction of building shadows. For better result, classification technique is applied. Customized shifted regions are categorized into different classes. With the recognized objects from image building samples can be altered.

For footprint extraction, shadow of image can be shifted. For the purpose of building footprint extraction, radio metric, edge detection based, and object detection-based techniques are already proposed. But without the user involvement, extraction is not proper in some complicated objects. To solve this problem, Roopa et al. [8] proposed the approach for automatic extraction of building footprint. The proposed technique integrates morphological top-hat filter and K-means algorithm to extract buildings footprint. To overcome the difficulties of aforementioned approaches, proposed method is an attempt to detect building footprint at pixel level. Shape is the most convenient factor to extract building. The proposed method comprises of a combination of superpixel, eccentricity, and color of superpixel segments used for building footprint extraction. Intensity value is the basic parameter used for this complicated task. Due to high spatial resolution of satellite images, small objects are also to be extracted.

3 Principle of the Proposed Approach

A new integrated technique for building footprint extraction is proposed here to extract footprint from satellite images. The process involved in the proposed building footprint extraction algorithm is shown in Fig. 1. The approach is centered on superpixel-based segmentation and extraction using colorization. The use of eccentricity factor extracts the exact shape of building footprint.

3.1 Mathematical Background

Shape is the most important feature of an image, since it is a reflection of human perception. Flawless extraction of shape makes footprint detection effortless. For extraction of shape, different shape descriptor operators are used. In the proposed method, eccentricity is a parameter used to describe shape of the object. Eccentricity is most suitable for determining the shape of elliptically structured objects [9]. When the value of eccentricity is zero, then it is considered as a circle. In an elliptical structure, eccentricity is ranging between 0 and 1. Scope of proposed work

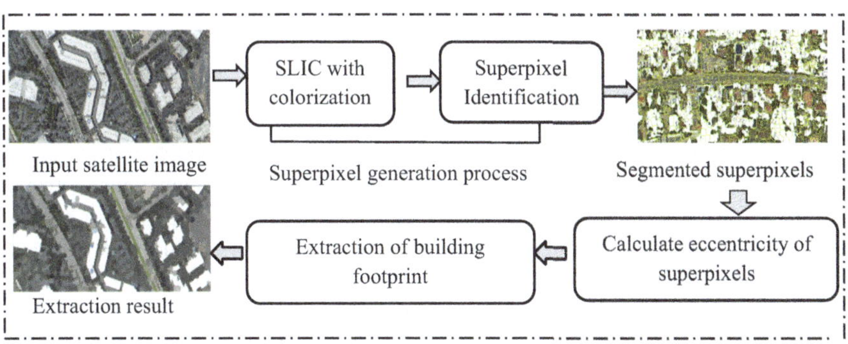

Fig. 1 Steps of proposed building footprint extraction algorithm

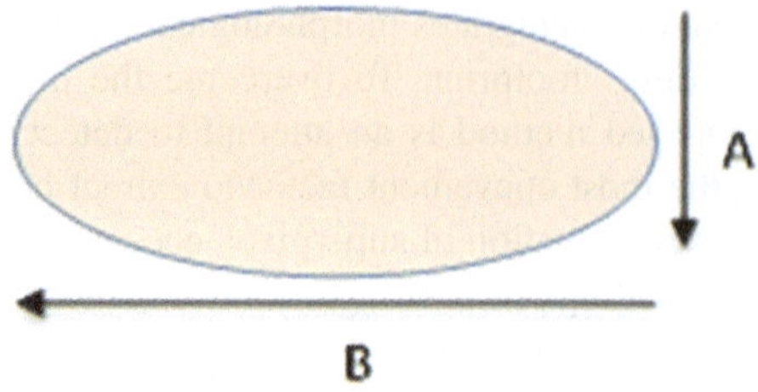

Fig. 2 Representation of eccentricity factor

lies in eccentricity values for elliptical objects. It focuses on shape descriptor parameter eccentricity and color features for achieving efficient and effective retrieval of footprint. The diagrammatic representation of eccentricity factor is shown in Fig. 2, where *A* represents the minor axis and *B* is the major axis of an ellipse.

The proposed algorithm generates superpixels from satellite image initially. Later, eccentricity values are calculated based on superpixel dimensions. For each superpixel, a maximum number of pixels along horizontal or vertical direction and minimum number along horizontal or vertical direction are computed. Maximum number is taken as major axis and minimum number is taken as minor axis. The eccentricity is computed as:

$$\text{Eccentricity} = \frac{\text{Minor Axis}}{\text{Major Axis}} \tag{1}$$

where ratio of minor axis to the major axis is called eccentricity of the region. Minor axis is a straight line segment joining the two points on the boundary least away from each other. Major axis is the straight line segment joining the two points on the boundary farthest away from each other [10]. Major axis and minor axis are perpendicular to each other. In superpixels, value of minor axis and major axis is represented by number of pixels. Maximum number of pixels along row or column in a particular superpixel is major axis. Minimum number of pixels along row or column in a particular superpixel is minor axis. Equation 2 is used to represent calculation of eccentricity parameter with respect to superpixels.

$$\text{Eccentricity} = \frac{\text{Minumum number of pixels in } x \text{ or } y \text{ direction}}{\text{Maximimum number of pixels in } x \text{ or } y \text{ direction}} \tag{2}$$

The value of eccentricity always ranges from 0 to 1. By analysis it is proved that, values greater than 0.5 always indicate superpixels or segments with footprint. In some cases it fails, at that situation, color of region is considered for perfect result. This technique is applied very easily in the superpixel-based image segmentation for detecting building footprint. Flawless segmentation leads to extraction of perfect building footprint. In the proposed work, input image is subjected for superpixel generation. From superpixels eccentricity factor is estimated for footprint extraction. Eccentricity factor reflects the shape of superpixels and some complicated superpixels, eccentricity factor is tending to 0. In such cases, color is considered for extraction. Combined approach of superpixel, eccentricity, and color can generate better results.

3.2 *Building Footprint Extraction Algorithm*

The proposed algorithm is evolved from the concept of superpixel, which represents the homogeneous information of collective pixels. For the purpose of superpixel generation, simple linear iterative clustering (SLIC) algorithm is used [11–14]. Exact segmentation of superpixels maps the small objects individually. Superpixels are most suited for satellite image processing, since they have minute buildings to extract. Eccentricity factor is calculated for superpixels and it ranges from 0 to 1. From experimental result it is clear that, superpixels with eccentricity value greater than 0.5 will accurately track the buildings. For some complex superpixels, eccentricity less than 0.5 cannot track proper shape, so in order to validate whether it is a building color feature is used. Figure 3 represents the building footprint algorithm. The building footprint extraction using the superpixel-based approach is a novel approach. This method concatenates eccentricity and color features with superpixels.

4 Implementation Details

We conducted various experiments to evaluate the effectiveness of the proposed building footprint extraction algorithm. The algorithm has been applied on images from the Google maps (satellite version) at Santa Ana, California [1, 15]. Ten test

```
ALGORITHM: Proposed building footprint extraction
algorithm
-------------------------------------------------
Read Input Image
Generate Superpixels
For each Superpixel
        Calculate Eccentricity Factor
          If eccentricity >=0.5
            If R>=190 And G>=190And B>=180
              Redefine pixels by white color
          Else If eccentricity <0.5
            If R>=190 And G>=190 And B>=180
            Redefine pixels by white color
        Else
        Redefine by black
End For
```

Fig. 3 Proposed algorithm for building footprint extraction

images are taken for the evaluation purpose from this area because the density of building in this city is high. The input images are of size 512 × 1024, which are usually high-dimensional images, in which building footprint covers 80% of each test image. All the experiments are executed on Anaconda (Python) with Intel(R) Core(TM) i7-7820EQ CPU with 3.0 GHz and 8 GB memory PC. For our experiments, superpixel-based segmentation has been used to extract various image segments, while colorization approach provides uniqueness of intensities, after that eccentricity of each superpixel is computed to extract building footprint. Maximum of 1200 superpixels are generated while processing. After determining eccentricity factor in Eq. (2) a value between 0 and 1, the threshold value for the extraction is set as 0.5. Eccentricity factor, which is higher than the threshold value, is chosen as the superpixels with building footprint. Following the influence of color parameter leads to extraction of better building footprint empirically. Here, the integrated features such as superpixels, color, and eccentricity are utilized for processing the high-resolution satellite image. The effectiveness of the extraction approach is evaluated with the correctness, completeness, F-measure [16], and Dice similarity measurements [11], which are evaluated in Sect. 4.1. The performance of the proposed algorithm has been evaluated by measuring the building footprint segments extraction through correctness, completeness, and F-measure. Prior to applying these metrics, extracted building footprint has been grouped as: |TP|, |FP| and |FN|, where |TP| is the true positive (extracted building are true building footprint), |FP| is false positive (extracted building are false building footprint), and |FN| is false negative (not extracted building are true building footprint). Three performance measurements have been applied after classification:

1. **Correctness**: It is the ratio of the number of correctly extracted building footprint to the total number of extracted building footprint in an image with respect to the ground truth image, as shown in Eq. 3.

$$\text{Correctness} = \frac{\text{TP}}{\text{TP} + \text{FP}} \tag{3}$$

2. **Completeness**: It is the ratio of the number of correctly extracted building footprint to the total number of building footprint in the ground truth, as shown in Eq. 4.

$$\text{Completeness} = \frac{\text{TP}}{\text{TP} + \text{FN}} \tag{4}$$

3. **F-measure**: It combines both correctness and completeness as a weighted average score, as shown in Eq. 5.

$$\text{F-measure} = 2 \times \frac{\text{Correctness} \times \text{Completeness}}{\text{Correctness} + \text{Completeness}} \tag{5}$$

For more comprehensive evaluation, Dice similarity coefficient (DSC) measuring the accuracy of the spatial overlap of ground truth image with the segmented image is calculated in Eq. 6.

$$D(A, B) = \frac{2|A \cap B|}{(|A| + |B|)} \tag{6}$$

where A is ground truth image and B is segmented image, DSC is in [0–1] range. The DSC of 1 corresponds to extraction of all building footprints successfully.

4.1 Qualitative and Quantitative Performance

(A) *Qualitative Evaluation*
We first illustrate the qualitative results in Fig. 4a1–a6 to explore the building footprint extraction effectively. First column represents the test patches from Santa Ana region, second column lists the ground truth images, and the extracted building footprint are shown in third column. From the visual results,

a1)
a2)
a3)
a4)
a5)
a6)

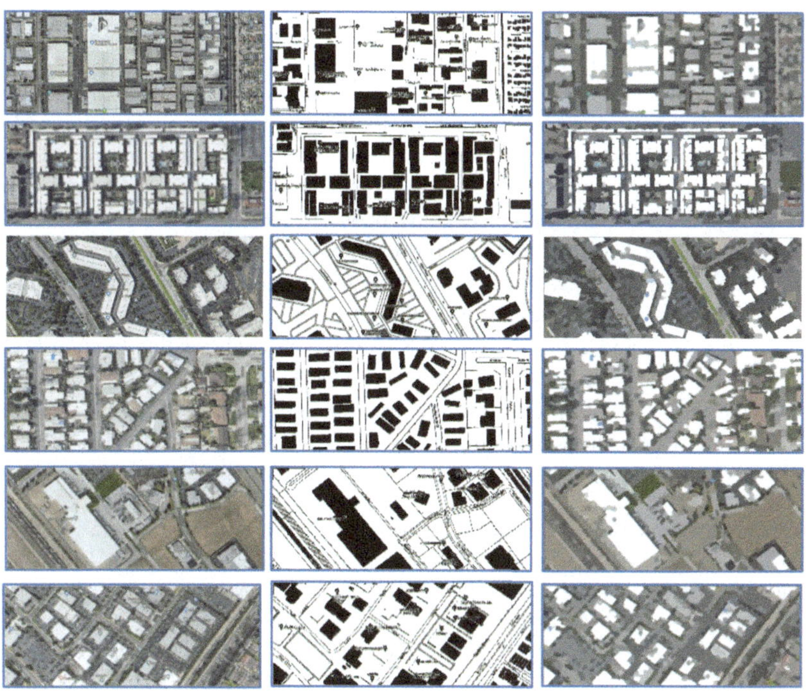

Fig. 4 (First column) Test images from Santa Ana (a1–a6). (Second column) Ground truth images. (Third column) Extracted outputs

b1)

b2)

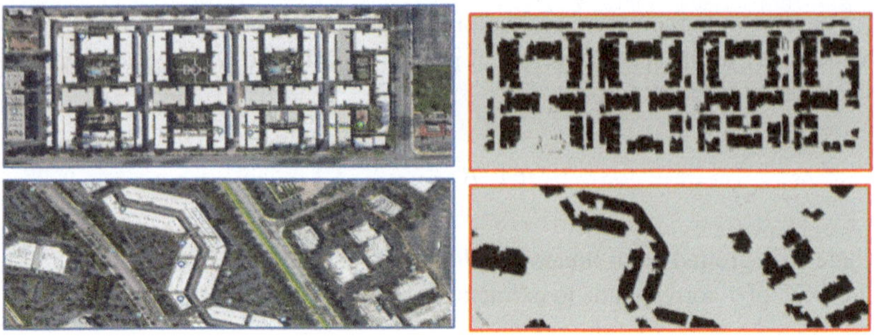

Fig. 5 (First column) Input images. (Second column) Extracted building footprint, where black squares indicate true positives detection (TPs)

extraction efficacy of the proposed algorithm is clearly evident with respect to ground truth image. We can observe that the extraction of exact shape of building is sometimes failed. Moreover, the eccentricity feature in the proposed algorithm can preserve the shape of footprint as shown in Fig. 4a2–a4. We further provide extracted footprint map as illustrated in Fig. 5b1, b2, and the second column displays the buildings which are truly extracted footprints (TPs). Our algorithm generates more exact building footprint for several applications.

(B) *Quantitative Evaluation*

For quantitative evaluation, the algorithm have been applied on a4 and a6 test patches from Fig. 4 to extract the footprints classified as TP, FP, and FN as illustrated in Table 1. From the table, it can be seen that the proposed method extracts the true building footprint precisely. For example, on the test patch a4, nine non-building objects have been extracted in building class, but at the same time, seven building footprints have been excluded. After classification, completeness, correctness, and F-measure have been computed, the high score of 0.89 (Completeness) confirms that the entire true building footprint has been extracted, while score of 0.86 (correctness) shows that the non-building footprint has been extracted falsely. Then the overall performance score (F-measure) of building footprint extraction achieves (0.87) higher quantitative value which shows the competency of the proposed algorithm. In Table 2, we

Table 1 Visual comparison of proposed algorithm with TP, FP, and FN

Test patch	Total building footprint in the image	Extracted true building footprint (TP)	Not extracted building footprint (FN)	Non-building objects extracted (FP)	Correctness	Completeness	F-measure
a4	65	58	7	9	0.86	0.89	0.87
a6	35	29	6	10	0.74	0.82	0.77

Table 2 Quantitative comparison of proposed algorithm with two state-of-the-art algorithms which shows the extraction effectiveness

Test images	Nitin et al. [1]			Gao et al. [7]			Our integrated approach		
.	Correctness	Completeness	F-measure	Correctness	Completeness	F-measure	Correctness	Completeness	F-measure
a1	0.76	0.71	0.73	0.67	0.85	0.74	0.70	0.93	0.79
a2	0.75	0.70	0.73	0.80	0.70	0.75	0.79	0.86	0.82
a3	0.45	0.94	0.61	0.76	0.71	**0.73**	0.56	0.91	0.70
a4	0.82	0.88	0.85	0.96	0.80	0.87	0.86	0.89	0.87
a5	0.97	0.89	0.92	0.97	0.89	0.92	0.96	0.89	0.93
a6	0.67	0.85	0.74	0.83	0.75	0.79	0.74	0.82	0.77

Table 3 Quantitative effectiveness of our algorithm with influence of color parameter

Images	Dice similarity coefficients without colorization	Dice similarity coefficients with colorization
a1	0.91	0.93
a2	0.83	0.84
a3	0.88	0.90
a4	0.93	0.93
a5	0.90	0.92
a6	0.93	0.95

also illustrate the quantitative comparison with other existing methods carried out by the aforementioned metrics. From the experimental evaluation, one can observe that our integrated algorithm outperforms existing methods proposed by Nitin et al. [1] and Gao et al. [7], with the best F-measure. In addition, Dice similarity score of the proposed algorithm have also been evaluated as shown in Table 3. On analysis, it is evident that the color parameter of the segmented area influences the effective extraction with an enhanced Dice similarity value. Benefiting from exploiting the color factor, our algorithm extracts complete building footprint as desired. Note that, use of the eccentricity factor (in Eq. 2) boosts the extraction performance significantly and preserves the shape of building footprint.

5 Conclusion

We have proposed a building footprint extraction method from satellite imaging. The algorithm uses an integrated approach to determine superpixels and to compute the eccentricity of each superpixel and colorization terms to extract exact shape of footprints. Experimental evaluation demonstrated the ability of proposed method in building footprint extraction. The approach yields high quantitative scores. It has shown significant enhancement in completeness and F-measure when compared to other existing approaches. Furthermore, the proposed algorithm also has been proved to achieve higher Dice similarity coefficient with respect to the color parameter. Future work will concentrate on improving the approach by deep learning concepts.

Acknowledgements Authors acknowledge the support extended by DST-PURSE Phase II, Govt of India.

References

1. Gavankar, N.L., Ghosh, S.K.: Automatic building footprint extraction from high-resolution satellite image using mathematical morphology. Eur. J. Remote Sens. **51**(1), 182–193 (2018)
2. Bhateja, V., Tripathi, A., Sharma, A., Le, B.N., Satapathy, S.C., Nguyen,G.N.: Ant colony optimization based anisotropic diffusion approach for despeckling of SAR images. Integrated Uncertainty in Knowledge Modelling and Decision Making. IUKM. Lecture Notes in Computer Science, vol. 9978. Springer, Cham (2016)
3. Bhateja, V., Tripathi, A., Gupta, A., Lay-Ekuakille, A.: Speckle suppression in SAR images employing modified anisotropic diffusion filtering in wavelet domain for environment monitoring. Measurement **74**, 246–254 (2015)
4. Schuegraf, P., Bittner, K.: Automatic building footprint extraction from multi-resolution remote sensing images using a hybrid FCN. ISPRS Int. J. Geo-Inf. **8**, 191 (2019)
5. Saeedi, P., Zwick, H.: Automatic building detection in aerial and satellite images. IEEE Trans. Geosci. Remote Sens. (2008)
6. Dahiya, S., Garg, P.K., Jat, M.K.: Object oriented approach for building extraction from high resolution satellite images. IEEE Trans. Remote Sens. (2012)
7. Gao, X., Wang, M., Yang, Y., Li, G.: Building extraction from RGB VHR images using shifted shadow algorithm. IEEE Access **6**, 22034–22045 (2018)
8. Roopa, E., Hebbar, R., Shubha, P.: Buildings detection from very high resolution satellite images using segmentation and morphological operations. In: International Conference on Design Innovations for 3Cs Compute Communicate Control (2018)
9. Luhmann, T.: Eccentricity in images of circular and spherical targets and its impact to 3D object reconstruction. Int. Arch. Photogramm. Remote Sens. Spatial Inf. Sci. **XL-5** (2014)
10. Mente, R., Dhandra, B.V., Mukarambi, G.: Image recognition using shape descriptor: eccentricity and color, IBMRD's J. Manag. Res. Print ISSN: 2277-7830, Online ISSN: 2348-5922 (2014)
11. Jyothish, V.R., Bindu, V.R., Greeshma, M.S.: An efficient image segmentation approach using superpixels with colorization. Third International Conference on Computing and Network Communications (CoCoNet'19). Procedia Comput. Sci. **17**, 837–846 (2020)
12. Wang, M., Dong, Z., Cheng, Y., Li, D.: Optimal segmentation of high-resolution remote sensing image by combining super pixels with the minimum spanning tree. IEEE Trans. Geosci. Remote Sens. (2018)
13. Tu, B., Kuang, W., Zhao, G., Fei, H.: Hyperspectral image classification via superpixel spectral metrics representation. IEEE Sig. Process. Lett. (2018)
14. Mehra, J., Neeru, N.: A brief review: super-pixel based image segmentation methods. Imperial J. Interdisc. Res. (IJIR) (2016)
15. Satellite Images of Santa Ana. https://www.maplandia.com/united-states/california/orange-county/santa-ana/
16. Xiao, J., Gerke, M., Vosselman, G.: Building extraction from oblique airborne imagery based on robust Façade detection. ISPRS J. Photogramm. Remote Sens. **68**, 56–68 (2012)

A Novel Low-Voltage High-Performance Current Mirror

Ashok D. Vidhate and Shruti Suman

Abstract Current mirror is utilized to produce replica of current flowing through active devices, from one end to other end of the circuit irrespective of loading effect. It is an indispensable part of an analog and mixed signal system and is popular due to its wide range of applications. In this paper, new CM circuit which consists of biasing amplifier provides sustainable voltage is proposed. This CM provides low input impedance, low power consumption, high output impedance with high accuracy and bandwidth. The proposed circuit is designed at 0.18 μm technology having high output resistance of 54 GΩ with 2 GHz BW at 100 μA input current.

Keywords Self-biased current mirror · MOS current mirror · Nano-scale MOSFET · Power dissipation · CMOS integrated circuit · Amplifier · Resistance · Cascode current mirror

1 Introduction

Current mirror is the basic building block of analog/mixed mode ICs. It is extensively used for manufacturing operational transconductance amplifiers (OTA), analog-to-digital converters, digital-to-analog converters, amplifiers and filters. CM provides constant output current without dependency of loading. In CM, the output current is the mirror image of input current. The characteristic parameters of the CM are accuracy, input resistance, output resistance, input/output compliance voltage, and operating frequency range of CM. Nowadays in circuit design MOS channel length is scaling down. This adversely affects CM parameters, hence improvement required in the design of CM for low-power high-performance operation.

The remaining part of the paper is subdivided in the following sections: Sect. 2 deals with basic MOS CM. CCM are discussed in Sect. 3. All the current mirrors that have been reviewed are discussed in Sect. 4 with proposed circuit in Sect. 5.

A. D. Vidhate (✉) · S. Suman
ECE Department, Koneru Lakshmaiah Education Foundation (K L University), Guntur, Andhra Pradesh, India

V. Bhateja et al. (eds.), *Computer Communication, Networking and IoT*, Lecture Notes in Networks and Systems 197, https://doi.org/10.1007/978-981-16-0980-0_30

Based on this performance parameters, simulation outcomes are added in Sect. 6 and finally the conclusion is derived in Sect. 7.

2 Basic Current Mirror

The current mirror circuit is the basic building block of analog circuit design. The circuit which is forced, output current equal to input current. The designs which can be formed by current mirror are analog converters, oscillators, and amplifiers [1–3]. Important characteristic of integrated circuit design is low-voltage operation, low power consumption, wide bandwidth, and minimum are requirements; therefore, there is a growing need for new low-voltage analog circuit design [4, 5]. The VCO and delay locked loops are important circuits operated at low power with current mirror [6–8]. The basic block diagram of current mirror is as shown in Fig. 1.

Ideal current mirror has zero input resistance and infinite output resistance, which implies that the input voltage does vary with the input currents and the output currents are independent of applied voltage [9]. In reality, a CM requires minimum voltage at the output to ensure that the device operate in saturation [10]. This called the output compliance voltage. Accurate mirroring of the signal requires perfect matching of the mirroring transistor M_1 and M_2.

From Fig. 2 it is observed that if a transistor is biased at I_{REF}, then it produces $\text{VGS} = f^{-1}(I_{\text{REF}})$. Thus, if this voltage is applied to the gate and source terminals of second MOSFET, the resulting current is $I_{\text{OUT}} = f \cdot f^{-1}(I_{\text{REF}}) = I_{\text{REF}}$ as shown in Fig. 2. From another point of view, two identical MOS devices that have equal gate source voltage and operate in saturation carry equal current (if $\lambda = 0$). From the figure having M_1 and M_2, neglecting channel length modulation, the current relations can be expressed as,

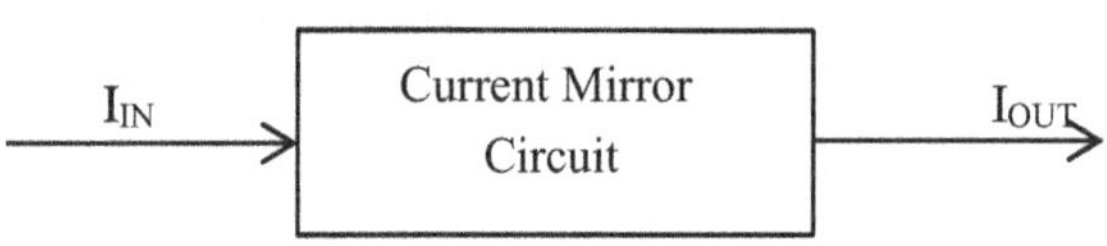

Fig. 1 Block diagram of current mirror

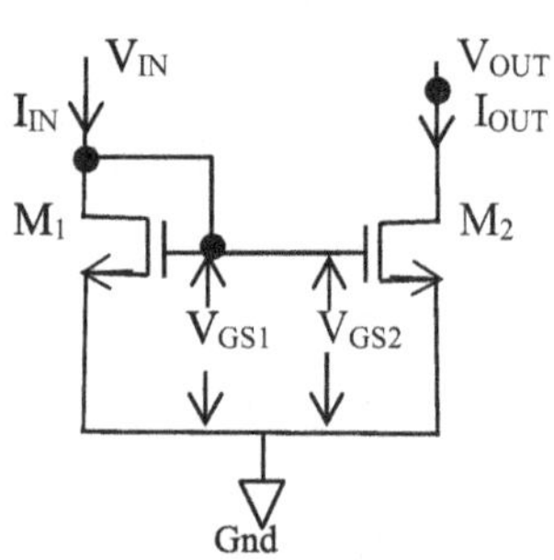

Fig. 2 Basic MOS current mirror

$$I_{\mathrm{REF}} = (1/2)\mu_n C_{\mathrm{ox}}(W/L)_1\left(V_{\mathrm{gs}} - V_{\mathrm{th}}\right)^2 \tag{1}$$

$$I_{\mathrm{OUT}} = (1/2)\mu_n C_{\mathrm{ox}}(W/L)_2\left(V_{\mathrm{gs}} - V_{\mathrm{th}}\right)^2 \tag{2}$$

Dividing Eq. (2) by Eq. (1), the relation between I_{OUT} and I_{REF} is obtained as,

$$I_{\mathrm{OUT}} = I_{\mathrm{REF}}\left[(W/L)_2/(W/L)_1\right] \tag{3}$$

If width-to-length ratio of both transistors is same mirroring of current from input to output takes place.

3 Cascode Current Mirror

To remove out the drawback of basic current mirror, the new structure is cascode current mirror. In cascode current mirror, the two basic CM circuits are connected one above one, i.e., in stacked form. This structure increases output resistance as well as increases gain of the CM. The cascade CM is as shown in Fig. 3.

The cascade structure is formed by using n-p-n MOSFET. In the previous circuit, the effect of channel length modulation is not considered due to that accuracy of basic CM is less. This problem is overcome by cascode CM. For cascode CM, small-signal output resistance is given as,

$$r_{\mathrm{OUT}} = r_{03} + r_{02}(1 + r_{03}(g_{m3} + g_{mb3})) \approx g_{m3} r_{03} r_{02} \tag{4}$$

Hence, the output resistance of cascode CM is equal to output resistance of simple CM multiplied by gain of MOSFET M_3. This means by increasing cascode levels the

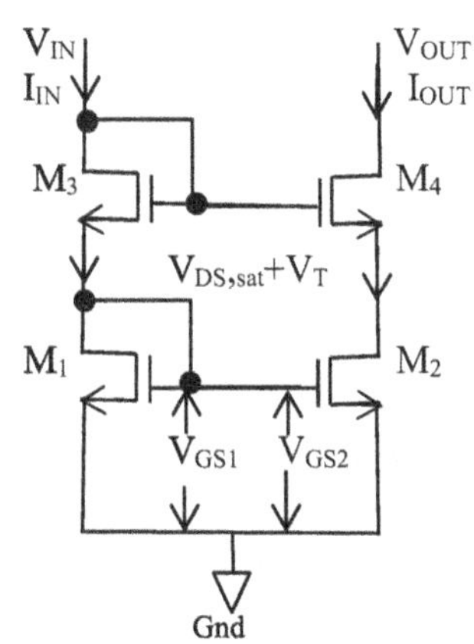

Fig. 3 Cascode current mirror

output resistance of CM increases. But the main drawback is that this also increases the voltage headroom which not applicable for power saving structures.

4 Literature Survey

In 2012, Bradely Minch worked on low-voltage current mirror. The proposed work implementation in a stacked MOS formation by making use of quasi-floating gate circuit, the capacitor functions as control gate to operate MOS in saturation boundaries. The dynamic performance gained due to stacked mirror. It also provides much higher output resistance than basic current mirror circuit but the voltage headroom increases by $V_{\text{diode}} + V_{\text{DSat}}$ which is slightly higher irrespective of current levels [11].

In August 2016, Yasin Bastan, Elach Hamzehil, Parniz Amiri worked on body-driven low-voltage current mirror. This body-driven technique gives higher output impedance by modifying the effective g_m of MOSFET. This also changes circuit parameters. This technique is useful in low-power application as low voltage requirement. The parameter improvements achieved are accuracy, bandwidth, low input resistance, and high output resistance. As compared to basic current mirror, it has just extra signal path having bulk MOSFET [12, 13].

In 2014, N. Raj, A. K. Singh and A. K. Gupta worked on self-biased bulk-driven low-voltage current mirror. This quasi-floating gate technique is combined with bulk-driven technique. This reduces threshold voltage at input side. The input resistance is less but the method suffers low transconductance. This gives limited bandwidth, complex twin-well fabrication process, and low gain [14, 15].

In 2014, Nidhi Tarun, Shruti Suman, and P. K. Ghosh worked on low-voltage current mirror. In this, CM biasing amplifiers are used to achieve low input resistance and high output resistance. This also proposes low compliance voltage requirement for CM [16, 17].

In December 2012, Maneesha Gupta and Bhawna Aggarwal proposed self-biased cascade current mirror. In this CM due to supercascode formation, the output resistance and BW get increased. The leakage current limits output resistance [18, 19].

In 2019, M. Doreyatim et al. proposed low voltage gain boosting CM which gives high voltage swing and better accuracy as mismatch between the devices is minimized by using internal amplifiers. The boosting structure is used to increase output resistance without limitation of input–output swing [20–22].

In 2020, K. Monfaredi and H. F. Baghtash extremely low voltage and high compliance CM in this current compensation scheme utilized which provides positive and negative feedback. The negative feedback is utilized to provide constant current at output side irrespective of increase in output resistance, whereas positive feedback is utilized to provide constant current at low voltage. But stability is the major concern as positive feedback is used [23, 24]. Due to use of cascaded inverters and inductive peaking, the bandwidth of the circuit increased [25].

5 Proposed Current Mirror

The proposed amplifier-biased current mirror (ACM) consists of biasing amplifier which provides sustainable voltage to keep transistor M_3 and M_4 in saturation. To obtain low input resistance along with basic current mirror, the transistor M_3 is connected in series. An amplifier having gain of '$-A_1$' is connected to transistor M_3 to control the gate voltage. Also, if input voltage is increased then due to amplifier combination input voltage to m_3 is decreased by 'A_1' results decrement in input resistance by some value of amplifier gain. Hence, supply voltage requirement gets decreased [26, 27].

Figure 4 shows amplifier-biased CM. In the output stage, supercascade configuration consists of amplifier having gain of ($-A_3$), transistor M_4 and M_5. The inverting amplifier A_3 provides biasing current I_{B1} to transistor M_5. This combination provides high output resistance as negative feedback loop forms in the circuit through transistor M_4, M_5 and M_6 [28]. This improves output resistance, accuracy, and gain of the current mirror.

5.1 Small-Signal Analysis of Amplifier-Biased CM

5.1.1 Input Resistance

Figure 5 shows transistor-level implementation of ACM. In this, the amplifiers are replaced by transistors which act as simple inverters. In the implementation M_3 is

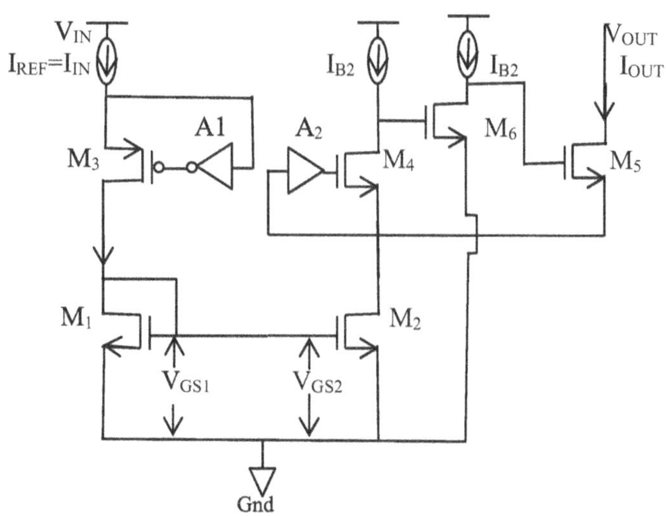

Fig. 4 Proposed amplifier-biased current mirror (ACM)

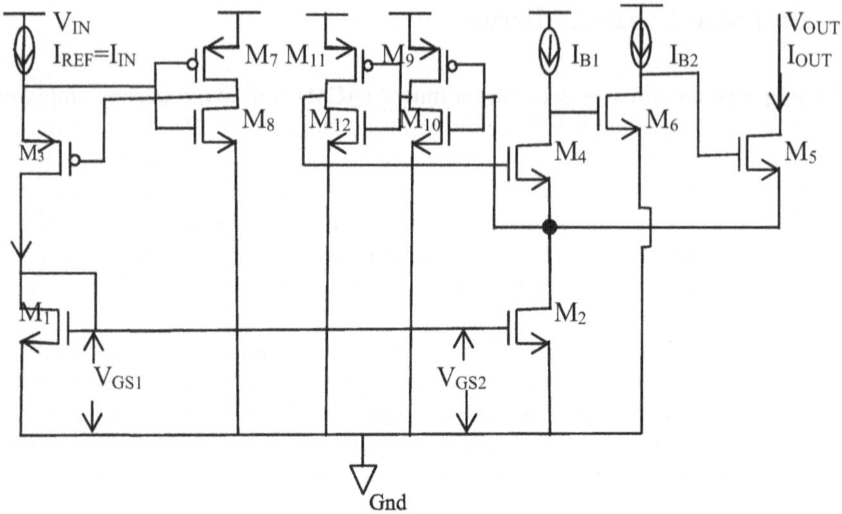

Fig. 5 Transistor-level implementation of proposed ACM circuit

p-type for this input is from inverting amplifier so only two transistors are enough for inverter input. But transistor M_4 is n-type so back to back two inverter structures are connected to produce required input for transistor M_4. To operate transistor circuit perfectly, the transistors M_7 and M_8 should work in saturation region. If any one of these transistors not work in saturation, then gain of the circuit gets reduced significantly.

Figure 6 shows small-signal equivalent model for input port of amplifier-biased CM.

By Kirchhoff's current law at differential node gives,

Fig. 6 Small-Signal equivalent model to calculate input resistance of ACM

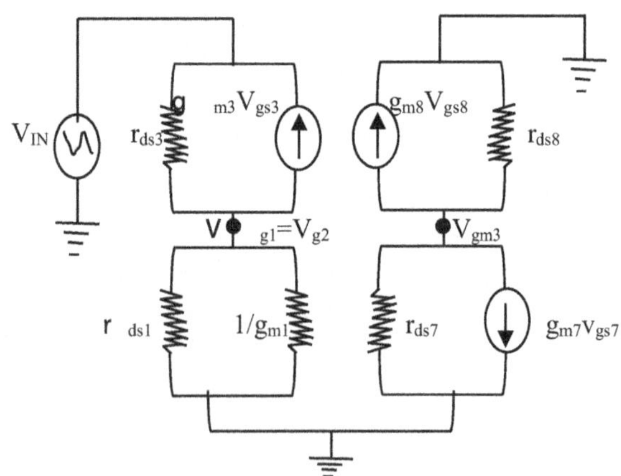

$$I_{\mathrm{IN}} = (V_{\mathrm{IN}} - V_1)/R = (V_1 - V_2)/r_{ds3} - g_{m3}V_{gs3} = \left(V_{g1}/r_{ds1}\right) + \left(V_{g1}/r_{gm1}\right) \quad (5)$$

The resistance r_{IN} is derived as

$$R_{\mathrm{IN}} = V_{\mathrm{IN}}/I_{\mathrm{IN}} \quad (6)$$

$$r_{IN} = (1 + r_{ds3})/(g_{m1} + g_{m3}g_{m1}r_{ds3}(A_1 + 1)) \quad (7)$$

From the above equation, it is observed that input resistance is inversely proportional to amplifier gain A_1. Hence, as value of amplifier gain is high the input resistance is low [29].

5.1.2 Output Resistance

Figure 7 shows the small-signal equivalent circuit to calculate output resistance. In the output side, negative feedback loop gives $V_{ds1} = V_{ds2}$, hence decreased effect of channel length modulation and accuracy maintains high [30].

By KCL at output side

$$V_{\mathrm{OUT}} = I_{\mathrm{OUT}} \cdot r_{ds5} + g_{m5}V_{gs5} + \left(V_{\mathrm{OUT}} - V_{gs4}\right)/r_{ds2} \quad (8)$$

$$I_{\mathrm{OUT}} = V_{gs5} \cdot g_{m5} + \left(V_{\mathrm{OUT}} - V_{gs4}\right)/r_{ds5} = V_{gs4}/r_{ds2} \quad (9)$$

$$r_{\mathrm{OUT}} = V_{\mathrm{OUT}}/I_{\mathrm{OUT}} = r_{ds5}[1 + r_{ds2}(g_{m5}r_{ds6}(1 + g_{m4}r_{ds4}) + (1/r_{ds5}) + g_{m5})] \quad (10)$$

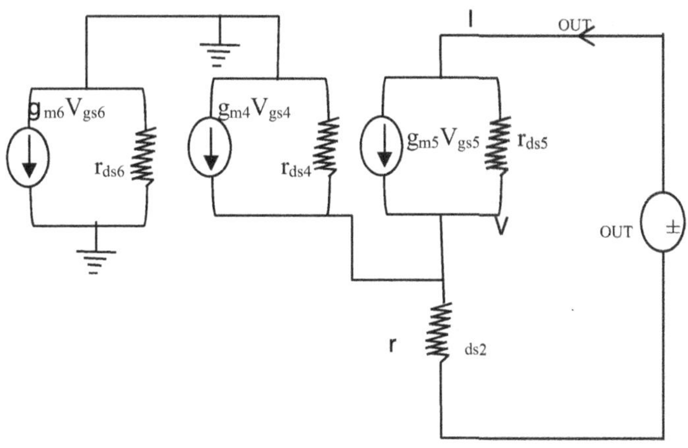

Fig. 7 Small-signal equivalent model to calculate output resistance of ACM

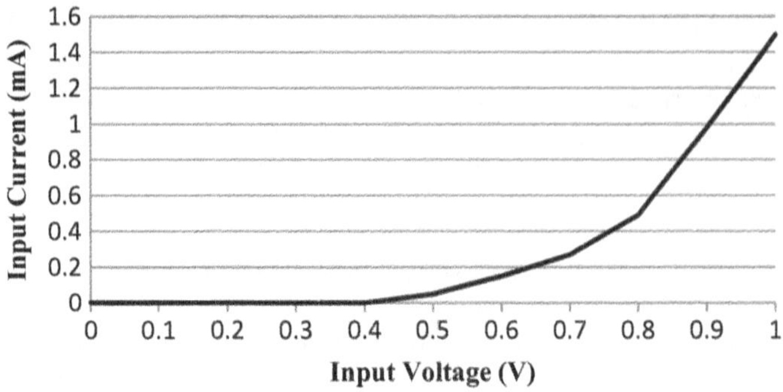

Fig. 8 Input characteristics of ACM

As g_m is large $g_m r_{ds} \gg 1$

$$r_{\mathrm{OUT}} \approx g_{m4} r_{ds4} g_{m5} r_{ds5} g_{m6} r_{ds2} \tag{11}$$

r_{OUT} is proportional to g_m of the transistor. If g_m is large then r_{OUT} is also large.

6 Simulation Results for Proposed Current Mirror

To find out results, the Tanner EDA tool is utilized. The 1 V input power supply is used. Simulations are carried out to find out input characteristic, output characteristic, and frequency response.

6.1 Input/Output Characteristic of ACM

Figure 8 shows input characteristics of ACM. Due to application of input voltage current starts to flow in the ACM. The current value is zero up to 0.5 V as input beyond this current starts increasing exponentially. The minimum compliance voltage obtained at the input side is 0.6 V; this implies that ACM has low input resistance.

6.2 Output Characteristics of ASCCM

Figure 9 shows output characteristics of ACM. In VLSI output resistance should be high. The minimum compliance voltage obtained for the ACM is 0.2 V.

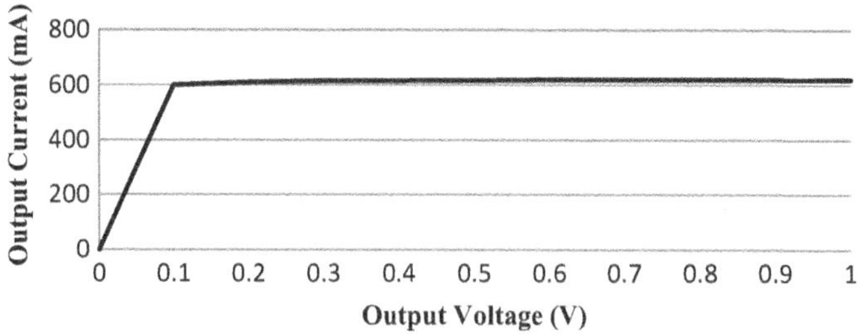

Fig. 9 Output characteristics of ACM

6.3 *Frequency Response of ACM*

The bandwidth enhancement obtained in ACM by adding super cascode configuration to primary current mirror. This gives bandwidth of 2 GHz at 100 μA (Fig. 10).

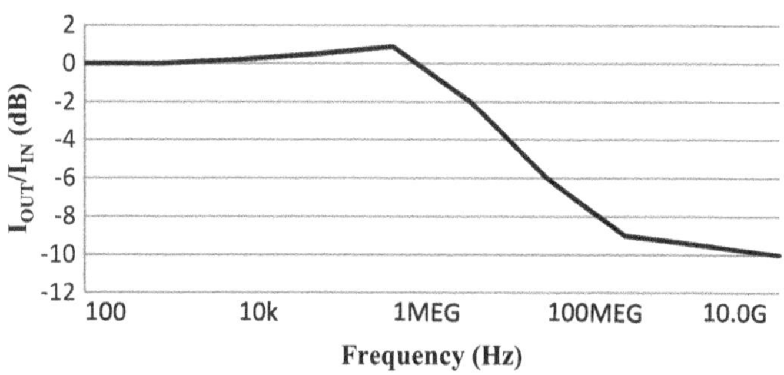

Fig. 10 Frequency response of proposed current mirror circuit

Table 1 Summary of current mirrors with ACM

References	[12]	[13]	[14]	[16]	Proposed ACM
Input resistance, R_{in}	13 GΩ	240 kΩ	675 kΩ	0.18 kΩ	180 Ω
Output resistance, R_{out}	39.5 GΩ	115 kΩ	485 kΩ	6.25 kΩ	54.68 GΩ
Input current, I_{in} (μA)	15 μA	447 μA	500 μA	–	100 μA
Bandwidth (MHz)	216 MHz	655 MHz	2.5 MHz	98.50 MHz	1.15 GHZ
Power	42.5 μW	0.825 mW	–	46 μW	340 μW
Supply voltage (Volt)	1 V	± 5 V	−1 V	0.8 V	1 V
Number of transistors used	7	11	10	10	12
Technology	0.18 μm	0.13 μm	0.18 μm	90 nm	0.18 μm

The specification of proposed ACM is given in Table 1 and the values are compared with related CM circuit. It can be seen the output resistance significantly increased with minimum input resistance. In addition, its bandwidth is increased without sacrificing any performance.

7 Conclusion

A novel self-biased CM with very high-performance characteristics has been introduced and verified with the help of simulations. Using this technique, input resistance and input compliance voltage get reduced to 180 Ω and 0.6 V, respectively. The frequency response is stable in nature and relatively high. Simulation result supports the utility of ACM for low voltage, high output resistance, and high performance. Improvement in output resistance has been carried out by implementation of transistor configuration at the output side. The active realization of the resistances, to overcome the drawbacks of passive components, has also been discussed. The proposed saturation region CM, where all required resistances have been replaced by active components, shows high degree of accuracy over a wide input current range. Small-signal analysis has been carried out at each step and the results have been validated using Mentor Graphics Eldospice based on TSMC 0.18 μm CMOS technology using 1 V supply voltage.

References

1. Kang, S.-M., Leblebici, Y.: CMOS Digital Integrated Circuits Analysis and Design, 3rd edn. McGraw Hill (2003)
2. Allen, P., Holbers, D.: CMOS Analog Circuit Design, 3rd edn. Oxford University Press (2011)
3. Mitwong, H., Kasemsuwan, V.: A 0.5 V Quasi-floating gate self-cascade DTMOS current-mode precision full-wave rectifier. In: 9th International Conference on Electrical Engineering/Electronics, Computer, Telecommunications and Information Technology. IEEE (2012)
4. Suman, S.: Theory and Design of Advanced MOS Current Mirrors, 1st edn. Lambert Academic Publishing (2018)
5. Kalyana Chakravarthy, G., Laskar, N.M., Nath, S., Chanda, S., Baishnab, K.L.: Flipped voltage follower based high dynamic range current mirror. In: Devices for Integrated Circuit (DevIC), pp. 311–315. IEEE (2017)
6. Suman, S.: Design of efficient ring VCO using nano scale double gate MOSFET. Mody Univ. Int. J. Comput. Eng. Res. **2**(1), 05–10. ISSN: 2456-9607 (2018)
7. Suman, S., Sharma, K.G., Ghosh, P.K.: 250 MHz multiphase delay locked loop for low power applications. Int. J. Electr. Comput. Eng. **7**(6), 3323–3331. ISSN: 2088-8708 (2017)
8. Suman, S., Tarun, N., Ghosh, P.K.: Design of comparator for ADC's using CMOS current mirror: Design of CMOS Comparator, 1st edn. Lambert Academic Publishing (2014)
9. Hosseini, S., Saberi, M., Lotfi, R.: A high-speed and power-efficient voltage level shifter for dual-supply applications. IEEE Trans. Very Large Scale Integr. (VLSI) Syst. **25**, 1154–1158 (2016)

10. Vajpayee, P., Srivastava, A., Rajput, S.S., et al.: Low voltage regulated cascode current mirrors suitable for sub-1V operation. In: IEEE Asia Pacific Conference on Circuits and Systems. IEEE (2010)
11. Minch, B.: A simple low-voltage cascode current mirror with enhanced dynamic performance. In: IEEE Subthreshold Microelectronics Conference, pp. 1–3. IEEE Digital Library, Waltham, USA (2012)
12. Bastan, Y., Hamazehil, E., Amiri, P.: Output impedance improvement of a low voltage low power current mirror based on body driven technique. Microelectron. J. Elsevier, 163–170 (2016)
13. Gupta, R., Sharma, S.: Quasi-floating gate MOSFET based low voltage current mirror. Microelectron. J. Elsevier, 439–443 (2012)
14. Raj, N., Singh, A.K., Gupta, A.K.: Low-voltage bulk-driven self-biased cascade current mirror with bandwidth enhancement. Electron. Lett. IEEE **50**, 23–25 (2014)
15. Young, Z., et al.: An energy-efficient and wide-range voltage level shifter with dual current mirror. IEEE Trans. Very Large Scale Integr. (VLSI) Syst., 1–5 (2017)
16. Tarun, N., Suman, S, Ghosh, P.K.: Design of low voltage improved performance current mirror. Control Theory Inf. **4**, 26–38 (2014)
17. Suman, S.: Design of two stage CMOS comparator with improved accuracy in terms of different parameters. Mody Univ. Int. J. Comput. Eng. Res. **2**, 64–67 (2018)
18. Torralba, A., Munoz, F., Ramirez, J., Carvajal, R.G.: Output stage for low supply voltage, high performance CMOS current mirrors. Electron. Lett. **38**, 1528–1529 (2002)
19. Ramirez-Angulo, J., Torrabla, A., Carvajal, R.G.: Low supply voltage high performance CMOS current mirror with low input and output voltage requirement. In: IEEE Midwest Symposium on Circuits and Systems, vol. 1, pp. 510–513 (2000)
20. Zhou, T., et al.: A high-precision and high-linearity readout integrated circuit for infrared focal plane array applications. Optik-Int. J. Light Electron Opt. **185**, 168–177. ISSN 0030-4026 (2019)
21. Doreyatim, M., et al.: A low-voltage gain boosting-based current mirror with high input/output dynamic range. Microelectron. J. **90**, 88–95 (2019)
22. Umar, L., et al.: Biosensor signal improvement using current mirror topology for dissolved oxygen measurement. Meas. Sci. Technol. **30**(6), 65–102 (2019)
23. Monfaredi, K., Baghtash, H.F.: An extremely low-voltage and high-compliance current mirror. Circuits Syst. Sig. Process. **39**(1), 30–53 (2020)
24. Julien, M., et al.: Breaking the speed-power-accuracy trade-off in current mirror with non-linear CCII feedback. Microelectron. J. **83**, 77–85 (2019)
25. Zohoori, S., et al.: A CMOS, low-power current-mirror-based transimpedance amplifier for 10 Gbps optical communications. Microelectron. J. **80**, 18–27 (2018)
26. Sahani, J.K., Suman, S., Ghosh, P.K.: Design of operational transconductance amplifier using double gate MOSFET. Innovative Syst. Des. Eng. IISTE **5**, 42–56 (2014)
27. Saini, H., Suman, S.: Design and analysis of nanoscale double gate MOSFET based current mirrors. Int. J. Appl. Eng. Res. **13**, 112–116 (2018)
28. Yang, H., Allstot, D.: An active-feedback cascode current source. IEEE Trans. Circuits Syst. **37**, 644–664 (1990)
29. Zhang, X.: A regulated body-driven CMOS current mirror for low voltage applications. IEEE Trans. Circuits Syst.-II: Express Briefs **51**, 571–577 (2004)
30. Azhari, S., Baghtash, H., Monfaredi, K.: A novel ultra-high compliance, high output impedance low power very accurate high performance current mirror. Microelectron. J. **42**, 432–439 (2011)

Self-automated Fault Diagnosis System for Internal Combustion Engines

Nitla Stanley Ebenezer, Abdul khurshid, K. Anjani Devi, Chodisetti Naga Sandeep, Penke Pragnana Manipal, Gorthi Siva Vijay, and Thotakura Sri Satya Dhanesh

Abstract A heat engine often coined as a system transmutes thermic and chemic energies to mechanical energy. The current review employs a traditional heat engine, i.e., an internal combustion engine, where a self-automated optimization technique is incorporated for determining best optimal parameters and for diagnosing the flaws thereby enhancing the overall efficacy. Several difficulties are witnessed during the effective functioning of an IC engine which eventually roots to multiple energy losses leading to the fatigue failure of the entire system. The current review engages a self-acting conceptualization for characterizing the several faults which deteriorates the overall performance traits. In contrast to the prevailing methodologies, viz. radiated sound signal, wavelet carton disintegration, the novel existent artificial neural network (ANN) methodology engages an extension pattern recognition technique encompassing an input skin and output skin. This simple, novel extension neural network (ENN) renders amplified optimization results and much easier data incorporation in contrast to the conventional neural networks. The current study is successful in devising a novel, self-automated, efficient diagnostic system. The inferences drawn from this novel technique tend to have superior optimization and fault identification rates.

Keywords Internal combustion engines · Fault diagnosis · Vibrations · Neural networks · Extension neural network · Optimization · Efficacy

N. S. Ebenezer (✉) · C. N. Sandeep · P. P. Manipal · G. S. Vijay · T. S. S. Dhanesh
Department of Mechanical Engineering, Aditya College of Engineering and Technology Surampalem, Kakinada, Andhra Pradesh 533437, India

A. khurshid
Department of Mechanical Engineering, JNTUK University College of Engineering, Vizianagaram, Andhra Pradesh 535003, India

K. A. Devi
Department of Physics, Computer Science and Engineering, Aditya Engineering College Surampalem, Kakinada, Andhra Pradesh 533437, India

V. Bhateja et al. (eds.), *Computer Communication, Networking and IoT*, Lecture Notes in Networks and Systems 197, https://doi.org/10.1007/978-981-16-0980-0_31

1 Introduction

Heat engine which is often abridged as H.E. is broadly engaged in innumerable vehicular configurations. The heat engine employed in the existent review is an internal combustion engine which is typically abbreviated as ICE. Several research enthusiasts investigating in the domains of diagnosis, optimization, and fault identification proposed several efficient methodologies for enhancing the overall efficacy. However, most of the prior proposed diagnostic tools rely mostly on the data assembled from former practices; henceforth, the cyclic improvement of the concerned dataset is often desirable. From the previous literature, a tallying diagnostic technology referred to as scan tool aids in enhancing the efficacy of the various internal combustion engines. Besides, it can only recognize the electrical faults which therefore urges several research fanatics the need for developing a novel optimization system that forecasts all sorts of inherent flaws. The prevalent structure for any diagnostic methods encompasses multiple loop phases. The initial phase comprises of characteristic datasets where the fault is to be gathered. Next phase involves depicting the relevant set of conditions by employing a precise tool. The succeeding phases retrieve the beneficial traits by engaging a specified tool. The final phase involves recognition of the flaws by array identification method [1, 2]. Detection of the flaws is established basing the formerly availed and gathered data [3–5]. This accumulated data reviews the working condition, processing parameters of the "ICE". For facilitating the effectiveness of the flaw detection, numerous sensors are mounted on the ICE for vindicating the pressure, temperature, voltage, and current. Vibrational energy losses pose a severe threat for the efficient functioning of most IC engines and material systems [6, 7]. The prior literature pertaining to vibrational loses is regarded as the utmost dominant information that leads to recognition of the fault as detailed in review by [8–10]. This investigation technique engages the sound response system in the absence of superior ambient medium equipped with multiple sensing devices for the diagnosis of the flaws. Several vibrational signal setups have been postulated for the flaw determination and the extent damage witnessed by the mechanical system [11–14]

2 Experimentation

In the existent paper, MATLAB is employed for designing the perfect optimal solution for envisaging the flaw detection. The entire experimentation encompasses the succeeding steps investigation of the sound response, development of various neutral networks, incorporating the intake patterns, teaching for the neural network (NN), eventually cropping the images/patterns for evaluating and obtaining the solution.

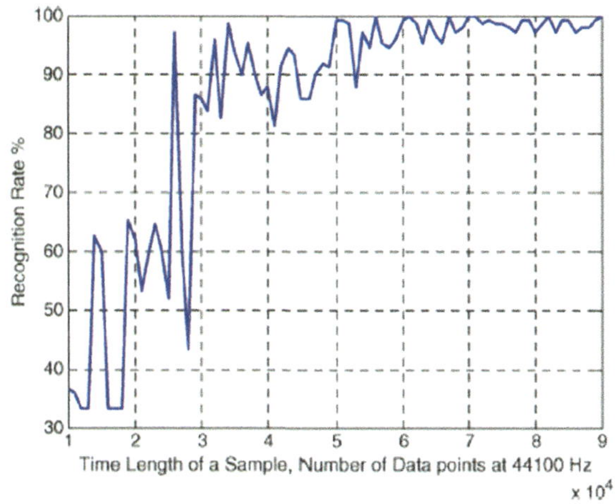

Fig. 1 Selection of subdomain length

2.1 *Pulse Period Response of the Test Sample*

The pulse period encompasses recording the vibrational pulse domain responses at a time interval rate of every two minutes. The experimentation follows the subsequent steps obtaining the various intake patterns followed execution and eventually to conduct test through ENN, for this interruption gap is divided into 100 subdomain signals for every 1 s. 50 subdomain responses are utilized for the evaluation of the neural network. The sound of a manually driven four-cylinder 1500 CC engine is recorded at three categories, i.e., no fault state, with and without any cylinder misfiring. Thereafter, the sound responses were categorized into multiple phases basing on the subdomain lengths. The wavelet packet decomposition "Wavelet PD" is engaged for deriving the diagnostic characteristics from the subdomain response and then sent into "ENN" for training followed by evaluating. For the flaw diagnosis in the existent system, the recognition rate is initiated by using irregular subdomain length as is shown in Fig. 1. From the figure, it is apparent that the identification rate is very low when the time length of the test sampling is around 49,000 data points. This acquired range is identical to 2.13 s of the registered time at a specimen rate of 43,100 Hz. Later the identification rate is stabilized at much higher values.

2.2 *Recording Method*

A sound recorder (model number ICD-UX5137) is employed for recording the intensity of the engine sound. The recorder is of dual channel equipped with (L and R) channels. The existent review employs the engine of Hyundai Elantra with an engine capacity of 1600 CC with four cylinders considering the subsequent parameters: no

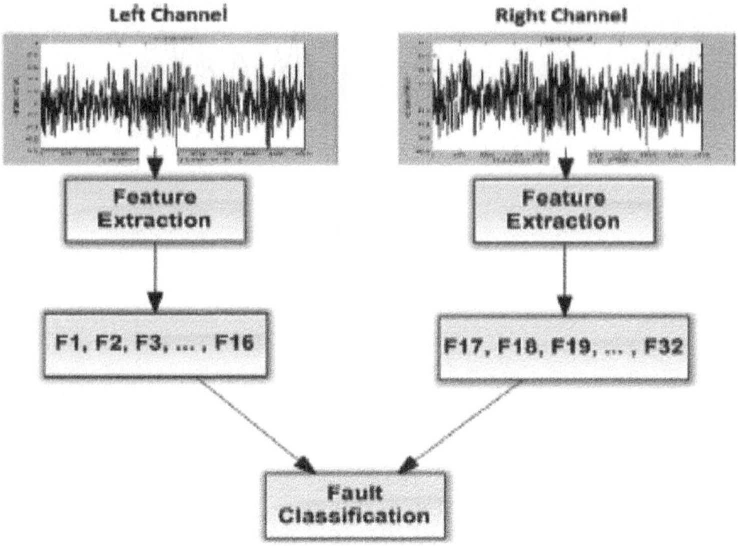

Fig. 2 Recording using dual mode

fault, single-cylinder backfire, defective mass flow sensor, air spillages, and disturbance in CAM shaft sensor state. The identification pace is evaluated with respect to the dual channels, i.e., *L* and channels with each 16, 16 traits (F1… F32) and depicted in Fig. 2.

3 Results and Discussion

3.1 Deriving the Characteristic Condition

The sound response measured in decibels of a manually driven four-cylinder 1500-CC engine is noted at various conditions such as no fault, having one-cylinder misfire, dual-cylinder misfire, and air spillages are documented thoroughly. As the sound signals do not characterize the time domain responses, the processing of the existent data is to be made for abstracting the features. Each sound signal is subjected to WPD, undergoes through multiple analyzing phases where the energy and entropy conditions are calculated at all nodes. Figure 3 represents that feature vector (FV) which is obtained from the recorder channel from the investigation, the images are divided by the help of pattern recognition tool. Later its standard deviation and variation are evaluated. It is observed that the first three principal components account of 90% of unevenness from the prior dataset. Later the patterns are arranged in three-dimensional space as represented in Fig. 4, where the visualization and evaluation of the faults are made clear.

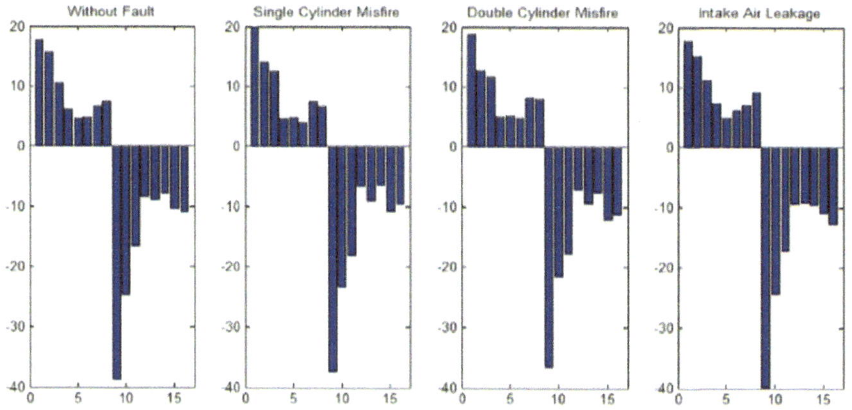

Fig. 3 FV for various faults showing the energy values and entropy

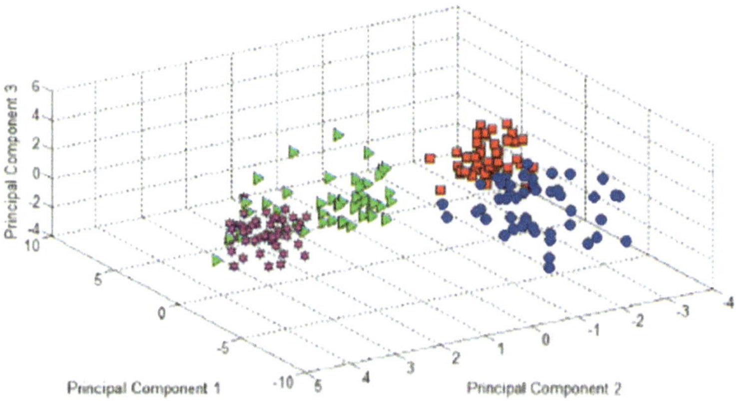

Fig. 4 Faults cluster. Triangle: No fault, square: single-cylinder misfire, circle dual-cylinder misfire, star: air spillage

3.2 *Classification Phase*

In this phase, backpropagation network (BPN) which is a multilayered neural network is engaged. The input and targeted vectors are categorized into no fault, single-cylinder backfire, damaged mass flow sensor, air spillages, and disturbance in CAM shaft position sensor and are test trained using ANN. Using the specified tool corresponding to NN in MATLAB, image recognition is constructed. Different functions show different performance characteristics of the ANN in Table 1; flaunts the execution of the ANN using three different training categories: steepest descent with momentum (GDM) function, CGD, and Levenberg–Marquardt training algorithm. The training is done for several trails; hence the variation of performance takes place at different runs (Fig. 5).

Table 1 Performance of BPN utilizing various algorithm and structure

Training function	Tree of the ANN	Average recognition rate (%)	Average number of epochs	Average training time (s)
Steepest descent with momentum	11-6	77.7	1000	16
	6-5-6	48.4	1000	31
Scaled conjugate gradient descent	11-6	99.8	81	3
	6-5-6	88.6	100	3
	5-7-6	95.5	121	4
	11-16-6	99.8	131	6
Levenberg–Marquardt	11-6	99.7	16	2
	6-5-6	98.8	35	4
	5-7-6	99.6	111	16

3.3 Correlation Among ENN and ANN

The resemblance among ENN and BPN is done in different aspects for scrutinizing the ENN and BPN at a constant distributed noise and added to the time domain signal response. Table 2 displays that ENN renders high resistance to environmental noise in contrast to the BPN.

3.4 Reliance on Flaw Recognition

The engine of Hyundai Creta 2200 CC is evaluated in various conditions and it is found that the identification pace in some flaws is higher than other flaws as presented in Table 4. Employing db2 mother wavelet the recognition rate obtained is 86%. Table 3 represents the faults of Hyundai Creta and Table 4 represents the faults of the Hyundai Elantra.

3.5 Feedback of Wavelet Assortment on Identification Rate

Attributes derived from the wavelet decomposition need selection of the specific mother wavelet features for the necessary evaluation. Three different categories of parent wavelets, "db5," "db9," and "db22," are reused for flaw identification engaging "ENN." Table 5 represents the inferences for all conditions when in contrast to "db5" used in the prior case. The implementation of the respective novel techniques mostly involves the overview of added monitoring practices which is not possible because of the time bound limitations and other resource accessibility. The lack of resources

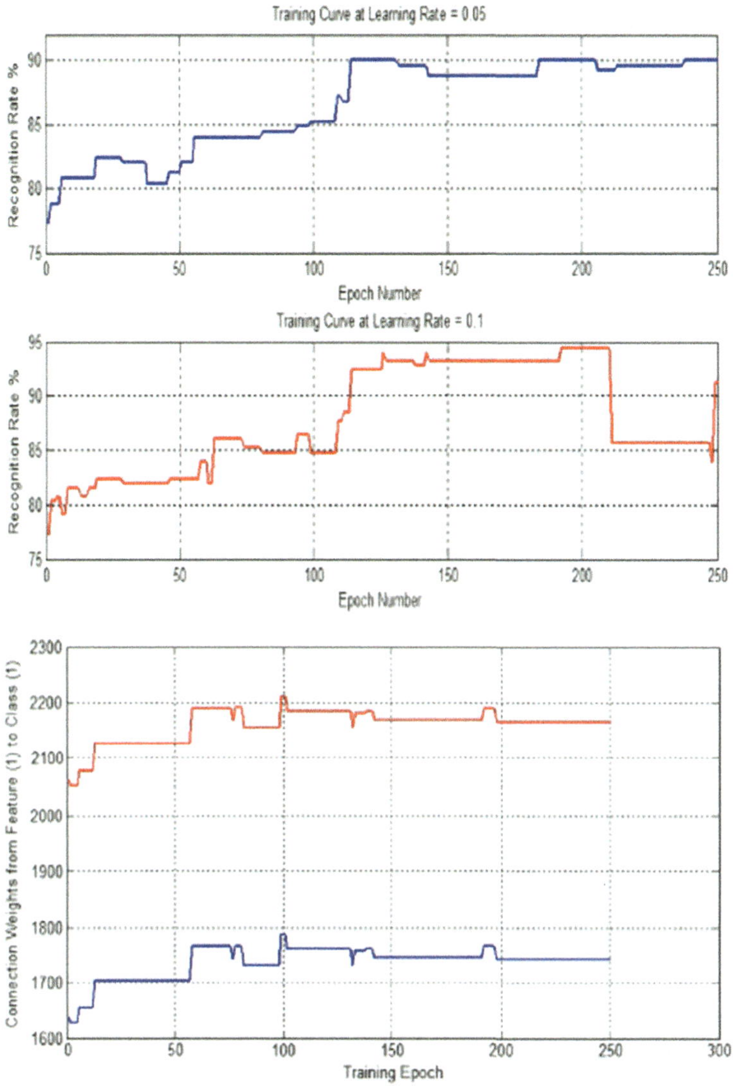

Fig. 5 Training progress

Table 2 Liberality of ENN versus BPN to environmental changes

Fault	Recognition rate %
No fault	91
Single cylinder	95
MAF sensor fault	99
Inhalation air leakage	99
Cam sensor fault	100

Table 3 Hyundai Elantra

Noise percentage	Recognition rate of ENN %	Recognition rate of BPN %
±0	98.9	91.6
±5	98.5	88.6
±15	96.5	87.6

Table 4 Hyundai Creta

Fault	Recognition rate %
MAF sensor fault	87
Inhalation air leakage	99
Cam sensor fault	61
Temperature coolant sensor fault	71

Table 5 Identification rate

	Identification rate %	
Primary wavelet	Hyundai Elantra	Hyundai Creta
db2	93.2	85
db5	95.8	89.3
db9	97.2	89.2
db22	97.7	89.6

and complexity leads to incompetent generation algorithm for proficient diagnosis employing neural networks.

4 Conclusion

The flaw diagnosis and flaw detection of combustion engines played a crucial part in effective functioning and enhancing the overall efficacy of several automobiles. The current study is successful in devising a novel, self-automated, efficient diagnostic system. The intended diagnostic system is sound-based system which entirely relies on the persistent sounds emitting from the engine during the dynamic conditions. ENN along with WPD as a specified is successful in fault identification. The postulated ENN added multiple advantages like precise flaw detection, categorization of flaws, elevating the overall efficacy by envisaging the damage due to the undesired vibrations in contrast to the conventional multifaceted ANN. Innovative diagnostic techniques like principal component analysis (PCA) can be regarded as prime potential techniques for attaining and monitoring the results in near future.

References

1. Velon, M., Malan, S.A., Giorelli, M., Irilli, A.: Genetic algorithm parameters estimation applied to vehicle powertrain dynamics. In: 18th European Control Conference (ECC), pp. 4168–4173. Naples, Italy (2019)
2. Shang, W., Zhou, X., Yuan J.: An intelligent fault diagnosis system for newly assembled transmission. Expert Systems with Applications, vol. 41(9), pp. 4060–4072 (2014)
3. Li, Y., Lu, X., Kar, N.C.: Rule-based control strategy with novel parameters optimization using NSGA-II for power-split PHEV operation cost minimization. IEEE Trans. Veh. Technol. **63**(7), 3051–3061 (2014)
4. Fischer, D.: Fault detection for lateral and vertical vehicle dynamics. Control. Eng. Pract. **15**, 315–324 (2007)
5. Sotudeh-Chafi, M.: Critical cases of a CNC drive system—fault diagnosis via a novel architecture. In: Proceedings of the NAFIPS FLINT 2002 International Conference, pp. 494–499. New Orleans, LA (2002)
6. Ebenezer, N.S., Srihari, P.S., Prasad, C.R., Appalaraju, P., Tatahrishikesh, A., Teja, B.S.: Experimental studies on damping behaviour of nickel electroplated A356. 2 alloy, Mater. Today: Proc. **27**(2), 1038–1044 (2020)
7. Ebenezer, N.S., Narayana, P.S., Ramakrishna, A.: Mechanical and micro-structural characterization nickel electroplated metal matrix composites. Mater. Today: Proc. **27**(2), 1278–1281 (2020)
8. Gebraeel, N., Lawley, M., Liu, R., Parmeshwaran, V.: Residual life predictions from vibration-based degradation signals: a neural network approach. IEEE Trans. Ind. Electron. **51**(3), 694–700 (2004)
9. Liu, S.: Detection of engine valve faults by vibration signals measured on the cylinder head. Proc. Inst. Mech. Eng., Part D: J. Automobile Eng. **220**(3), 379–386 (2006)
10. Ekici, S.Y.: Energy and entropy-based feature extraction for locating fault on transmission lines by using neural network and wavelet packet decomposition. Expert Syst. Appl. **34**(4), 2937–2944 (2008)
11. Soualhi, A., Clerc, G., Razik, H.: Detection and diagnosis of faults in induction motor using an improved ant clustering technique. IEEE Trans. Ind. Electron. **60**(9), 4053–4062 (2013)
12. Yen, G.G., Lin, K.C.: Wavelet packet feature extraction for vibration monitoring. IEEE Trans. Ind. Electron. **47**(3), 650–667 (2000)
13. Soualhi, A., Razik, H., Clerc, G., Doan, D.D.: Prognosis of bearing failures using hidden Markov models and the adaptive neuro-fuzzy inference system. IEEE Trans. Ind. Electron. **61**(6), 2864–2874 (2014)
14. Soualhi, A., Clerc, G., Razik, H., Badaoui, M., Guillet, F.: Hidden Markov models for the prediction of impending faults. IEEE Trans. Ind. Electron. **63**(5), 3271–3281 (2016)

References

[illegible]

A Comprehensive Analysis over Controller Placement and Network Expansion in SD-WAN

W. Rose Varuna and R. Vadivel

Abstract The primary factors related to software-defined networks (SDN) deployment are the placement of the controller. The baseline issue encountered by network operators is to expand the network with reduced cost to fulfill the needs of substantial network users and to expand their services. To handle these issues, this work concentrates on implementing three different ideas for controller placement. The controller placement problem is dealt with the use of the Markov decision process, the implementation of multiagent aggregators for balancing network loads, and finally, game theory is performed to show effectual controller placement. The current network model is reorganized with metrics like delay, packet loss, and throughput. These metrics are measured before and after the installation of controllers. As this network expansion problem is more generalized, these approaches can be applied to perform newer network functionalities by updating the existing network model.

Keywords Software-defined network (SDN) · Controller placement · Markov decision process · Multiagent aggregator · Game theory

1 Introduction

The primary cause behind the software-defined network (SDN) establishment is to partition the data plane and the control plane to diminish cost and to enhance the innovative competency [1]. In the control plane, one or more controllers make precise decisions to establish a global view toward the network. This is considered as the intelligence behind the network model. In the data plane, some cheaper commodity switches are used. When switches have no idea toward the incoming packets, it will transfer it to the controller and waits for further instructions. The communication establishment among the controllers and the switches are generally carried out with OpenFlow.

W. Rose Varuna (✉) · R. Vadivel
Department of Information Technology, Bharathiar University, Tamil Nadu, Coimbatore 641046, India

V. Bhateja et al. (eds.), *Computer Communication, Networking and IoT*, Lecture Notes in Networks and Systems 197, https://doi.org/10.1007/978-981-16-0980-0_32

The critical concern in SDN is the placement of controllers as it may show a particular influence on cost and network performance. For instance, to describe how and where the controllers have to be deployed based on its delay factor the brute-force model is applied [2]. Similarly, the management framework to regulate connectivity among the controllers and the switches is based on its utilization that anticipated in [3]. This problem has been handled with the use of heuristic approaches and brute-force models [4]. Despite on concentrating delay factor, controller load has utilized and anticipates a model to deal with this issue [5]. Recently, an optimization approach has been predicted to provide newer SDN while dealing with various added constraints [6].

The above-mentioned approaches deal with planning issues that depict that there is no prevailing infrastructure model defined yet [7]. As an outcome, the network model is eventually updated to include more number of switches [8]. This problem is depicted as follows: Assume a set of current SDN infrastructure, how it can be expanded such that financial needs are reduced [9]. As an outcome, the placement model anticipated in this work is competent to either plan some newer network from expanding or scratch a prevailing one [10].

In this investigation, various mathematical models have been used to handle the network expansion problem and controller placement problem. Initially, the Markov decision process is applied for a node clustering-based partitioner algorithm. Next, a multiagent-based multiaggregator is used for splitting the network into subnetworks and for diminishing end-to-end delay. Finally, the gaming model is used to determine the placement of players in some appropriate locations. This model attempts to reduce the cost for reorganization and to offer a feasible solution (various constraints like latency, capacity, traffic, and so on have to be fulfilled). The remainder of the work is structured as trails: Sect. 2 is problem formulation to handle the network issues, Sect. 3 is the various proposed model applied to control the controller placement issues and network expansion model. Section 4 is the numerical results and discussion section. Section 5 is the conclusion of the anticipated model.

2 Problem Formulations

To formulate the controller placement and network expansion issues, the following are the factors that have to be analyzed. (1) Consider two different positions of switches and the total amount of traffic generated toward a controller; (2) providing various link types (bandwidth and cost) that are applied for expanding and designing the network model; (3) different controller types have to be installed (value, capacity, and a number of ports); (4) maximal tolerance toward data flow and latency encountered between the controllers and switches to assist various services and applications; (5) the general SDN topology has to be maintained for data transmission.

3 Methodologies

This section discusses the methods used for handling the controller placement problem.

3.1 *Partitioner Algorithm Using Markov Decision Process*

The objective is to reduce delay identified in switches and controllers. Initially, the controller placement problem is dealt with the use of the Markov decision process. Secondly, the implementation of multiagent aggregators for balancing network loads and finally gaming theory is performed to show effectual controller placement.

(a) **Node Clustering Network Partition Algorithm (NB-PAMDP)**

The network partitioner demonstrates the clustering problem from general clustering algorithms. Every cluster center is provided with clustering algorithms. (1) Initiate k-clusters and assign center for every cluster with random sampling; (2) assign node with Euclidean distance; (3) recompute cluster centroid; (4) repeat 2 and 3 till there is no modification in every cluster.

Initially, a random selection of centers cannot reduce latency among nodes. Next, Euclidean distance cannot be utilized in distance computation among nodes. Thirdly, the controller is positioned to ensure physical connection among centroid and nodes. Here, the partitioner algorithm with Markov decision process is modeled to address this crisis.

Algorithm 1: NB-PAMDP

Input: Network topology $N = (V, E)$.

Output: Cluster centroid.

Step 1: Switch allocation to 'K' clusters.

Step 2: Compute E2E latency among two nodes $E2E\ (u, v)$.

Step 3: Randomly chose one node as the initial center.

Step 4: Compute the actual center of the network. Node with least sum E2E latency to other nodes is chosen as a centroid.

Step 5: Compute the second center of the network—node with the highest sum of E2E latency to chosen centroid as the second centroid.

Step 6: Compute vertex to the initial cluster with equation below

$$V \in cluster; \quad for\ all\ cluster \in \{1, 2, \ldots, k\}$$

Step 7: Update centroid $C' = \{C^1, C^2, \ldots, C^K\}$ where E2E latency of nodes in the cluster to new centroid C^I is reduced.

$$C_i' = V_m; if\ E2E(V_m, V) = minimum,$$

$$for\ all \in size(cluster_i), v \in cluster, i = \{1, 2, \ldots, k\}$$

Step 8: Repeat step 4, 5, and 6 till network is partitioned into 'k' clusters.

(b) **Multicontroller Placement Algorithm**

Latency among controllers and switches is composed of E2E and queuing latency. As the density of switches in diverse subnetworks changes, controllers of subnetwork possess massive switches and encounter higher traffic load.

Algorithm 2: Multicontroller Placement Algorithm

Step 1: Choose subnetwork and evaluate all simultaneous packets transmitting rate

$$\lambda = \sum_{n=i} \lambda_i$$

Step 2: Set network topology = 1 and evaluate delay/latency among nodes.
Step 3: Compute the maximum total latency.
Step 4: Increase node until the total delay is smaller than the threshold value.
Step 5: Set K = 'm' and carry out proposed network partitioning that is associated with subnetwork into 'm' subnetwork and discover its centroids.
Step 6: Place 'm' controllers into centroids.
Step 7: Repeat steps 1–6 until subnetworks are placed with suitable controllers.

The objective is to compute the number of controllers by examining total delay (max { D_{total})} and the position of the controller to the centroid. The number of controllers is continuously increased until latency is lesser than the delay threshold D_T. If appropriate controllers are placed, it is used for partitioning and determines 'm' clusters.

(c) **Markov Decision Process for Controller Placement**

Here, the proposed model is infinite state, discrete and finite action Markov decision process as 4-tuple, where 's' is a limited set of states, 'A' is a finite set of actions, and 'P' is the probability of state transition to new state till action is initiated. 'R' is expected immediate reward received after state transition, based on the control activities carried out.

(d) **Multicontroller Placement Parameters**

There are three parameters to model a finite set of states: (1) flow queue size F_{qs}, (2) flow entry size (F_{es}), and (3) transmitted packets count $\left(T_{pc}\right)$.

$$C_p = F_{qs} * F_{es} * T_{pc} \tag{1}$$

$$A \in \begin{cases} 1 \text{ Broadcast} \\ 0 \quad \text{remain} \end{cases} \tag{2}$$

(e) **Placement Reward Function**

Immediate reward R_I is attained after state transition with $F_{qs}, \; F_{es}, T_{pc}$

$$R_I = R_{F_{qs}} + R_{F_{es}} + R_{T_{pc}} \tag{3}$$

$$R_I = w_f R_{F_{qs}} + w_q R_{F_{es}} + w_t R_{T_{pc}} \tag{4}$$

$$\begin{cases} 1 & 0 = f = F_{min} \\ F_{max} - f/(F_{max} - F_{min}) & F_{min} < f < F_{max} \\ 0 & f = F_{max} \end{cases} \tag{5}$$

where 'f' is available flow entry size. F_{min} is the minimum acceptability of unused flow. F_{max} is the maximum acceptable unused flow. Rewards with flow queue size F_{qs} and transmitted packet count T_{pc}:

$$R_{F_{qs}} = \begin{cases} 1 & 0 \le q \le Q\_\min \\ \frac{Q_{\max} - q}{Q_{\max} - Q_{\min}} & Q\_\min < Q < Q\_\max \\ 0 & q \ge Q\_\max \end{cases} \tag{6}$$

where 'q' is queue size. $Q_{\min}$ is a minimum queue size.

$$R_{T_{pc}} = \begin{cases} 0 & 0 \le t \le T\_\min \\ \frac{T\_\max - t}{T\_\max - T\_\min} & T\_\min < t < T\_\max \\ 1 & t \ge T\_\max \end{cases} \tag{7}$$

where 't' specifies transmitted packet count at the present state, $T_{\min}$ is a minimum acceptable transmitted packet count, $T_{\max}$ is the maximum fair transmitted packet count. The cost is based on the above-mentioned reward functions. With the data flow over the network and reduced queue size, the transmission process is reduced. Therefore, it influences the cost function of the model.

3.2 *Multiagent Multiaggregator (M-MA)*

Generally, SDN controller resources are pooled to serve as multiple switches dependent on multitenant models with diverse virtual and physical resources flexibly and dynamically allocated and reassigned to data plane requirements. This work utilizes a novel multiagent aggregator-based decision model as the primary material for this research.

Algorithm 3: Multiagent-Based Controller

Input: Nodes with connected network topology.

Output: Average delay computation.

Step 1: Allocate with appropriate clustering.
Step 2: Compute E2E delay computation between connected nodes.
Step 3: Randomize node for network initialization.
Step 4: Validate actual network centroid.
Step 5: Determine the route between connected nodes.
Step 6: Merge controllers agents without message loss and track package route.
Step 7: Deploy multicontrollers for core partitioning.
Step 8: Compute load of controllers with Eq. (8).
Step 9: Determine the objective function with Eq. (9)
Step 10: Determine link utilization rate.
Step 11: Compute path reliability among nodes '*i*' and '*j*'.
Step 12: Compute the shortest path between nodes '*i*', and '*j*'.
Step 13: Evaluate the centrality index of normalization degree.
Step 14: Compute multiobjective constrain with Eqs. (21), (22) and (23)
Step 15: Output delay based on constrained factors.

To fulfill the reproducibility of multiagent functionality, decision modeling is available for partitioning control plane. This model makes agents simulate the characteristics of SDN components. Network-based multiagents specify network components like a group of clusters, and these may hold varying characteristics. Moreover, some of the network agents may possess non-proper or malicious characteristics. These characteristics are intended for compromising privacy and considering non-authorized data. From this perspective, these kinds of network agents considered are termed as 'malware agents'. So as to differentiate type of malware agents from other agents, this work specifies it as 'normal network agents' in this work. Then, the 'controller agent' specifies the centralized controller of SDN. In multiagent-based aggregation, these agents primarily work to integrate isolation protocol. In multiagent-based aggregation, these agents may work primarily to merge the isolation protocol. Controller agents join the table with trust with every network agent as a 0–100% interval. This aggregation may simulate a massive amount of messages from sender to receiver. The receiver may establish that message was received or lost appropriately.

(a) **Markov Decision Process-Based Load Balancing**

In SD-WAN, the control plane is deployed by multiple controllers; these controllers monitor and handle core partitioning. Henceforth, load balancing during partition is considered as an essential performance to resolve controller deployment. Some investigations find balancing of a number of switches in every partitioning. The flow size is processed by all switches, as in Eq. (8):

$$L = \left(\sum_{i=1}^{k} \alpha\left(\frac{\sum_{j=1}^{N_i} \frac{\lambda_{i,j}}{k}}{\sum_{i=1}^{K}\sum_{j=1}^{N_i} \lambda_{i,j}} - \frac{1}{k}\right) + \beta\left(\frac{N_i}{N} - \frac{1}{k}\right)^2\right)^{\frac{1}{2}} \tag{8}$$

where $\lambda_{i,j}$ is the request rate of a switch, $\alpha \& \beta$ is weighting coefficient during load balancing. When $\alpha = 1$ is load balancing index, $\beta = 1$ is partitioned by the number of switches. The objective function is specified as in Eq. (9):

$$\min L \tag{9}$$

Load balancing constraints are as in Eqs. (10)–(12):

$$\overrightarrow{f_{vi}} = \overleftarrow{f_{vi}}, V_i \in V \tag{10}$$

$$l_{i,j} \leq 1, (i, j) \in E \tag{11}$$

$$\sum_{i \in N_i} \lambda_{i,k} \leq \phi_k, k \in K \tag{12}$$

Equations (10), (11), and (12) specify controller partition-based flow that is provided equally. The link utilization of every link is lesser than 1. Higher than reliability, the control plane works continuously without any failure. Failure results in interruption between the domain controller and switch. Henceforth, security is also an essential performance metrics to resolve controller deployment. The link includes communication between domain and switch controller, and between the root controller and the domain controller. Path reliability among 'i' and 'j' is depicted as in Eq. (13):

$$R(i, j) = \prod_{e \in E_{i,j}} (1 - p_e) \prod_{v \in V_{i,j}} (1 - p_v) \tag{13}$$

where $E_{i,j}$ is a link set of shortest path among node 'i' and 'j', '$V_{i,j}$ is the shortest path set between node 'i' and 'j'.

$$R_c(k) = \sum_{j \in C_k} R(cc(k), j) \tag{14}$$

where C_k is the kth partition where node 'j' is located, path reliability of the root controller is Eq. (15):

$$R_{rc} = \sum_{k \in K} R_c(k) \cdot R(cc(k), rc) \tag{15}$$

Various studies consider path reliability in resolving the reliability crisis; however, do not consider other node factors. Here, not only the safety of links and nodes are considered; however, request rate and node degrees are considered. Node degree is the number of edges related to the node. The centrality index of the normalization degree is provided as in Eqs. (16)–(18):

$$R_{\text{degree}} = \frac{\text{degree}(i)}{\max\limits_{i \in N}(\deg(i))} \tag{16}$$

$$R_{\lambda,i} = \frac{\lambda_i}{\max\limits_{i \in N}(\lambda_i)} \tag{17}$$

$$R_i^{rc} = R_{rc} \cdot R_{\text{degree}} R_{\lambda i} \tag{18}$$

When choosing the domain controller and root controller, objective function as in Eqs. (19)–(20):

$$\max R_i^c \tag{19}$$

$$\max R_i^{rc} \tag{20}$$

The multiobjective constraint is provided as in Eqs. (21)–(23):

$$\sum_{i \in N_i} \lambda_{i,k} \leq \phi_k,\ k \in K \tag{21}$$

$$\sum_{k \in K} \sum_{i \in N_i} \lambda_{i,k} \leq \phi \tag{22}$$

$$\sum y_{cc}^{rc} = K \tag{23}$$

Above-mentioned Eqs. (21), (22) and (23) are based on a total load of every domain lesser than maximum load tolerance of domain controller. The tolerance level of the controllers is higher with the Markov decision process and reduces the cost by improving the efficiency. The load distribution and cost are directly proportional to one another. When cost is high, the load to the available nodes is also higher. When the cost is low, the load to other nodes is also lesser.

3.3 Gaming Theory

The gaming theory model is discussed for controller placement in SD-WAN. Generally, the game theory concept facilitates analyzing and describing various placement problems between agents who have to make appropriate decisions to ensure their objectives. An extensive work comprises of building a tree with a combination of decisions with time notation in successive actions p_{ij} is payoff (payload) of every player. 'i' is a combination of judgments while it refers to players. For instance, p_{11} is a payoff combination of players' decisions, while p_{12} is payoff combination of player two choices. This model may provide a better solution by understanding the solution generated by the previous solution by players. Tree construction is a particular representation of the different combination of probable decisions provided to every player in an alternative manner. Therefore, it is essential to move from an extensive tree to a standard form to perform a Nash equilibrium. This may provide a global solution. The objective cost function of controller placement has been taken into account. The combination of both long-term and short-term decisions is considered here. The long-term decision is made only when clusters include a considerable amount of nodes.

Algorithm 4: Partitioning and Placement Algorithm

Input: Network topology; subnetworks.

Output: switches allocation and centroids; delay, throughput, packet loss.

Step 1: Chose random nodes as the initial center of topology.
Step 2: Evaluate the network center.
Step 3: Choose nodes with lesser end-to-end delay from the centroid.
Step 4: Determine the next center using the gaming model.
Step 5: Set next center with higher end-to-end delay from the second center.
Step 6: Provide vertex $v \in V$ as a cluster for placement.
Step 7: Update centroid with E2E delay for all nodes in the cluster.
Step 8: Repeat steps 3–7 until network controller placement is done correctly with partitioning to subnetworks.
Step 9: Place controller into centroids.
Step 10: Repeat till appropriate placement.

Next, short-term planning is based on the available amount of lesser nodes. For all nodes, investigators determine the number of nodes that may include or exclude nodes into the cluster (short-term prediction and power consumption of nodes). To make appropriate decisions, short-term tree is generated for all nodes connection. Equilibrium is evaluated for all connected nodes. To test the efficiency of the proposed gaming model, an analysis has been done with delay, packet loss, and throughput. Here, the testing scenario is performed, and experiments are carried out with various metrics that may facilitate insight toward WAN controller performance. The impact of propagation delay is made with Nash equilibrium without any time delay. Therefore, this may improve the number of responses per second. This may provide higher propagation latency. Moreover, consideration of these latencies is not appropriate as in Eqs. (24) and (25):

$$\min\{\max\{D_{\mathrm{E2E}}(c_n) + D_{\mathrm{queue}}\}\} \tag{24}$$

$$c_n \in \mathrm{SDN}_i(\forall i \in k) \tag{25}$$

where k is the total number of controller, c_n is nth controller, D_{E2E} end-to-end delay, and D_{queue} is delay of waiting. Controller placement has to fulfill reliability and performance constraints. Here, the evaluation of controllers is done by throughput computation. This is specifically to examine the average throughput of all controllers with a different amount of connected nodes. SDN controller placement is measured by two essential characteristics: throughput and latency. The objective is to acquire minimal latency and maximal throughput for all controllers. These performance metrics were analyzed in a below-given section. This approach uses SDN controller placement based on gaming theory with the core concept is to use the Nash equilibrium. Many SD-WAN providers charge is based on volume of traffic transiting services. This is done by assigning an additional cost based on circuit size connecting to the SD-WAN. 10 Mbps costs X, and a 25 Mbps line costs 2X. These costs add up for a large network. The cost savings is approximately $30–50/month/site.

4 Numerical Results and Discussions

This section demonstrates numerical values attained by performing computation with the above-mentioned models. Here, simulation is shown in the NS-3 simulator. Metrics like delay, packet loss, and throughput were considered.

Tables 1 and 2 depict a comparison of delay encountered before/after node placement in SD-WAN. The total number of nodes is 20 used for simulation.

Tables 3 and 4 depict a comparison of packet loss encountered before/after node placement in SD-WAN.

Tables 5 and 6 depict comparison of throughput encountered before/after node placement in SD-WAN.

Figure 1 depicts a graphical representation of the average delays in milliseconds. 'X'-axis shows the number of controllers, and the 'Y'-axis shows the average delay (ms). Figure 2 depicts graphical representation average packet loss. Figure 3

Table 1 Delay (ms) before node placement

S. no.	No. of nodes	Delay (ms) with Markov decision process	Delay (ms) computation with M-MA	Delay (ms) computation with gaming theory
1	0	0.042	0.035	0.030
2	5	0.087	0.080	0.075
3	10	0.096	0.090	0.085
4	15	0.112	0.110	0.105
5	20	0.134	0.125	0.120

Table 2 Delay (ms) after node placement

S. no.	No. of nodes	Delay (ms) with Markov decision process	Delay (ms) computation with M-MA	Delay (ms) computation with gaming theory
1	0	0.17	0.15	0.10
2	5	0.94	0.90	0.80
3	10	1.37	1.3	1.2
4	15	2.19	2	1.5
5	20	3.03	2.98	2.80

Table 3 Packet loss (bps) before node placement

S. no.	No. of nodes	Loss (bps) with Markov decision process	Loss (bps) computation with M-MA	Loss (bps) computation with gaming theory
1	0	1.20	1.15	1.10
2	5	3.0.82	3.0.80	3.75
3	10	8.56	8.50	8.45
4	15	10.47	10.40	10.35
5	20	12.98	12.90	12.85

Table 4 Packet loss (bps) after node placement

S. no.	No. of nodes	Loss (bps) with Markov decision process	Loss (bps) computation with M-MA	Loss (bps) computation with gaming theory
1	0	9.67	8.60	8.50
2	5	14.14	13.15	13.10
3	10	17.98	16.95	16.90
4	15	21.82	20.75	20.70
5	20	24.46	23.50	23.45

Table 5 Throughput (bps) before node placement

S. no.	No. of nodes	Throughput (bps) with Markov decision process	Throughput (bps) computation with M-MA	Throughput (bps) computation with gaming theory
1	0	14.45	14.40	13.30
2	5	13.18	13.10	12.05
3	10	10.17	10.10	8.05
4	15	6.21	6.15	6.10
5	20	3.31	3.25	2.20

Table 6 Throughput (bps) after node placement

S. no.	No. of nodes	Throughput (bps) with Markov decision process	Throughput (bps) computation with M-MA	Throughput (bps) computation with gaming theory
1	0	26.45	25.54	30.56
2	5	25.39	24.30	30.54
3	10	24.26	22.20	28.56
4	15	23.88	22.77	25.54
5	20	22.90	20.85	23.56

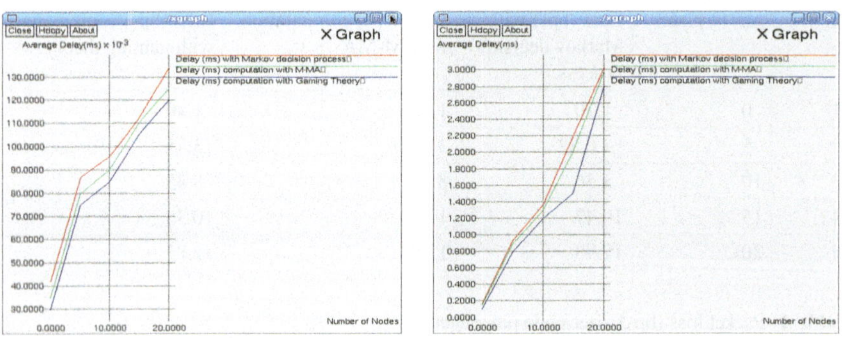

Fig. 1 Delay computation before and after node placement

illustrates graphical representation average throughput in milliseconds. From all the above-mentioned models, the gaming model is considered to work well during controller placement and network expansion.

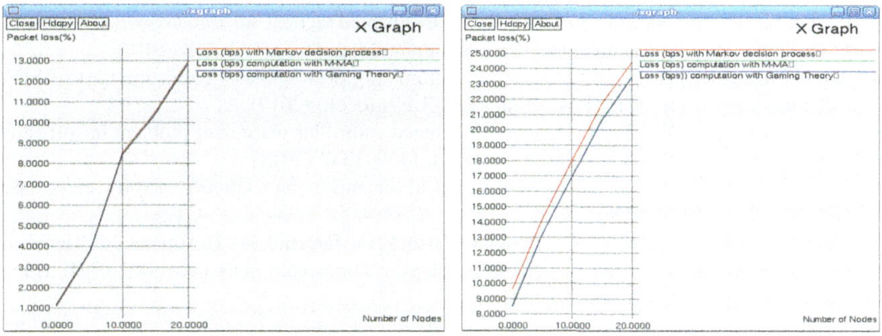

Fig. 2 Packet loss before and after node placement

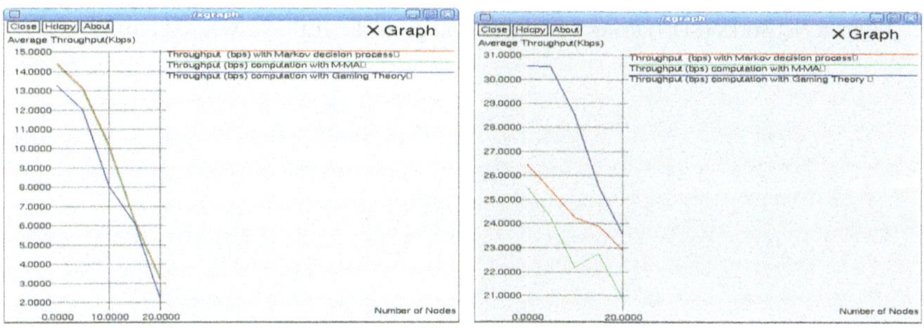

Fig. 3 Throughput before and after node placement

5 Conclusion

The controller placement is hugely significant in SDN as it provides a crucial impact on controller placement and network expansion. Here, a novel mathematical model has been initiated to assist operators in planning and upgrading the SDN infrastructure. As network expansion is considered as a generalized problem, the anticipated model can be applied for solving these above-mentioned problems. As a future research direction, some of the factors have to be analyzed. They are: (1) effectual cost multiplier; (2) valid network infrastructure quality and output of model expansion; and (3) implementation of heuristic approaches.

References

1. Qin, Q., Poularakis, K., Iosifidis, G., Tassiulas, L.: SDN controller placement at the edge: optimizing delay and overheads. In: IEEE Infocom (2018)
2. Wang, J., Miao, Y., Zhou, P., Hossain, M.S., Mizanur, S.: A software defined network routing in wireless multihop network. J. Netw. Comput. Appl. **85**, 76–83 (2017)

3. Poularakis, K., Iosifidis, G., Tassiulas, L.: SDN-enabled tactical Ad Hoc networks: extending programmable control to the edge. IEEE Commun. Mag. **56**(7), 132–138 (2018)
4. Xu, H., Yu, Z., Qian, C., Li, X., Liu, Z.: Minimizing flow statistics collection cost of SDN using wildcard requests. In: Proceedings of IEEE Infocom (2017)
5. Yao, G., Bi, J., Li, Y., Guo, L.: On the capacitated controller placement problem in software-defined networks. IEEE Commun. Lett. **18**(8), 1339–1342 (2014)
6. Guo, M., Bhattacharya, P.: Controller placement for improving resilience of software-defined networks. In: IEEE ICNDC (2013)
7. Lange, S., Gebert, S., Zinner, T., Tran-Gia, P., Hock, D., Jarschel, M., Hoffmann, M.: Heuristic approaches to the controller placement problem in large scale SDN networks. IEEE Trans. Netw. Serv. Manage. **12**(1), 4–17 (2015)
8. Ksentini, A., Bagaa, M., Taleb, T., Balasingham, I.: On using bargaining game for optimal placement of SDN controllers. In: Proceedings of IEEE ICC (2016)
9. Ul Huque, Md.T.I., Si, W., Jourjon, G., Gramoli, V.: Large-scale dynamic controller placement. IEEE Trans. Netw. Serv Manage. **14**(1), 63–76 (2017)
10. Rath, H.K., Revoori, V., Nadaf, S.M., Simha, A.: Optimal controller placement in Software Defined Networks (SDN) using a non-zero-sum game. In: IEEE WoWMoM (2014)

An Adaptive Data Dissemination in VANETs Using Dynamic Routing

R. V. S. Lalitha, A. Lakshmi, Haritha Darisi, Surya Kala Nagireddi, and K. Kavitha

Abstract VANET communication mechanisms mainly focus on improving the reliability of the links and finding energy-efficient routes. Due to high mobility characteristic of nodes and dynamic topology, VANET communication poses the problem of link failures. The frequent disconnections in the network restrain communication among the vehicular nodes. Vehicular nodes require frequent path maintenance in dynamic topology environment. In this paper, information of the VANETs is maintained dynamically in routing tables. The novelty in this research methodology is that when a link failure occurs routing tables are updated so that neighboring nodes will get detected. The change in transmission path takes place and message will get transmitted in the new path tracked. This exploits efficient transmission even during times of failure of links. The topology information of the VANET is maintained in routing tables and is updated time to time for facilitating accurate topology information and for tracing alternate path in case of link failures if any. In this paper, dynamic topology based algorithm is used to detect link failures and establishing alternate paths dynamically.

Keywords VANET · Routing table · Topology based routing · Link failures · Path maintenance and Wireless Ad hoc Networks

R. V. S. Lalitha (✉) · H. Darisi · S. K. Nagireddi
Department of CSE, Aditya College of Engineering and Technology, Surampalem, India

A. Lakshmi
HCL Technologies, Bangalore, India

K. Kavitha
Department of CSE, Gokaraju Rangaraju Institute of Engineering and Technology, Hyderabad, India

V. Bhateja et al. (eds.), *Computer Communication, Networking and IoT*, Lecture Notes in Networks and Systems 197, https://doi.org/10.1007/978-981-16-0980-0_33

1 Introduction

Vehicular Ad hoc Networks are formed with vehicles as nodes. The communication between nodes in the VANET improves road safety measures. The key challenge of Vehicular Ad hoc Network (VANET) is Vehicle-to-Vehicle (V2V) communication, Vehicle-to-Infrastructure (V2I), and Infrastructure-to-Vehicle (I2V). The routing mechanisms in VANETs are ad hoc routing, position-based routing, cluster-based routing, broadcast routing, geocast routing, etc. Ad hoc On Demand Distance Vector (AODV) protocol is used in Ad hoc routing. In the position-based routing, Greedy Perimeter Stateless Routing (GPSR) is used. In the cluster-based routing short-lived clusters are created for message transmission. In VANETs messaging may be of unicasting/multi-casting/broadcasting. The qualitative communication is achieved by improving the design factors of communication mechanism. VANETs use messaging through neighboring vehicles using multicasting. They also check whether the messages are already received or not using epidemic messaging algorithms [1]. The conventional mechanisms operate well with strong signal communication and without link failures. Since the vehicles travel with high speed, there will be frequent disconnections in the network. Hence, routing mechanisms need to be enhanced apparently for achieving effective communication. In this paper, initially, data dissemination between two nodes is performed based on the routing table information. In case of link failures routing table information is updated for establishing path. This improves reliable transmission capability across VANET nodes. In order to observe the current day statistics, Plymouth database is downloaded from the Internet which is free. https://plymouth.thedata.place/dataset/traffic-counts/resource/c99a8d79-a61b-4563-ae55-f2c4d54ae2d1. The traffic data of various vehicles moving from various junctions is shown in Fig. 1 and type of vehicles moving from one junction is shown in Fig. 2. It is evident from the diagram that more number of cars and taxis are moving from single junction, which infers that efficient VANET communication became essential.

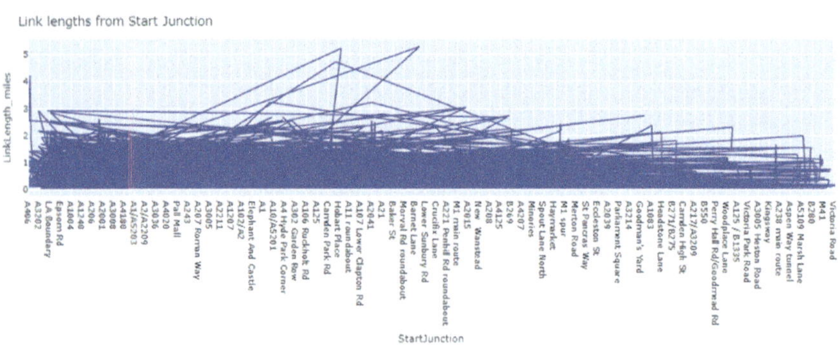

Fig. 1 Link lengths from each junction

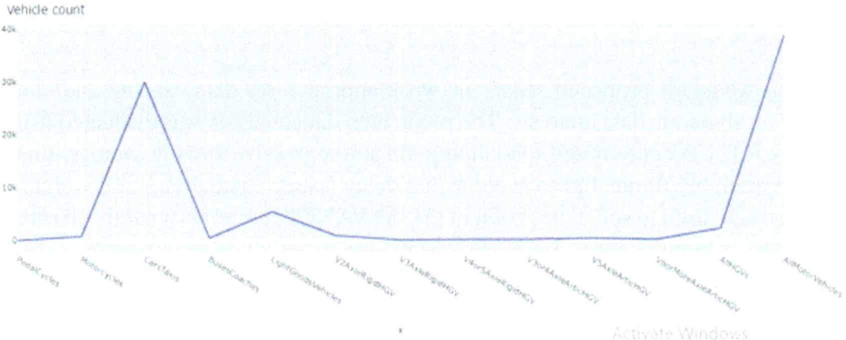

Fig. 2 Vehicle count from one junction

A comprehensive survey on dynamic topology challenges are elaborated in Sect. 2, and the state of the art of adaptive data dissemination using dynamic routing is explained in detail in Sect. 3 which includes Architecture of adapted data dissemination in VANET, Proposed algorithm to display topology information, Updating routing table information, Initiating Data dissemination, Path establishment between two nodes, Checking for link failures for reestablishing paths and embedding access list information in gateways is discussed to improve dynamic routing procedures. Finally, the concept is concluded by stating the improved efficiency of VANET communication with the new approach.

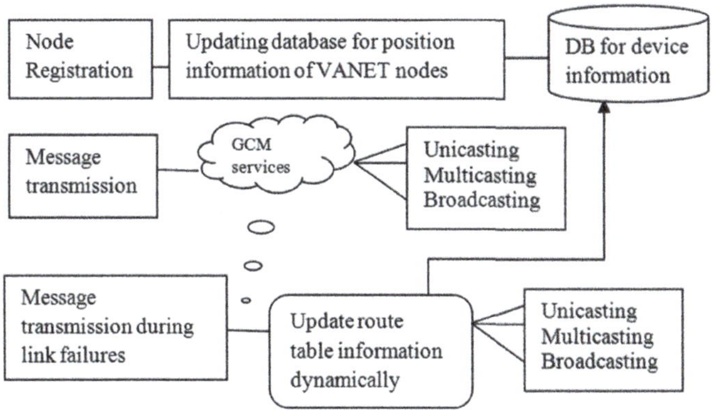

Fig. 3 Adapted data dissemination in VANETs

2 Related Work

Sahraoui Abdelatif proposed sensor network approach for data sensing and cloud storage for dynamic data analysis. The predicated data analysis was evaluated using fuzzy logic [2]. A delay-based spatialtemporal autoregressive moving average model was proposed by Yanjie Tao to avoid traffic delay issues. Spatialtemporal weighted matrices were used to solve the problem [3]. As VANETs are very dynamic, to establish robust communication Ankita Srivastava proposed a communication methodology using Ant Colony algorithm. In the first phase, neighboring nodes are predicted and in the next phase route objects will be tracked [4]. With the specific characteristic of VANETs and short-range communication issues, data dissemination will be based on decision-making of VANETs. Joahannes B. D. da Coasta proposed a traffic management system using DDRX. The tracking of nodes will be based on sub-graphs created using DDRX. Coverage issues can be minimized using this algorithm [5]. The research challenges and issues in VANETs were well explored by Muhammed Sameer Sheikh. In addition to that the security and location privacy problems in VANET communication mechanisms and Vehicular cloud computing were discussed [6]. Shujuan Wang proposed an algorithm named Instantly Decodable Network Coding during tough transmission times. This application assigns high priority to safely related applications and low priority for data requests in order to avoid congestion. The simulation results provided minimizes delay time and miss ratio [7]. Muhammad Ali discussed position-based emergency messaging using cluster-based and broadcasting techniques to reduce communication delays. In his paper, he clustered vehicles moving in the same direction and PDR was improved upto 14% with the proposed mechanism [8]. Xiaofeng Liu illustrated overview of performance metrics of VANETs for controlling congestion in VANETs. Event-driven, reactive and proactive congestion control mechanisms were elaborated based on CBR and transmission parameters. The typical driving scenarios like low vehicle density, high vehicle density, acceleration and lane changing, car crash were illustrated with respect to safety messaging [9]. Rakesh Shrestha proposed the application of edge computing in block chain to secure communication messaging. The positions of the vehicles are broadcasted using beacon messages [10].

3 Adaptive Data Dissemination in VANET Using Dynamic Routing

Initially, nodes that wish to communicate needs to get registered. They can adapt any conventional communication modes for transmitting messages. If the link failure occurs, and communication fails, then by updating routing table information, the message can be transmitted via newly established path. The adapted data dissemination in VANET is shown in Fig. 3.

The algorithms for displaying topology information and message transmission are given below.

3.1 Proposed Algorithm to Display Topology Information

```
Algorithm upload (nodeinf, latitude, longitude){
1. Register node information along with position information (lat,long)
2. Make an entry in the routing table
3. Display the topology of the VANET in a particular region.}
Algorithm Send (Message m, Src, Dst){
1. Check the link between Src and Dst.
2. If link is established between src and dst, message transmission follows through routing table entries.
3. Update routing table information dynamically to identify new paths if any.
4. Send message using entries of the updated route tables.}
```

3.2 Updating Routing Table Information

The topology of the VANET is maintained in the server by storing the details of their location and speed. The ecosystem is built using Net Beans IDE and Java. The topology of VANET information is displayed in Figs. 4 and 5.

3.3 Initiating Data Dissemination

The initial entries of the route tables will be based on the number of vehicles in the destination node, via node, and path cost. The adaptive mode of transmission in VANET is shown in Figs. 6, 7, 8, 9, 10 and 11.

3.4 Path Establishment Between Two Nodes

Due to mobility of the vehicles, the route tables will be updated in reactive mode. The details of source and destination vehicles will be based on the path, which follows

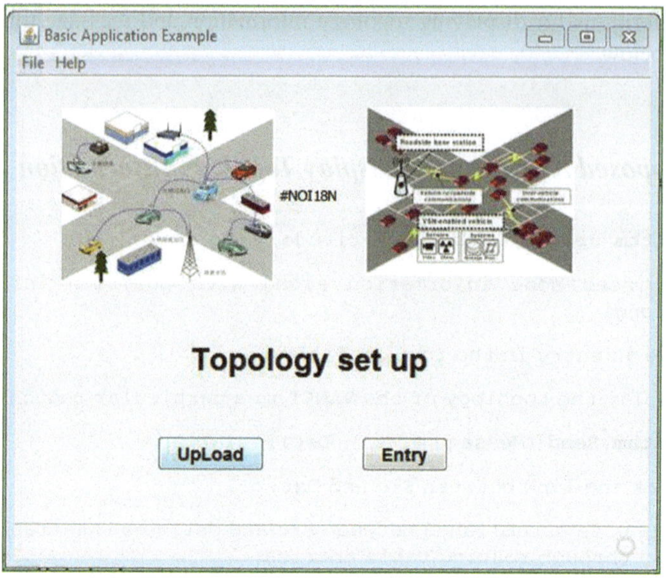

Fig. 4 Uploading position information to display VANET topology

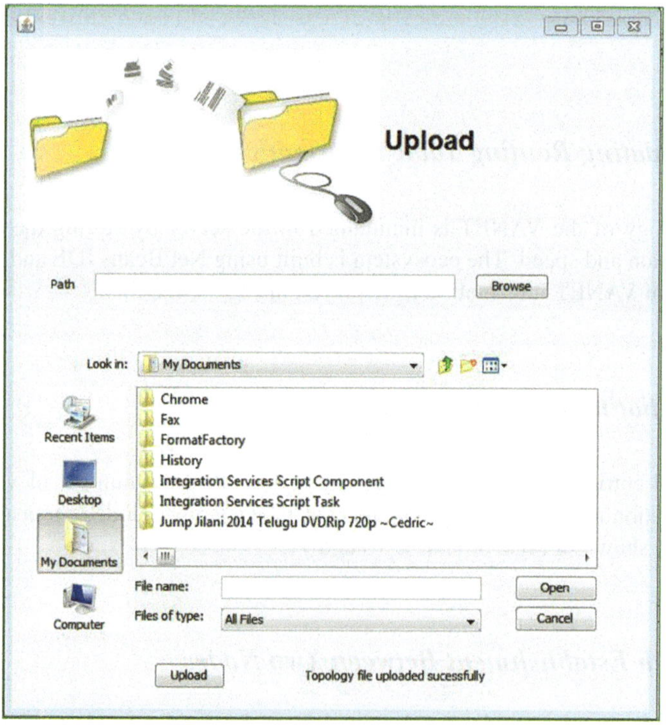

Fig. 5 Setting topology after successful upload of position information of VANET nodes

Fig. 6 Entering node information in VANET ecosystem

Fig. 7 Current topology of VANET nodes

the route guided using entries in the routing table. Before sending a message, it has to be checked for the availability of the route between the source and destination. If the route is not available, routing table needs to be updated. Also, the links from source node to destination node will be tracked.

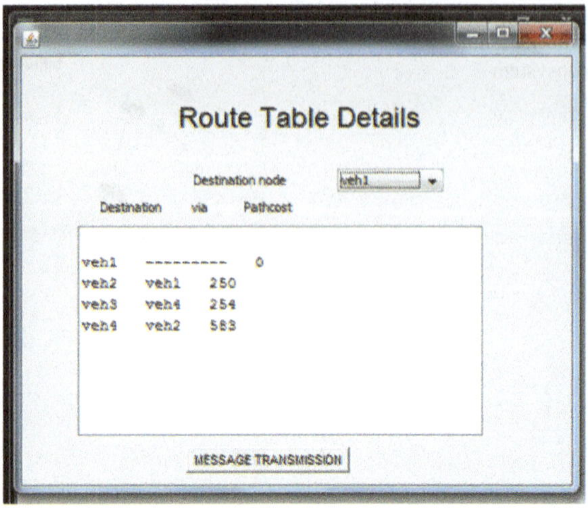

Fig. 8 Routing table information

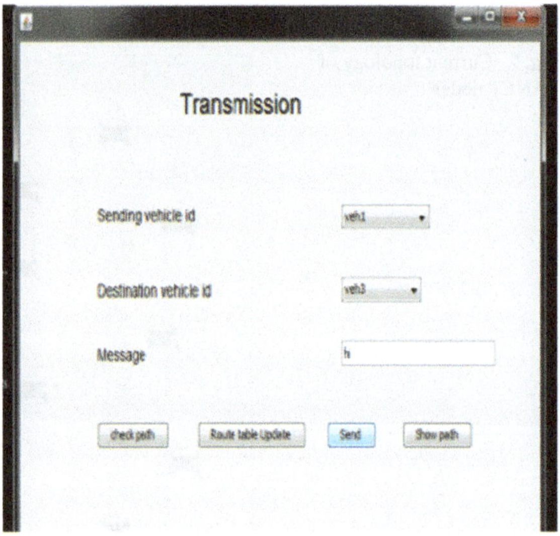

Fig. 9 Mode of message transmission in VANET

3.5 Checking for Link Failures and Reestablishing Paths

Due to the mobility of the vehicles, the existing link may become weak or fail. If the link failure occurs, it will discover the new path by calling updating method of route table. New route from source to destination needs to be identified due to the existing path failure. A new path will be initiated based on the routing table information in an optimized manner. The message transmission will then be based on recently updated route tables. A scenario is depicted from Figs. 12, 13, 14, 15 and 16.

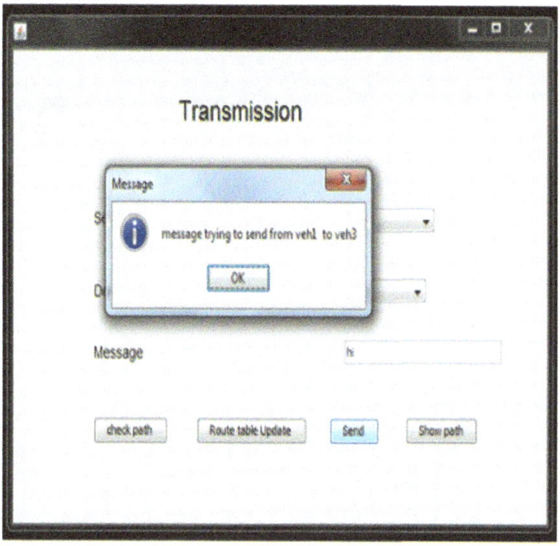

Fig. 10 Message transmission

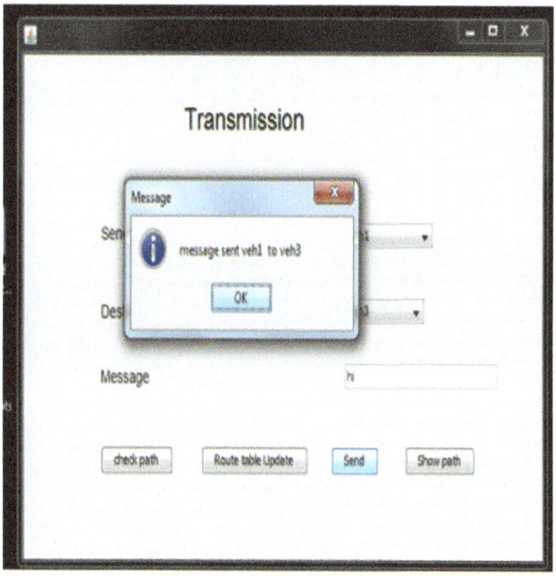

Fig. 11 Transmission of message from vehicle 1 to vehicle 3

3.6 Refining Routing Mechanism with Access Lists and Virtual Routers

Routing enables nodes to communicate with one another in the same network or on a subnet. In case of link failures, they communicate through neighborhood networks. Routers connect multiple networks or subnets by configuring IP. In static routing, the

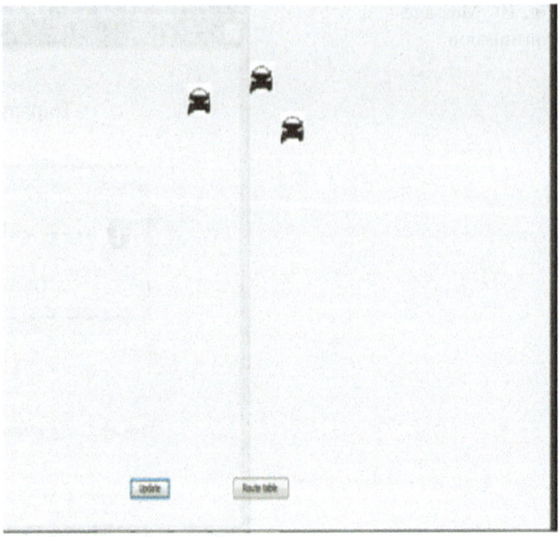

Fig. 12 Instance of link failure in VANETs

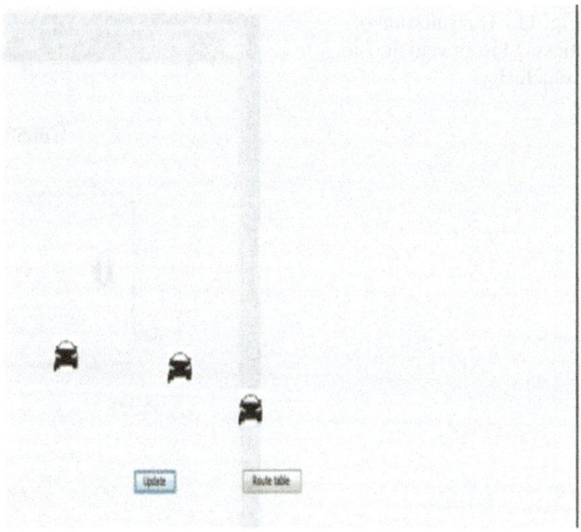

Fig. 13 Updating routing table information for re-routing

routing information is updated manually. Static routing maps IP network addresses to destination address. Dynamic routing relies on incoming information and reachability of the destination nodes. Multicast routing enables routers to disseminate data to one or more destination nodes. If the routing information changes frequently due to mobility in nature Routing Information Protocol (RIP) is used to update the database dynamically. Since the data is very big and time-varying, communication between nodes will definitely falls in one of the static/dynamic/multicast routing mechanisms. One solution to maintain constantly varying data is by maintaining

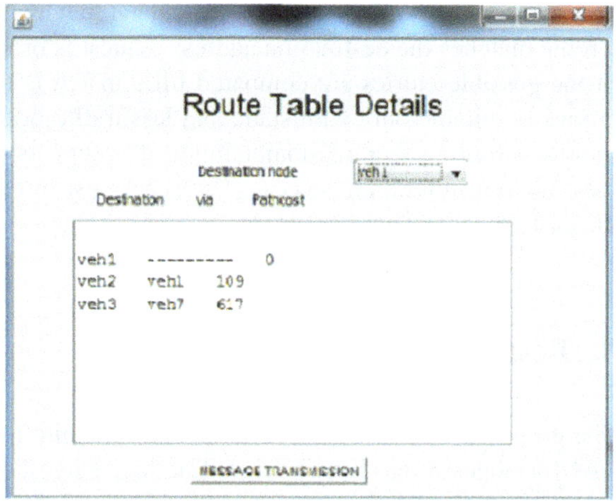

Fig. 14 Updated routing table information

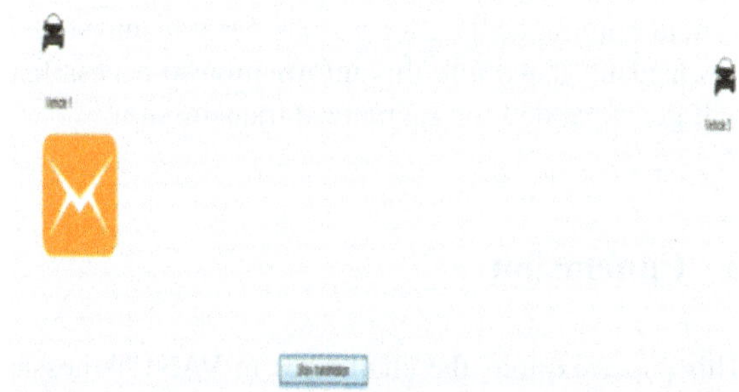

Fig. 15 Re-route is established after updating routing table information

Fig. 16 Message transmission after tracking path

access lists. Access list checks for the subnet to which the destination IP belongs. If a route matches the destination address request is processed. The access list entries of the possible entries are compared till a match is found. Updating routing table entries is suitable only with static and less traffic-oriented networks. Dynamic and multicast routing requires maintaining of access lists and virtual routers to make message transmission a success. The access list information is to be embedded in the customized gateway software.

4 Results and Discussion

The proposed research method establishes definite message transmission between VANET nodes. It has been analyzed in three stages. At first, message transmission between two VANET nodes when there is no signal interference or connection to destination node is clear. In second phase, mode of message transmission when the destination node is not visible or path between the two nodes is disconnected. In this scenario, by updating the routing table entries new paths will be traced and system automatically chooses one for transmission. To enrich the communication mechanism, if we link this information to access list the scope of communication will get expanded for guaranteed transmission.

5 Conclusion

This paper exploits the challenges in VANET message communication by considering the chances of link failures. By updating the routing table, new route will be established dynamically between the source and destination nodes. Apart from that if this software is integrated with access list information the chances of traversal paths will get enhanced and effective communication takes place. Since the primary requirement of VANET communication is to transmit emergency messages in case of traffic jams/accidents, this idea provides preferential benefits.

References

1. Spadaccino, P., Cuomo, F., Baiocchi, A.: Epidemic and timer-based message dissemination in VANETs: a performance comparison. Electronics **9**, 595 (2020). https://doi.org/10.3390/electronics9040595
2. Abdelatif, S., Derdour, M., Ghoualmi-Zine, N., Marzak, B.: VANET: a novel service for predicting and disseminating vehicle traffic information. Int. J. Commun. Syst. (2020). https://doi.org/10.1002/dac.4288
3. Tao, Y.,Sun, P., Boukerche, A.: A novel travel-delay aware short-term vehicular traffic flow prediction scheme for VANET. In: 2019 IEEE Wireless Communications and Networking

Conference (WCNC) Marrakesh, Morocco. IEEE, Morocco (2019). https://doi.org/10.1109/WCNC.2019.8885921
4. Srivastava, A., Prakash, A., Tripathi, R.: An adaptive intersection selection mechanism using Ant colony optimization for efficient data dissemination in urban VANET. Peer to Peer Netw. Appl. (2020). https://doi.org/10.1007/s12083-020-00892-8
5. da Costa, J.B.D., de Souza, A.M., Rosario, D., Cerqueira, E., Villas, L.A.: Efficient data dissemination protocol based on complex networks' metrics for urban vechicular networks. J. Internet Serv. Appl. **10**, Article no. 10:15 (2019). https://doi.org/10.1186/s13174-019-0114-y
6. Sheikh, M.S., Liang, J., Wang, W.: Security and privacy in vehicular ad hoc network and vehicle cloud computing: a survey. Wireless Commun. Mob. Comput. **2020**, 25, Article id 6758536 (2020). https://doi.org/10.1155/2020/5129620
7. Wang, S., Zhang, Q., Lu, S.: NCaAC: network coding aware admission control in prioritized data dissemination in vehicular ad hoc networks. Wirel. Commun. Mob. Comput. **2020**, 14 (2020). https://doi.org/10.1155/2020/6758536
8. Malik, M.A.A.W., Rahman, A.U., Iqbal, S., Hamayun, M.M.: Position based emergency message dissemination for internet of vehicles. Int. J. Distrib. Sens. Netw. (2019). https://doi.org/10.1177/1550147719861585
9. Liu, X., Jaekel, A.: Congestion control in V2V safety communication: problem, analysis, approaches. Electronics **8**(5), 540 (2019). https://doi.org/10.3390/electronics8050540
10. Shrestha, R., Bajracharya, R., Shrestha, A.P., Nam, S.Y.: A new type of block chain for secure message exchange in VANET. Digit. Commun. Netw. **6**(2), 177–186 (2020). https://doi.org/10.1016/j.dcan.2019.04.003

High Utility Pattern Mining Distributed Algorithm Based on Spark RDD

Sunil Kumar and Krishna Kumar Mohbey

Abstract High utility pattern identification is a key pattern mining method that extracts items with more utility than the mentioned utility threshold limit. The demand for Mining of high utility items has grown in the last few years. It addresses frequent pattern identification barriers, which considers the only frequency of itemset as a measure to find patterns. Several techniques for high utility pattern identification were designed. However, such methods are effective for smaller data sizes and operate on an Independent system. Individual machines can not process massive transaction data files. To deal with that problem, a distributed approach is proposed for High utility mining and represented as Dis-EFIM. It performs partitioned mining tasks on a cluster of multiple nodes. Extensive experiments on various real-time datasets are performed with the proposed Dis-EFIM and found that Dis-EFIM is more efficient than current cutting-edge techniques EFIM and PHUI-Miner.

Keywords High utility pattern identification · Distributed computing · Spark · RDD · Big data

1 Introduction

The immense study is of Frequent Itemset Mining (FIM) has been performed in the context of data analytics. Its task is to determine groups of items that are frequently purchased by the customer. However, Frequent itemset mining does not promise the maximum profit provided by the frequent products discovered. For maximum potential from the extraction procedure an additional measure is defined, known as utility. Utility belongs to the significance of an itemset, e.g., price, quantity of items, and so on. This extension of pattern identification using utility metric is classified as high utility pattern identification, that offers patterns with great business benefits. HUI mining has been widely popular in recent years due to its high relevance. For example, supermarkets will be able not only to recommend products that people usually purchase but also sell items that lead to greater profits. The utility is the

S. Kumar (✉) · K. K. Mohbey
Department of Computer Science, Central University of Rajasthan, Ajmer, India

V. Bhateja et al. (eds.), *Computer Communication, Networking and IoT*, Lecture Notes in Networks and Systems 197, https://doi.org/10.1007/978-981-16-0980-0_34

combination of internal and external utilities. The utility corresponds to the combination of external and internal utilities. External utility depends on the factor of usability/profitability for the business and internal utility contains information of frequency of an itemset in a set of transactional data. A group of items containing utility more than mentioned threshold is selected as a high utility pattern. However, the HUI mining process is considerably more difficult The utility corresponds to the combination of external and internal utilities. External utility depends on the factor of usability/profitability for the business and internal utility contains information of frequency of an itemset in a set of transactional data. A group of items containing utility more than mentioned threshold is selected as a high utility pattern [1]. Several High utility mining (HUIM) algorithms [2, 3] emphasizes the numbers of products and benefits (external utilities). After the emergence of Big Data, several studies have been conducted with parallel computing to compute the large datasets. Only few parallel and distributed algorithms have been proposed so far for the frequent itemset mining as well as for high utility item mining. PHUI-Growth, a parallel HUI mining algorithm, using pattern-Growth on the Hadoop framework [4], was proposed by Lin et al. [5]. PHUI-Miner [6], P-FHM+ [7] uses more efficient distributed Spark framework [8]. Since Spark is able to perform in-memory computation, it is much more efficient. In recent, EFIM [9] technique is proposed to enhance execution time and memory of the HUIM process. In most instances, EFIM is two to three times quicker than the state of the art UP-Growth+ [10], HUP-Miner [11], d2HUP [12], HUI-Miner [3], FHM [13] algorithms and uses up to eight times less memory. However, EFIM is designed for an independent system that is not suitable for processing large datasets. In this paper, a distributed version of the EFIM algorithm is being proposed to address the Big data problem. Dis-EFIM is a Spark RDD based distributed approach that performs on the Spark cluster effectively. The paper is divided into the following parts. Section 1 sets out essential condition for high utility pattern identification. The proposed algorithm is outlined in Sect. 3. Section 4 looks at tests and outcomes, followed by a conclusion in Sect. 5.

2 Preliminaries

Let $S = \{e_1, e_2, \ldots, e_k\}$, be a group of elements in set of transaction of dataset $D = \{t_1, t_2, \ldots, t_l\}$. A finite set of items V such that $V \subseteq S$. In the HUI mining process, several definitions are used as follows. Table 1 is a transaction database and Table 2 reprsents external utility.

Table 1 Set of transactions

TID	Items: freq
t_1	$g:3, h:2, l:1, m:3$
t_2	$h:2, i:2, k:1, l:3, m:1$
t_3	$h:1, j:3, m:2$
t_4	$g:4, h:2, j:3, l:2$
t_5	$g:1, h:2, l:3, m:3$

Table 2 External utilities

Item	g	h	i	j	k	l	m
Value	3	2	3	3	1	2	2

Definition 1 (*Type of Utility*) Count of an element in a transaction is internal utility. The external utility of a element is the usefulness and profitability for the business, which is same for each transaction in dataset unlike internal utility. Internal and external utilities are formulated as $I_U(i, t_l)$ and $E_U(i)$.

Definition 2 (*Utility*) The utility value of a element e in transaction t_l is expressed as $U(e, t_l) = I_U(e, t_l) \times E_U(e)$ The utility of a set of elements V in transaction t_l, i.e., $U(V, t_l)$, is measured as aggregate of all containing elements utility. $U(V, t_l) = \sum_{e \in V \wedge V \subseteq t_l} U(e, t_l)$. In dataset D utility of X, i.e. $U(D)$, is expressed as: $U(X) = \sum_{X \subseteq t_l \wedge t_l \in D} U(X, t_l)$.

Definition 3 (*Transaction Weighted Utility*) TWU is the aggregate of utility of all trasactions containing itemset V. TWU$(V) = \sum_{V \subseteq t_l} TU((t_l))$. Here, TU of t_l is the total of each elements utility in transaction t_l.

Definition 4 (*Local Utility*) Let be an group of element α and an single element $z \in E(\alpha)$. The Local Utility of z w.r.t. α is $lu(\alpha, z) = \sum_{\alpha \cup \{z\} \subseteq t_l \wedge t_l \in D}[U(\alpha, t_l) + re(\alpha, t_l)]$.

Definition 5 (*Sub-Tree Utility*) Let be a set of items α and an item z which can extend α ($z \in E(\alpha)$). The Sub-tree Utility of item z w.r.t. α is $su(\alpha, z) = \sum_{\alpha \cup \{z\} \subseteq t_l \wedge t_l \in D}[U(\alpha, t_l) + U(z, t_l) + \sum_{i \in t_l \wedge i \in E(\alpha \cup \{z\})} U(i, t_l)]$.

3 Dis-EFIM: Distributed High Utility Itemset Mining

In this section, we elaborate the distributed proposed algorithm which identify high utility itemset, named Dis-EFIM that is a distributed extention the of state-of-the-art HUI mining algorithm EFIM.

EFIM uses High-Utility Database Projection (HDP) strategy in order to reduce transaction scanning costs. HDP based on observation that all the items not belong to extensions of α can be overlooked in the process of utility calculation of a group of items α in the $\alpha - subtree$ and max limits on its utility. A set of transactions after removing overlooked unnecessary items is projected set of transactions.

Definition 6 (*Projected Database*) The dataset D projection with an itemset α is described as $\alpha - D = \{\alpha - t_l | t_l \in D \wedge \alpha - t_l = \phi\}$.

An efficient approach which is based on arrays being used to evaluate the upper limits in linear space and time called Fast Utility Counting (FUC). It is based on utility-bin. This method is highly efficient in measuring upper bounds. Utility-bins are an extremely compact data structure.

Definition 7 (*Utility-Bin*) A array size is of length same as total number of elements in dataset D is being used for utility-bin U_b, referred as $U_b[z]$ for each item $z \in S$. Each location of utility-bin is used to store a utility value.

TWU computation for all items. Firstly, a utility-bin array U_b is initialized. Then, the utility-bin is updated as $U_b[z] = U_b[z] + TU(t_l)$ for each item $z \in S$ in every transaction. At the end of the database scan, the utility-bin $U_b[z]$ contains TWU for each item.

Sub-tree utility computation for an itemset α. Utility-bin array is used to calculate the sub-tree utility. Firstly, a utility-bin array U_b is initialized zeros. Then, the utility-bin is updated $U_b[z] = U_b[z] + U(\alpha, t_l) + U(z, t_l) + \sum_{e \in t_l \wedge e \succ z \wedge e \in Secondary(\alpha)} U(e, t_l)$ as for each item $z \in t_l \cap E(\alpha)$.

Local utility computation for an itemset α. A utility-bin array U_b has first initialized zeros. Then, the utility-bin $U_b[z]$ is updated as $U_b[z] = U_b[z] + U(\alpha, t_l) + re(\alpha, t_l)$ for each item $z \in t_l \cup E(\alpha)$. After the last transaction scan, we have $U_b[k] = lu(\alpha, k) \forall k \in E(\alpha)$.

A distributed algorithm for efficient identification of high utility set of items has been proposed in this paper. It provides quick and efficient execution for large-scale transaction data for the Apache Spark platform. Data is spread through several cluster nodes and processed on individual nodes separately. It follows the divide and conquers strategy, in which the entire task is divided into smaller-jobs and processed locally on each worker node. Every node calculates 1-itemset's TWU and sorts every element to TWU in the upward order. Then, transactions are readjusted by sort through items in ascending order of TWU and dropping non-HUI 1-itemset. The search space is split equally between the nodes by a sorted 1-itemsets. Transaction set of data to process locally on each node are produced according to the search space. From the allocated partition data, each node generates HUI individually. The complete operation of Dis-EFIM algorithm is included in Fig. 1.

A substantial strategy is used for the uniform division of the tasks across each worker node [14], in the proposed algorithm, the same approach is implemented. Let, cluster have N number of nodes as worker and $ItmSet = \{it_1, it_2, \ldots, it_k\}$ ordered as per TWU values. Nodes are granted a item one by one as node $1, 2, 3, \ldots, N$, then remaining items are assigned to $N, N-1, \ldots, 1$, and so on. A portion of the search space is allocated to every worker in order to extract HUIs. Every node has given a part of transaction data to process set of items in a search space. A specific disjoint set of items is generated and evaluated locally at a distinct node which helps to mitigate the workload and data communication across the cluster nodes. The algorithm used to generate specific projected transactions for each node in cluster is shown in Algorithm 1.

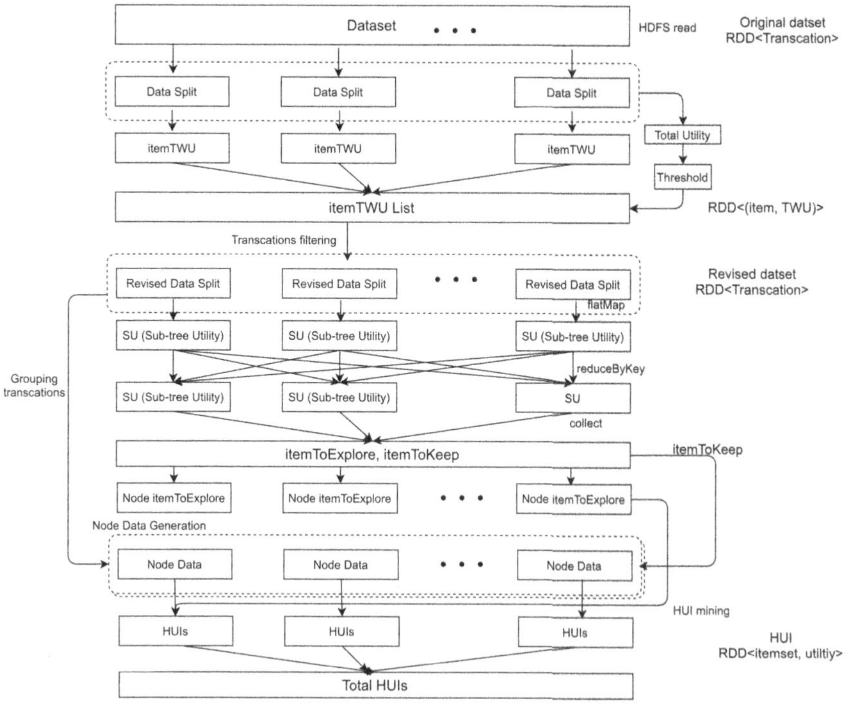

Fig. 1 Process of proposed Dis-EFIM approach

Algorithm 1: Generating node data

Input: D:dataset, itemToExplore, nlst
Output: $< node\_id, transaction >$

for all transaction $t \in D$ **do**
  added ← empty set
  for i ← 0 *to* $|t| - 1$ **do**
    find item t[i] in mapItmToNode and get node id nid
    if nid is not in added **then**
      add nid to added
      st ← subset of t from i to end
      output $< nid, t >$
    end if
  end for
end for
for i ← 0 *to* $|nodes|$ **do**
  nodeitem[i] ← itemToExplore ∩ nodeData.groupBy(i)
end for

After producing the node data, each node mines local HUIs for assigned transactions. Algorithm 2 demonstrates the mining process.

Algorithm 2: The Mining procedure (Miner)

Input: α: prefix itemset, $\alpha - D$: the α projected Node Data, itemToKeep: the primary (α) , itemToExplore: the secondary (α), minthres: minimum utility limit

Output: the set of HUI's

for all *item i* $\in$ *itemToExplore* **do**
  $\beta \leftarrow \alpha \cup \{i\}$
  Calculate $U(\beta)$ and create $\beta - D$
  if $U(\beta) \geq minthres$ **then**
    output β
  end if
  Calculate $su(\beta, w)$ and $lu(\beta, w)$ for all item w $\in$ itemToKeep
  itemToExplore $\leftarrow \{w \in itemToKeep \mid su(\beta, w) \geq$ minthres$\}$
  itemToKeep $\leftarrow \{w \in itemToKeep \mid lu(\beta, w) \geq$ minthres$\}$
  Miner (β, $\beta - D$, itemToExplore, itemToKeep, minthres)
end for

4 Experiment Results

To evaluate the results we have created 4 node cluster on spark framework which include three nodes as worker node and one master node. The configuration of nodes in the cluster are: Ubuntu 18.04.3 LTS OS, 8 VCPU, Memory 8GB, Disk 1TB, Spark framework V-2.3.0, Java V-8, and Scala V-2.13.0. Each node in the cluster are linked using a switch and Ethernet of speed 100 MBPS. The proposed algorithm employs HDFS to store and process information and Scala the programming language selected to develop the programs. Dis-EFIM was compared with PHUI-Miner [6] and EFIM [9].

Two real-time datasets chainstore and kosarak [15] are used for experiments. Chainstore includes real-life customers purchase details with real utility values of a big grocery store chain in USA. Chainstore containing 1,112,949 transactions and a total of 46,086 items. Kosark includes hungarian online news portal click-stream data with synthetic utility values. Kosarak contains 990,002 transactions and 41,270 items in total.

Dis-EFIM and PHUI-Miner both algorithms are based on spark framework, to evaluate the results both are executed on a Spark cluster using a chainstore and kosarak dataset. EFIM in a single machine algorithm is executed on the single system setting using the same parameters. All the algorithms were executed with different minimum utility threshold values. Dis-EFIM running time is significantly lesser than PHUI-Miner and EFIM for different sets of minimum utility threshold. When the minimum threshold declines, the overheads of the techniques begin to rise, and

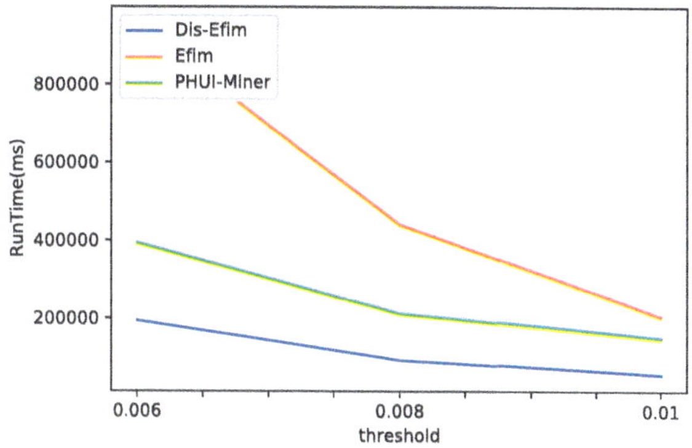

Fig. 2 Running time of the proposed algorithm on kosarak dataset

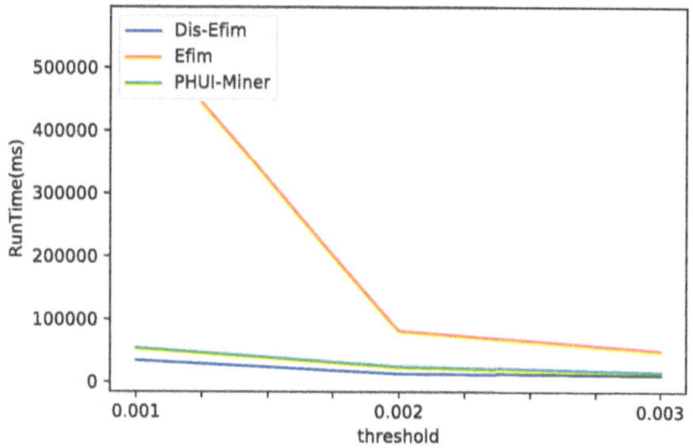

Fig. 3 Running time of the algorithm on chainstore dataset

the benefits of the distributed framework significantly reduce the impact of rising overheads. In particular, Dis-EFIM efficiency is less than that of EFIM on small datasets, notably when the threshold is high. Dis-EFIM however always outperforms PHUI-Miner. The Dis-EFIM is capable of handling dense data sets and achieving a Better execution. On the dataset chainstore and kosark, as shown in Figs. 2 and 3 when the threshold value is large, we can observe that the three algorithms execution time is closer to each other. However, the execution times of the proposed alogorithm are mlutiple time significantly smaller than other algorithms, when the threshold decreases.

5 Conclusion

The proposed algorithm Dis-EFIM is an extension of the advanced algorithm EFIM to handle big data in the Spark parallel and distributed platform. EFIM introduces a new utility counting method based on array structure termed as Fast Utility Counting and employs a database projection HDP (High utility database projection) to reduce the database scans cost. Traditional algorithms for HUI mining are not suitable for Big

Data. Hence, the proposed algorithm is efficient for handling and processing big data. For iterative algorithms, Spark is one of the most efficient distributed frameworks. The entire Dis-EFIM cycle can be divided into three stages: search space split, generation of node data, and data mining at nodes. In order to assess the efficiency of Dis-EFIM, comprehensive tests have been performed on different datasets. However, Dis-EFIM lacks in memory efficiency. Therefore, better techniques can be further explored to improve memory use as future work.

References

1. Liu, Y., Liao, W., Choudhary, A.: A two-phase algorithm for fast discovery of high utility itemsets. In: Pacific-Asia Conference on Knowledge Discovery and Data Mining, pp. 689–695 (Springer, 2005)
2. Kumar, S., Mohbey, K.K.: A review on big data based parallel and distributed approaches of pattern mining. J. King Saud Univ. Inf. Sci. (2019)
3. Liu, M., Qu, J.: Mining high utility itemsets without candidate generation. In: Proceedings of the 21st ACM International Conference on Information and Knowledge Management, pp. 55-64 (ACM, 2012)
4. Shvachko, K., Kuang, H., Radia, S., Chansler, R.: The Hadoop distributed file system. In: MSST, pp. 1–10 (2010)
5. Lin, Y.C., Wu, C.-W., Tseng, V.S.: Mining high utility itemsets in big data. In: Pacific-Asia Conference on Knowledge Discovery and Data Mining, pp. 649–661 (Springer, 2015)
6. Chen, Y., An, A.: Approximate parallel high utility itemset mining. Big Data Res. **6**, 26–42 (2016)
7. Sethi, K.K., Ramesh, D., Edla, D.R.: P-FHM+: parallel high utility itemset mining algorithm for big data processing. Proc. Comput. Sci. **132**, 918–927 (2018)
8. Zaharia, M., Chowdhury, M., Das, T., et al.: Resilient distributed datasets: a fault-tolerant abstraction for in-memory cluster computing. In: Proceedings of the 9th USENIX Conference on Networked Systems Design and Implementation, p. 2 (USENIX Association, 2012)
9. Zida, S., Fournier-Viger, P., Lin, J.C.-W., et al.: EFIM: a fast and memory efficient algorithm for high-utility itemset mining. Knowl. Inf. Syst. **51**, 595–625 (2017)
10. Tseng, V.S., Shie, B.-E., Wu, C.-W., Philip, S.Y.: Efficient algorithms for mining high utility itemsets from transactional databases. IEEE Trans. Knowl. Data Eng. **25**, 1772–1786 (2012)
11. Krishnamoorthy, S.: Pruning strategies for mining high utility itemsets. Expert Syst. Appl. **42**, 2371–2381 (2015)
12. Liu, J., Wang, K., Fung, B.C.M.: Direct discovery of high utility itemsets without candidate generation. In: 2012 IEEE 12th International Conference on Data Mining, pp. 984-989 (IEEE, 2012)
13. Fournier-Viger, P., Wu, C.-W., Zida, S., Tseng, V.S.: FHM: faster high-utility itemset mining using estimated utility co-occurrence pruning. In: International Symposium on Methodologies for Intelligent Systems, pp. 83–92 (Springer, 2014)
14. Li, H., Wang, Y., Zhang, D., et al.: PFP: parallel FP-growth for query recommendation. In: Proceedings of the 2008 ACM Conference on Recommender Systems, pp. 107–114 (ACM, 2008)
15. Fournier-Viger, P., Gomariz, A., Gueniche, T., et al.: SPMF: a Java open-source pattern mining library. J. Mach. Learn. Res. **15**, 3389–3393 (2014)

Evaluation of Power Efficient Hybrid Multicast Routing Protocol for MANET in Wireless Communications

Madas Pushpalatha, M. Srinivasan, and P. Ramadevi

Abstract A mobile ad hoc network is a rambling gathering of nodes, multi-hop and distributed wireless ad hoc network in which nodes are mobile in general, and hosts in such networks communicate with each other without the existing of fixed infrastructure and without central control. Routing is a critical issue in mobile ad hoc network. Here, the clustering method is used for adapting, which will be more energy efficient, clusters are made among the nodes as a shared self-leader system, and hence, every node will be a member of a few clusters and inside a cluster. Every node could arrive at any other node by significant power. Every cluster maintains head node that behaves like transmitting head for its nodes, for easy communication, cluster uses accessible bandwidth, and with that, clusters face energy or power utilization problems. Since load on the clusters and addition of clusters. In this work, we have proposed IP-ERMP to give power efficient multicasting for high mobility utilizations in the multi-source multicast environment, clustering also helps to overcome energy/power utilization problem, the main aim of intelligent multicast power efficient multicast routing protocol (IP-EMRP) is to overcome the drawbacks of proactive and reactive routing protocols so that to reduce the bandwidth and power problems to provide efficient communication in multicasting, simulation shows comparing new protocol with ZRP, DSR and AODV. We have simulated protocols by taking the number of nodes $N = 50$ in MATLAB software and obtained better results the network parameters in terms of throughput and packet delivery rate compared with existing ZRP and DSR protocols.

Keywords Ad hoc network · HRP · ZRP · DSR clustering · MATLAB

M. Pushpalatha (✉) · M. Srinivasan
SSSUTMS, Sehore, Bhopal, Madhya Pradesh, India

P. Ramadevi
Vardhaman College of Engineering, Hyderabad, Telangana, India

V. Bhateja et al. (eds.), *Computer Communication, Networking and IoT*, Lecture Notes in Networks and Systems 197, https://doi.org/10.1007/978-981-16-0980-0_35

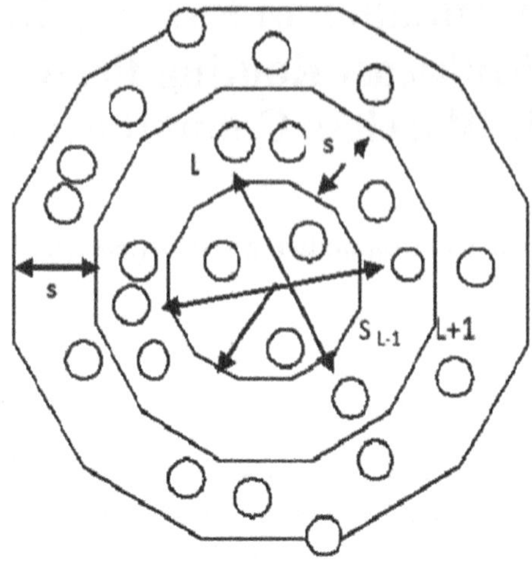

Fig. 1 Various layers in the cluster

1 Introduction

Ad hoc network is a group of a variety of devices such as PDAs, laptops and sensors that are jointly known as node of ad hoc networks. These nodes are prearranged in also heterogeneous or homogeneous manner. Each node of this network is communicated to other wireless inside a limited area. These nodes are able to move in whichever direction it wants due to its dynamic nature [10, 11]. Consequently, every node must carry out the functions of a router. Throughout the network lifetime, a few nodes may become unavailable due to a variety of reasons such as physical damage and lack of power resources. The main goal of ad hoc network is not simply to transmit data but also to enlarge the lifetime of the network (Fig. 1).

2 Related Work

Lot of research work carried out in energy efficient multicast routing for wireless ad hoc and sensor networks. The important routing protocols are discussed below, and multicast routing protocols in MANETare shown in Fig. 2.

Table-driven or Proactive Protocols: Proactive routing protocols will maintain consistent, up to date routing information between each node in the network. Proactive protocols are WRP, vector (DSDV) routing, OLSR, clustered gateway switch routing (CGSR) and [15] FSR.

On-demand or Reactive Protocols: Reactive protocols are not like the proactive ones. These protocols will establish the routes to the destination. Reactive routing protocols are AODV & DSR [2] routing, TORA and [12] ABR.

Hybrid Protocols: Hybrid routing protocol [4] is a combination of the proactive and reactive approaches [2]. Hybrid protocols are ZRP and ZHLS.

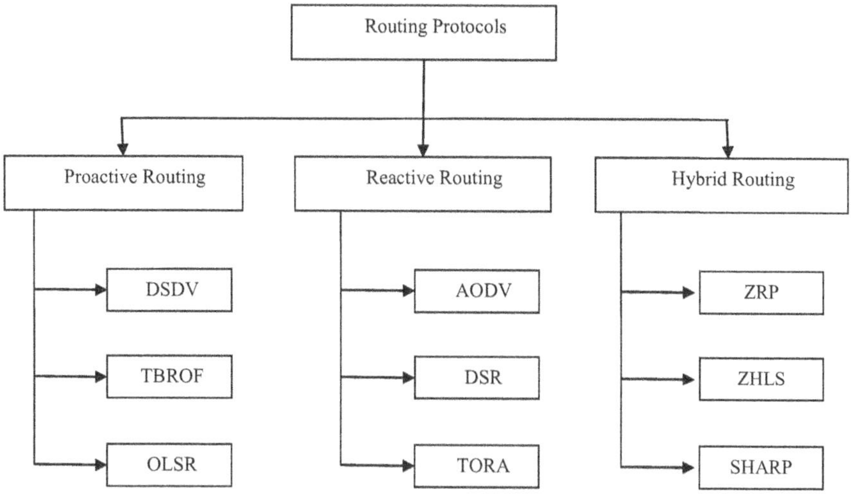

Fig. 2 Multicast routing protocols in MANET

Power/Energy Absorption in Clustering: Whenever the packet moves between the pairs of mobile nodes, in that the energy and cost of variable are constant among the transmitter and receiver devices. Now, consider the energy absorbed by the single hop represented as

$$E = \text{Energy}_t + \text{Energy}_r + \text{Energy}_{\text{oh}}$$

Energy_t = energy absorbed by the transmitter, Energy_r = energy absorbed by the receiver and overhead from the media access control into the transmitter and receiver. We consider four parameters *u*, *v*, *w* and *x*

$$U * w^{v} + x,\ w = \text{distance among the pair of nodes},$$

$$x = \text{transmitter and receiver cost of electronics}$$

v = varies from 2 to 4, U = coefficient cost of energy based on gain of antenna.

Basically if cluster size is increased/large, it will be helpful for the head of cluster to transmit more multicast information for all the receivers at a time, and we can see in the case of normal cluster also.

$$p_j = \text{total nodes, form } j \text{ to } j+1,\ H_j = \text{sum of receivers},\ j = \text{levels}$$
$$S_j = \text{Diameter of area},\ s = \text{length of two sucessive level, equal to constant},$$

Consider the size of the cluster that will be $L - 1$, and total receivers of multicast are given as

$$\frac{H_{L+1} \sum_{j=1}^{j=L-1} P_j}{\sum_{j=1}^{j=L+1} P_j}$$

Whenever we want to rise the power by 1 to L, we could include P_L nodes to the cluster.

Step 1: Select $L + 1$ power range to cover every node in the cluster. Energy in this step is given as

$$\text{Energy}_1 = u(S_{L-1} + 2s)^v + x + H_{L+1}x$$

Step 2: Select L power only to cover L level

$$\text{Energy}_2 = \text{Energy}_{\text{oh}} + \overline{\text{Energy}}$$

$$\text{Energy}_{\text{oh}} = u(S_{L-1} + 2s)^v + x$$

$$\overline{\text{Energy}} = \frac{H_{L+1 P_{L+1}}}{\sum_{J=1}^{L+1} P_J}(us^v + 3x) + H_{L+1}x$$

Accordingly, head of the cluster accepts and resends the information for including nodes into the layer

$$\text{Energy}_1 - \text{Energy}_2 = b_L(u(S_{L-1} + s)^v + 3x) - 3\delta$$

$\delta = (S_{L-1} + 2s)^v - (S_{L-1} + s)^V$, $b_L = H_{L+1}\frac{P_{L+1}}{3(P_L + P_{L+1})}$, L = coefficient adaptation.

if $P_{L+1} > P_L$ will benefits using step 1, if $P_{L+1} < P_L$ will loss by utilizing big cluster.

2.1 Node Clustering

To accomplish efficient multipath routing, we make the accompanying suppositions that are equivalent to LID and different algorithms in the writing do. Each node has a remarkable ID and knows the ID of its 1-jump neighbors. A message sent by a node is got accurately inside a limited time by the entirety of its 1-jump neighbors. (a) The topology of the network does not change during the algorithm execution. Above all else, it is essential to have at the top of the priority list the prerequisites of the protocol we propose to manufacture, which are the accompanying. (b) The algorithm should bring about an exceptionally uniform cluster [15] arrangement that can accomplish a pressing proficiency nearest to hexagonal with decreased cover.

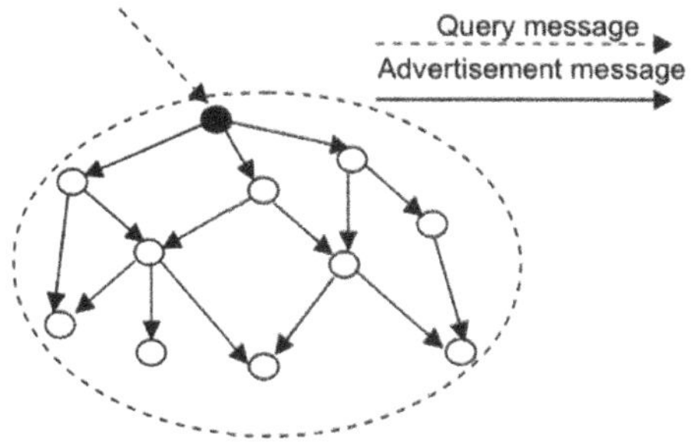

Fig. 3 Example of cluster formation

(c) A cluster head cannot meddle with another cluster head, as the objective of this system is to shape zones that do not meddle with one another.

At the point when the clustering protocol completes, all the nodes are clustered (have a place with a cluster) or are cluster heads (cluster leaders). As far as we could possibly know, there is no past protocol in the writing that meets all the prerequisites recorded previously. Along these lines, we adapted the cluster-based [17] algorithm, which gives a decent cluster dissemination yet does not guarantee that cluster heads do not meddle. The way toward clustering can be portrayed as a mix of two phases, cluster arrangement and cluster upkeep. The cluster arrangement stage manages the sensible parcel of the mobile nodes into groups and the appointment of suitable nodes to go about as leaders in each group. The goal of cluster upkeep is to protect the current clustering structure however much as could reasonably be expected. The example of cluster formation is shown in Fig. 3.

3 Contribution of the Proposed Work

In this section, we propose a new multicast routing protocol which improves power efficiency in MANET. We have introduced techniques in before sessions, which are using to generate a new multicast routing protocol IP-EMRP. This overcomes the problem of energy consumption and improves the power in multicasting by managing and electing the cluster.

1. To improve, the vitality proficient correspondence utilizes group supervisor-based bunch head.
2. But our work is entirely unexpected from past approaches.
3. In our work, the group administrator is answerable for picking bunch head choice. For the proposed approach astute force effective multicast steering convention (IP-EMRP) plan and its working standard, the algorithms are quickly talked about in this segment.
4. We look at a few existing conventions and portray the upsides and downsides.
5. Bunch director jobs are significant, and furthermore, it exercises as a reinforcement hub just as an observing hub.

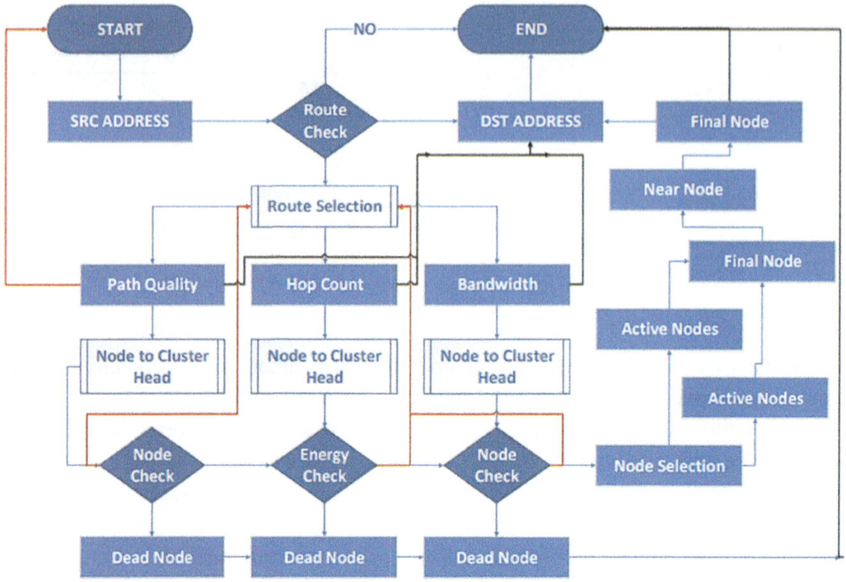

Fig. 4 Proposed methodology flow

6. To decrease the vitality utilization, information traffic and improve the throughput rate by means of stable directing.
7. The fundamental proverb of the proposed convention is a spontaneous creation of correspondence execution and system lifetime by methods for vitality consummate and least correspondence delay (Fig. 4).

The above flow chart explains how our EMRP works and its efficiency and quality of service. EMRP starts with the source address of the node which is going to have a network connectivity to the destination address. Our system is the combination of proactive and reactive network topology, and hence, routes will be created on demand or it will use the existing routes to travel. When routed is validated, our algorithm will choose the best available path to reach the destination. This selection includes path quality, hop count and available bandwidth. Here, we are calculating and analyzing multiple parameters to make the path effective and energy saving. Once path selection is completed means, node to cluster handover will take place. Cluster is the combinations of multiple nodes, and hence, cluster finds out the active and dead nodes. If there is no active nodes to transfer the data means, the packets will be dropped off or else will be forwarded to the active nodes. All active nodes will work together to create a path to the destination. Once the packet crosses all active nodes, then it will reach the final node which is one hop ahead of the destination address. The last node will deliver the packets to the destination address. Following steps are outlook flow of the present work

Step 1: START

Step 2: Source address verification
Step 3: Route validation
Step 4: Route selection
Step 5: Path quality check
Step 6: Node to cluster handover
Step 7: Cluster node election process
Step 8: Node pass through
Step 9: Final node acknowledgement
Step 10: END.

The IP-EMRP is an adaptive multipath routing algorithm that helps in power efficient and improves speed of packet travel than other existing protocols.

3.1 Cluster Maintenance

In WMNs, the topology can change because of nodes coming up short, leaving, joining or basically moving. Subsequently, the cluster arrangement may change after some time, and an upkeep system is required. The cluster-based algorithm protocol that was utilized as a base to the clustering protocol we use runs when the setup of the network happens, and afterward, the data gave by the algorithm is static and does not change. For refreshing data about clusters, there are two alternatives concerning when the data is refreshed: (i) the algorithm can be continually running in the foundation, creating the relocation of the clusters when arriving at a specific edge or condition; (ii) clustering data can be refreshed intermittently when a blunder in a way happens because of a disappointment or node developments. In one end, the principal choice is an adaptive procedure, thus, in every second, the clusters are refreshed and structure groups with the best attributes. On the opposite end, such a methodology creates an excess of traffic and is continually searching for the immaculate setup of the network, which is not essential for the protocol to perform well. Having this as a main priority, the subsequent choice is considerably more suitable for picking the best time to adjust clusters, diminishing the overhead connected with clustering.

Since these progressions are required to be rare in wireless mesh networks, we picked to re-apply the cluster development algorithm at whatever point a change is distinguished. The conditions under which it is done are expressed as cry. To distinguish changes in the topology, we utilize a responsive methodology, where broken connections are recognized during the proliferation of use information, subsequently saving money on the flagging expense. The last node in the way that got however could not advance the information message is answerable for starting the cluster development algorithm. To this end, it floods the network with a re-cluster message that, when got, restores each node to its underlying (unclustered) state and restarts the cluster development strategy.

Cluster Bead Election Algorithm

1. Step-1
2. Start
3. Group of mobile node deployment $MN_{i,j}$
4. Request to Hi message to all neighboring nodes in the cluster
5. Head of the cluster selecting among all nodes ($CH1$, $CH2$, … CHn)
6. For ($i = 0$; $i > n$; i++)
7. Step-2
8. Find energy level of node
9. $N_E = N_{tr}$ (p) *N_{pt} (*processing time of node*
10. If ($MN_{i,j} < N_E$) //compare the node energy to node mobile
11. The cluster node is defined as high energy of the mobile node (CH_1)
12. Else
13. That mobile node participates in the network communication
14. Current information is updated and stored in cluster manager C_{mr}
15. For $C_{mr} = (CH_1)$// performance of cluster head and energy level observed by cluster manager C_{mr}
16. Step 3
17. Find the energy level of CH_1
18. If ($CH_1 > N_E$);
19. CH_1 acts as mobile node and elects other cluster head CH_2 to follow step 2
20. Else
21. ($CH_1 N_E$);
22. Continue CH_2 cluster head
23. End if
24. End process ().

Cluster Manager Selecting Algorithm

1. Start
2. Group of Mobile 11 node deployment $N_{i,j}$
3. Manager of Cluster selection C_{mr}
4. Request to Hi message to all neighboring nodes in the cluster
5. Find the node distance
6. Distance of Node (N_d) → $\left(N_{x_i} - N_{y_i}\right)^2 + \left(N_{x_j} - N_{y_j}\right)^2$
7. After selecting cluster manager, will find the energy level of the node, i.e.,
8. Find the energy level of the node
9. Energy node (N_E) (*Transmission power of node* N_{tr} (p) * $N_{camm\ time}$
10. for (i = 0; i)
11. If (Find energy and distance of node N_E and N_d)
12. If N_d *is low and* N_E *high*
13. Then, the Cluster manager is equal to the node
14. Else
15. If N_d *is high and* N_E *low*, then it acts as a mobile node
16. End

3.2 Clustering Improvements on IP-EMRP

The protocol we propose utilizes connect state for the nodes to have a review of the whole network. It is realized that this sort of protocols produces a lot of overhead to accomplish this objective. Be that as it may, for our situation, by utilizing clustering, we decrease the overhead by (i) limiting flooding; (ii) limiting the length of the information messages; (iii) limiting the preparing exertion expected to locate the most extreme number of non-meddling ways. Flooding minimization information flooding happens when a node needs to make an impression on all nodes in the network. As between two cluster-heads there are all things considered two jumps, the source data can be decreased by one third, contrasting and an unadulterated connection state approach. This happens, in particular when a node plays out a course disclosure by sending a message or a Topology message. As all nodes in a cluster are neighbors of the cluster head, the accompanying methodology can be applied to diminish the quantity of messages expected to flood data along the network. (a) If the packet shows up at a cluster head, additional data will be added to the message and afterward broadcasts the packet including the address of the doors it knows are an approach to arrive at different clusters and the clusters that are reached through that entryways. (b) If a clustered node gets a topology message, it spares the data it advertises and afterward disposes of it. On the off chance that it gets a Rreq message, it is dropped right away. (c) For the situation, a node gets a broadcast packet (topology or Rreq), its address is remembered for the door show, and it recovers the following cluster from the message and looks for its cluster head or the following entryway to arrive at the cluster.

If the packet shows up at a cluster head node and it had already got the packet, it is dropped. Minimization of messages length source routing has the disadvantage to build packet size as the nodes need to remember for its header the way through which the information should pass. In any case, for this situation when the source node concludes that a way is legitimate as per the protocol administers, the information needs to follow a particular way, and a bounce by hop approach can lead to an inappropriate activity as a middle of the road node can have in excess of a way to a similar destination node. So that, source routing must be utilized to ensure that ways utilized are the right ones.

Clustering additionally assists with limiting the way data remembered for the messages as notwithstanding all the nodes that have a place with the way just the cluster heads are put away in message header. In between two cluster heads data forwarding and receiving considered within two jumps to its neighbors. Minimization of the processing effort by having the chart that speaks to the network (data about the nodes and the connections that associate them), it is conceivable to figure all the accessible ways between a source and destination and decide the non-meddling ways. Obviously, this methodology is not common sense as the data required is excessive and the resources to figure ways would be high. Clustering can likewise lessen the preparing exertion that is important to ascertain these ways as the network is spoken to by an overlay network characterized by cluster heads, and it is more affordable to investigate them instead of breaking down all the nodes.

4 Performance Evaluation

Evaluated the performance of the existing algorithm using MATLAB. We compare our method IP-EMRP (intelligent power efficient multicast routing protocol with zone routing protocol (ZRP), AODV, DSR. Here, we are taking only few parameters for showing simulation results, and still work is going on, to bring simulation results for some other parameters.

4.1 Throughput and Packet Delivery Rate Performance Versus Channel Bandwidth

Figures 5 and 6 separately show the throughput and the packet delivery rate (PDR) as an element of the channel transmission bandwidth (BW). Note that the channel BW builds the throughput and packet conveyance rate increment. This is on the grounds that the channel limit (i.e., information rate) is relative to the channel BW. Be that as it may, the EMRP conspire outflanks AODV, ZRP and DSR plots in terms of the packet conveyance rate by up to 13%, 84% and 142%, separately. Likewise, the

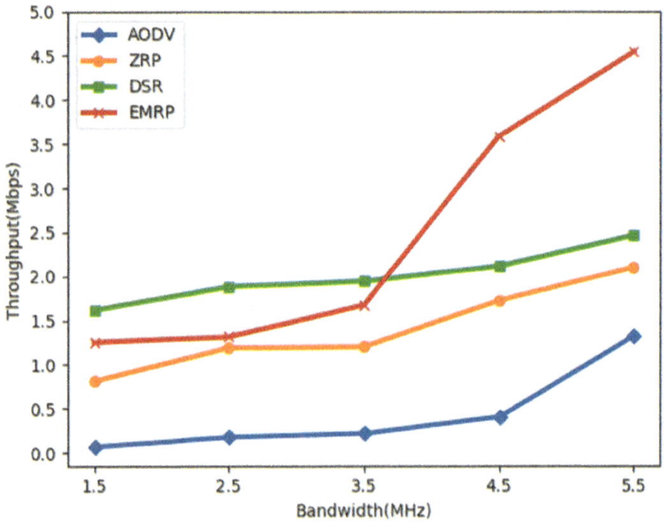

Fig. 5 Throughput versus channel bandwidth under PI = 0.8 ($N = 50$, $Nr = 18$, $M = 25$, $P_t =$ 0.1 W, $D = 5$ KB)

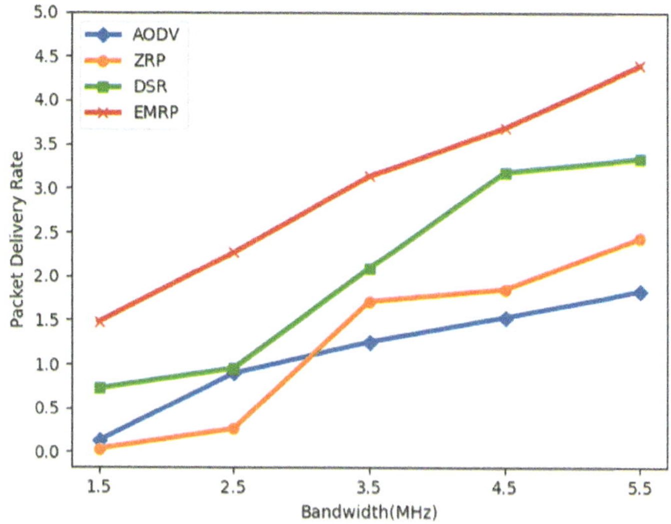

Fig. 6 PDR versus channel bandwidth under PI = 0.8 ($N = 50$, $Nr = 18$, $M = 25$, $P_t = 0.1$ W, $D = 5$ KB)

EMRP conspire beats AODV, ZRP, and DSR plots in terms of throughput by 11%, 46% and 67%, separately.

4.2 *Throughput and Packet Delivery Rate Performance Versus the Packet Size*

Figures 7 and 8 plot the throughput and the packet delivery rate (PDR) for all the schemes versus the packet size (D). Figure 7 shows the critical improvement gain in term of throughput. Figure 8 shows the improvement gain in term of the packet delivery rate. The proposed plot beats AODV, ZRP and DSR plots in terms of the packet delivery rate by up to 9.3%, 110% and 162%, individually. Additionally, the proposed plot beats AODV, ZRP and DSR conspires in terms of the throughput by up to 11.18%, 66.5% and 84%, individually.

4.3 *Throughput and Packet Delivery Rate Performance Versus the Transmission Power*

It is well known that increasing the transmission power can significantly improve the achieved data rate, decrease the required transmission time and result in improving network performance as shown. However, the maximum transmission power for

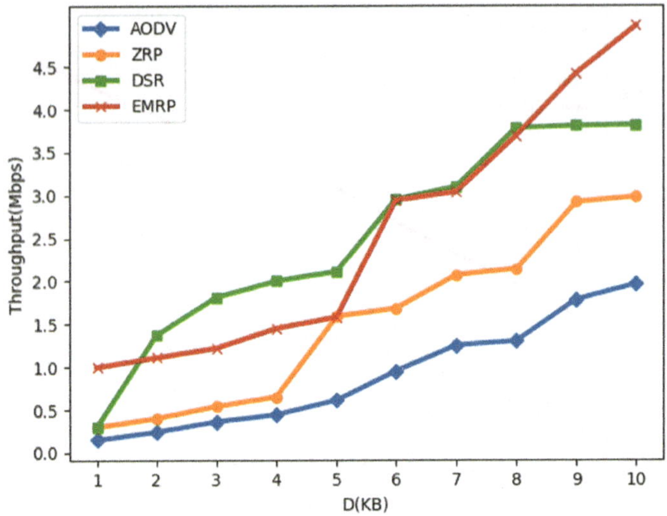

Fig. 7 Throughput versus the packet size under PI = 0.8 ($N = 50$, $Nr = 18$, $M = 25$, $P_t = 0.1$ W, BW = 1 MHz)

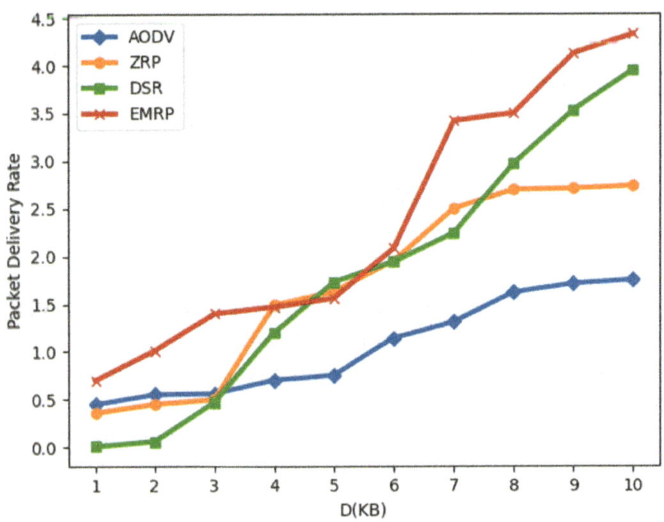

Fig. 8 PDR versus the packet size under PI = 0.8 ($N = 50$, $Nr = 18$, $M = 25$, $P_t = 0.1$ W, BW = 1 MHz)

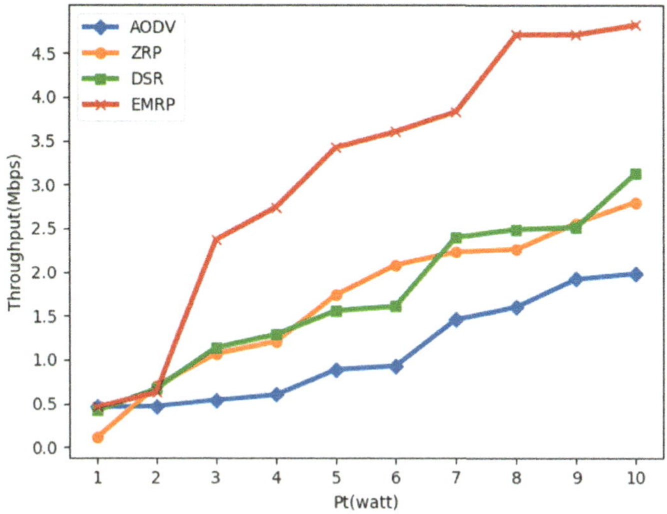

Fig. 9 Throughput versus the transmission power under PI = 0.8 ($N = 50$, $Nr = 18$, $M = 25$, BW = 1 MHz, $D = 5$ KB)

CR users is very limited, compared to the PU transmission power. One watt is too high for such applications because this will increase interference significantly. The proposed scheme outperforms AODV, ZRP and DSR schemes in terms of the packet delivery rate by up to 8.7%, 84% and 152%, respectively. Also, the proposed scheme outperforms AODV, ZRP and DSR schemes in term of the throughput by up to 11%, 59% and 67.6%, respectively (Figs. 9 and 10).

5 Conclusion

In this work, we have proposed evaluation power efficient hybrid multicast routing protocol IP-ERMP to give power efficient multicasting in the multi-source multicast environment. It is enhanced clustering method, and simulation shows by comparing new protocol with ZRP and DSR. We have simulated protocols in MATLAB software and obtained better results in terms of throughput and packet delivery rate compared with existing ZRP and DSR protocols. The simulation results show that selection of node speed parameter value affects the performance of multicast routing protocols in MANET. Future work will show some more simulations for IP-EMRP algorithm and data security for transmission.

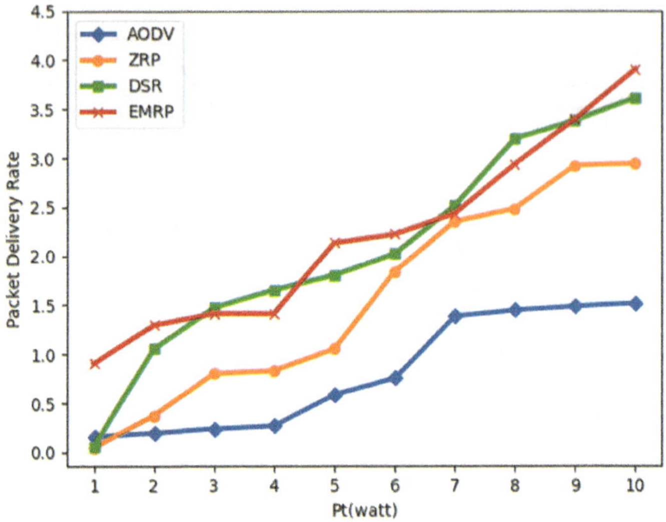

Fig. 10 PDR versus the transmission power under PI = 0.8 ($N = 50$, $Nr = 18$, $M = 25$, BW = 1 MHz, $D = 5$ KB)

References

1. Adhvaryu, K.U., Kamboj, P.: Performance comparison between multicast routing protocols in MANET. In: 2017 4th IEEE Uttar Pradesh Section International Conference on Electrical, Computer and Electronics (UPCON), GLA University, Mathura, 26–28 Oct (20170
2. Pushpalatha, M., Srinivasan, M., Ramadevi, P.: Analization of multicaste routing protocols for mobile ad-hoc networks in wireless communication. J. Adv. Res. Dyn. Control Syst. **12**(02) (2020)
3. Hu, S.-C., Tsai, C.-H., Lu, Y.-C., Pan, M.-S., Tseng, Y.-C.: An energy-efficient multicast protocol for ZigBee-based networks. In: IEEE Wireless Communications and Networking Conference (WCNC 2016)—Track 3—Mobile and Wireless Networks
4. Vidhyalakshmi, R., Sankardayal, P., Sankareeswari, G.: Design of hybrid multicasting network for authentication. Int. J. Innovative Res. Comput. Commun. Eng. (Tirupur, Tamil Nadu, India on 6th & 7th Mar 2014) (2014)
5. Koyama, A., Nishie, T., Arai, J., Barolli, L.: A new quality of service multicast routing protocol based on genetic algorithm. In: Proceedings of the 2005 11th International Conference on Parallel and Distributed Systems (ICPADS'05) 0-7695-2281-5/05 $20.00 © 2005 IEEE
6. Beijar, N.: Zone routing protocol (ZRP). Networking Laboratory, Helsinki University of Technology, Finland (2002)
7. Sethi, S., Udgata, S.K.: An efficient multicast hybrid routing protocol for MANETs. In: Das, V.V., et al. (eds.) BAIP 2010, CCIS 70, pp. 22–27, 2010. © Springer-Verlag, Berlin, Heidelberg (2010)
8. Hoebeke, J., Moerman, I., Dhoedt, B., Demeester, P.: An overview of mobile ad hoc networks: applications and challenges. J. Commun. Netw. **3**(3), 60–66 (2004)
9. Panda, I.: A survey on routing protocols of MANETS by using QOS metrics. Int. J. Adv. Res. Comput. Sci. Softw. Eng. **2** (2012)
10. Bansal, A., Kaur, N., Khullar, S., Bedi, R.P.S.: Performance analysis of ZHLS-GF routing protocol for MANETs through simulations. [6] Theofanis, K.: A presentation on hybrid routing protocols (2011)

11. www.scribd.com/Manets [13] Hemagowri, J., Baranikumari, C., Brindha, B.: A study on proactive routing protocol in Ad-hoc network. Int. J. Mod. Eng. Res. (IJMER), 01–04.
12. Bouhafs, F., Merabti, M., Mokhtar, H.: A semantic clustering routing protocol for wireless sensor networks. In: IEEE Communications Society subject matter experts for publication in the IEEE CCNC 2006 proceedings
13. Sharma, M., Singh, G.: Evaluation of proactive, reactive and hybrid Ad hoc routing protocol for various battery models in VANET using Qual. Int. J. Smart Sens. Ad Hoc Netw. (IJSSAN) **1**(2). ISSN No. 2248-9738 (Print) (2011)
14. Pushpalatha, M., Srinivasan, M., Ramadevi, P.: Different types of an energy efficient multicast routing protocols for QOS in wireless communication. Int. J. Recent Technol. Eng. (IJRTE) **8**(6S). ISSN: 2277-3878 (2020)
15. Rappaport, T.S.: Wireless Communications: Principles and Practice. Prentice Hall (1995)
16. Raghavendra, C.S., Prasanna, V.: Energy efficient adaptation of multicast protocols in power controlled wireless Ad hoc networks
17. Chan, H., Perrig, A.: ACE: An emergent algorithm for highly uniform cluster formation. In Conference of European Workshop on Wireless Sensor Networks, 19 January 2004

Ensuring Fault Tolerant Connectivity in IoT Networks

N. N. Srinidhi, E. Nagarjun, J. Shreyas, S. M. Dilip Kumar, and Dharamendra Chouhan

Abstract The Internet of things (IoT) is the new paradigm which extends the traditional Internet from laptops and mobile phones to different set of devices with own processing and communicating capability. Mobility among IoT devices challenges devices energy efficiency as these devices are battery powered. Most of the existing work lacks to provide efficient connectivity when devices are mobile. Hence, in this work, proactive mobility support for IoT devices has been propounded. Here, Kalman filter is used to predict the point of attachment of a mobile node and trajectory. Fault tolerant characteristics are being considered in the proposed algorithm to prevent the network from failing and for efficient routing of data. Proposed approach performs well in terms of handoff delay, movement prediction and prediction accuracy when compared to existing approach.

Keywords Fault tolerance · Handoff delay · Internet of things · Kalman filter · Seamless connectivity · Self-healing

1 Introduction

Evolution of the next-generation technologies has resulted in a huge scope for Internet of things (IoT) and its application across a myriad range of domains [1]. The IoT establishes a communication network among the various physical devices (which are integrated with actuators and sensors) and the cyber-world constrained under some communication protocol. These devices may be mobile or static [2]. IoT applications have a very low tolerance for data loss and require the reliable transfer of critical data. Another requirement required to facilitate different tasks is mobility.

N. N. Srinidhi (✉)
Department of Computer Science and Engineering, Sri Krishna Institute of Technology, Bengaluru, India

E. Nagarjun · J. Shreyas · S. M. Dilip Kumar · D. Chouhan
Department of Computer Science and Engineering, University Visvesvaraya College of Engineering, Bengaluru, India

V. Bhateja et al. (eds.), *Computer Communication, Networking and IoT*, Lecture Notes in Networks and Systems 197, https://doi.org/10.1007/978-981-16-0980-0_36

However, mobility of devices poses problems such as lack of mobility support, high energy usage, low connectivity and communication delay which degrade the network efficiency and performance [3]. The aim of this work is to negate aforementioned issues. In the proposed work, the aim is to reduce the handoff delay and to reduce the signaling cost. Reducing the signaling cost increases the power efficiency and overloading at different links. Reduced handoff delay results in seamless connectivity among mobile nodes and provides better routing. Proposed method optimizes signaling costs by fewer and more efficient signal, message transmission and accurately predicting the next PP (preferred parent) for mobile nodes in their trajectory before handoff to reduce handoff delay. So, the main objective is to (i) establish mobility among mobile nodes to provide faster and seamless access to access points (parents), avoid disconnection and to find new viable parents and (ii) reducing the signal overhead. The proposed method has been coupled with the movement prediction of mobile nodes to deal with the issues caused by mobility on the network performance. Movement prediction of an object or a person in many real applications has many advantages such as faster communication, reduced energy consumption and so on. The application where the movement prediction can be used is the IoT-based healthcare system which requires a movement prediction scheme to predict the next movement of IP-based mobile sensor nodes and to reduce the hand off cost of the network. If there are ineffective communications between the nodes, then it will be recognized as critical safety problems for the patients in IoT healthcare application. Movement prediction in mobile IP-based sensor networks is mainly carried out to address the continuous connectivity and to reduce hand off cost. In the proposed approach, movement prediction is the key requirement to provide seamless connectivity. The Kalman filter is an efficient recursive filter that estimates the state of a linear dynamic system from a series of noise measurement [4]. This provides high prediction accuracy based on a small amount of information. Kalman filter has two phases, the first is the prediction phase, wherein the next state of the system is predicted given the previous measurements, and the second is update phase, where the current state of the system is estimated based on the actual measurement at that instance. The Kalman filter is essentially a set of mathematical equations that implement a predictor-corrector type estimator that is optimal in the sense that it minimizes the estimated error covariance when some presumed conditions are met [5]. In this work, self-healing of mobile nodes is considered, when mobile nodes fail due to draining of battery or other network issues. Due to the failure of mobile nodes, the connection between the nodes will be lost and the data loss occurs. In order to prevent this data loss, whenever nodes fail, Dijkstra's algorithm is used to find the shortest path between the mobile nodes, whenever the nodes fail, by not considering the failed nodes. When static nodes do not receive the acknowledgement from the root node, one or more nodes may be failing in the path of sending the data. At this time, the shortest path between the root node and static node will be recalculated by excluding the failed nodes. Dijkstra's algorithm is a routing algorithm, where we calculate the shortest path between the source node and the root node.

Proposed work contributions are:

1. Supports mobility and provides seamless connectivity
2. Handles mobility via accurate mobility prediction using Kalman Filter
3. Is self-healing takes alternative steps in case of node failure
4. Reduces energy consumption and increases efficiency in the WSN as a result of the above three measures.

Paper is furnished as follows: Existing works summary is presented in Sect. 3. Section 3 provides problem statement of the work. Section 4 presents proposed algorithm. Section 5 presents evaluation of the work. Summary of the work is presented in Sect. 6.

2 Related Work

Authors in [6] have proposed method to predict users next movement in the future time interval. This model uses ensembling technique during training phase to predict users next movement location along with the pattern during movement. Method also uses Markovian model to predict the trajectories followed during the movement. This work has been evaluated using real-time datasets and performs well in respect to prediction of user movement. This method performs low in terms of prediction when the high variance dataset is used in training the model. In [7], authors have proposed prediction method to determine mobile and static users using Markovian predictor. This work performs well in predicting user movement in 110 locations. But the accuracy of predicting the user movement is only 60%, and this method uses transition matrix while modeling which considers visited logs as inputs during modeling which ignores other content information. In [8], authors have proposed mobility scheme for 6LoWPAN. In this method, location of the nodes is included by the cluster head in order to reduce the delay in route discovery process and the handover happens in parallel rather than in serial in different layers. Method discards the necessity of accounting participants address during handover which result in packet loss. Method performs well in handover process but it lacks in providing seamless connectivity happens due to mobility of the nodes in the network. Authors in [9] proposed an predicting method to determine future location of infrastructure less network. This method uses artificial neural network to predict continuous coordinates. But this method fails to consider temporal variable that means user movement during different time in the day.

3 Problem Statement of the Proposed Work

The problem is to provide seamless connectivity in IoT networks in case of mobile nodes to ensure faster communication and avoid preferred parent search for each mobile node after disconnecting from its current parent.

4 Proposed Algorithm

This section provides in detail explanation of the proposed *Kalman Filter* algorithm.

4.1 Terminology

- Q: It is the vertex set of nodes and distance of each node from the source node.
- A_k: Estimated mean of the state for the time stamp k after measurement.
- O_k: State covariance for the time stamp k after measurement.
- B_k: Measurement mean.
- R_k: Innovation or the measurement residual on time stamp k.
- Q_k: Measurement prediction covariance on the time stamp k.
- KG_k: It is the filter gain, which tells how much the predictions should be corrected on time stamp k.
- A_k^-: Predicted mean of the state on the time stamp k before seeing the measurement.
- O_k^-: Covariance of the state on the time stamp k before seeing the measurement.

4.2 Motion Identification Stage

Motion Identification Stage: Node which is in motion is connected to static favored parent (FP) in the communication range. FP is meant to monitor the quality of link of mobile node (MON) to check the motion of MON and to initiate second phase. Received signal strength indicator (RSSI) value of MON is continuously monitored by FP through received packets. If the average RSSI value falls below the threshold value, then FP presumes that MON is going to leave the communication range. Before interruption of link, FP informs MON through unicast disconnection started message (DNSM) and flag is set to 1. After receiving the DNSM message, MON will search for new node for the association in order to prevent network disconnection and to continue its connection with current FP through sending and receiving of packets. Associated FP monitors RSSI value of the MON until signal quality degrades than the threshold value. Associated FP notifies MON to halt packets sending whenever quality of the signal degrades and new FP is considered based on neighbor RSSI value and Kalman prediction.

4.2.1 Reactive Stage

This stage is initiated whenever MON receives DNSM message (flag = 1) from its present FP. After receiving DNSM message, MON starts to collect information about new FP in its vicinity to select and predict new FP. The selection process is based on the direction and new neighbor list of MON. Here, the main purpose is to predict further MON movement and location in order to maintain new connection for longer period of time and to minimize signaling cost and maximize network parameters. Movement of the MON is based on the present position and present FP. Static nodes position is known to MON from the starting, whereas present MN is determined through prediction. Prediction is based on the RSSI values of static nodes in the new position and trajectory movement prediction through Kalman filter. Whenever reactive stage is initiated, MON collects necessary data about nearby static nodes and their associated RSSI table values. MON broadcasts DNSM message (flag = 2) in response to received disconnection starts order (DISO) (flag = 1) message from static nodes. Static nodes neighboring waits for random amount of time before DISO message to prevent collision among them. Whenever DISO collection period is expired, MON calculates RSSI value of each link, its position and movement direction to select new connection. Otherwise, whenever MON receives message of reaching threshold, it halts packet sending to prevent packet loss as it is in the verge of reaching border of communication area.

4.2.2 Notifying Stage

MON is connected to new FP whenever, it gets disconnected from previous FP. This is different from reactive stage wherein MON will search for new node for association only when it gets disconnected from present FP. Whenever new FP is selected, routing path should be updated. MON will update its parameters like FP, default route and ranking information on the basis of received DSO message from static nodes. MON sends connection order (CNO) message to selected node for establishing new connection. Once the connection is initiated, MON notifies previous FP through disconnection order (DSO) for disconnection.

5 Experimental Evaluation

In this section, proposed work along with discussion about results obtained is explained.

5.1 Simulation Settings

The implementation of Kalman filter in the network system has a very huge impact on both minimizing in the amount of packet loss, reducing the handoff delay, efficiency in energy consumption, etc. The implementation is done using Python 3.7.3 and the measurement data is stored in numpy arrays. The measured data has been obtained from Kaggle datasets, wherein human mobility data is considered.

Algorithm 1: Prediction Algorithm

Begin ;
Step 1: Initially setup the movement of MN
Step 2: Collect the signal strength B_k as the input variables.
Step 3: Let A_{k-1} and O_{k-1} are both the initial and recursive matrices of mean and covariance respectively at time $k - 1$.
Step 4: **if** A_k *does not have same value as* A_k^- **then**
    // prediction at time stamp k
    $A_k = C_{k-1}\ A_{k-1} + D_k\ W_k$
    $O_k^- = C_{k-1}\ O_{k-1}\ C_{k-1}^T + I_{k-1}$
// updation at time step k
$(KG)_k = O_k^-\ F_k^T\ Q_k^{-1}$ // calculating kalman gain $(KG)_k$
// variables used for estimation
$R_k = B_k - F_k\ A_k^-$
$R_k = F_k\ O_k^-\ F_k^T + J_k$
// updation
$A_k = A_k^- + (KG)_k\ R_k$
$O_k = O_k^- - (KG)_k\ Q_k\ (KG)_k^T$
Step 5: **if** A_k *has same value as* A_k^- **then**
    Go to Step 6 and return values of A_k and O_k
else
    Go to Step 4
Step 6: End

5.2 Simulation Results and Performance Analysis

In Figs. 1 and 2, the graphs are plotted for the packet loss versus the movement prediction for 50 nodes and 100 nodes, respectively. Graphs show that as the prediction accuracy increases, the data packet loss decreases. So, from the graphs, it is understood that as the prediction accuracy of mobile nodes increases, the data packet loss decreases due to lowered handoff delay. In case the estimated next movement of the mobile node is equal to the actual movement, then packet loss will decrease to a large extent.

In Fig. 3, the graph illustrates that the prediction made by Kalman-filter for a random movement of nodes where the node moves from grid to grid and it has been compared with observed values. Here, next movement of each node in the grid, Kalman filter prediction, is made and estimated based on the measurements obtained from the sensor.

This work is compared with existing MLOWPAN [8], and in Fig. 4, the graph shows that handoff delay is less as compared with MLOWPAN method due to the movement prediction of mobile nodes in prior to the next movement.

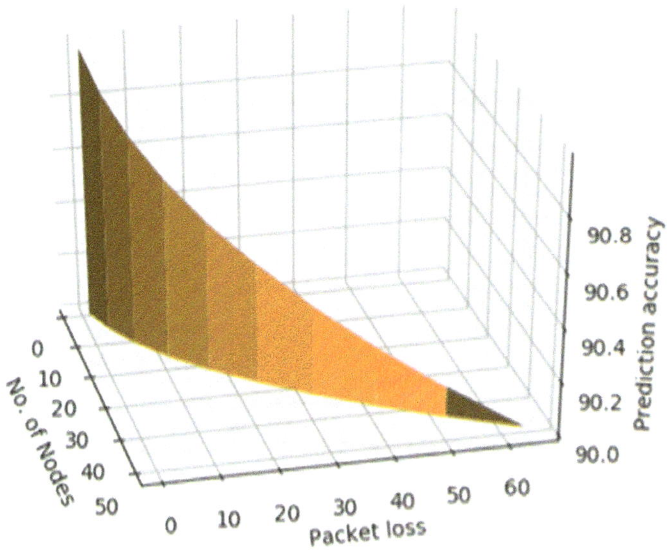

Fig. 1 Packet loss versus prediction accuracy for 50 nodes

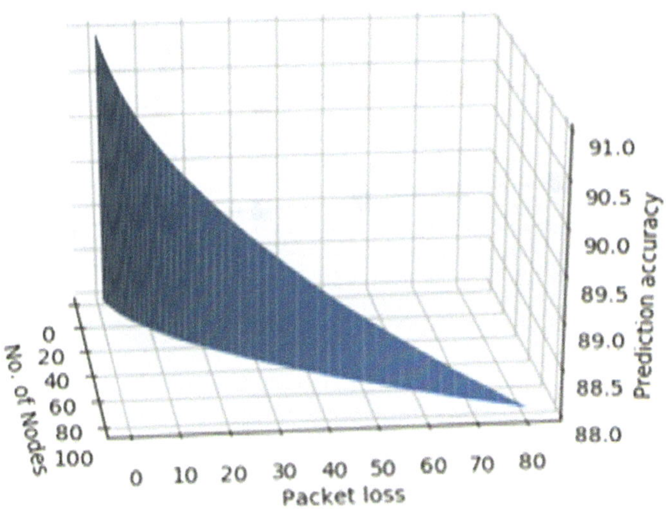

Fig. 2 Packet loss versus prediction accuracy for 100 nodes

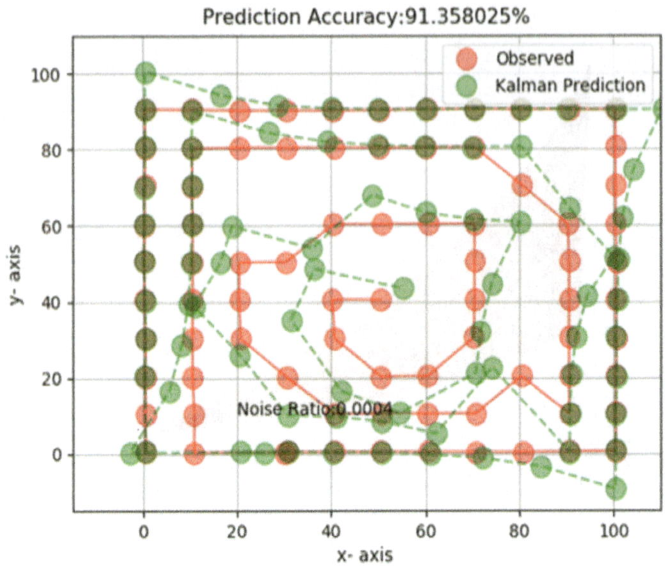

Fig. 3 Prediction accuracy for 100 nodes

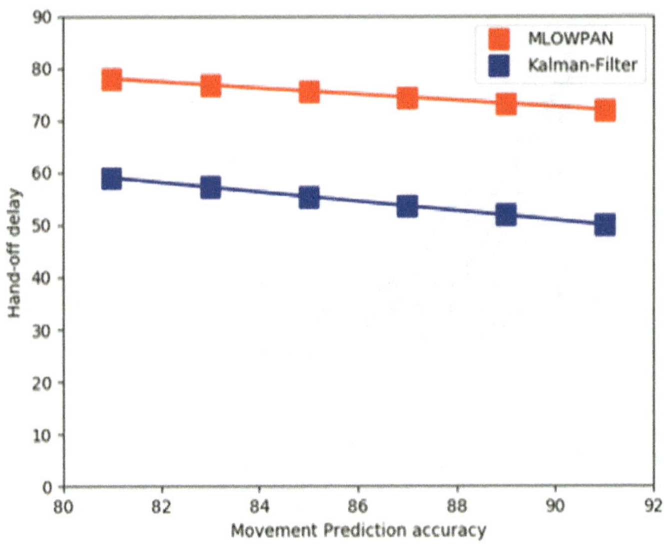

Fig. 4 Movement prediction accuracy versus handoff delay

6 Conclusion and Future Work

In the proposed work, seamless connectivity aids in reduced handoff delay and signaling cost. The energy consumption of the nodes is minimized and overall network performance is maximized. The goal of negating the issues caused in the network due to mobility has been achieved. The use of Kalman filter in the proposed method helped in achieving movement prediction and thereby reducing the handoff delay and the energy consumption. The Kalman filter uses a series of noisy measurements to predict the movement of the object accurately. Kalman filter also refines its outputs as the dataset value increases. The handoff trigger mechanism uses a real-time mobility data obtained from Kaggle datasets and it will be applied proposed prediction mechanism. The mobility information is used to predict a future handoff event, based on the value of the relative signal strength between the current and target base stations at a future time. The handoff trigger is set to be active if the signal strength of the static node reduces below a threshold for a given mobile node. As the mobile node disconnects from its parent only after it has found a new parent, loss of critical data is avoided.

The work can be further improved by using extended Kalman filter that can work with nonlinear trajectory of mobile nodes or unscented Kalman filter that can work with highly ambiguous data. Deep learning algorithms can also be used to achieve more complex and accurate movement prediction results.

Acknowledgements This research work has been funded by the Science and Engineering Research Board (SERB-DST) Project File No: EEQ/2017/000681.

References

1. Srinidhi, N.N., Nagarjun, E., Dilip Kumar, S.M.: HMCRA: hybrid multi-copy routing algorithm for opportunistic IoT network. In: 2019 International Conference on Smart Systems and Inventive Technology (ICSSIT), pp. 370–375 (IEEE, 2019)
2. Srinidhi, N.N., Sunitha, G.P., Nagarjun, E., Shreyas, J., Dilip Kumar, S.M.: Lifetime maximization of IoT network by optimizing routing energy. In: 2019 IEEE International WIE Conference on Electrical and Computer Engineering (WIECON-ECE), pp. 1–4 (IEEE, 2019)
3. Srinidhi, N.N., Sagar, C.S., Deepak Chethan S., Shreyas, J., Dilip Kumar, S.M.: Machine learning based efficient multi-copy routing for OPPIOT networks. In: International Conference on Computational Intelligence, Security and Internet of Things, pp. 288–302 (Springer, 2019)
4. Emami, A., Sarvi, M., Bagloee, S.: Using Kalman Filter algorithm for short-term traffic flow prediction in a connected vehicle environment. J. Mod. Transp. **27**(3), 222–232 (2019)
5. Srinidhi, N.N., Sunitha, G.P., Raghavendra, S., Dilip Kumar, S.M., Chang, V.: Hybrid energy-efficient and QoS-aware algorithm for intelligent transportation system in IoT. Int. J. Grid Util. Comput. **11**(6), 815–826 (2020)
6. Zhao, Z., Karimzadeh, M., Gerber, F., Braun, T.: Mobile crowd location prediction with hybrid features using ensemble learning. Future Gener. Comput. Syst. **110**, 556–571 (2018)
7. He, H., Qiao, Y., Gao, S., Yang, J., Guo, J.: Prediction of user mobility pattern on a network traffic analysis platform. In: Proceedings of the 10th International Workshop on Mobility in the Evolving Internet Architecture, pp. 39–44 (ACM, 2015)

8. Wang, X., Le, D., Yao, Y., Xie, C.: Location-based mobility support for 6LoWPAN wireless sensor networks. J. Netw. Comput. Appl. **49**, 68–77 (2015)
9. Cadger, F., Kevin, K., Santos, J., Moffet, S.: Location and mobility-aware routing for improving multimedia streaming performance in MANETs. Wirel. Pers. Commun. **86**(3), 1653–1672 (2016)

Intelligent Invigilation Using Video Surveillance

Gayatri Sakya, Shubhanshu Singh, Shreya Mishra, and Shashank Mani Tiwari

Abstract At the examination center, examinees daily face difficulties in finding their seating position that consumes a lot of time. Also in COVID-19 pandemic, we need to maintain social distancing while taking examination offline. So, this paper proposes system that can help examinees in knowing their seat number in their respective examination hall with the help of online Web portal. It can help in eliminating the crowd at the notice board and can also save their time. With the help of an admit card generated to which a barcode be attached, an examinee can scan that code each day and know their seating location as per a daily basis. Another system is also proposed for face recognition that can help the examiner in identifying the identity of candidates that can also alert them if they are not found in the database using deep learning. Within the examination hall, the orientation of the head and movement of the mouth of an examinee provide us the clue of suspicious behaviors. Nowadays, most of the suspicious behaviors monitoring procedures are done manually that involves a lot of invigilators in each examination hall. In the pandemic situation, we proposed an intelligent invigilation using video surveillance that can autonomously detect and track examinee's eye gaze, head orientation, and mouth movement to robustly detect their cheating activities. Algorithms which are implemented independently include eigen-face, fisherface, and linear binary pattern histograms. With the help of webcam installed at each examinee's desk, if any suspicious activity is detected, the system will generate an alarm indicating such behaviors.

Keywords Web development · Faces recognition · Deep learning · Intelligent invigilation · Video surveillance

1 Introduction

The paper aims at providing aids to the task of examination invigilation. This project is divided into three parts. The first step of invigilation is to maintain the seating

---

G. Sakya (✉) · S. Singh · S. Mishra · S. M. Tiwari
JSS Academy of Technical Education, Noida, India
e-mail: gayatrisakya@jssaten.ac.in

V. Bhateja et al. (eds.), *Computer Communication, Networking and IoT*, Lecture Notes in Networks and Systems 197, https://doi.org/10.1007/978-981-16-0980-0_37

arrangement and sticking it to the notice board which leads to chaos if the examination is on a large scale.

To simplify this, the project has "Smart Admit Card Generation" which is incorporated by a QR code. The QR code has a roll number of the student fed in it. The seating arrangement is to be done in a database. The QR code when scanned gets linked to the database and the seating arrangement is displayed to the student. This will simplify the task of invigilators; they do not have to check it manually as well as the students who find it difficult to find the seating arrangement from a long list of roll numbers on the notice board. The second task would be face verification by capturing the face of students and matching it to the pictures already saved in the database. There exists a database containing 2–3 images of the candidate, and the live photograph is tested against those images. If the characteristics match, the student is verified, otherwise not verified. The last step is to keep an eye on the students and their movements. If the face movements of any student defy the angles and positions already fed into the database, then an alarm would be set at the system through which an eye is kept. This task helps in proctoring the examination to a significant extent. The process is explained in Fig. 1.

2 Various Techniques for Intelligent Invigilation

In recent years, the techniques based on the biological properties of human beings having the much significant in the identification of individuals where the other techniques like passcodes, OTP generation, and other types of security modes, having the possibility of getting stolen, misused, forged, and many more. Hence, the biological properties like identification of face, fingerprints, palm, ear, iris, retina, and signature can be used which are not easily accessed by anyone. Here are the following methods which developed for feature screening and matching. These are the development phases for the biological checking of the face.

The eigenface method uses the vanguard eigenvectors of the sample covariance matrix to characterize the lower dimensional. The fisherface method and block LDA algorithm segment the entire image into several blocks and structures each block as a row vector. Linear discrimination analysis is performed on the row vectors for the block from the two-dimensional matrices. Support vector machines to improve the classification performance of the supervised learning techniques are used to train SVM generally. Admit cards have been an inseparable part of our education system for a very long time. They came into existence to verify the identity of the student going to give the exam. Various columns like roll number and the face of the student are manually verified by the invigilator, and then, the seating arrangement is verified. Face verification has also been a difficult task as a single image on the admit card can be a little or more different from the real candidate. Invigilation for cheating is a really important part since the inception of examinations to not tamper the integrity of examinations but still, the students tend to cheat in some or the other way.

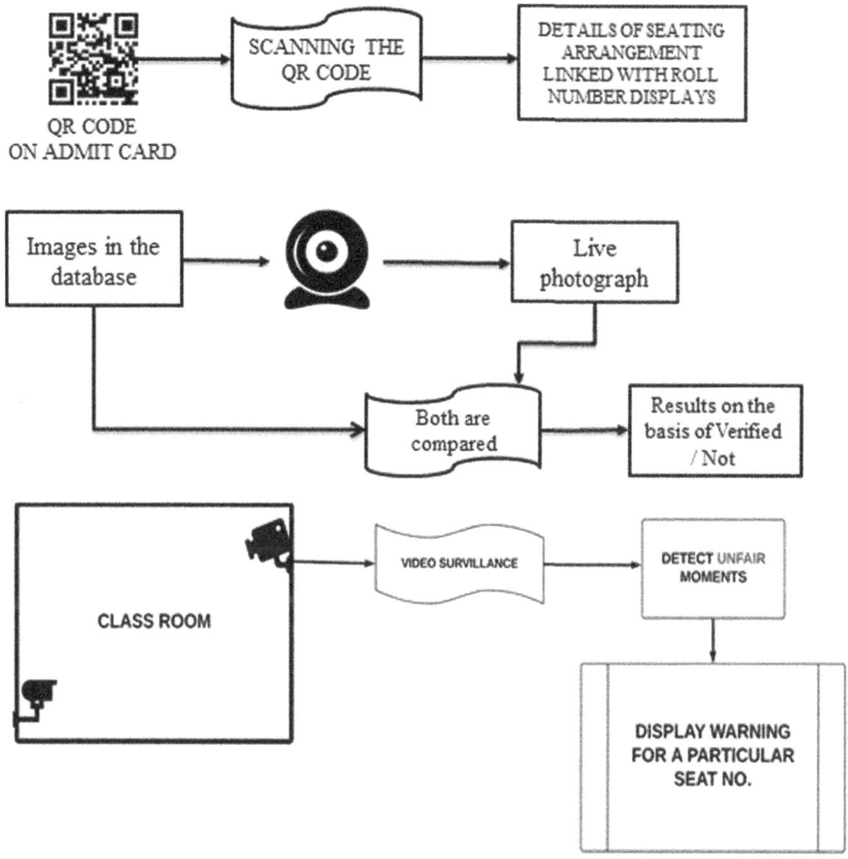

Fig. 1 Block diagram of workflow

In [1] described the use of barcodes in supermarkets and give a brief idea about barcode generation, and identification software development is discussed. Also in [2], the alertness of driver is monitored using the position of eye gazes and head movement which motivated to perform this work using head movement. In this big data era, management of database is a big challenge and also discussed in [3] the various database management systems and methods which may be useful in this work. In [4], the cascade Haar-like features are explained, and also method to improve the scanning of images and quality is discussed. So for our work, we have chosen cascade Haar. In [5], the author analyzes and obtains the 3D head pose. In our work, also, we need to check the head movement for ant cheating activity. In [6], the author discusses a detailed survey of various for face detection. In [7], the author discusses the vision-based human activity recognition and latest methods mainly the activity presentation and classification. The abnormal behavior analysis methods are discussed in [8], and an algorithm is proposed using Markov model for observing video frames. In [9] and [10], deep learning methods are discussed for abnormal behavior detection and image classification. By doing the literature survey, the work is motivated to find the exact flow of performing the intelligent invigilation. The survey motivated for developing the Web portal, for using the facial recognition algorithm to do automatic verification of person and also to use deep learning methods for detecting the abnormal behavior like cheating.

3 Objectives and Methodology of the Proposed Work

Admit cards have been an inseparable part of our education system for a very long time. They came into existence to verify the identity of the student going to give the exam. Various columns like roll number and the face of the student are manually verified by the invigilator, and then, the seating arrangement is verified. Face verification has also been a difficult task as a single image on the admit card can be a little or more different from the real candidate. Invigilation for cheating is a really important part since the inception of examinations to not tamper the integrity of examinations but still, the students tend to cheat in some or the other way. So our work includes three main objectives.

- Phase 1: Development of Web portal to generate each day seating plan by scanning the QR code on admit card.
- Phase 2: Check the identity of the examinee using deep learning algorithm without manual check.
- Phase 3: To check any cheating activity by capturing the neck movement at desk.

Phase 1: It includes admit card generation portal. The admit card has been generated by using the concepts of Web development. The portal is made upon a hosting site by HTML and CSS. The portal requires the roll number and name of the student and redirects it to the database, and a admit card is generated for the candidate which has a QR code on it developed by using PHP. The QR code has roll number fed in it, and as soon as the candidate scans it by using the camera, the seating arrangement is displayed in front of student and invigilator so there is no need to verify the admit card by the teacher for roll number and name. Following image shows the steps involved in Fig. 2.

Meanwhile, the data is stored in the database which is created in phpMyAdmin which works with MySql and PHP languages. The database has to be linked with the admit card portal to do so. The following images show the structure and table of the database, and the administration dealing with seating arrangement can modulate it easily according to the requirements. The images of Web portal are shown below in Fig. 3.

Phase 2 Face Recognition:

It includes face verification. So we apply deep learning and OpenCV along (with no libraries apart from scikit-learn) to detect faces, compute 128D face embedding's to reckon a face, train associate degree SVM on high of embedding's, and recognize

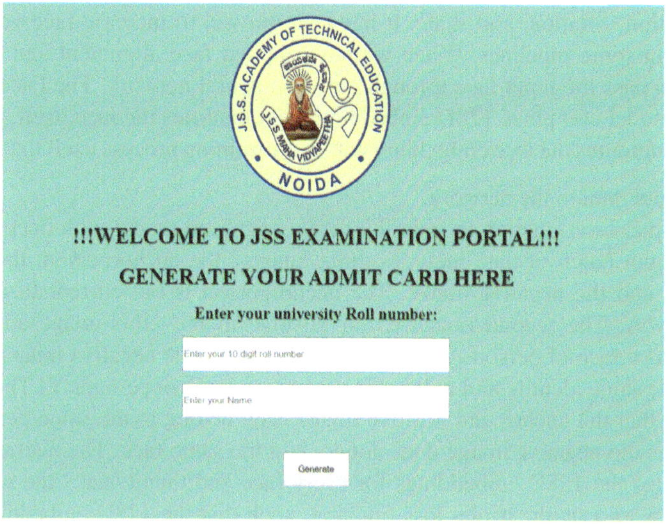

Fig. 2 Admit card generations

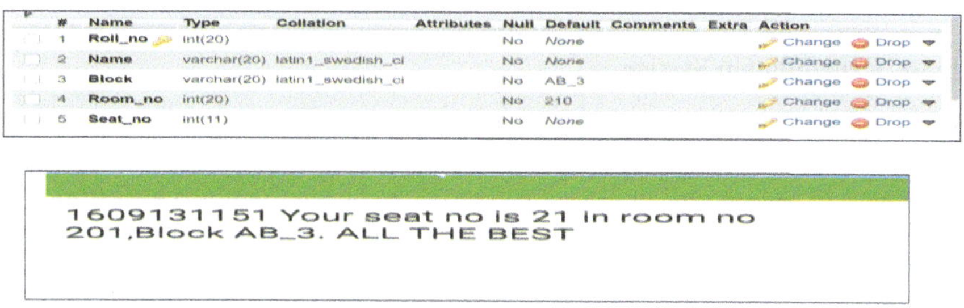

Fig. 3 Web portal images

faces in live video streams. All these tasks are accomplished with OpenCV, enabling us to get a pure OpenCV to face recognition pipeline.

- **OpenCV's face recognition**: To build OpenCV face recognition pipeline, we will apply deep learning in two key steps:
- To apply face detection that detects the presence and placement of a face in a picture, however, does not determine it.
- To extract the 128D feature vectors (called "embeddings") that quantify every face in a picture.

First, we input the video frame to our face recognition pipeline. Given the input, we apply face detection to detect the placement of a face within the image. We are then able to cipher the facial landmarks, enabling us to preprocess and align the face. Face alignment is a method of distinguishing the geometric structure of the faces and making an attempt to get a canonical alignment of the face based on the interpretation, rotation, and scale. It has been proven to increase face verification accuracy in some pipelines. When we have applied face alignment and cropping, we tend to pass the input face through our deep neural network. The FaceNet deep learning model computes 128D embedding that quantifies the face itself. Then, the network computes the face embedding within the training process itself that includes:

- The input data to the network.
- The triplet loss function to train the face recognition model with deep learning, each input batch of data includes three images: the anchorperson, the positive image, and the negative image. The anchorperson is our current face and has identity A. The second image is our positive image—this image additionally contains a face of person A. On the opposite hand, the negative image does not have the same identity and will belong to person B, C, or perhaps Y! The purpose here is that the anchor and positive image each belong to the same person/face, whereas the negative image does not contain the same face. The neural network computes the 128D embeddings for every face and then tweaks the weights of the network (via the triplet loss function) such that the 128D embedding of the anchor and positive image lie nearer along.
- While at the same time, pushing the embeddings for the negative image further away.

Phase 3: Invigilation System

A webcam is installed on each of the examination desks. This will automatically detect and track the examinee's eye gaze and mouth movement. As soon as the orientation of the eye goes out of a certain range or the examinee tries to talk to the neighboring examinee, the system will alert a message indicating the suspicious behavior of the examinee.

4 Algorithms Developed

4.1 Face Recognition Algorithm

For this work, three algorithms are implemented independently that are eigenface, fisherface, and linear binary pattern histograms, respectively. All three can be implemented using OpenCV libraries. There are three stages for face recognition as follows:

- Collecting images IDs.
- Extracting unique features, classifying them, and storing in XML files.
- Matching features of an input image to the features in the saved XML files and predict identity.

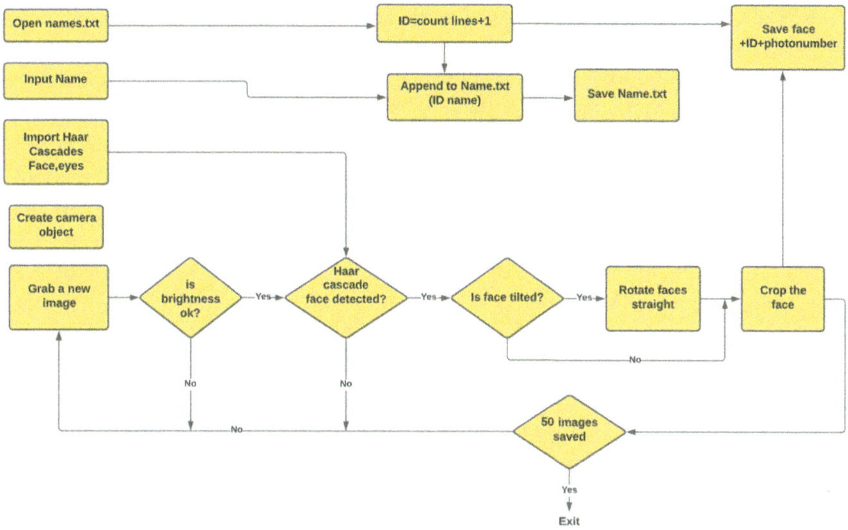

Fig. 4 Process for face recognition

Collecting classification images is usually done manually using photo editing software to crop and resize photos. Furthermore, PCA and LDA require the same number of pixels in all the images for the correct operation. This is time consuming, and the laborious task is automated through an application to collect 50 images with different expressions. The application detects suitable expressions between 300 ms, straightens any existing tilt, and saves them.

The application starts with a request for a name to be entered to be stored with the ID in a text file. The face detection system starts in the first half. However, before the capturing begins, the application checks for the brightness levels and will capture only if the face is well illuminated. Furthermore, after the face is detected, the position of the eyes is analyzed. If the head is tilted, the application automatically corrects the orientation. These two additions were made considering the requirements for the eigenface algorithm. The image is then gathered and saved using the ID as a filename to be known later. A loop runs this program until 50 worthwhile pictures are collected from the person. This application made data collection efficient (Fig. 4).

4.2 *Training the Classifiers*

OpenCV facilitates the creation of XML files to accumulate the features extracted from datasets using the face recognizer class. The stored images are imported, converted to grayscale, and saved with IDs in two lists with the same indexes. Face recognizer objects are created using face recognizer class. Each recognizer can take in parameters that are described further.

cv2.face.createEigenFaceRecognizer(): Takes in the number of components for the PCA for crating the eigenfaces. OpenCV documentation indicates that 80 can provide adequate reconstruction capabilities.

cv2.face.createFisherfaceRecognize(): The first argument is the number of components for the LDA for creating fisherfaces. OpenCV references it to be kept zero if uncertain. Similar to the eigenface threshold, −1 if the threshold is passed.

cv2.face.createLBPHFaceRecognizer(): The radius from the center pixel to build the local binary pattern. The number of sample points to shape the pattern. Having a considerable number will slow down the computer. The number of cells needs to be created on *X*-axis and *Y*-axis. A threshold value analogous to eigenface and fisherface is decided. If the threshold is passed, the object will return −1. Recognizer objects are created, and images are imported, resized, converted into Numpy arrays, and stored in a vector. The ID of the image is gathered from splitting the file name and stored in another vector. All three of the objects are trained by using FaceRecognizer.train(NumpyImage, ID). It is important to note that the images were resized only for eigenface and fisherface, not for LBPH. The configuration model is then saved as a XML file using FaceRecognizer.save(FileName).

4.3 The Face Recognition

The face recognizer object is created using the desired parameters. The face detector is used to detect faces in the image, cropped, and transferred to be recognized. This is done using the same technique used for the image capture application. For each face detected, a prediction is made using FaceRecognizer.predict() which returns the ID of the class and confidence. The process is the same for all algorithms, and if the confidence is higher than the set threshold, ID is −1. At last, names from the text file with IDs are used to display the name and confidence on the screen. If the ID is −1, the application will print unknown faces without the confidence level. The algorithm is shown in Fig. 5.

4.4 Phase 3: To check any cheating activity by capturing the neck movement at desk the algorithm is discussed in Fig. 6.

Algorithm

1. Load OpenCV's Haar cascade for detecting face.
2. Initialize the video stream by turning on the surveillance camera.
3. Loop over the video stream frame-by-frame.
4. Capture a frame and convert it to grayscale.
5. Using the detectMultiScale method, face is detected in the captured frame.

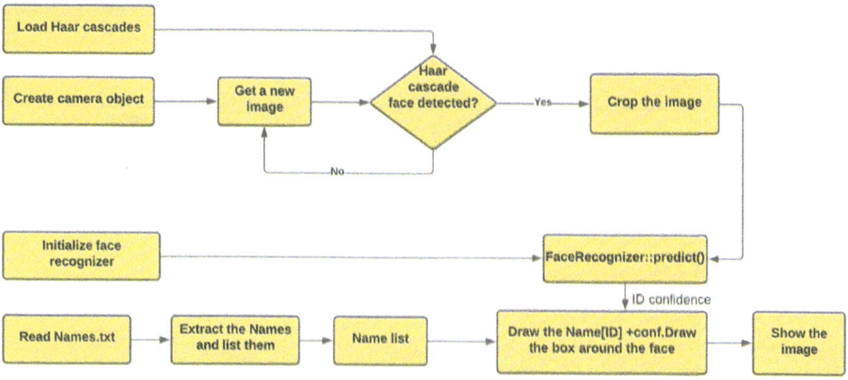

Fig. 5 Algorithm for face recognition

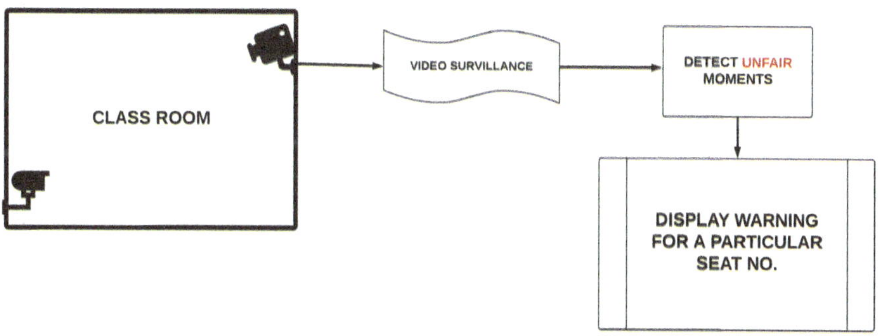

Fig. 6 Process flow

6. The result of face detection model is a list of bounding box rectangle.
7. If any suspicious activity is detected, then an alarm is triggered and seat number is displayed on the screen.

5 Observation and Results

Database of candidate is maintained with ample of photographs with different orientations. These will be used against the live picture of candidate for verification.

As a result of the work admit card with QR code was generated. QR code was successfully linked to the database of seating arrangement. **Seating arrangement displayed** upon scanning. **Face verification with live photograph was done.** Eyes and head **alignment detection for unfair means** was successful. Figures 7 and 8 discuss the example database.

Fig. 7 Example database for face recognition

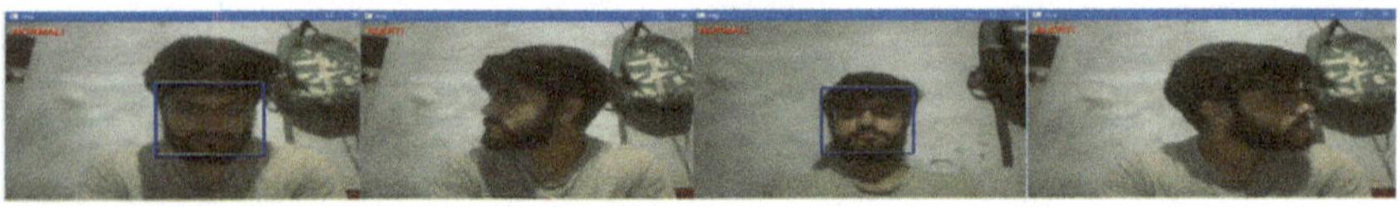

Fig. 8 Example database for checking cheating activity

6 Conclusion

The work helps in automatically letting three teachers and the students know the seating arrangement without any chaos by using QR-based admit card which is linked to the database and verifies the roll number and seat number along with building name. Then, the second part verifies the face of the candidate by using image processing methodology in which there is a database of several images of the candidate which is checked against the live photograph of the candidate. If a characteristic matching percentage is high, the identity is verified. The last is the video surveillance which checks for unfair movement by the candidate by considering the angles by which the face and eyes are moving. This has a large future scope in coming examinations of companies and institutions and can be a great aid to the invigilation process.

References

1. Weng, D., Yang, L.: Design and implementation of barcode management information system. In: Zhu, R., Ma, Y. (eds.) Information engineering and applications. Lecture Notes in Electrical Engineering, vol. 154. Springer, London (2012). https://doi.org/10.1007/978-1-4471-2386-6_158
2. Debnath, P.P., Hasan, A.F.M.R., Das, D.: Detection and controlling of drivers' visual focus of attention. In: 2017 International Conference on Electrical, Computer and Communication Engineering (ECCE), Cox's Bazar, pp. 301–307 (2017). https://doi.org/10.1109/ECACE.2017.7912922
3. Mapanga, I., Kadebu, P.: Database management systems: a NoSQL analysis. Int. J. Mod. Commun. Technol. Res. (IJMCTR) **1**(7). ISSN: 2321-0850 (2013)
4. Malikovich, K.M., Akhmatovich, T.K., Ugli, I.S.Z., Nizomovich, M.O.: Triangle method for fast face detection on the wild. J. Multimedia Inf. Syst. **5**(1), 15–20 (2018)

5. Pan, Y., Zhu, H., Ji, R.: 3-D Head pose estimation for monocular image. In: Wang, L., Jin, Y. (eds.) Fuzzy systems and knowledge discovery. FSKD 2005. Lecture Notes in Computer Science, vol. 3614. Springer, Berlin, Heidelberg (2005). https://doi.org/10.1007/11540007_35
6. Kumar, A., Kaur, A., Kumar, M.: Face detection techniques: a review. Artif. Intell. Rev. **52**, 927–948 (2019). https://doi.org/10.1007/s10462-018-9650-2
7. Zhang, S., Wei, Z., Nie, J., Huang, L., Wang, S., Li, Z.: A review on human activity recognition using vision-based method. J. Healthc. Eng. (2017)
8. Wanga, T., Qia, M., Deng, Y., Zhouc, Y., Wang, H, Lyua, Q., et al.: Abnormal event detection based on analysis of movement information of video sequence. Optik **152** (2018)
9. Ko, K.-E., Sim, K.-B.: Deep convolutional framework for abnormal behavior detection in smart surveillance system. Eng. Appl. Artif. Intell. **67** (2018)
10. Sudarshana, T., Jyotsna, C., Amudha, J.: Deep learning model for image classification. In: International Conference on Computational Vision and Bio-Inspired Computing (ICCVBIC) (2019)

Detection of Anomaly in Streaming Dataset

Shashikumar G. Totad, K. G. Karibasappa, R. B. Geeta, T. Nithish, Prasanna J. Hiremath, Rahul Elagoudanavar, and Nisha Rao

Abstract Data is composed of actual data that translates to a sequence that can be processed by the machine. Data can be analysed and used to enhance the revenue of the company, to create a new chemical formulation, to make unique applications and to aid in the establishment of innovative modes of communication. The key element of work for any industry is to analyse the data acquired and adapt trends or achieve economic insights to tackle any given problem. The analysis of data is as qualitative as the accuracy of the information. The data obtained should be accurate and consistent. Improper data collection may lead to incorrect data visualizations resulting to inaccurate analysis and erroneous results. Most industries aim for accuracy and better performance where a tiny mistake may lead to heavy losses. In this condition, the incorrect decisions taken due to improper data analysis may lead to crucial circumstances. In this work, we propose a hybrid model which is the combination of long short-term memory (LSTM) and recurrent neural network (RNN) for detecting the anomalies or abnormalities in the data during the request raised by the customer. The proposed model is found to be even more efficient in identifying an anomaly in the data collection, taking into consideration of date and time constraints. The proposed model delivered enhanced performance compared to other traditional approaches.

Keywords Anomaly · LSTM · RNN · Sarima · Decision tree · Isolation forest

1 Introduction

Data is a set of accurate information that transforms to form in such a fashion that the system can easily process it. Data is biggest assets for industry to make strategic decision in dynamic nature of the world. Data can be analysed and used to boost the company's revenue, to design a new chemical formulation, to allow innovative

S. G. Totad · K. G. Karibasappa · T. Nithish (✉) · P. J. Hiremath · R. Elagoudanavar · N. Rao
School of Computer Science and Engineering, KLE Technological University, Hubballi, Karnataka, India

R. B. Geeta
Department of CSE, KLE Institute of Technology, Hubballi, Karnataka, India

V. Bhateja et al. (eds.), *Computer Communication, Networking and IoT*, Lecture Notes in Networks and Systems 197, https://doi.org/10.1007/978-981-16-0980-0_38

solutions and to help establish innovative forms of communication. Without adequate information, it is impractical to serve the globe with innovative and trending technologies to boost the economy. It is indeed essential that data collection carries out effectively and efficiently. Detection of anomalies is the method of finding patterns in the data which do not strictly adhere to the anticipated behaviour [1]. We recognize an anomaly as a time point where the system's action is unusual and considerably different from previous, acceptable behaviour. This problem has important applications on identifying the patterns on different aspects such as identifying the patterns of customer's login and interest of purchases, orders issued from the specific customer at regular interval of time and so on [2]. In this paper, we have considered a hybrid model to identify the anomaly in the text of customers purchase orders. The process which has been considered here in this work is to identify the abnormalities and classify the data based on the need of the user which will help the online vendors to predict the requirements of the customers which have been ordered and identifies whether the customer order is an normal or abnormal. In this work the method was called to define the abnormality and classify the data based on the user's need to help the online vendors predict the customer's requirements that were ordered and determine the whether the customer's order is an anomaly or un-anomaly [3]. Previous approaches to anomaly detection from text mostly constructed lexical features and employed a classifier [4]. The main aim of this work is to find anomaly in inbound request of the customer, to detect abnormalities and to study its impact analysis. In a data driven industry, where every decision is taken based on historical and current data. This point drives as main motivation to propose new models for detecting abnormalities. During data collection, data cannot be used for meaningful analysis because of presence of abnormalities in the data. Thus, it is essential to create a system to detect abnormalities and alter the users. The proposed method gave accuracy of 93% and LSTM and RNN model is the best promising model in comparison and various conventional approaches. The remaining part of the paper is organized as given below. Section 2 offers an description of related work. Section 3 provides information more about LSTM (LSTM) neural network anomaly detection approach. Section 4 depicts experimental results and discussion. Finally, Section 5 conclusion.

2 Related Work

Federico Giannoni et al. have used univariate Gaussian predictor and S-ESD algorithm to identify abnormalities and outliers [5]. The paper proposed a novel anomaly detection approach in a time series based on LSTM network that uses Kernel estimator of quantile functions [6]. This paper proposes novel anomaly detection algorithm that meets these constrictions. The strategy takes advantage of an algorithm online storage of the series is called hierarchical temporary memory (HTM), and present study provides results by using a baseline containing abnormalities recognized in streaming data in the actual world [7–9]. The identification of irregularities in streaming data in real time has practical and important implications across several sectors. Use cases like maintenance work, fraud protection, identification of faults, and surveillance can be found across multiple industries, including finance, IT, defence, health, electricity, e-commerce, farming and social networks. For certain cases, the system's metrics can shift over time, which is an issue recognized as idea drift. In these instances, models must adjust in an unsupervised, automated fashion to a new phenomenon of standard [10]. Statistical analysis is the most widely used data form in existence, as well as the statistical analysis is hot research topic that gathers data for analysis. It involves a series of chronologically ordered over across from observational objects. Time series is widespread and widely used in numerous industry sectors. Zhao et al. [11] anomalous series data sub-string identification is a really complex area. Most data series with unusual occurrences have seemed as normalization, and unusual trends seem to quite very seldom, but rare unusual shapes encompass very vital information [12, 13]. For diagnose the irregularity of the time-series data, yet another-class SVM has been used as it can identify the abnormalities mostly on set of data of the vector. After the transformation of time-series data into phase space [14]. In LSTM framework, it is to tackle the question of time-series data outlier detection. The layered LSTM network educated only on training dataset has been used to simulate over a variety of time levels. They presumed the consequent error terms with a Gaussian distribution which has been used to predict the significance of anomaly behaviour and attitude. Their model had been shown to perform well on four datasets.

3 Long Short-Term Memory (LSTM) Neural Network

They have [15] introduced LSTM, an evolution of RNN, to fix the concerns of layer RNN results primarily downsides by providing new encounters per subsystem (or cell). LSTMs are really a particular type of RNN worthy of learning long-term dependencies, and then data collection is a default method for longer periods of time. The LSTM framework is arranged in a chain structure format. The recurring component though has a different composition. This has communications infrastructure layers with a special medium of interaction instead of a neural network [16] classifier like a traditional RNN. The system proposed is made of multiple components. The very first component, time-series data determinant, estimates times and dates for an accept or, and the second subsystem, anomaly detector is liable for the normal or abnormal labelling of the specified time-series analysis items. Deep learning was used in a large range of applications mainly owing to its capacity to quickly explore unique patterns [17]. This automated analysis function uses neural network a viable contender for anomaly-based issue in time-stamped data. Variants in contrast with other neural networks and mathematical models are stable, too. Because of its capabilities to extract lengthy-term patterns in the time-series data experienced, the LSTM produces well enough on statistical features. However, we have shown in this study that due to its parameter efficiency RNN can be a good alternative for both univariate and MV time-series data. RNN and LSTM are commonly used in the literature for question of time-series classification [18–20].

3.1 Time-Series Predictor

LSTM's predictor module was indeed based on RNN. RNN is a neural network that has been commonly used in different domain names such as pattern recognition and machine learning processing because of its parameter efficiency in a variety of different capabilities. This system developed a statistical procedure, joint structures, as the name indicates. The function applies the convolution layer in first point which leads to linear detection's. At linear stimulation, a nonlinear activation function is applied in the next step. The convolution operation is normally denoted as asterisk:

$$s(t) = (x * w)(t) \tag{1}$$

This new function s can be defined at the time-stamp t, as a smooth approximation or a weighted average of the $x(\pi)$ function, where the weighting is given by the amount t. In (1), function x is referred to as the input, and function w is referred to as the kernel. The output is called the function map one-dimensional convolution which is defined as:

$$s(t) = \sum_{\tau=-\infty}^{\infty} x(\tau)w(t-\tau) \tag{2}$$

The final coupling phase in RNN is a convolution layer. Every neuronal from either a preceding network is connected to all output neurons in this zone. The activation for convolution and fully connected layers is given in (4), (5) and (6), respectively, where k is defined as

$$z^l_{ji} = \sum_{-k}^{k} W^l_{jk} a^{l-1}_{i-k} + b^l_j \tag{3}$$

$$a^l_{ji} = \max\left(z^l_{ji}, 0\right) \tag{4}$$

$$z^l_j = \sum_{k=1}^{e} W^l_{jk} a^{l-1}_k + b^l_j \tag{5}$$

$$a^l_j = \max\left(z^l_j, 0\right) \tag{6}$$

3.2 Anomaly Detector

When the time-series predictor allows the prediction of next time stamp $xt + 1$, this module detects abnormalities in a given time sequence. The predicted value is passed to this module by the predictor module [21], and the calculation of the difference between the observed and the estimated value. Euclidean distance given in (7) is used as a measure of the discrepancy.

$$(y_t, y_t') = \sqrt{(y_t - y_t')^2} \tag{7}$$

where y_t is actual value and y_t' is predicted value.

The distance to the Euclidean has been used as anomaly value. A high anomaly score indicates a major anomaly in the time stamp in question. A threshold for this module must be specified based on the type of time series which is required for most anomaly detection algorithms.

4 Experimental Results

We have assessed LSTM on informational index and given an in depth evaluation client request demand anomaly detection strategies which incorporate a few best in class strategies. Both real and unreal time-series information from various spaces are utilized for tests. Exploratory arrangement is categorized into two parts: in first part, data is segregated into week days. In second part, an LSTM and SARIMA models are deployed in segregated dataset [22].

4.1 Experimental Settings

4.1.1 Data Collection Module

In this process, data stored in the CVS format is accessed. This dataset contains 874 actual and logical time-series anomaly instances.

4.1.2 Data Preparation Module

In this process, the data is first segregated into weekday datasets and various visualizations and graphical representations are created to under the trends, seasonality and spikes in the data has shown in Fig. 1. The main aim is to find anomaly in inbound request of the customer. The order placed by customers, abnormalities detection and its impact is shown in Figs. 1, 3 and 4, respectively.

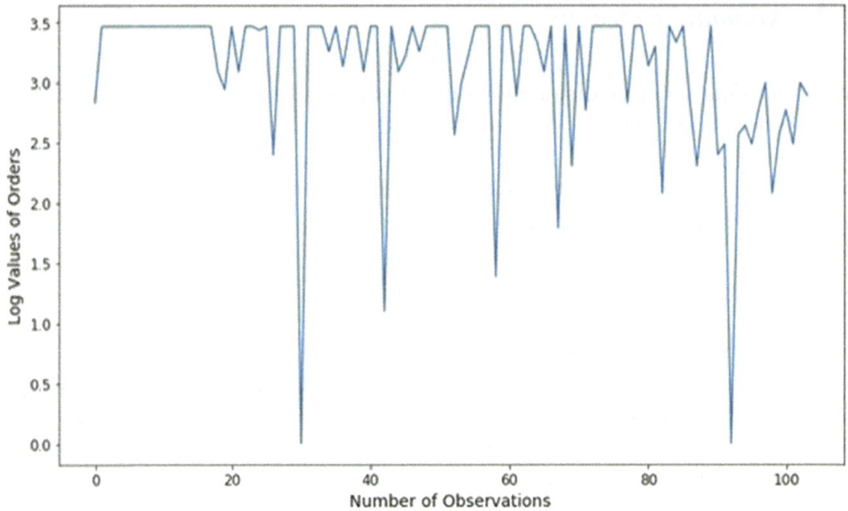

Fig. 1 Graphical representation of orders

To address this issue, hybrid model was proposed in this work which is integration of long short-term memory (LSTM) and recurrent neural network (RNN) for detecting the abnormalities in the data during the request raised by the customer. Our model combines both packages of time-series predictor and anomaly detector. The model is found more effective in order to identify anomalies on the marketing sales dataset by considering the constraints of time and date. The model gave an enhanced performance compared with the traditional approaches [23].

4.2 *Evaluation Metric and Experimental Setup*

$F - score$ is most normally utilized singleton metric which fills in as a marker of the models performance. Therefore, we utilized $F - score$ as the assessment metric for our models. All the peculiarity location strategies in this exploratory setting are applied on each time arrangement independently. Normal $F - scores$ per benchmark are accounted for every technique.

$$F{-}score = 2 \cdot \frac{\text{Precision} \cdot \text{Recall}}{\text{Precision} + \text{Recall}} \tag{8}$$

LSTM performance is compared with SARIMA-based model, and evaluation results show that LSTM performance outweighs the performance of SARIMA model. The abnormalities detected on segregated Monday dataset by LSTM model is shown in Fig. 3.

	precision	recall	f1-score	support
Anamoly	1.00	0.50	0.67	2
UnAnamoly	0.93	1.00	0.96	13
accuracy			0.93	15
macro avg	0.96	0.75	0.81	15
weighted avg	0.94	0.93	0.92	15

0.9333333333333333

Fig. 2 Confusion matrix of the classifier model

Fig. 3 Abnormalities detected in dataset

It is vividly visible that LSTM model provides better anomaly detection than SARIMA model for Monday dataset as shown in Fig. 4.

The confusion matrix of the classifier model is shown in Fig. 2. The clustering classification approach with respect to accuracy and RMSE score is shown in Fig. 5.

The random forest is applied on the preprocessed dataset and is obtained an accuracy of 93%, and the result of average LSTM RMSE score is 7.25 compared to SARIMA RMSE Scores 8.96. The weekdays score of LSTM and RMSE scores is shown in Fig. 6.

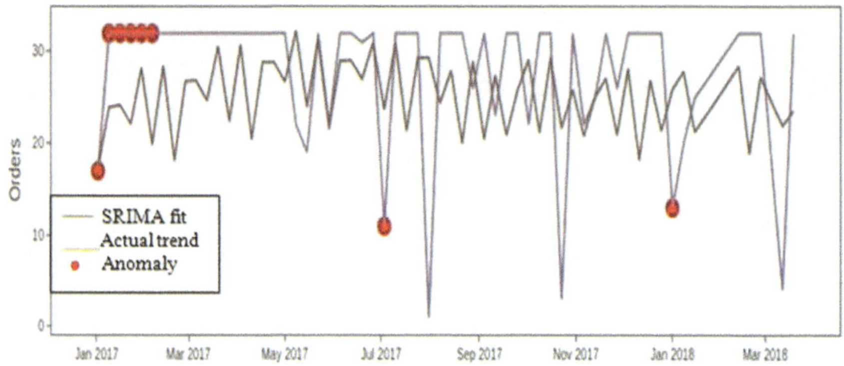

Fig. 4 Impact on abnormalities detection

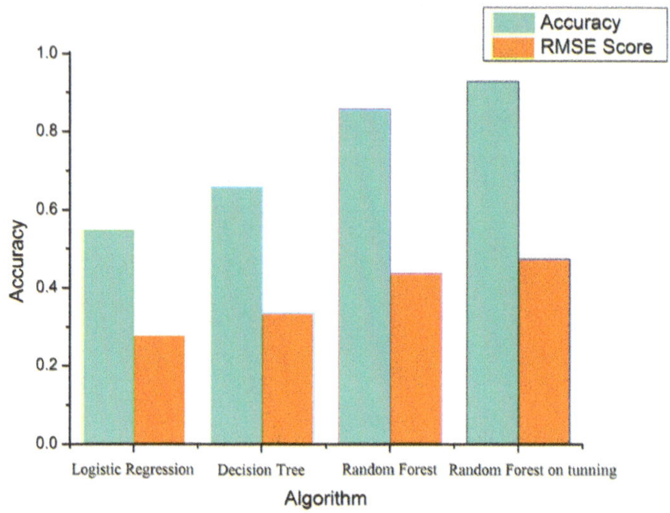

Fig. 5 Accuracy versus RMSE score of clustering classification

5 Conclusion

Detection of abnormalities is a significant aspect for the industry. Detection of abnormalities plays a vital role in the decision-making and the development of the sector. It can be unambiguously concluded from the analysis that the time-series approach is better than the clustering and classification model because it provides a broader understanding of data trends, seasonality and spikes, and the time-series approach provides and results in the form of meaningful segregation on weekdays that help in analysing errors and to identify abnormalities. The root mean square errors for time-series approaches are an approaches that outweigh performance of clustering and classification models. From the observations, it can be concluded that LSTM

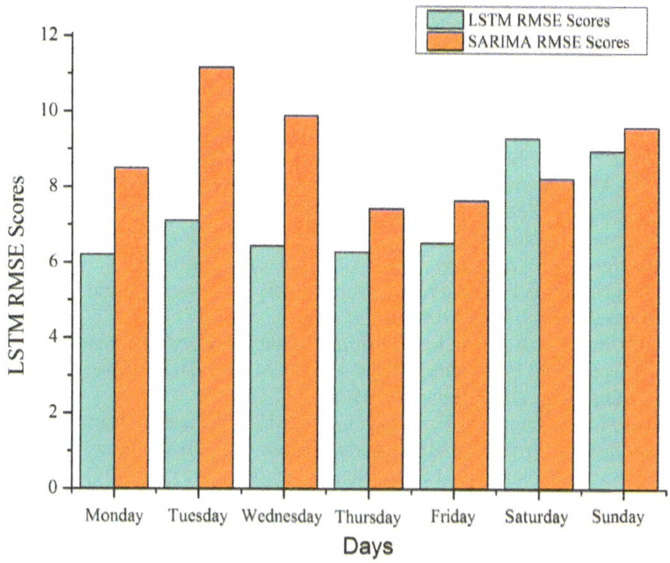

Fig. 6 Weekdays score

provides better detection of abnormalities than SARIMA as LSTM's average RMSE score is lower than SARIMA's score. It is indeed concluded that time-series LSTM technique performs superior at presenting the results for the anomaly detection task for the given dataset. It is analysed that strengthened pre-processing and model furnishing can provide better results. Users will use the other time-series techniques to analyse various datasets to get higher performance.

References

1. Chandola, V., Banerjee, A., Kumar, V.: Anomaly detection: a survey. ACM Comput. Surv. (CSUR) **41**(3), 1–58 (2009)
2. Geeta, R.B., Totad, S.G., Reddy, P., Shobha, R.B.: Big data structures and usage mining coaliation. Int. J. Serv. Technol. Manag. **21**(4–6):252–271 (IF:0.21) (Inderscience Enterprises Ltd.)
3. Guthrie, D.: Unsupervised detection of anomalous text. Dissertation. University of Sheffield (2008)
4. Manevitz, L., Yousef, M.: One-class document classification via neural networks. Neurocomputing **70**(7–9), 1466–1481 (2007)
5. Giannoni, F., Mancini, M., Marinelli, F.: Anomaly detection models for IoT time series data. ArXiv abs/1812.00890 (2018)
6. Malhotra, P., Vig, L., Shroff, G., Agarwal, P.: Long short term memory networks for anomaly detection in time series. In: European Symposium. Artificial. Neural Network, pp. 22–24 (2015)
7. Ahmad, S., Lavin, A., Purdy, S., Agha, Z.: Unsupervised real-time anomaly detection for streaming data. In: Proceedings of the 15th International Conference on the Machine Learning Application. IEEE, Miami, FL (2017)

8. Lavin, A., Ahmad, S.: Evaluating real-time anomaly detection algorithms—the Numenta anomaly benchmark. In: Proceedings of the 14th International Conference on Machine Learning Application. Miami, FL (2015)
9. Gama, J., Žliobaitė, I., Bifet, A., Pechenizkiy, M., Bouchachia, A.: A survey on concept drift adaptation. ACM Comput. Surv. **46**, 1–37 (2014)
10. Bianco, A.M., García Ben, M., Martínez, E.J., Yohai, V.J.: Outlier detection in regression models with ARIMA errors using robust estimates. J. Forecast. **20**, 565–579 (2001)
11. Zhao, Z., Zhang, Y., Zhu, X., Zuo, J.: Research on time series anomaly detection algorithm and application. In: IEEE 4th Advanced Information Technology, Electronic and Automation Control Conference (IAEAC), pp. 16–20. Chengdu, China (2019)
12. Akouemo, H.N., Povinelli, R.J.: Probabilistic anomaly detection in natural gas time series data. Int. J. Forecast. **32**, 948–956 (2015)
13. Laptev, N., Amizadeh, S., Flint, I.: Generic and scalable framework for automated time-series anomaly detection. In: Proceedings of the 21th ACM SIGKDD International Conference on Knowledge Discovery Data Mining, pp. 1939–1947 (2015)
14. Pratama, M., Lu, J., Lughofer, E., Zhang, G., Er, M.J.: Incremental learning of concept drift using evolving type-2 recurrent fuzzy neural network. IEEE Trans. Fuzzy Syst., 1
15. Hochreiter, S., Schmidhuber, J.: LSTM. Neural Comput. **9**, 1735–1780 (1997)
16. Deshapande, A.S., Giraddi, S.G., Karibasappa, K.G., Desai, S.D.: Fungal disease detection in maize leaves using haar wavelet features. In: Information and Communication Technology for Intelligent Systems, pp. 275–286
17. Rumelhart, D.E., Hinton, G.E., Williams, R.J.: Learning representations by back-propagating errors. Nature **323**, 533–536 (1986)
18. Athani, S.S., Kodli, S.A., Banavasi, M.N., Sunitha Hiremath, P.G.: Student performance predictor using multiclass support vector classification algorithm. In: 2017 International Conference on Signal Processing and Communication (ICSPC). IEEE (2017)
19. Joshi, Y., Geeta, R.B., Totad, S.G., Prasad Reddy, P.V.G.D.: Mobile agent based frequent pattern mining for distributed databases. In: International Conference on Intelligent Computing and Communication (ICICC-2017). MAEER's MIT College of Engineering, Pune, 2–4 Aug 2017; Intelligent Computing and Information and Communication, pp. 77–85. Springer, 20 Jan 2018
20. Sung, J., Lee, J., Chung, I.-M., Heo, J.-H.: Hourly water level forecasting at tributary affected by main river condition. Water **9**, 644 (2017)
21. Athani, S.S., Kodli, S.A., Banavasi, M.N., Sunitha Hiremath, P.G.: Student academic performance and social behavior predictor using data mining techniques. In: 2017 International Conference on Computing, Communication and Automation (ICCCA). IEEE (2017)
22. Kothapalli, S., Totad, S.G.: A real-time weather forecasting and analysis. In: IEEE International Conference on Power, Control, Signals and Instrumentation Engineering (ICPCSI-2017). Saveetha Engineering College, Thandalam, Chennai, Tamil Nadu, 21–22 Sept 2017
23. Dhulavvagol, P.M., Totad, S.G., Sourabh, S.: Performance analysis of job scheduling algorithms on Hadoop multi-cluster environment. In: Emerging Research in Electronics, Computer Science and Technology(ICERECT-2018). Lecture Notes in Electrical Engineering, vol. 545, pp. 457–470. Springer, Singapore. First Online 24 Apr 2019

IOT-Based Occlusion Invariant Face Recognition System

Ch. Rathna Jyothi and K. Ramanjaneyulu

Abstract Nowadays, one of the most important challenges of any face recognition system is partial occlusion. The paper addresses face recognition in the presence of sunglasses and scarf occlusion. The method that proposed is a robust approach which consists of detecting the face region that is not occluded and uses this region to obtain the face recognition. Face images are captured from the USB web camera on the Raspberry Pi. To segment the occluded and non-occluded parts, fuzzy C-means clustering is used, and for recognition, block-based mean weighted local binary patterns (MWLBP) are used. The input image is divided into number of sub-blocks, and each block is checked if occlusion present or not, and only from non-occluded blocks MWLBP features are extracted and are used for classification. Experiment results show that our method is giving promising results when compared to the other conventional techniques.

Keywords Face recognition · Occlusion · Fuzzy segmentation · SVM · IOT · Mean weighted LBP

1 Introduction

Due to the rapid growth in the technology, there is a high demand for high level of security for persons and organization to sustain their privacy in the business world. So in order to secure assets and privacy, one needs more secure system called biometrics. If the number of images exceeds certain limit, humans fail to identify correctly, but today the face recognition system is achieving good accuracy of 92% in 1.6 millions of faces as stated by National Institute of Standards and Technology (NIST) [1]. But it is observed that when the database size increases, its accuracy decreases linearly.

Ch. Rathna Jyothi (✉)
Computer Science and Engineering Department, Andhra Loyola Institute of Engineering and Technology, Vijayawada, India

K. Ramanjaneyulu
Electronics and Communication Engineering Department, Prasad V. Potluri Siddhartha Institute of Technology, Vijayawada, India

V. Bhateja et al. (eds.), *Computer Communication, Networking and IoT*, Lecture Notes in Networks and Systems 197, https://doi.org/10.1007/978-981-16-0980-0_39

There are several biometrics like iris, retina, finger print, face, etc., and out of several biometrics, face biometric is the one that is mostly used because as stated above it is the natural human acting way of recognition, it does not need any cooperation of the user, and it is more reliable biometric since it possess complex nonlinear patterns. The reliability of face biometric is also a big issue because the accuracy of face biometric depreciates when face images are added with different variations like pose, illumination, and expression. There are occasions where face biometric fails are aging, resolution, and occlusion.

2 Related Work

The original image of the face greatly influenced by the amount of occlusion. The literature [2–15] focuses on finding occlusion-tolerant features or classifiers to reduce the effect of partial occlusions in face recognition. Though, the hidden features from the occluded parts can still deteriorate the recognition performance. Recently, the studies [16–19] demonstrated that prior information about the occlusion (e.g., type, location, and size) can be used to exclude the occluded parts to achieve significant improvement in the recognition rate. Hence, explicit occlusion analysis is become a necessary step in occlusion-robust face recognition.

3 Partitioning the Image into Sub-blocks

Image partitioning is an important factor to determine the functionality and the efficiency of face recognition system. By breaking images into smaller and more manageable units, it usually becomes easier to compress, store, access, and retrieve the image data.

The coarse segmentation framework partitions the image into fixed sized sub-blocks. The grid structure used to partition the image is shown in Fig. 1. In general, the aggregate number of sub-blocks [20] for an image is calculated by Eq. (1).

Fig. 1 Partitioned image

$$\text{Total no. of Sub-blocks} = (\text{No. of row pixels})/p \times (\text{No. of Column Pixels})/q \quad (1)$$

here, 'p' is the number of rows required in the image sub-block, and 'q' is the number of columns required in the image sub-block. The size of the image block is chosen in such a way that it should improve the effectiveness at a cost of computation time. Smaller the block size may contain more details but increase the computation time as well. Conversely, we increase the block size in order to reduce the computation time; in that case, we may lose the finer information. In the coarse segmentation framework, each image can be represented formally as in Eq. (2).

$$I = \#B_{ij} \quad \text{where} \quad i = 1\ldots4 \; j = 1\ldots6 \tag{2}$$

The image contains 24 sub-blocks subscripted from B_{11} to B_{46} as mentioned in Eq. (2). These sub-blocks will serve as basic building blocks to recognize the images.

SVM is used to find the optimal separating hyper plane $\{\alpha_i, i \in [1, N]\}$ and predict the label of an unknown face x by Eq. (3)

$$f(x) = \text{sign}\left(\sum_{j=1}^{N} \alpha_j y_j K(x_i x) + b\right) \tag{3}$$

where $\{x_j, j \in [1, N]\}$ indicates support vectors.

4 Occlusion Segmentation

The standard fuzzy C-means algorithmic rule (FCM) is used for segmentation of occluded part; by excluding occluded portion, the recognition performance can be greatly improved.

Let us consider the face image contains N pixels, and let us suppose this N should be segmented into $K(K = 2)$ clusters in fuzzy fashion. That means a point i does not necessarily belonging in one of the two classes (one is occluded portion and the other is non-occluded portion) but can partially belong to two classes as well. For each point $i \in N$, the sum of memberships with respect to k classes must be one, i.e.,$\sum_{c=1}^{k} u_{ij=1}$ and $u_{ij} \in [0,1]$. In FCM approach, the segmentation is carried out to minimize an objective function shown in Eq. (4).

$$F = \sum_{c=1}^{k} \sum_{i=1}^{m} u_{ic}^{m} \| y_i - y_c \|^2 \tag{4}$$

where y_c represents the cluster center, and m denotes the fuzziness parameter typically set to 2.

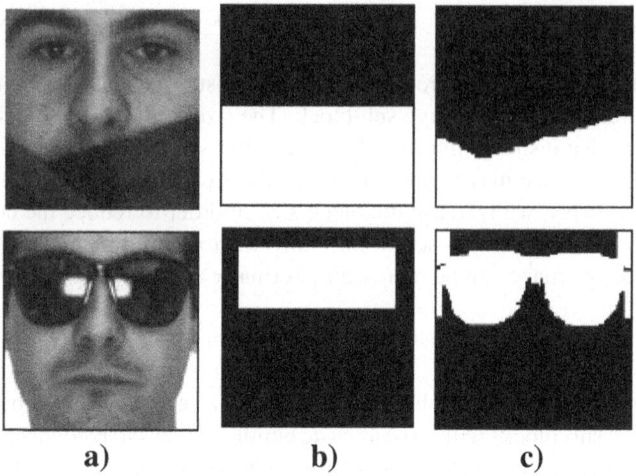

Fig. 2 Illustration of occlusion segmentation **a** Face images occluded by scarf and sunglass. **b** Initial guess of occlusion through thresholding. **c** Visualization after applying fuzzy segmentation

The fuzzy segmentation is applied on a sample image, and the corresponding segmentation results are presented in Fig. 2.

5 The Face Recognition Procedure

The main component of our system is a face recognition procedure that identifies the user based on the captured facial image. In general, biometric identification consists of two distinct phases: enrollment and recognition [5]. During the enrollment phase, we acquire biometric data, process it, and then store it in the system's database along with the users identity.

Enrollment: An overview of the enrollment procedure is shown in Fig. 3.

In the enrollment, images are acquired from the USB web camera on the Raspberry Pi, which by default, captures ten images from the USB web camera on the Raspberry Pi. Users then upload the images to the web application running on a back-end server where they are further processed. For the enrollment procedure, face detection is first performed on the captured images to exclude the image background from the

Fig. 3 Enrollment phase

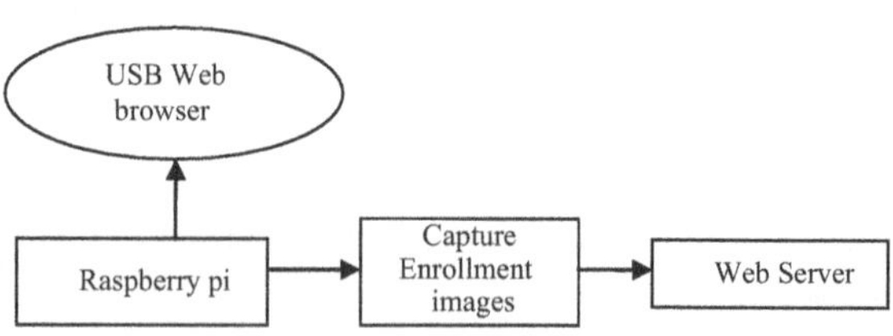

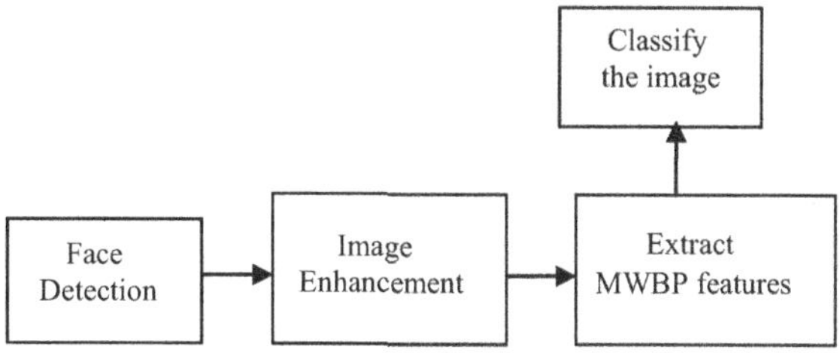

Fig. 4 Recognition phase

processing and focus solely on the region of interest. and it is resized to predefined width and height, i.e., 200 × 200 pixels. Next, the resized image is enhanced by applying Gaussian blurring and histogram equalization. This step is needed to reduce noise and mitigate potential effect of the external lighting. Each image is represented as a 1D feature vector. The feature vectors of all enrollment images are collected in a 2D matrix and stored in the system's database.

Recognition: An overview of the recognition procedure is shown in Fig. 4. During the recognition phase, motion detection is employed to detect whether the user would like to start using the system. When a sufficient level of motion is detected, the frame that triggers the recognition procedure is captured. Each frame is pre-processed by converting it to gray-scale and applying Gaussian blur to remove high-frequency noise; so, we can focus on the structural objects of the image.

6 Experimental Results and Evaluation

Dataset

To test the robustness of the proposed method, experiments are conducted on different bench mark datasets. Images with partial occlusion(AR) dataset are used. The experiments are also conducted on real-time dataset collected.

Experiment #1

The AR dataset contains images with sunglass and scarf. Each set of partially occluded images of sunglass and scarf is considered separately to evaluate the performance of the proposed feature extraction technique, and results are given in Table 1. The performance chart of the same is presented in Fig. 5.

Experiment #2

Table 1 Recognition performance on AR database with partial occlusion

Method		Partial Occlusion
	Sunglasses	Scarf
PCA	5.42	12.5
ICA	69	37
LBP	60.8	50.8
MWLBP	**92.8**	**94.8**

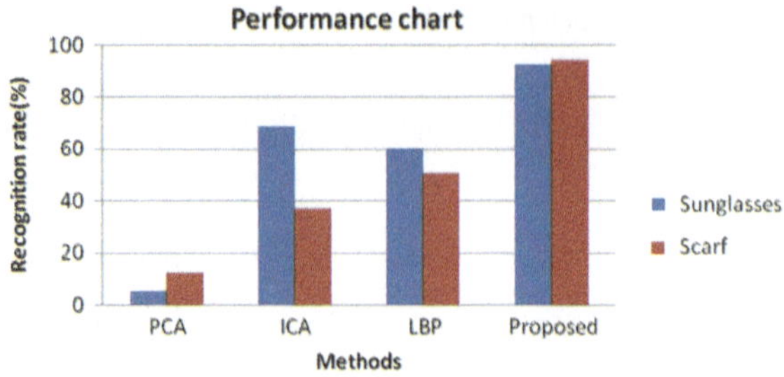

Fig. 5 Recognition performance on AR face database with partial occlusion

Table 2 Robustness to different facial variations

Method	Recognition rate (%)
PCA	51
LBP	86.25
MWLBP	**96.25**

To further evaluate the recognition performance of the proposed technique, the algorithm is tested on images with varying facial expressions of the proposed real-time dataset (RTD). Three expressions are considered—smiling, angry, and sad. RTD dataset is used to check the functionality in which the neutral expression is used for training the model, while expression faces are for the investigation set. Table 2 displays the performance of the proposed technique on RTD.

7 Discussion

The proposed feature extraction technique is comparatively performing well on different benchmark face image datasets considered along with real-image dataset. On AR dataset, it is achieving remarkable performance while compared with other state-of-the-art techniques that are used by many authors. On the dataset proposed in this work, it is giving 96.25% accuracy.

8 Conclusion and Future Research Issues

The work addressed the problem of face recognition under occlusions caused by scarves and sunglasses and also addressed the problems of face recognition under illumination, pose, and scale variations.

The experimental results show the significantly improved performance of the proposed algorithm under severe illumination, partial occlusion, and low-quality images in uncontrolled imaging conditions. The best recognition rate indicated approximately **28.15%** improvement with respect to accuracy and nearly **20%**

improvement in terms of speed over the state-of-the-art techniques in this area. The proposed face recognition system can also be applied to recognize the face images in general that are occluded with hat, beard, and so on, and also, it can be extended to security applications.

References

1. Li, S.Z., Jain, A.K.: Handbook of Face Recognition, 2nd edn. Springer (2011)
2. Grother, P.J., Quinn, G.W., Phillips P.J.: Multiple-Biometric Evaluation (MBE) 2010, Report on the Evaluation of 2D Still-Image Face Recognition Algorithms. NIST Interagency Report 7709 (2010)
3. Abate, A.F., Nappi, M., Riccio, D., Sabatino, G.: 2D and 3D face recognition: a survey. Pattern Recogn. Lett. **28**(14), 885–906 (2007)
4. Zhao, W., Chellappa, R., Phillips, P.J., Rosenfield, A.: Face recognition: a literature survey. ACM Comput. Surv. **35**(4), 399–458 (2003)
5. Crawford, M.: Facial Recognition Progress Report. Defense & Security, SPIE Newsroom (2011)
6. www.nist.gov
7. Jain, A.K., Klare, B., Park U.: Face matching and retrieval in forensics applications. IEEE Multimedia Multimedia Forensics Secur. Intell. 1–9 (2012)
8. Rao, M.V.G.: A survey of recent advances in face detection. Int. J. Control Theory Appl. **8**(5), 2145 (2015)
9. Hernández-Aceituno, J., Acosta, L., Piñeiro, J.D.: Pedestrian detection in crowded environments through Bayesian prediction of sequential probability matrices. J. Sens. **2016**, 8 pages (2016) (Technical report, Hindawi Publishing Corporation, Article ID 4697260)
10. Klare, B.F., Burge, M.J., Klontz, J.C., Bruegge, R.W.V., Jain, A.K.: Face recognition performance: role of demographic information. IEEE Trans. Inf. Forensics Secur. **7**(6), 1789–1801 (2012)
11. Spaun, N.A.: Facial comparisons by subject matter experts: their role in biometrics and their training. Adv. Biometr. Lect. Notes Comput. Sci. **5558**, 161–168 (2009)
12. Ahonen, T., Hadid, A., Pietikainen, M.: Face description with local binary patterns: application to face recognition. Pattern Anal. Mach. Intell. IEEE Trans. **28**(12), 2037–2041 (2006)
13. Li, S.Z., Jain, A.K.: Hand Book of Face Recognition, 2nd edn. Springer (2011)
14. Indiveri, G., Liu, S.C.: Memory and information processing in neuromorphic systems. Proc. IEEE (2015)
15. Siswanto, A.R.S.: Implementation of face recognition algorithm for biometrics based time attendance system. In: IEEE Conference on ICT for Smart Cities, pp. 24–25 (2014)
16. Wang, J., Jia, L.: Eigenfaces vs. Fisherfaces: recognition using class specific linear projection. Pervasive Computing Research Center, Institute of Computing Technology, Chinese Academy of Sciences (2016)
17. He, X., Yan, S., Hu, Y., Niyogi, P., Zhang, H.: Face recognition using Laplacianfaces. IEEE Trans. Pattern Anal. Mach. Intell. **27**(3), 328–340 (2005)
18. Kim, T., Kittler, J.: Locally linear discriminant analysis for multimodally distributed classes for face recognition with a single model image. IEEE Trans. Pattern Anal. Mach. Intell. **27**(3), 318–327 (2005)
19. Chen, H., Chang, H., Liu, T.: Local discriminant embedding and its variants. In: Proceedings of the IEEE International Conference on Computer Vision and Pattern Recognition (2005)
20. ParisaBeham, M.: Robust face recognition using automatic pose clustering and pose estimation. In: Fifth IEEE Conference on Advanced Computing (2013)

21. Yan, S., Xu, D., Zhang, B., Zhang, H., Yang, Q., Lin, S.: Graph embedding and extension: a general framework for dimensionality reduction. IEEE Trans. Pattern Anal. Mach. Intell. **29**, 40–51 (2007)
22. Qiao, L., Chen, S., Tan, X.: Sparsity preserving projections with applications to face recognition. Pattern Recogn. **43**(1), 331–341 (2010)
23. Fidler, S., Skocaj, D., Leonardis, A.: Combining reconstructive and discriminative subspace methods for robust classification and regression by subsampling. IEEE Trans. Pattern Anal. Mach. Intell. **28**(3), 337–350 (2006)
24. Khadatkar, A., Khedgaonkar, R.: Occlusion invariant face recognition system. In: IEEE World Conference on Futuristic Trends in Research and Innovation for Social Welfare (Startup Conclave) (2016)
25. Kim, J., Choi, J., Yi, J., Turk, M.: Effective representation using ica for face recognition robust to local distortion and partial occlusion. IEEE Trans. Pattern Anal. Mach. Intell. **27**, 1977–1981 (2005)
26. Liao, S., Jain, A.K.: Partial face recognition: an alignment free approach. In: Biometrics, International Joint Conference on, 0, pp. 1–8 (2011)
27. Martinez, A.M.: Recognizing imprecisely localized, partially occluded and expression variant faces from a single sample per class (2002)
28. Oh, H.J., Lee, K.M., Lee, S.U.: Occlusion invariant face recognition using selective local non-negative matrix factorization basis images. Image Vision Comput. **26**(11), 1515–1523 (2008)
29. Park, B.-G., Lee, K.-M., Lee, S.-U.: Face recognition using face-arg matching. IEEE Trans. Pattern Anal. Mach. Intell. **27**(12), 1982–1988 (2005)

Adaptive Digital Image Watermarking for Color Images using Wavelet

Sarita P. Ambadekar and Jayashree Khanapuri

Abstract Internet and social media have become an inseparable part of the human life which has led to extensive sharing of all kinds of digital information including but not limited to photos and videos. The risk of plagiarism and counterfeiting follows with the technology of information sharing. Digital watermarking deals in encrypting useful information within various forms of media for copyright protection and authentication. In this paper, an algorithm is used to embed a colored watermark image within colored cover image in a robust fashion. In the proposed technique, principal component analysis (PCA) is applied to get compressed watermark image. By using Bhattacharyya distance and kurtosis, the embedding and scaling factors are computed to create an adaptive digital watermarking mechanism. The embedding and extraction algorithms are developed by considering image segmentation in RGB space. Performance of the proposed method was tested using PSNR and NCC parameters on multiple cover and watermark images and also by considering common image processing attacks like salt and pepper, cropping, rotation.

Keywords Digital image watermarking · Principal component analysis · Discrete wavelet transform · Kurtosis · Bhattacharyya distance

1 Introduction

In recent times due to data sharing and continuous exchange of digital information over the Internet, ensuring authentication of digital content ownership is a major concern. Digital image watermarking is the method of embedding a watermark image into a cover or host image without distorting its visual form. It also maintains the authenticity and integrity of the cover image. One of the key criteria of a watermarked image is to be robust against multiple attacks. Imperceptibility means the

S. P. Ambadekar (✉) · J. Khanapuri
K. J. Somiaya Institute of Engineering and Information Technology, Mumbai, India
e-mail: sarita.ambadekar@somaiya.edu

J. Khanapuri
e-mail: jayashreek@somaiya.edu

V. Bhateja et al. (eds.), *Computer Communication, Networking and IoT*, Lecture Notes in Networks and Systems 197, https://doi.org/10.1007/978-981-16-0980-0_40

visual quality of the original image is not affected by watermark embedding. A robust watermark is the one that remains detectable even after common image processing operations as well as the one which is intended to merely to remove the watermark [1]. Every watermarking algorithm should possess these properties to make it consistent and effective for its purpose. The technique is reliable only if it satisfies properties of robustness, imperceptibility, and embedding capacity. These techniques are classified as spatial domain and transform domain [2]. Frequency domain techniques are considered better as it increase imperceptibility. Discrete cosine transform (DCT) is a lossy compression technique in frequency domain where data is lost when the original image is reconstructed from compressed image as described by Arvind Kumar [3]. Watermark embedding and extraction algorithm using DWT coefficients, distance measurement and encryption using Arnold transform are demonstrated [4]. Prasanth Vaidya [5] suggested an adaptive digital watermarking scheme in the wavelet domain, where scaling and embedding factors are calculated using Bhattacharya distance and kurtosis. Authors presented watermarking through texturization, statistical parameters (mean, standard deviation, skew, kurtosis, and variance) and Bhattacharyya distance [6]. Kamal N. Kaur proposed algorithm for image watermarking using visual cryptography that generates two shares with DWT-SVD [7]. As the watermark image is embedded in the cover image, it tends to affect the visual quality of the cover image. Hence, dimensionality reduction is required so that only the significant components required to generate an image (in this case, watermark image) are selected, and all other insignificant components are removed. This process is known as principal component analysis. In this process, only significant components are embedded in the cover image to form the embedded watermarked image by maintaining its visual quality [8–10]. Author [11] reviews the significant work addressing the need for a formal generic watermarking model. The paper presents a formal generic model for digital image watermarking and key-based watermarking and various attacks. Chen and Zhaob [12] used redundant discrete wavelet transformation and SVD-based watermarking scheme for copyright protection of videos. The proposed method was tested for variety of attacks like adding noise, rotation, and compression. Most of the watermarking algorithms developed earlier have a static embedding factor. In this paper, a digital watermarking algorithm with an adaptive embedding factor is proposed. Due to this, any image can be used as a cover image and any image can be used as a watermark image. Adaptive embedding factor makes the system more flexible, robust, and imperceptible against attacks. The proposed algorithm also uses RGB color splitting in order to adapt watermarking of color images.

In this paper, Sect. 2 provides complete information about algorithms proposed for watermark embedding and extraction process. Section 3 present results obtained and observations. Section 4 concludes the paper.

2 Watermark Embedding and Extraction Procedure

The proposed algorithm mainly focuses on color images by implementing DWT-PCA with Bhattacharyya distance and kurtosis, to create robust and adaptive digital watermarking. Kurtosis and Bhattacharyya distance are combined for the calculation of scaling and embedding factors. These two factors have an impact on the shape of the probability distribution. The similarity between the input images determines the strength of the watermark image, which is embedded in the cover image; as a result, the perceptibility of the cover image is retained to the maximum. Using Kurtosis and Bhattacharyya distance, we are able to determine correct values for the scaling and embedding factors, thus making the task simple rather than random guessing, which may not result in better output. Further, in the embedding process, the colored cover image and watermark image are divided into RGB, that is, red, blue, and green channels/components. Here each component is treated as a separate monochromatic cover image and watermark image, respectively. Further, DWT is applied to the red (R) component of the cover image. The red component of watermark image is embedded into the approximation sub-band, i.e., the LL sub-band of the red component of the cover image and then combined with the remaining sub-bands to form the final red component of the image embedded with watermark. Here the LL sub-band is chosen for embedding as it contains the maximum information of the image. Similarly, the above procedure is repeated for blue (B) and green (G) components of the cover image in which blue and green components of watermark image are embedded, respectively. Now all three obtained RGB components are merged to form the embedded image. During the extraction process, the embedded image is separated into RGB components. Then the LL sub-band for the red component of extracted watermark image is obtained and further modified by adding the scaling and embedding factors. Similarly, the procedure is repeated for the remaining two components, which are green and blue. Finally, the obtained RGB components are merged to form the extracted watermark image. PCA is a technique that is used for various application like dimension reduction, feature extraction, and visualization. Use of PCA in digital watermarking compresses the watermark image to get principle components which reduce the number of pixels to be embedded. As a result, only the principal components of the watermark image get embedded into the cover image, which further helps to recover the watermark image more precisely. Figure 1 shows the watermark embedding process, and Fig. 2 shows the watermark extraction process.

2.1 Watermark Embedding Algorithm

Input: Colored cover image (I), Colored watermark image (W)
Output: Watermark Embedded Image (W´)

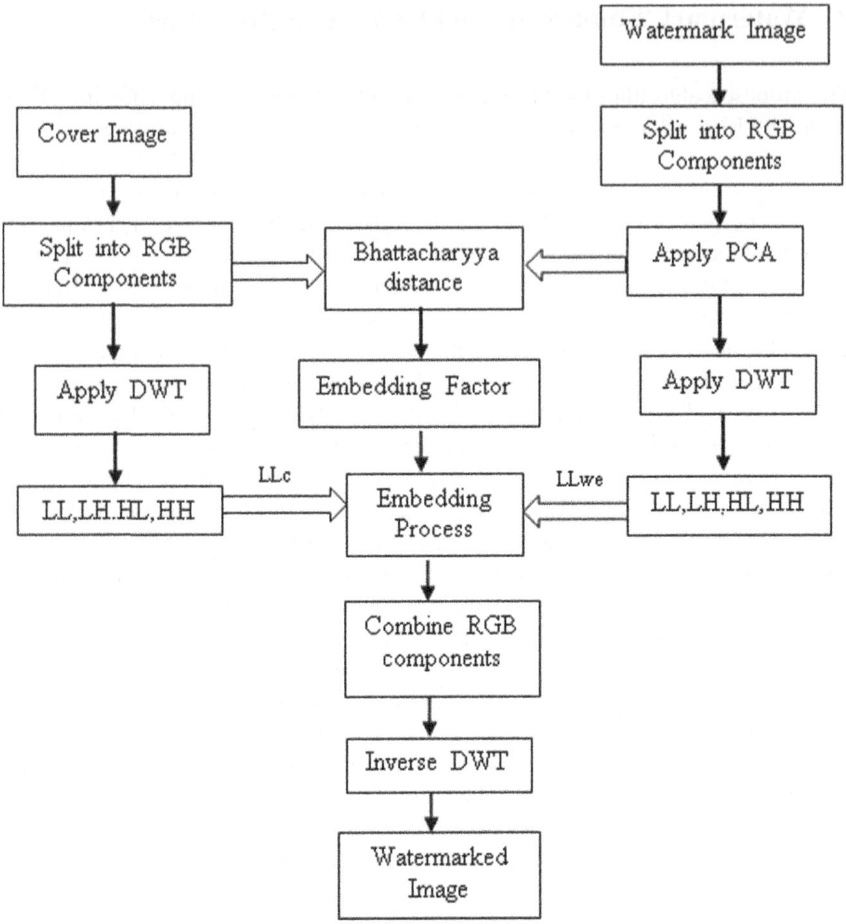

Fig. 1 Watermark embedding process

1. Cover and watermark images are divided into their respective red, green, and blue (RGB) components.
2. Perform PCA on the R component of watermark image and compress the image to r.
3. Calculate the scaling and embedding factors (α, β) using Bhattacharyya distance and kurtosis.
4. Apply 2-level DWT using Haar wavelet on the cover image R component.
5. Extract LL band.
6. Insert the watermark using the following equation:

$$\mathrm{LL}' = \mathrm{LL} + (\beta/\alpha) \times r \quad (1)$$

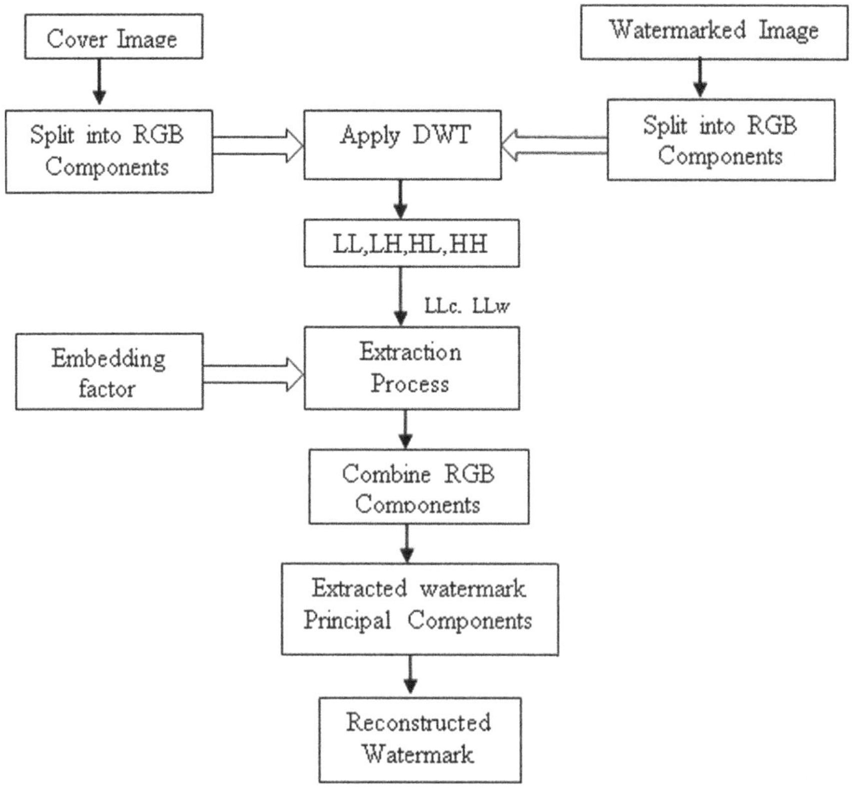

Fig. 2 Watermark extraction process

where α = Bhattacharyya Distance and β = Kurtosis

7. Combine the modified LL sub-frequency band (LL´) with other LH, HL, and HH bands of the cover image R component.
8. Apply inverse DWT to get *R* component of watermark embedded image.
9. Repeat steps 1–8 for G and B Components.
10. Combine the obtained RGB components to form watermark embedded image.

2.2 *Watermark Extraction Algorithm*

Input: Cover image (I) and watermark embedded image (I´).
Output: Extracted Watermark Image (wm).

1. Divide watermark embedded image, cover image and watermark image into RGB components.
2. Apply 2-level DWT using Haar wavelet on the R component of watermark embedded image and R component of cover image.

3. Extract low-frequency bands of both images.
4. Obtain scaling and embedding factors (α, β) calculated during the embedding process.
5. Extract R component of watermark image using following equation:

$$\text{LLwm} = \left(\text{LL}' - \text{LL}\right) \times \alpha/\beta \tag{2}$$

6. Reconstruct the R component of extracted watermark image by applying inverse DWT.
7. Repeat steps 1–6 for G and B components.
8. Combine these RGB components to form extracted watermark image.

3 Simulation Results

Watermark embedding and extraction algorithm was implemented on different images which is presented below. The entire processing was done using MATLAB software and executed on Intel i5 processor with 4 GB RAM and 1.19 GHz processing speed.

Case-I

Figure 3a shows the original cover images, and Fig. 3b shows the watermark image of Cameraman. (Images from internet and from C:\Program Files\MATLAB\R2020a\toolbox\images\imdemos).

Three general attacks such as salt-and-pepper attack, geometric attack through rotation, and cropping attack are considered. Figure 4 shows the three types of attacks on watermark embedded image used for extraction algorithm.

The two important metrics were determined to evaluate the results of the proposed method which are:

A. Peak signal-to-noise ratio (**PSNR**): It is a quality metric used to measure distortion in the image after the embedding process.

(a) **(b)**

(a) Cover Images (Fruits, Lena, flamingos and strawberries) (b) Watermark Image (Cameraman)

Fig. 3 Cover and watermark images (Case-I)

(a) Embedded and Extracted Images – (i) no attack (ii) salt and paper attack (iii) cropping attack (iv) rotation attack

Fig. 4 Results (Cameraman) (Case-I)

$$\text{MSE} = \frac{1}{m \times n}\left(\sum_{x=1}^{m}\sum_{y=1}^{n}|I(x, y) - O(x, y)|^2\right) \quad (3)$$

$$\text{PSNR} = 10 * \log_{10}\left(\frac{255^2}{\text{mse}}\right) \quad (4)$$

where $I(x, y)$ and $O(x, y)$ are the input carrier and watermarked output image, respectively. $m \times n$ are the dimensions of carrier image.

Table 1 shows PSNR (dB) values, and Table 2 shows the watermarking parameters NCC evaluated between original watermark image and extracted watermark image at various attacks and without attacks.

B. Normalized correlation coefficient (**NCC**): It is used to find correlation coefficient between watermark and extracted watermark with and without attack.

$$\text{NCC} = \frac{\sum_{x=1}^{m}\sum_{y=1}^{n} w(x, y) * we(x, y)}{\sum_{x=1}^{m}\sum_{y=1}^{n} |w(x, y)|^2} \quad (5)$$

Table 1 PSNR ratio (peak signal-to-noise ratio) (in dB) (Case-I)

Mode used	Images used			
	Fruits	Lena	Flamingos	Strawberries
DWT + PCA	45.797	40.245	45.795	44.851
DWT	41.148	40.143	42.169	40.176

Table 2 NCC (between extracted watermark and watermark) (Case-I)

Mode used	Types of attacks	Images used			
		Fruits	Lena	Flamingos	Strawberries
DWT + PCA	No attack	0.999	0.999	0.999	0.999
	Salt and pepper	0.363	0.243	0.360	0.326
	Cropping	0.709	0.714	0.747	0.694
	Rotation	0.314	0.506	0.598	0.558
DWT	No attack	0.999	0.999	0.999	0.999
	Salt and pepper	0.306	0.234	0.324	0.345
	Cropping	0.701	0.703	0.697	0.674
	Rotation	0.225	0.491	0.406	0.508

where $w(x, y)$ is the input watermark and $we(x, y)$ is the extracted watermark image.

Case-II

Figure 5 shows the original cover images and watermark image (boat image). Figure 6 shows the three types of attacks on watermark embedded image used for extraction algorithms.

(a) Cover Images (Fruits,Lena,Flamingos and Strawberries) (b) Watermark Image (Boat)

Fig. 5 Cover and watermark images (Case-II)

Embedded and Extracted Images
(i) no attack, (ii) salt and paper attack, (iii) cropping attack, (iv) rotation attack

Fig. 6 Results color (Boat) (Case-II)

Table 3 PSNR ratio (peak signal-to-noise ratio) (in dB) (Case-II)

Mode used	Images used			
	Fruits	Lena	Flamingos	Strawberries
DWT + PCA	50.329	45.746	42.855	48.309
DWT	48.078	44.034	40.105	48.135

Table 4 NCC (between extracted watermark and watermark) (Case-II)

Mode used	Types of attacks	Images used			
		Fruits	Lena	Flamingos	Strawberries
DWT + PCA	No attack	0.999	0.999	0.999	0.999
	Salt and pepper	0.211	0.306	0.289	0.260
	Cropping	0. 318	0.598	0.498	0.500
	Rotation	0.314	0.506	0.598	0.558
DWT	No attack	0.999	0.999	0.999	0.999
	Salt and pepper	0.183	0.413	0.295	0.268
	Cropping	0.544	0.599	0.530	0.496
	Rotation	0.550	0.478	0.526	0.521

Table 3 shows PSNR ratio, and Table 4 shows the watermarking parameters NCC evaluated between original watermark image and extracted watermark image at various attacks and without attacks.

4 Conclusion

In this paper, a digital image watermarking through PCA and DWT on image segments in RGB space is presented. Statistical parameters like kurtosis and Bhattacharyya distance are used to create an adaptive digital watermarking mechanism for colored images. Performance of the technique was tested on multiple carrier images like fruits, lena, flamingos, strawberries and watermark images like cameraman and boat, under common image processing attacks, which yield better results. The PSNR and NCC values of both test cases are sufficiently high to imply that the algorithm ensures robustness and imperceptibility. In the future work, blind digital watermarking method can be used where host image is not required to extract the watermark.

References

1. Tao, H., Chongmin, L., Zain, J.M., Abdalla, A.N.: Robust image watermarking theories and techniques: a review. J. Appl. Res. Technol. **12**, 122–138 (2014)
2. Kulkarni, A.S., Lokhande, S.S.: Imperceptible and robust digital image watermarking techniques in frequency domain. Int. J. Comput. Technol. Electron. Eng. (IJCTEE) **3**, 33–36 (2013)
3. Kumar, A., Agarwal, P., Choudhary, A.: A Digital Image Watermarking Technique Using Cascading of DCT and Biorthogonal Wavelet Transform. Springer India (2016)
4. Ambadekar, S.P., Jain, J., Khanapuri, J.: Digital image watermarking through encryption and DWT for copyright protection. Adv. Intell. Syst. Comput. Book Ser. (AISC) **727**, 187–195 (2019)
5. Prasanth Vaidya, S., Chandra Mouli, P.V.S.S.R.: Adaptive Digital Watermarking for Copyright Protection of Digital Images In Wavelet Domain, 1877-0509 © 2015 The Authors. Published by Elsevier B.V. Procedia Computer Science, vol. 58, pp. 233–240 (2015)
6. Ambadekar, S.P., Jain, J., Khanapuri, J.: Adaptive Digital Image Watermarking Technique Through Wavelet Texture Features, Advances in Intelligent Systems and Computing, vol. 1155. Springer, Singapore (2020)
7. Kaur, K.N., Gupta, D.I., Singh, A.K.: Digital Image Watermarking Using (2, 2) Visual Cryptography With DWT-SVD Based Watermarking. Springer Nature Singapore Pte Ltd. (2019)
8. Ayesha, S.K., Masilamani, V.: An Imperceptible Digital Image Watermarking Technique by Compressed Watermark Using PCA. Springer International Publishing, Switzerland (2015)
9. Shinde, G., Mulani, A.: A robust digital image watermarking using DWT-PCA, Novateur Publications. Int. J. Innov. Eng. Res. Technol. **6**(4) (2019) [ijiert]. Issn: 2394-3696
10. Rajasekhar Reddy, M., Sai Sandeep, P., Mounika, Ch., Gopi, J., Suganya, S.D.: Optimized watermarking technique using DWT-PCA and log filter. In: 2017 International Conference on Networks & Advances in Computational Technologies (NetACT). https://doi.org/10.1109/netact.2017.8076754
11. Nyeem, H., Boles, W., Boyd, C.: Digital image watermarking: its formal model, fundamental properties and possible attacks. EURASIP J. Adv. Signal Process. **2014**, 135 (2014) (Springer Open Journal)
12. Chen, L., Zhaob, J.: Adaptive digital watermarking using RDWT and SVD. In: 2015 IEEE International Symposium on Haptic, Audio and Visual Environments and Games (HAVE). https://doi.org/10.1109/have.2015.7359451

Effect of LTE Handover on TCP Performance Over Optical Networks

Konstantinos G. Tsiknas and Kyriakos E. Zoiros

Abstract High-speed mobile networks have rapidly evolved over the last years. The unique environment they operate, however, sets new challenges for the transmission control protocol (TCP). In this paper, we study the impact of long-term evolution (LTE) handover on TCP performance in a wide range of networks settings including different handover algorithms, random channel errors and support of error correction at lower layers. First, results indicate a severe TCP goodput reduction in most considered scenarios.

Keywords LTE · Handover · TCP · Optical networks · Ns-3

1 Introduction

The tight relationship between mobile and optical networking has already been established in 4G and 5G mobile networks, by emerging fiber-optic links to/from remote cell sites, as the leading high-speed, low-latency connectivity solution [1–4]. LTE, in particular, was designed to provide higher throughput and lower latency than high-speed downlink packet access (HSDPA) [5]. LTE also provides all-IP network and services to the user end (UE).

Handover is a key parameter for the proper and effective support of user mobility function in LTE networks, whereas LTE supports mobility at high speeds up to 500 km/h. In these speeds, handover should be executed analogously more frequently and faster. Since handover failures cause packet drops, temporal disconnections and poor user satisfaction, it is vital to ensure that the LTE handover mechanism is reliable and maintains high performance levels.

K. G. Tsiknas (✉) · K. E. Zoiros
Department of Electrical and Computer Engineering, Democritus University of Thrace, Vas. Sofias 12, 67100 Xanthi, Greece
e-mail: ktsiknas@ee.duth.gr

K. E. Zoiros
e-mail: kzoiros@ee.duth.gr

V. Bhateja et al. (eds.), *Computer Communication, Networking and IoT*, Lecture Notes in Networks and Systems 197, https://doi.org/10.1007/978-981-16-0980-0_41

Issues that degrade TCP performance during handover may include the interference between source and target eNodeB (eNB). Another cause of system performance degradation during handover is the link interruption due to the fact that LTE supports only hard handover [6]. Although the loss or out-of-order delivery of packets is avoided by forwarding them via the X2 interface (used for inter-eNB communication over a wired link) [7], the interruption causes delays on high level protocols, such as TCP, because no data can be sent between eNB and the user end (UE) [8]. Even though some losses may be mitigated at lower layers, yet a portion of them can remain at the transport layer when packet loss rates are high. Standard TCP Reno congestion control algorithm misinterprets these packet segment losses as congestion and acts by reducing its congestion window (cwnd) to half of its previous size, which leads to severe throughput deterioration [9].

Several promising protocols have been developed to deal with this issue: scalable TCP (STCP) [10], high-speed TCP (HSTCP) [11], FAST-TCP [12], H-TCP [13], binary increase congestion control (TCP-BIC) [14], TCP CUBIC [15], TCP BIAD [16].

Moreover, several comparative studies of these protocols in terms of goodput, packet loss rate, TCP friendliness and fairness [17–20] have been conducted in high-speed long-distance networks. Orozco et al. [21] assessed the impact of GPON MAC layer on TCP performance. Ramjee et al. [22] described various performance issues of TCP over 3G networks with high rate and delay variations. Winsteing et al. [23] showed that the error control mechanisms provided by the low level protocols often result in additional delays on TCP congestion control mechanism. Similar conclusions were drawn by Erman et al. [24] who identified that the absence of synchronization between TCP and the underlying protocols degrades application performance. Huang et al. [25] reported various drawbacks in TCP over LTE networks, like undesired timeouts. Nguyen et al. [8] investigated the impact of LTE handover on TCP performance and concluded that lossless handovers results in higher TCP segments delays, whereas seamless handovers cause packet losses at TCP.

Other research works analyzed the impacts of handover algorithms and radio link control protocols and parameters on TCP performance. Lin et al. [26] proposed a new handover algorithm which aims at minimizing the number of handovers and the system delay. Dimou et al. [27] considered the setting of handover triggers of primary importance for the design of handover mechanism and suggested their adaptation on the basis of speed, propagation conditions and cell sizes. Anas et al. [6] proposed linear and dB domain L3-filter performance improvements in terms of global number of handovers.

Despite the long investigation, however, most of these studies have not simulated networks in these particular conditions applied during the handover. As a consequence, the root causes of the observed performance problems have not been largely investigated. Therefore, it is important to assess the benefits of increased bandwidth for popular mobile applications and essential network protocols, such as TCP, to identify their limitations and suggest the necessary improvements [25].

In this paper, we investigate through simulation results the interaction of TCP with the underlying protocols. In particular, we investigate TCP behavior subject to different handover types and with or without the presence of error correction mechanisms provided by the lower layers of LTE protocol stack in order to analyze and deduce how these factors affect TCP performance. We do not experiment with TCP variants to overcome the performance hits, but we rather study the interaction between TCP and the underlying layers of the LTE network. For this purpose, we use the ns-3 simulation environment to investigate the impact of LTE handover on TCP performance.

The rest of this paper is organized as follows. Section 2 gives an overview of LTE handover functionality, Sect. 3 describes the simulation testing environment and reports the results, and finally, Sect. 4 contains the conclusions drawn from our work.

2 Overview of LTE Handover

In this section, a brief overview of the basic handover mechanisms for LTE network is given, focusing on the details relevant to our performance study. For a more detailed description to the LTE mechanisms and standards, we refer the reader to [28].

2.1 Seamless and Lossless Handover

Figure 1 displays the protocol stack of the LTE radio interface between eNB and UE.

The three protocols that constitute Layer 2 of the stack are the medium access control (MAC), the radio link control (RLC) and the packet data convergence protocol (PDCP). The MAC protocol provides the multiplexing and the scheduling capabilities between the user end (UE) and eNB. The RLC protocol is responsible for in order packet delivery, segmentation–reassembly and packet retransmission when errors are detected. PDCP maintains a retransmission queue for forwarding the untransmitted packets to the eNB in case of handover (see Fig. 2).

Depending on how these data are handled, there are two types of handover in LTE:

Seamless handover: Only untransmitted PDCP service data units (SDUs) are forwarded to the target eNB during handover with the use of X2 interface. All data in the PDCP retransmission queue are dropped. This type of handover is used for applications which are reasonably tolerant to losses but less tolerant to delays (as, for instance, in voice services).

Lossless handover: In addition to the untransmitted PDCP SDUs, all data in the PDCP retransmission queue are forwarded to the target eNode. Lossless handover is appropriate for delay-tolerant applications, like, for instance, in file transfers and email services.

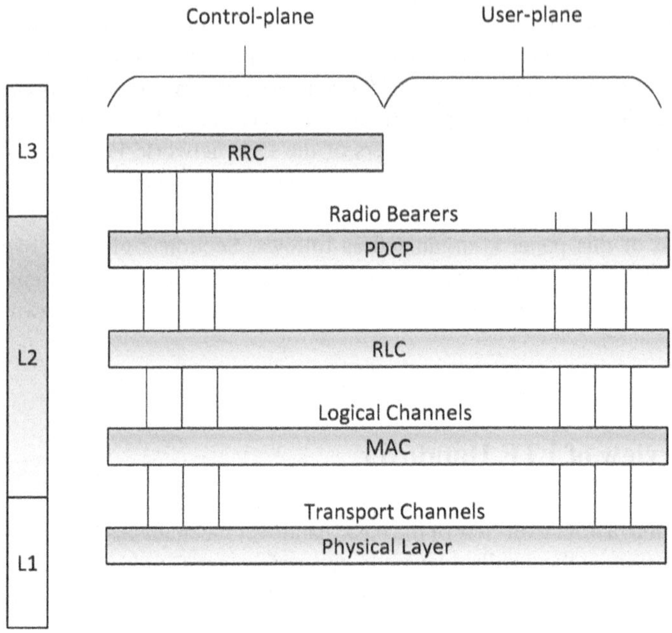

Fig. 1 LTE radio interface [29]

Fig. 2 PDCP retransmission queue [8]

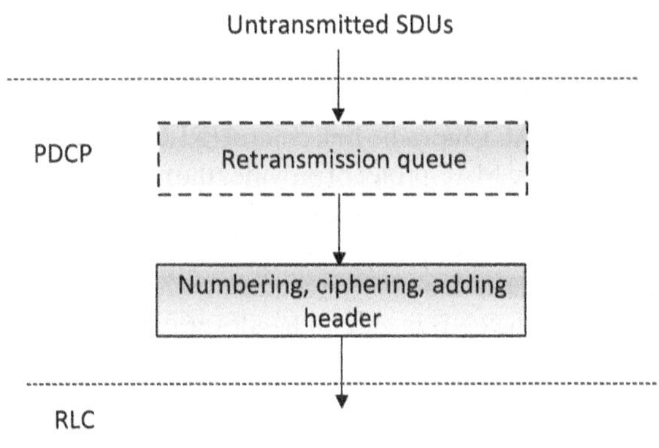

2.2 Handover Signaling Events

The 3GPP LTE standards define measurements as a sequence of events. Two main types of events are specified, the A3 event and the A2-A4 event pairs. A3 event is triggered when the reference signal received power (RSRP) of the neighboring cell is higher than an offset value of the RSRP serving cell. The A2-A4 event is triggered when the reference signal received quality (RSRQ) of the neighboring cell exceeds the RSRQ value of the serving cell [30].

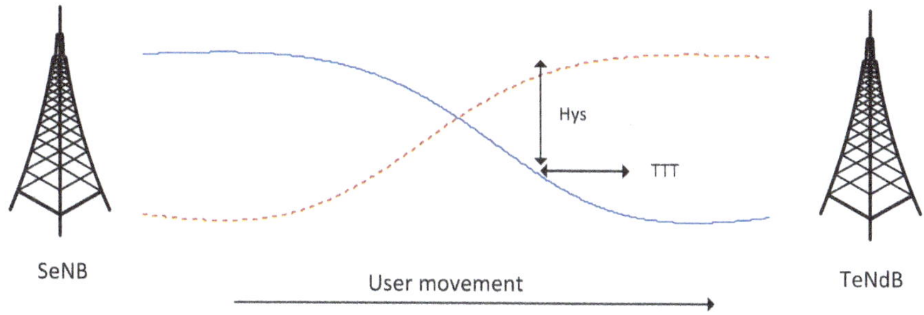

Fig. 3 Handover based on A3 event measurements [31]

Figure 3 displays an A3 event-based handover for a user movement from source eNB (SeNB) to target eNB (TeNB). The handover is triggered when hysteresis (Hys) and time-to-trigger (TTT) pair values (referred to as handover operation point or HOP) are both satisfied when UE moves toward a neighboring cell.

A3 handover events are simpler in comparison to the A2-A4 pair of events, since they rely only on one measurement. However, the A2-A4 pair of events is usually more reliable in making estimations about the handover owing to the exploitation of additional information, like, for instance, the channel quality, the noise and interference levels.

3 Performance Evaluation

In this section, the simulation setup environment is presented, and the obtained experimental results are discussed. The setup environment uses the ns-3 network simulator version 3.29 [32] and the build-in LTE module, which also known as LENA [33, 34].

3.1 Generic Simulation Scenario

Our simulation setup presented in Fig. 4 considers a simple LTE-based broadband network. The fixed part of the test setup consists of a file server and the notional evolved packet core (EPC) node (S-GW and P-GW) [28] connected with a 100 Gbps, 10 ms link optical broadband network (Gi interface). The EPC node and the eNBs are connected with a 10 Gbps, 1 ms Optical Network (S1-U interfaces).

The mobile part of the network consists of a notional UE configured to move along the x-dimension, at a distance 'yDistance' meters from the x-axis. The default value of 'yDistance' is 1000 m, so the initial position of the US is defined by the Cartesian coordinates (0, 1000). Two eNBs are configured, the source eNB (SeNB)

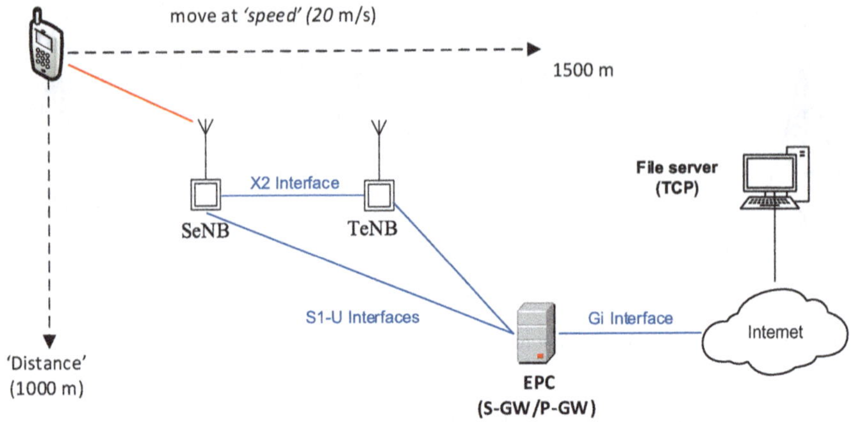

Fig. 4 Notional LTE–EPC network topology

to target eNB (TeNB), which are separated by a distance of 500 m from each other and from the x-axis. The SeNB position is at (500, 500), and the TeNB position is at (1000, 500). While UE is moving horizontally at 'Speed,' a long file transfer session is established over TCP during the simulation between file server and the UE. The TCP protocol variant used in simulation was TCP Reno.

The setup uses the Round Robin scheduler in MAC layer of LTE Radio protocol stack, whereas the RLC protocol is set to Acknowledged Mode (default value). The simulation time is configured to correspond to the time it takes for the UE to move to the halfway point between the SeNB and the TeNB, around the halfway point of the simulation (default value is 75 s). Table 1 summarizes the critical model parameters and their default values in the simulation.

Table 1 Simulation parameters

Parameter	Value
Network dimension	1500 m × 1000 m
Distance between eNBs	500 m
Speed of Gi interface	100 Gbps
Delay of Gi interface	10 ms
Tx power used by eNBs	46 dBm
UE speed	20 m/s
Handover types	A2A4 (default), A3Rsrp
RLC acknowledged mode	Enabled (default), Disabled
Simulation time	75 s
Traffic type	FTP application
File size	200 MB
TCP variant	Reno

3.2 TCP Performance with Different Handover Algorithms

In first simulation setup, the handover algorithm initiates handover at time 38.2 s, based on A2-A4 pair of events. At this time, the reference signal received quality (RSRQ) of the neighboring cell (TeNB) overcomes the RSRQ level of the serving cell (SeNB) (see Fig. 5).

TCP goodput variations are plotted in Fig. 6 for the whole duration of simulation. Note that goodput was calculated as the total number of bits successfully received by the UE over a small time interval (100 ms).

As results demonstrate, TCP seizes quickly the maximum available channel capacity (14.688 Mbps) thanks to slow-start algorithm which employs at the beginning of the transmission (see also cwnd variations in Fig. 7). During the handover, however, TCP goodput drops sharply to 3.684 Mbps and then increases slowly right afterward, eventually reaching the available channel capacity after 4.8 s.

As simulation results depict, the handoff mechanism is not 'hitless' as there are some packet drops that the RLC layer is not able to recover. Since the standard TCP Reno congestion control algorithm reacts to a single packet loss by reducing cwnd to half of its previous value, it becomes clear that the multiple losses cause consecutive reductions of cwnd to very low values (close to 40 segments from 64.000 segments before the handover).

Furthermore, the protocol employs an additive increase algorithm in congestion avoidance phase by increasing cwnd by 1 segment per receiving acknowledgment (ACK) (see Fig. 8). However, this low increase rate leads to a very slow convergence of the transmission rate at the available channel capacity after handover. Figure 9 plots goodput variations during A3Rsrp event handover.

As Fig. 9 demonstrates, the handover takes place later (at 50 s, approximately, 12 s after the A2-A4 event handover) due to the hysteresis and TTT margins imposed by

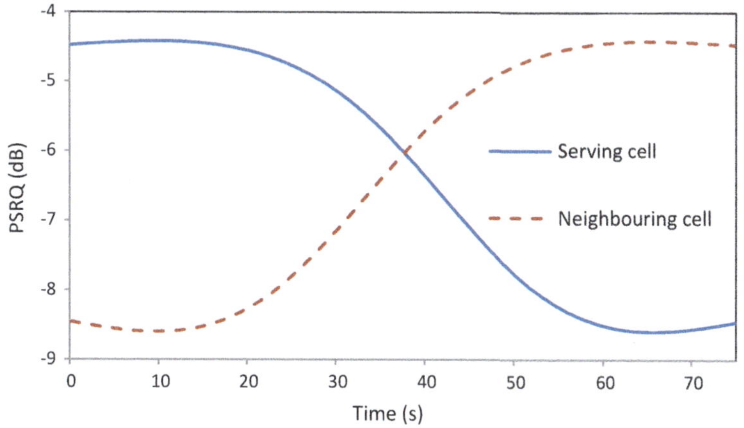

Fig. 5 Reference signal received quality versus time

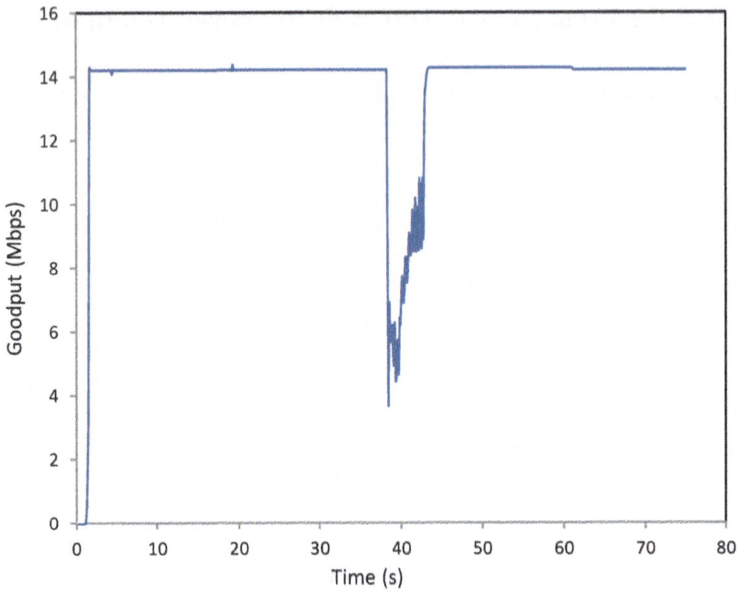

Fig. 6 TCP goodput in A2-A4 event handover

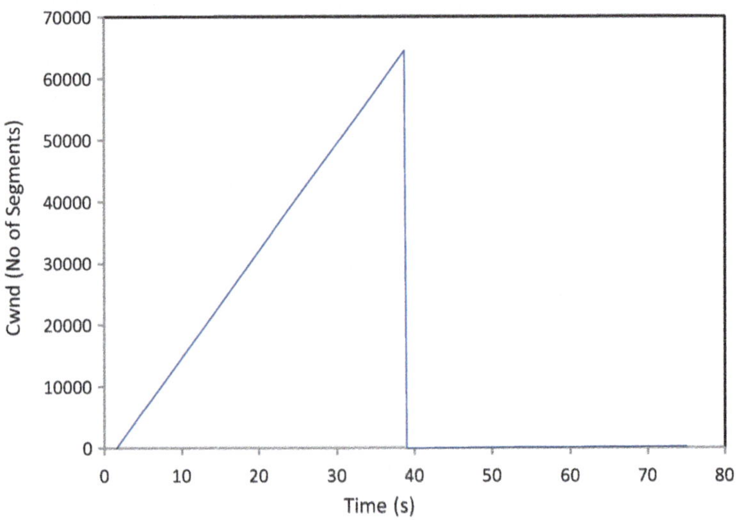

Fig. 7 Congestion window dynamics for A2-A4 event handover

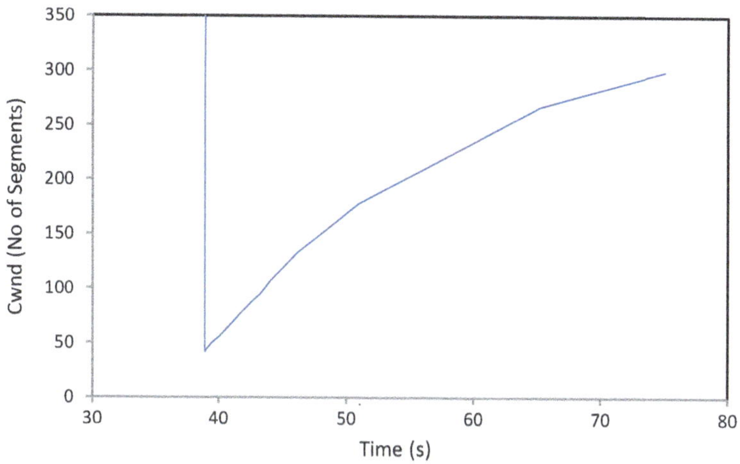

Fig. 8 Additive increase of TCP after A2-A4 event handover

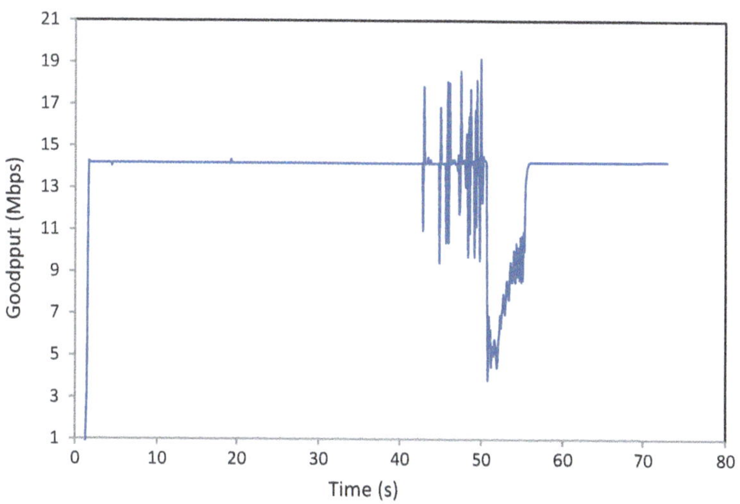

Fig. 9 TCP goodput in A3Rsrp handover

the algorithm (see also Fig. 3). The algorithm causes some delay variations (jitters) on TCP performance for a time interval of approximately 10 s, before handover occurs, perhaps due to low signal coverage during UE movement toward cell 2. This may cause negative effects on delay-sensitive packets, such as VoIP or real-time audio/video streams. Apart from this fact, there are no other implications on goodput, since the extent and the duration of performance degradation remain the same as with the A2-A4 handover event.

3.3 *TCP Performance in RLC Unacknowledged Mode*

LTE RLC unacknowledged mode (RLC UM) is mainly utilized by delay-sensitive and error-tolerant real-time applications. In this mode of operation, packet retransmission services (automatic repeat request or ARQ) are not provided by RLC.

As Fig. 10 demonstrates, TCP is very inefficient in RLC UM. Single packet losses result in frequent reductions in transmission rate and subsequent low performance and channel utilization levels of TCP (see Fig. 11).

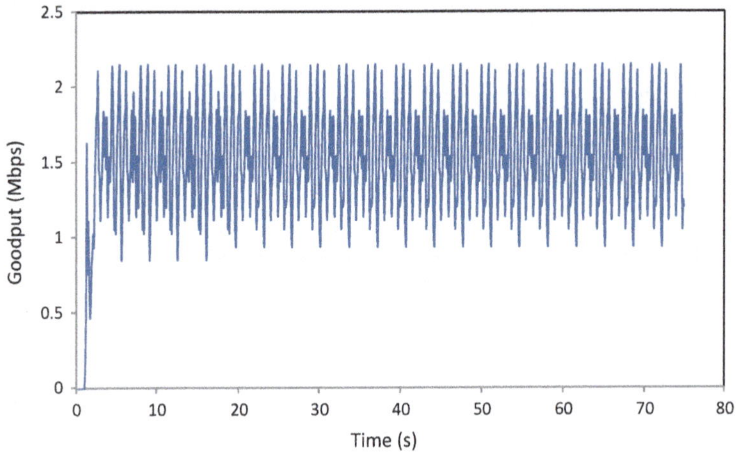

Fig. 10 TCP goodput in RLC UM

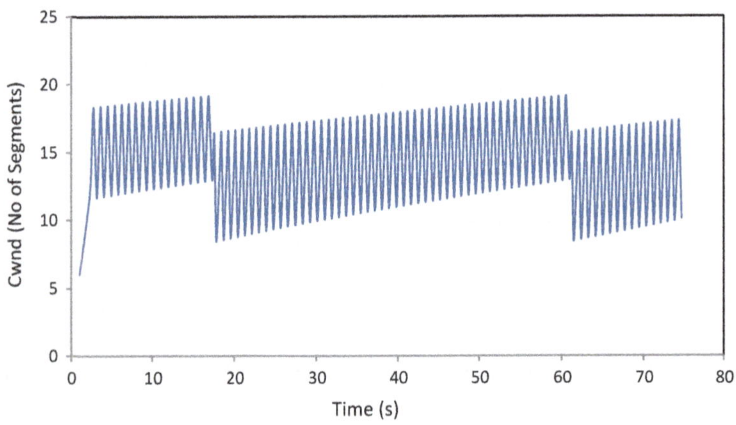

Fig. 11 Congestion window dynamics in RLC UM

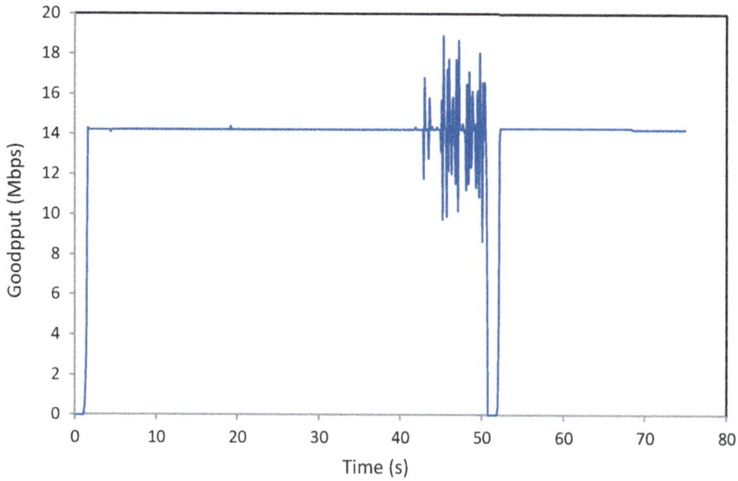

Fig. 12 TCP goodput in the presence of random losses (BLER = 0.37)

3.4 Effect of Random Losses on TCP Performance

In order to test the effect of random losses on TCP performance, errors following the AWGN distribution were inserted at PHY layer of LTE radio protocol stack with different block error rates (BLER) [35]. Our results indicate that low BLER values does not cause any performance degradation during LTE handover, since these errors can be corrected by RLC layer. However, higher BLERs result in spurious timeouts [36] at TCP sender as shown in Fig. 12.

As results demonstrate, random losses may have significant implications on TCP performance. In particular, TCP performance is reduced at very low levels (zero value) for about 1 s after the handover. However, TCP recovers quickly due to the slow-start algorithm it employs, immediately after the request timeout event (see Fig. 13).

4 Conclusions

In this paper, we have studied the effect of LTE handover on TCP performance with the use of ns-3 network simulator. As results demonstrate, handover can degrade TCP throughput by 40% on average for a duration of approximately 5 s. TCP performance degradation is even higher when packet drops are not corrected by RLC, due to the frequent reductions of the cwnd of TCP sender. Finally, scenarios involving single errors with high probability may incur loss of communication for short intervals after LTE handover.

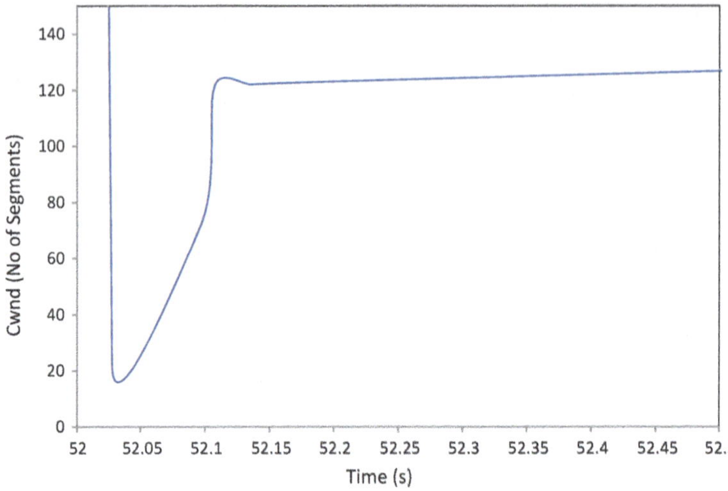

Fig. 13 Congestion window dynamics in the presence of random losses (BLER = 0.37)

TCP performance degradations have a significant impact on users' experience for a broad set of applications and services for which TCP is used as the transport layer [37]. Taking this into account, our future research direction is to extend the obtained results, which can form as the basis, and investigate TCP in a wider range of operating environments involving concurrent TCP flows and wireless channel errors and to suggest improvements in TCP congestion control algorithm.

References

1. Itu-T.: G.984.1 Gigabit-capable passive optical networks (GPON): General characteristics. Networks (2008)
2. Terada J.: Optical Network Technologies for 5G Mobile Network. In: Optical Fiber Communications Conference and Exhibition, (OFC) (2019). https://doi.org/10.1364/ofc.2019.w4h.1
3. Jaber, M., Imran, M.A., Tafazolli, R., Tukmanov, A.: 5G backhaul challenges and emerging research directions: a survey. IEEE Access. **4**, 1743–1766 (2016). https://doi.org/10.1109/ACCESS.2016.2556011
4. Kani, J.I., Terada, J., Suzuki, K.I., Otaka, A.: Solutions for future mobile Fronthaul and access-network convergence. J. Lightwave Technol. **35**(3), 527–534 (2017)
5. Toskala, A., Holma, H., Kolding, T., Mogensen, P., Pedersen, K., Reunanen, J.: High-Speed Downlink Packet Access. In: WCDMA for UMTS: HSPA Evolution and LTE, Fourth Edition (2008). https://doi.org/10.1002/9780470512531.ch12
6. Anas, M., Calabrese, F.D., Östling, P.E, Pedersen, K.I., Mogensen, P.E.: Performance analysis of handover measurements and layer 3 filtering for UTRAN LTE. In: IEEE International Symposium on Personal, Indoor and Mobile Radio Communications, PIMRC (2007). https://doi.org/10.1109/PIMRC.2007.4394671
7. Alexandris, K., Nikaein, N., Knopp, R., Bonnet, C.: Analyzing X2 handover in LTE/LTE-A. In: 14th International Symposium on Modeling and Optimization in Mobile, Ad Hoc, and Wireless Networks, WiOpt (2016). https://doi.org/10.1109/WIOPT.2016.7492906

8. Nguyen, B., Banerjee, A., Gopalakrishnan, V., et al.: Towards understanding TCP performance on LTE/EPC mobile networks. In: All Things Cellular 2014—Proceedings of the 4th ACM Workshop on All Things Cellular: Operations, Applications, and Challenges (2014). https://doi.org/10.1145/2627585.2627594
9. Jacobson, V.: Congestion avoidance and control. Comput. Com. Rev. **18**(4), 314–329 (1988)
10. Kelly, T.: Scalable TCP: improving performance in highspeed wide area networks. SIGCOMM Comput. Commun. Rev. **33**(2), 83–91 (2003)
11. Floyd S.: HighSpeed TCP for large congestion windows. RFC3649, pp. 1–34, December (2003). https://doi.org/10.17487/rfc3742
12. Jin, C., Wei, D.X., Low, S.H.: FAST TCP: motivation, architecture, algorithms, performance. In: Proceedings of IEEE INFOCOM, vol. 4, pp. 2490–2501 (2004). https://doi.org/10.1109/INFCOM.2004.1354670
13. Leith, D., Shorten, R.: H-TCP protocol for high-speed long-distance networks. Proc PFLDnet, Argonne, pp. 1–16 (2004). http://www.hamilton.ie/net/htcp3.pdf
14. Xu, L., Harfoush, K., Rhee, I.: Binary increase congestion control (BIC) for fast long-distance networks. In: Proceedings of IEEE INFOCOM, vol. 4, pp 2514–2524 (2004). https://doi.org/10.1109/INFCOM.2004.1354672
15. Ha, S., Rhee, I.: CUBIC : a new TCP-friendly high-speed TCP variant. ACM SIGOPS Oper. Syst. Rev. Res Dev Linux kernel **42**(5), 64–74 (2008). https://doi.org/10.1145/1400097.1400105
16. Tsiknas, K., Rantos, K., Schinas, C.J., Soilemes, A.: Performance evaluation of TCP-BIAD in high-speed, long-distance networks. Computing **101**(4), 319–337 (2019)
17. Tsiknas, K., Stamatelos, G.: Comparative performance evaluation of TCP variants in WiMAX (and WLANs) network configurations. J. Comput. Netw. Commun. **2012** (2012). https://doi.org/10.1155/2012/806272
18. Alrshah, M.A., Othman, M., Ali, B., Mohd, Hanapi Z.: Comparative study of high-speed Linux TCP variants over high-BDP networks. J. Netw. Comput. Appl. **43**, 44–75 (2014). https://doi.org/10.1016/j.jnca.2014.04.007
19. Abdeljaouad, I., Rachidi, H., Fernandes, S., Karmouch, A.: Performance analysis of modern TCP Variants: a comparison of cubic, compound and new Reno. In: 25th Biennial Symposium on Communications (QBSC) (2010). https://doi.org/10.1109/BSC.2010.5472999
20. Taruk, M., Budiman, E., Setyadi, H.J.: Comparison of TCP variants in long term evolution (LTE). In: 5th International Conference on Electrical, Electronics and Information Engineering: Smart Innovations for Bridging Future Technologies, ICEEIE (2018). https://doi.org/10.1109/ICEEIE.2017.8328776
21. Orozco, J., Ros, D.: TCP performance over gigabit-capable passive optical networks. In: Third International Conference on Access Networks, vol. 6, pp. 264–279 (2009)
22. Chan, M.C., Ramjee, R.: TCP/IP performance over 3G wireless links with rate and delay variation. Wirel. Netw. **11**, 81–97 (2005)
23. Winstein, K., Sivaraman, A., Balakrishnan, H.: Stochastic forecasts achieve high throughput and low delay over cellular networks. In: 10th USENIX Symp Networked Syst Des Implement (NSDI), pp. 459–471 (2013)
24. Erman, J., Gopalakrishnan, V., Jana, R., Ramakrishnan, K.K.: Towards a SPDY'ier mobile web? In: Proceedings of the 9th International Conference on Emerging Networking Experiments and Technologies (CoNEXT) (2013). https://doi.org/10.1145/2535372.2535399
25. Huang, J., et al.: An in-depth study of LTE: Effect of network protocol and application behavior on performance. Comput. Commun. Rev. (2013)
26. Lin, C.C., Sandrasegaran, K., Mohd Raml, H.A., Basukala, R.: Optimized performance evaluation of LTE hard handover algorithm with average RSRP constraint. Int. J. Wirel. Mob. Netw. **2**(3), 1–16 (2011)
27. Dimou, K., Wang, M., Yang, Y., et al.: Handover within 3GPP LTE: design principles and performance. In: IEEE Vehicular Technology Conference (VTC) (2009). https://doi.org/10.1109/VETECF.2009.5378909

28. ETSI. TS 136 331—V15.3.0—LTE; Evolved Universal Terrestrial Radio Access (E-UTRA); Radio Resource Control (RRC); Protocol specification. 3GPP TS 36331 version 1371 Release 13 (2018)
29. Sesia, S., Toufik, I., Baker, M.: LTE—The UMTS Long Term Evolution: From Theory to Practice, p. 752. Wiley, New York (2011)
30. Hu, H., Zhang, J., Zheng, X., Yang, Y., Wu, P.: Self-configuration and self-optimization for LTE networks. IEEE Commun. Mag. **48**(2), 94–100 (2010)
31. Bălan, I.M., Moerman, I., Sas, B., Demeester, P.: Signalling minimizing handover parameter optimization algorithm for LTE networks. Wirel. Netw. **18**(3), 295–305 (2012)
32. ns-3 Consortium. ns-3. ns3 homepage. https://www.nsnam.org/
33. Gupta, R., Bachmann, B., Ford, R., et al.: Ns-3-based real-time emulation of LTE testbed using LabVIEW platform for software defined networking (SDN) in CROWD project. In: ACM International Conference Proceeding Series (2015). https://doi.org/10.1145/2756509.2756516
34. Piro, G., Baldo, N., Miozzo, M.: An LTE module for the ns-3 network simulator (2012). https://doi.org/10.4108/icst.simutools.2011.245571
35. Ahmad, I., Kaleem, Z., Chang,K.: Block error rate and UE throughput performance evaluation using LLS and SLS in 3GPP LTE downlink. Proc. Korean Inst. Commun. Inf. Sci. Daegwallyeong-myeon, South Korea: Yongpyong Resort (2013). https://arxiv.org/abs/1810.01162
36. Gurtov, A., Ludwig, R.: Responding to spurious timeouts in TCP. In: Proceedings of IEEE INFOCOM (2003)
37. Davey, R.P., Grossman, D.B., Rasztovits-Wiech, M., et al.: Long-reach passive optical networks. J. Lightwave Technol. **27**(3), 273–291 (2009)

Comparative Analysis of Watermarking Methods on CFRP Sample Thermal Images

B. Kalivaraprasad, M. V. D. Prasad, K. Rajesh Babu, Syed Shameem, S. Mohan, and Vasudevan Vani

Abstract Digital watermarking became indispensable in today's scenarios, and a lot of digital data have become distributed worldwide. Internet connectivity has made it very convenient and inexpensive. Due to rapid technological progress in the past 15 years, current users. There has been a rise in the likelihood of various digital attacks, such as forgery, damaging digital data, downloading, copying and altering digital knowledge. The use of a privacy mechanism that can protect digital data from the owner of digital data and its rights has been relevant in these situations. In this paper, on comparing the various watermarking techniques such as LSB, DWT-SVD hybrid, and reversible blind watermarking on CFRP samples to identify the efficacy of each method. The reversible bind watermarking is better for recovering the layout information of the CFRP sample than the LSB and hybrid method. The watermarking strategies of thermal images layout can become more secure.

Keywords Steganography · Watermarking · DWT · SVD · CFRP sample

1 Introduction

Nowadays, watermarking is a technique used in the use of protecting the copyrights of digital assets. It has used to embed data and information with digital multimedia such as digital images, audio, digital video and various digital documents that are used for watermarked to improve the protection of the respective owner's digital assets [1]. Watermarking is the method of embedding a watermark with the original object, such as image, video and audio, without sacrificing quality. In 1282, in Italy, the first watermarking technique was used in paper mills. This approach can mainly be used for authentication, security of data, telemedicine, broadcasting, aircraft and military applications [2]. The watermarking process is divided into two sections, such as the

B. Kalivaraprasad (✉) · M. V. D. Prasad (✉) · K. Rajesh Babu · S. Shameem
Department of Electronics and Communications Engineering, K.L. University, Guntur- DT, Green Fields, Vaddeswaram, Andhra Pradesh, India

S. Mohan · V. Vani
Department of Computer Science Engineering, University of South Pacific, Suva, Fiji

V. Bhateja et al. (eds.), *Computer Communication, Networking and IoT*, Lecture Notes in Networks and Systems 197, https://doi.org/10.1007/978-981-16-0980-0_42

watermark being inserted into the host image and the watermark being extracted from the image. By many scholars, digital watermarking has been suggested as one of the most promising methods for providing sample protection, reliability, authenticity and mechanical application knowledge (MAK). The watermarking technique provides additional security beyond the existing security technology [3]. According to human perception, the watermarks are of four types such as visible watermark, invisible-robust watermark, invisible-fragile watermark and dual watermark. A secondary translucent overlay is a watermark in the primary image. The reversible bind watermarking is more appropriate for recovering the CFRP sample's layout details than the LSB and the process hybrid system. And the DWT-SVD process gives better embedding, extraction and retrieving of the layout information with minimum loss of data and time than LSB and reversible blind process. Moreover, with robust watermarking techniques, the sharing of the defective part information of thermal images layout can become more secure. In this paper, the comparison of the different watermarking techniques is used for mechanical applications. In the aircraft's sensitive materials, i.e., CFRP samples are used to manufacture so many parts. Thermal images are taken with an IR camera to identify the damaged parts of CFRP material for repair and diagnosis purpose. In order to unlawfully assert ownership or prevent the transfer of information to intended recipients, the primary motive behind attacks might be to alter, adjust or even delete the document watermark. It was an interesting topic for top researchers in this field, because if CFRP material is not protected properly, an unauthorized person can copy the original defective information. So, the enemies would find it much easier to attack. Then, it will be easy to attack by the enemies. So, watermarking techniques are required to protect the authorized knowledge of the owner. The paper has an arrangement as follows: In part 2, the description of watermarking algorithms is explained in depth, Sect. 3 presented some experimental results, and Sect. 4 concludes this work.

2 Methodology

The techniques for image watermarking can be defined as 'spatial domain' and 'transform domain' techniques [4]. The spatial domain techniques are straightforward and computationally simple with the LSB substitutions, correlation-based and spread spectrum. Watermark data is specifically embedded in spatial domain watermarking by manipulating the pixel, bitstream or code values of the host signal (cover media) [5]. However, the spatial domain techniques offer less robustness against the signal processing attacks. The data is processed by modulating the coefficients of a transform, using a discrete Fourier transform (DFT), discrete cosine transform (DCT), discrete wavelet transform (DWT) and singular value decomposition (SVD). The techniques of transforming domain watermarking are computationally complex and offer greater robustness to watermarked data [6].

2.1 LSB Technique

For a simple operation, the LSB technique is used to insert information into a cover image. The simplest algorithm is least significant bit insertion, in which each 8-bit pixel's least significant bit is overwritten with a bit from the watermark [7]. In a digital image, information can be inserted directly into every bit of image information, or the busier areas of an image can be calculated to hide such messages in less perceptible parts of an image. This technique is based on LSB modifications of the pixel value. It is simple to understand, easy to implement and high visual fidelity [8]. Incorporating the watermark into the least significant bits of the cover image [9] would be the easiest form of watermark embedding. Despite its simplicity, LSB substitution suffers from many disadvantages, such as cropping, and the watermark is likely to be overcome by adding noise or lossy compression.

2.2 DWT-SVD (Hybrid) Method

The technique named LSB is easy to embed secret information but vulnerable for small cover modifications [10]. It is easy to apply signal processing techniques to destroy entire secret information. The minor change will result from lossy compression to a complete loss of information in certain instances. The insertion (embedding) of data in the domain provides greater robustness for watermarked data [11]. In important areas of the cover, transform domain techniques mask confidential information, which makes them extremely resilient than spatial domain techniques to signal processing attacks. The next subsections [12] present essential transformation domain techniques. The DWT, SVD and DCT-based watermarking approach enhances robustness against signal processing attacks without major degradation of the cover image quality. For multiple images and text watermarks, the efficiency of the hybrid approach has also been evaluated. The technique has four parts, namely the image, watermark, embedding and extraction processes. The method proposed in [13] was modified to embed and extract text watermark.

2.3 Reversible Blind Watermarking

This data nonetheless needs to be decoded to extract the watermarked image. The original cover data is needed in the extraction process [14]. This process is, therefore, called a blind technique. In a non-blind technique, the original cover data is unknown to the recipient; hence, the decoding process will have just relied on the watermark. Reversibility property in watermarking systems ensures the restoration of the host image after watermark extraction. It is an indispensable requirement in sensitive medical, military, forensic imagery and mechanical applications [15]. Triangular numbers (Peterson 2002) are figurate numbers, which are represented as triangular patterns of dots. They can be generated with the TNG function given in Eq. (1). A TNG function uniquely encodes a pair of integers (x, y) into an integer T, which can be factored back without any overhead [16]

$$T = f(x, y) = [(x + y)^2 + 3x + y]/2 \tag{1}$$

An integer pair is seen to be uniquely coded, i.e., f (x, y) and f (x, y) are distinct. The sequence of triangular numbers appears in the first row of the table. However, any number coded with TNG function is called triangular in this research for generality. Any integer encoded with TNG function can be factored back into (x, y) pair, as shown in Eq. 2.

$$C = \left[\text{sqrt}(8T + 1) - 1\right]/2 \tag{2}$$

where $C = x + y$, $x = T - C(C + 1)/2$ and $y = C/(C + 3)/2 - T$. From the above equations, it is seen that x and y recovered from T without any side information. TNG also exhibits blind extraction property [17]. By using some techniques that include extraction of the watermark, reversible watermarking prevents irreversible distortions in the host cover picture. In order to reduce the possibility of incorrect sample content, for image watermarking, these techniques are suggested [18, 19].

3 Results and Discussion

For the experiment procedure, the images of equal size with the same number of rows and columns are considered as the square matrix. The original image size is 512 × 512, and the same is taken for the watermark image. Here, the CFRP sample is the original image, and the watermark is the thermal image and its style. The cover image and the image of the watermark used to enforce the method of LSB, DWT-SVD and reversible blind watermarking are shown in Fig. 1. The proposed watermarking technique is implemented in MATLAB (2018a). As per the algorithm mentioned, watermarking scheme is executed. The watermark extracted and recovered the defective layout information, respectively, after the implementation of the watermarking scheme. Table 1 demonstrates the outcome of the three forms of watermarking schemes. As The research in this field face unauthorized problems in materials. If the defective information is protected by different watermarking techniques, then easy to avoid the authorization.

Three performance measurement metrics, such as mean square error (MSE), peak signal-to-noise ratio (PSNR), and normalized cross-correlation (NCC), have been measured. The MSE parameter is used to approximate or calculate the average of the "error" squares between the host image and the image with the watermark, and the PSNR factor is the imperceptibility of the image. More the PSNR shows that the watermarked image is perceptible, or watermark is not recognized by naked eyes.

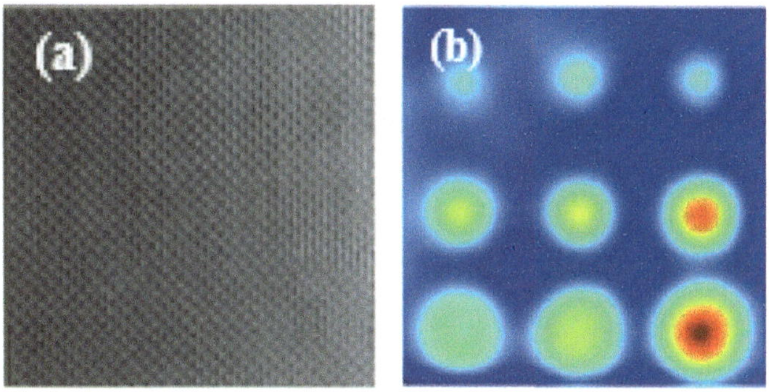

Fig. 1 **a** Cover image (CFRP Sample). **b** Watermark image (thermal image and layout)

Table 1 Results of watermarking methods

Parameter	Cover Image	Watermark	Watermarked	Extracted	Recovered
LSB		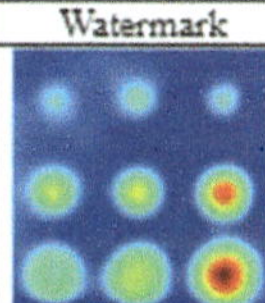	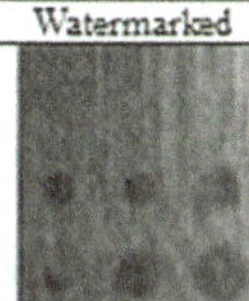	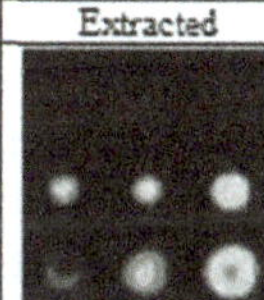	
DWT-SVD		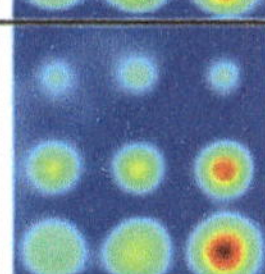		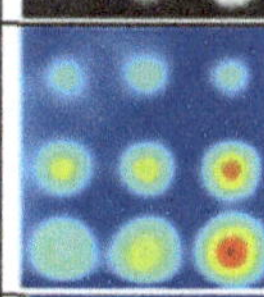	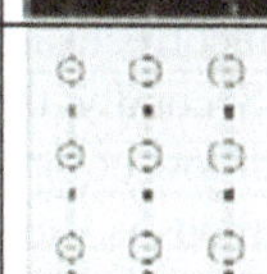
BLIND		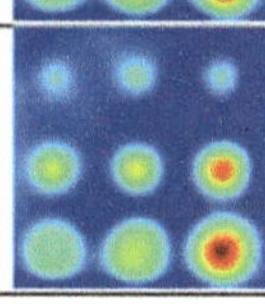		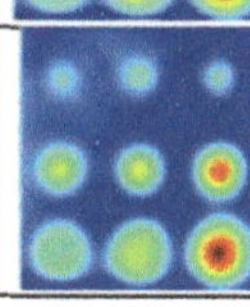	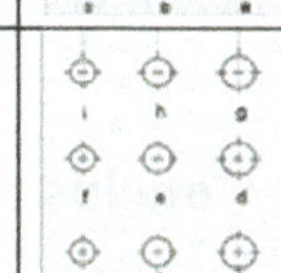

The NCC is mainly preferred to calculate the correlation between the watermarked and extracted images.

First, according to the performance evaluation, the reversible blind watermarking method has better results of MSE, PSNR and NCC of 48.2615, 37.12 and 0.9998, respectively. The performance evaluation of three methods is presented in Table 2.

Table 2 Performance evaluation of watermarking algorithms

Parameter	MSE	PSNR (dB)	NCC	Elapsed time (s)
LSB	52.3890	34.61	0.9743	7.04
DWT-SVD	51.9928	35.98	0.9960	**3.10**
Reversible	**48.2615**	**37.12**	**0.9998**	23.20

The bold values represent the low MSE, high PSNR and NCC attained by reversible watermarking technique as compared with other techniques.

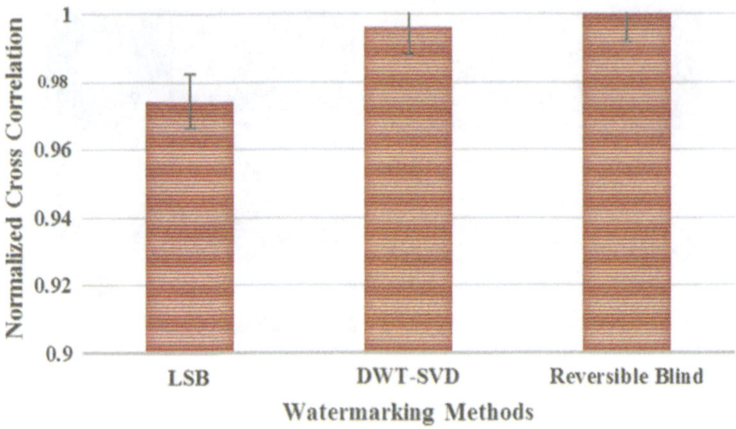

Fig. 2 Comparison of NCC for watermarking algorithms

And the effective extraction of the defective layout information is compared with the spatial domain (LSB) and frequency domain methods (DWT-SVD).

In image watermarking, these techniques are preferred to decrease the probability of incorrect sample information.

Secondly, the hybrid method performs very well as compared with the LSB and with low computational time, i.e., 3.12 s. The comparison of NCC among the three methods is shown in Fig. 2.

4 Conclusion

Digital watermarking has become an important way of preserving digital content that has become widespread on the Internet, as well as preserving the copyrights of the owner. In this paper, the watermarking methods and the usage of watermarking techniques in thermal images of CFRP sample of airplanes and aircrafts application, besides, discussed some watermarking techniques for authentication, integrity control, data hiding and an understanding of how each technique has several advantages and disadvantages. The CFRP sample is used as a host image embedding with the thermal image, in which layout defective information is hidden, of the host image. The reversible bind watermarking is therefore ideal for recovering the CFRP sample layout information as compared with LSB and hybrid system. The DWT-SVD approach offers improved embedding, extraction and recovery of the layout information with minimal information and time loss compared to LSB and reversible blind process. Moreover, with the usage of robust watermarking techniques, the sharing of the defective information of thermal images layout can become more secure.

5 Future Work

In future, the defective information of the CFRP sample may be calculated accurately by using the advanced watermarking techniques along with pulse compression technique by measuring the full width half maximum (FWHM) parameter.

Acknowledgements Thankfully acknowledge the management of KL (Deemed to be University) to provide every source and required facilities for completion of this work.

References

1. Zain, J.M., Fauzi, A.R.M., et al.: Clinical evaluation of watermarked medical images. In: Proceedings of the 28th IEEE EMBS Annual International Conference, New York City, USA, 30 Aug–3 Sept 2006
2. Su, Q., Niu, Y., et al.: A blind dual-color images watermarking based on IWT and state coding. Opt. Commun. 1717–1724 (2012)
3. Song, C., Sudirman, S., Merabti, M.: Recent Advances and Classification of Watermarking Techniques in Digital Images. ISBN: 978-1-902560-22-9 (2009)
4. Suwadi, W., et al.: A recent survey of a self-embedding fragile watermarking scheme for image authentication with recovery capability. EURASIP J. Image Video Process. 1–22 (2019)
5. Arsalan, M., Qureshi, A.S.: Protection of medical images and patient-related information in healthcare using an intelligent and reversible watermarking technique. Appl. Soft Comput. J. (City, University of London Institutional Repository) **51**, 168–179 (2017)
6. Mohanarathinam, A., Kamalraj, S., et al.: Digital watermarking techniques for image security: a review. J. Ambient Intell. Humanized Comput. (2019)
7. Kelkar, V., Tuckley, K., et al.: Novel variants of a histogram shift-based reversible watermarking technique for medical images to improve hiding capacity. J. Healthc. Eng. (Hindawi) **2017**, 1–7. Article ID 3538979 (2017)
8. Begum, M., Uddin, M.S.: Digital image watermarking techniques: a review. Information (MDPI) 1–38 (2020)
9. Ibrahim, M.M., Abdel Kader, N.S., et al.: Video multiple watermarking technique based on image interlacing using DWT, Hindawi Publishing Corporation. e Sci. World J. **2014**, 1–12. Article ID 634828 (2014)
10. Laskar, T.A., Hemachandran, K.: Digital image watermarking techniques and its applications. Int. J. Eng. Res. Technol. (IJERT) **2**(3), 1–8 (2013)
11. Bamatraf, A., Hussein, T., et al.: Digital watermarking algorithm using LSB. In: International Conference on Computer Applications and Industrial Electronics (ICCAIE 2010), pp. 155–159, 5–7 Dec 2010
12. Chammem, A., Mitrea, M., et al.: DWT-based stereoscopic image watermarking. In: Stereoscopic Displays and Applications XXII, Proceedings of SPIE-IS&T Electronic Imaging, SPIE, vol. 7863, pp. 786326 (2011)
13. Badshah, G., Liew, S.-C., et al.: Watermarking of ultrasound medical images in teleradiology using compressed watermark. J. Med. Imaging **3**(1), 017001 (2016)
14. Manupriya, Kumar, T.: Analysis of wavelets with watermarking through wavelet transformation. Int. J. Comput. Appl. **96**(22), pp. 17–22 (2014)
15. Singh, A.K.: Some New Techniques of Improved Wavelet Domain Watermarking for Medical Images. 2K-12/NITK/Ph.D./1425–1430 (2015)
16. Agrestea, S., Andalorob, G., et al.: An image adaptive, wavelet-based watermarking of digital images. J. Comput. Appl. Math. (2006)

17. Allaf, A.H., Kabir, M.A.: Usage of watermarking techniques in medical imaging. In: Association for Computing Machinery, SCA'18, 10–11 Oct 2018
18. Saini, L.K., Shrivastava, V.: A survey of digital watermarking techniques and its applications. Int. J. Comput. Sci. Trends Technol. (IJCST) **2**(3), 70–73 (2014)
19. Kalivaraprasad, B., Prasad, M.V.D., Raghava Prasad, C., Sasikala, N., Ravi, S.: Flower image segmentation using watershed and marker controlled led watershed transform domain algorithms. ARPN J. Eng. Appl. Sci. **13**(21) (2018)

Video Frame Rate Doubling Using Generative Adversarial Networks

Ashwin R. Bharadwaj, Hardik Gourisaria, and Hrishikesh Viswanath

Abstract This work explores a Deep Neural Network based approach to double the frame rate of a video by adding synthetically generated frames between two consecutive frames in the original video. The Neural Network is a Generative Adversarial Network that consumes two consecutive frames from the original video, interpolates them into one image, and generates a synthetic intermediate frame. Apart from the Neural Networks, Statistical and Image Processing techniques have been used to further enhance the generated frame. The final generated frame is an amalgamation of the interpolated frame and generated frame which are merged through masking. This has various applications including but not limited to remastering legacy videos with low frame rates to improve their quality. Finally, this work compares the results of the above technique against various pre-existing techniques.

Keywords Neural networks · Generative adversarial networks · Video frames · Frame interpolation · Frame duplication · Digital image processing · CycleGAN · Histogram matching · Structural similarity index

1 Introduction

Efficient video streaming is a requirement of many internet users to reduce network load and buffering time. On the server end, the streaming service could reduce the frame rate of the videos to improve streaming speed [1] and on the user end, boost the frame rate to get back the original quality. If the frame rates could be boosted on the user end, then ideally, each video would only need half as many frames, reducing the size of the video on the server end.

This work presents a method to double the frame rate of a video where lower frame rate videos can be converted into videos with twice the frame rate. The process involves the addition of synthetic frames into the original video to increase the frame

A. R. Bharadwaj · H. Gourisaria · H. Viswanath (✉)
Department of Computer Science and Engineering, People's Education Society University, Bangalore, India

V. Bhateja et al. (eds.), *Computer Communication, Networking and IoT*, Lecture Notes in Networks and Systems 197, https://doi.org/10.1007/978-981-16-0980-0_43

rate. Synthetic frames [2] are produced by extracting information from the adjacent frames in the original video.

There is some amount of noise present in the synthetic high frame rate video that has been generated from the original lower frame rate video [3]. Noise is introduced due to the synthetic frames not having the same color scheme of the original frames, which leads to flickering in the synthetic output video. However, with sufficient training of the Generative Adversarial Network (GAN) [4] followed by applying the image processing techniques mentioned ahead, to the GAN output, it is possible to minimize such noise.

Traditional methods of performing video frame rate boosting involve video frame duplication and video frame interpolation [5]. The former technique does not improve the visual quality of the video while the latter generates synthetic frames that have a high mean squared error compared to the desired frames, in highly dynamic videos. This work aims to address these issues.

2 Related Work

Xiao et al. [3] used a Variational Auto Encoder to boost the frame rate of videos. The problem of flickering persisted in their output. Flickering can be reduced by a significant amount using techniques described in the following sections. Goodfellow et al. [4] propose a method to estimate generative models through an adversarial process, essentially constructing a GAN. Sharma et al. [5] suggest a method to increase the frame rate of videos by interpolation. A spatio-temporal medial filtering method was implemented to improve the quality of the interpolated frames. McCarthy et al. [6] suggest that viewers were less susceptible to changes in frame rate but were more sensitive to changes in resolution. This insight was used to implement a model that processes videos of known resolution. Feng et al. [7] propose techniques for transmitting videos over the network by compressing them. A client-side buffer is utilized to smoothen the frame rate of the video.

Xu et al. [8] suggest a method to improve the frame rate of videos through an efficient sub-pixel convolution neural network. A higher resolution is realized by combining motion estimation between the adjacent frames with CNN. Further amplification of the frame rate was done by frame interpolation. This was done by calculating the image optical flow between the frames. The method was proven to have advantages in the quality of video restoration. Clark et al. [9] suggest a method to produce longer videos of higher resolution than the original using a Dual Video Discriminator GAN. The assumption is that the pixels of each frame don't depend directly on other pixels of the video. The model utilizes two discriminators - Spatial Discriminator and Temporal Discriminator to resolve the issue of generating large videos. The Spatial Discriminator discerns the contents of a single frame while the temporal discriminator signals movement. Ying et al. [10] propose a method to generate previously unseen frames of a video to predict the future to aid intelligent agents in prediction. The process of predicting is aided by synthetic inter-frame difference. Two paths were designed—a Coarse Frame Generator, to determine the coarse details of the future frame and a Difference Guide Generator, to generate the difference image that contains complementary details. Qi et al. [11] explore a method to reduce bandwidth consumption by reducing the frame rate of the videos on the server-side and further increase them through interpolation on the client-side. Wang et al. [12] have developed a deep learning model to denoise the images. The model performs operations such as contrast enhancement and motion awareness. The denoising model requires tuning parameters that have large variance across various settings.

Li et al. [13] have discussed a method to upscale the frame rate of 3D videos. The color values are encoded in parallel to increase the speed of interpolation. Aigner et al. [14] have used PGGAN to predict the future frames of the video. New layers are added while training to accommodate larger datasets. Chen et al. [15] have implemented a GAN based model to translate videos. The model processes each frame and translates it to match the style of the target video. Janzen et al. [16] performed a study to understand whether the frame rate was more important to viewers than latency. Frame Rate affects the ability of the viewers to recognize objects. Latency affects the viewers' perception only when the frame rate is abysmally low.

It can be concluded from these works that while GAN can be used successfully to generate synthetic frames, inserting the frames into the video might lead to flickering. This work attempts to address the same.

3 Proposed Method

Three methods are discussed in this paper to improve the frame rate of videos. The first one uses a simple Pix2Pix GAN [2] like Xiao et al. [3] to improve the frame rate of videos. The second method, which is supposed to provide an improvement over GAN, uses CycleGAN [17] to reduce flickering that occurs with the usage of a regular GAN. Our proposed method combines GAN and interpolation techniques [5] to fully exploit the two, each of which can work well in situations that the other model failed. GAN technique is shown to work better in videos with rapid movement while interpolation has better performance when applied to videos with limited movement.

3.1 Data Collection

The dataset of videos that are used to train the models is taken from the HMDB 51 dataset [18]. The videos used to train the models are required to have a high degree of variation to provide a dynamic set of frames. Failing to do so will cause the model to over-fit to those types of videos and will not generalize well to other types of videos.

3.2 Dataset Preparation

Three sets of frames were extracted from each video and stored separately with two labels—input and output. The input data set contained the endpoint frames while the output contained the intermediate frames. This was done to separate the input from the desired output of the model. The two input frames were then interpolated by pixel-wise averaging. To tackle the hardware limitations of training a GAN model, the images were resized to $256 \times 256 \times 3$.

3.3 Model Definition and Training

Pix2Pix Generative Adversarial Network [4] consists of a Generator network and a Discriminator network. The Generator is trained to generate a synthetic frame from the given input comprising two interpolated frames. The Discriminator is trained to discern if a frame is original or synthetically generated. Both models act as adversaries of each other and in turn, improve the performance of the other one. The generator aims to generate frames that are indistinguishable from the ground truth. The discriminator compares the features of the synthetic and original images to differentiate between the two.

The generator is a U-Net [2] model with 7 encoder layers and 7 decoder layers. The discriminator has 6 Convolution Layers in a sequence and takes the output of the generator and the target frame as the input. The output of the discriminator helps identify the degree of deviation of the generated synthetic frame from the ground truth frame. Each of the convolution layers performs 2D convolution between the image and the filter. A 4×4 filter is selected to mask the image padded with zeroes. Convolution operation between two 2D matrices is given by the following equation

$$y[i, j] = \sum_{m=-\infty}^{\infty} \sum_{n=-\infty}^{\infty} h[m, n] \times x[i - m, j - n] \tag{1}$$

where, y is the activation map matrix. $h$is the filter matrix and x is the input matrix. Convolution is done to extract important features of the image and reduce the dimensions of the said image. The resulting dimension of the activation map y after applying a single filter is given by the formula

$$A = \frac{I-F+2P}{S} + 1 \tag{2}$$

where I is the dimension of the input image, F is the filter size, P is the padding, and S represents the stride. The final dimension of the activation map will be (A, A, n), where n is the number of filters applied to the image. Each filter captures a specific feature of the image.

The output of each convolution layer, including the hidden layers, is batch normalized to reduce the co-variance shift and to speed up the learning process. The Co-variance shift is the shift in the mapping between input and output when the input distribution changes. By normalizing the values, higher learning rates can be used since the values don't fluctuate and conform to a distribution. It prevents overfitting by adding noise to the activation of hidden layers, like regularization.

Normalization is done by first determining the mean of a batch of input denoted by $I = \{x_1, x_2, \ldots, x_N\}$ where $x_1, x_2, \ldots, x_N$ are activation values.

$$\mu_I = \frac{1}{N} \sum_{k=1}^{N} x_k \tag{3}$$

The normalized value of x is given by

$$\widehat{x_I} = \frac{x_I - \mu_I}{\sqrt{\sigma_I^2 + \epsilon}} \tag{4}$$

ϵ is an arbitrary constant and σ is the standard deviation of the batch. The normalized output is finally scaled and shifted in the following manner

$$y_i = \gamma \widehat{x_I} + \beta \tag{5}$$

The process of learning happens through minimizing the loss, which is calculated using cross-entropy with logits. Logits are zero matrices for synthetic output and unit matrices for original output and are denoted by x. The image matrix is denoted by z.

$$\text{loss} = z(-\log \sigma(x)) + (1 - z)(-\log(1 - \sigma(x))) \tag{6}$$

The loss is minimized using the Adam Optimizer, which aims to dynamically alter the learning rate by using a weighted sum of previous gradients, momentum and, Nesterov Accelerated Gradient. Nesterov Accelerated Gradient uses a look ahead to slow down the momentum as the weights approach the target value and prevent the value from overshooting the optimum. Momentum does the exact opposite. It increases the rate of gradient descent with each epoch, thereby, speeding up the convergence.

$$W_{t+1} = W_t - \frac{\eta V_t}{\sqrt{S_{dw} + \epsilon}} \tag{7}$$

$$V_t = \beta V_{t-1} + (1 - \beta) g_t \tag{8}$$

$$S_{dw} = \beta S_{dw-1} + (1 - \beta) g_t^2 \tag{9}$$

In Eq. 7, V_t represents the momentum in the epoch t as a weighted average of the previous momentum and the previous gradient. ϵ is an arbitrary constant to prevent the denominator from becoming zero. In Eq. 8, β is a value between [0, 1], used to perform exponential smoothing. At $\beta = 1$, the momentum will be an average of all the data points. As β approaches 0, the momentum reduces. g_t represents the gradient in the epoch t.

In Eq. 9, S_{dw} represents the weighted average of the previous gradients, with the most recent gradient contributing the highest to the next gradient.

The model code, when written in Tensorflow, requires large amounts of training data and epochs to converge to the optimum. The model was trained for 200,000 steps before it started providing acceptable results with low mean squared error and structural similarity values closer to 1. The same model was implemented using Keras and due to the internal optimizations in Keras, the model converged after 15,000 steps.

The model was trained on a system with Intel i7, Hex-Core Processor, and an Nvidia GeForce 1050Ti Graphics Card.

3.4 Histogram Matching

As proposed by Rakwatin et al. [19], Histogram Matching is a technique that compares two Cumulative Distribution functions and aims to map the source histogram to the target histogram.

The distribution of the image generated by the generator is represented as a histogram. This output is modified to fit the distribution of the previous frame.

However, this method does not significantly improve the quality of the output because the autoencoder, which is the primary component of the Generator, is expected to match the distribution of the images. The generator works by minimizing a linear combination of mean squared error and Kullback-Leibler Divergence (KL Divergence) [20] between the input and the output.

KL Divergence is a representative of how different the two distributions are. Lower values of KL Divergence indicate higher similarity between the histograms.

3.5 Filtering Noise in Synthetic Frames

Various filters were implemented to remove the additional noise [21] introduced in the synthetic frame generated by GAN [3]. These include a Mean filter, Median Filter, and Histogram Equalization. Median Filter outperformed the other filters.

4 CycleGAN

CycleGAN is a model that enables unsupervised training of images. It is primarily used for the translation of images [17]. The network mainly learns the mapping between images. The motivation for using CycleGAN began with the flickering that was present in the output video of the Pix2Pix GAN. The initial idea was to use CycleGAN to translate noisy or rather flickering frames into expected frames. There was however no considerable improvement in the quality of the output as determined by mean squared error and structural similarity index.

5 Region of Interest

Another approach to resolve the issue of flickering in videos is to extract the region of interest from frames and use these regions to selectively pass to GAN and interpolation models.

Region of interest refers to the region that changes rapidly between frames. Any fast-moving object is in the region of interest while any slow-moving objects in the frames are discarded while training.

As mentioned by Wang et al. [22], if regions of interest are used instead of the entire frame, the number of bits required to encode the frame is lesser and consequently, the model has to learn fewer sets of weights, reducing the possibility of overfitting.

A mask is used to distinguish the region of interest in the frames from the other regions of the frame. The white sections of the mask correspond to the regions of interest of the frame which are obtained using GAN and the black sections of the mask correspond to the remaining regions of the frame which are obtained using interpolation.

Using the mask to distinguish between the region of interest and the remaining regions of the frame helps minimize the flickering that is introduced using GAN.

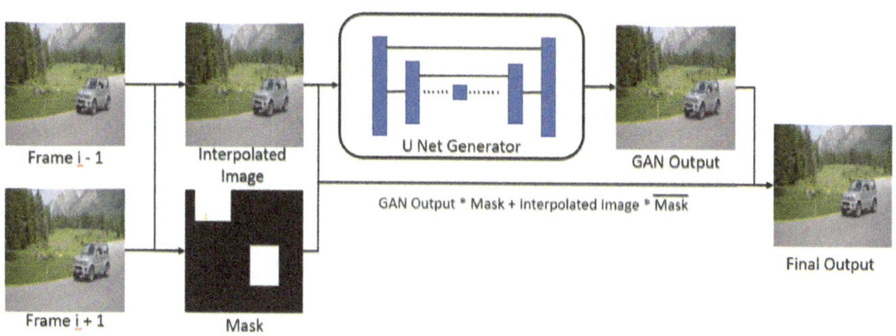

Fig. 1 Proposed model architecture

6 Evaluation

Setting aside the qualitative analysis of the model's functionality by the visual inspection of the output, it is important to compute and compare the results quantitatively.

High FPS videos are downgraded to half their FPS by removing alternate frames. The FPS of the downgraded video is doubled by inserting synthetically generated frames. The output video is compared against the original high FPS video by pixel-wise differencing of the generated and the expected frame. Mean Squared Error (MSE) is computed for the output generated by all the methods demonstrated. MSE is computed using the following equation

$$\mathrm{MSE} = \frac{1}{fnm} \sum_{i=1}^{n} \sum_{j=1}^{m} \sum_{k=1}^{f} \left(Y_{i,j,k} - \widehat{Y_{i,j,k}} \right)^2 \tag{10}$$

Structural Similarity Index (SSIM) values are determined to perform a two-fold evaluation of the output and is a measure of how similar the two frames in question are. If the SSIM value of two images is 1, that implies that the two frames are the same. By comparing the synthetic frames with original frames, we can conclude whether the two are indistinguishable. For multi-dimensional images (RGB), SSIM compares luminance (L), structure (S), and contrast (C) and is computed using the following equation

$$\mathrm{SSIM}(x, y) = [L(x, y)]^{\alpha} \times [C(x, y)]^{\beta} \times [S(x, y)]^{\gamma} \tag{11}$$

$$L(x, y) = \frac{2\mu_x\mu_y + C_1}{\mu_x^2 + \mu_y^2 + C_1} \tag{12}$$

$$C(x, y) = \frac{2\sigma_x\sigma_y + C_2}{\sigma_x^2 + \sigma_y^2 + C_2} \tag{13}$$

$$S(x, y) = \frac{\sigma_{xy} + C_3}{\sigma_x\sigma_y + C_3} \tag{14}$$

In the above equations, μ terms refer to mean and σ terms refer to standard deviation σ_{xy} is cross-covariance. SSIM is calculated between each synthetic and original frame of the video. The average value of SSIM of every synthetic frame is a representative of how similar the synthetic video is to the original one.

7 Performance and Results

Table 1 depicts three instances of output frames which are initially evaluated by visual inspection. Frame $i - 1$ and frame $i + 1$ are frames that are passed as input

Table 1 Sample results

Frame $i-1$	Frame $i+1$	Interpolation	Final Output

to the models. The output frames obtained through interpolation and using GAN are indicated in the rightmost columns. As it can be observed, the output frame of the proposed model is sharper than the output frame generated through simple image interpolation. This is particularly evident in the second example, where the leaves of the tree are more clearly visible in the output obtained through the proposed model.

In Table 2, lower values of MSE indicate a higher similarity between the generated frames and the ground truth frames. Higher values of SSIM indicate that the expected frames and the ground truth frames are structurally similar. In ideal cases, the MSE is zero and SSIM is one. The MSE between the ground truth frames and the frames generated by the proposed model is lesser than the MSE obtained by frame duplication. However, frame interpolation has a lower MSE than the GAN network.

Table 2 Performance of the models

Method used	Pixel-wise mean squared error	Average SSIM
Frame duplication	0.01372	0.75089
Frame interpolation	0.00918	0.88873
GAN model	0.00936	0.84048
CycleGAN	0.00966	0.86300
GAN + interpolation	0.00929	0.88250

This is because GAN introduces additional background noise through flickering, which can be reduced by further training and focusing on the region of interest.

CycleGAN has lower training time than Pix2Pix GAN but the performance is found to be slightly worse than GAN. This difference was not discernable in most of the videos. Occasionally, higher values of MSE are obtained, but it is visually not observable.

A combination of GAN and interpolation outperforms the other models when used individually. The generated videos are visually pleasing. The combined GAN interpolation model is tested by splitting the video into fast-moving scenes and slow-moving scenes. This is done by extracting the region of interest and determining the MSE between consecutive regions of interest. Fast-moving scenes are fed to the GAN while slow-moving scenes are fed to the interpolation model. The resulting frames are impossible to identify as being synthetically generated. The only situation where the synthetic frames are imperfect is when an extremely dark frame transitions to a very bright frame.

A final approach to remove flickering is to determine the difference in average RGB values of the color palette of the synthetic frame and the original frame. Based on this difference, the brightness of the synthetic frames is dynamically increased. This difference in average color palette values is consistent and is easy to scale. It is not necessary to focus on individual pixels because any minor discrepancies are taken care of with median filters.

8 Future Work

More advanced models can be implemented and tested for improving the accuracy of the synthetic frames generated. Due to hardware limitations, the input dimension of the image is restricted to $256 \times 256 \times 3$. Future work involves optimizing this model to handle high-resolution videos without resizing them.

9 Limitations

Hardware limitations restricted the training process to small datasets. The models were repeatedly retrained on batches of small datasets. The video resolution was restricted to $256 \times 256 \times 3$ for the same reason. The proposed method is limited to doubling the frame rate of videos and cannot be used to raise the frame rate to any arbitrary value.

Acknowledgements The authors would like to thank Dr. Natarajan Subramanyam who provided valuable insight and feedback. The work was done while the authors were at PES University.

Declaration We have taken permission from competent authorities to use the images/data as given in the paper. In case of any dispute in the future, we shall be wholly responsible.

References

1. Wu, J., Yuen, C., Cheung, N.M., Chen, J., Chen, C.W.: Enabling adaptive high-frame-rate video streaming in mobile cloud gaming applications. IEEE Trans. Circuits Syst. Video Technol. **25**(12), 1988–2001 (2015)
2. Isola, P., Zhu, J.-Y., Zhou, T., Efros, A.A.: Image-to-image translation with conditional adversarial networks. In: Proceedings of the IEEE Conference on Computer Vision and Pattern Recognition, pp. 1125–1134 (2017)
3. Xiao, T., Puri, R., Kesineni, G.: Frame rate upscaling with deep neural networks. Term Paper for CS294-129 Deep Neural Networks, Fall (2016)
4. Goodfellow, I.J., Pouget-Abadie, J., Mirza, M., Xu, B., Warde-Farley, D., Ozair, S., Courville, A., Bengio, Y.: Generative adversarial networks (2014)
5. Sharma, R.K., Hazra, R., Kasai, A.: Method and apparatus for increasing video framerate. US Patent 6,192,079, 20 Feb 2001
6. McCarthy, J.D., Sasse, M.A., Miras, D.: Sharp or smooth? comparing the effects of quantization vs. frame rate for streamed video. In: Proceedings of the SIGCHI Conference on Human Factors in Computing Systems, CHI '04, New York, NY, USA, pp. 535–542. Association for Computing Machinery (2004)
7. Feng, W.: Critical bandwidth allocation techniques for stored video delivery across best-effort networks. In: Proceedings 20th IEEE International Conference on Distributed Computing Systems, pp. 56–63. IEEE (2000)
8. Xu, M., Wang, D., Du, X.: A video frame resolution and frame rate amplification method with optical flow method and espcn model. In: Proceedings of the 2020 3rd International Conference on Image and Graphics Processing, pp. 91–95 (2020)
9. Clark, A., Donahue, J., Simonyan, K.: Adversarial video generation on complex datasets. arXiv preprint arXiv:1907.06571 (2019)
10. Ying, G., Zou, Y., Wan, L., Hu, Y., Feng, J.: Better guider predicts future better: difference guided generative adversarial networks. In: Asian Conference on Computer Vision, pp. 277–292. Springer (2018)
11. Qi, X., Yang, Q., Nguyen, D.T., Zhou, G.: Context-aware frame rate adaption for video chat on smartphones. In: Proceedings of the 2013 ACM Conference on Pervasive and Ubiquitous Computing Adjunct Publication, pp. 111–114 (2013)
12. Wang, Z.W., Jiang, W., He, K., Shi, B., Katsaggelos, A., Cossairt, O.: Event-driven video frame synthesis. In: Proceedings of the IEEE International Conference on Computer Vision Workshops, pp. 0–0 (2019)
13. Li, Y., Sun, L., Xue, T.: Fast frame-rate up-conversion of depth video via video coding. In: Proceedings of the 19th ACM International Conference on Multimedia, pp. 1317–1320 (2011)
14. Aigner, S., K¨orner, M.: Futuregan: anticipating the future frames of video sequences using spatio-temporal 3d convolutions in progressively growing gans. arXiv preprint arXiv:1810.01325 (2018)
15. Chen, J., Li, Y., Ma, K., Zheng, Y.: Generative adversarial networks for video-to-video domain adaptation. arXiv preprint arXiv:2004.08058 (2020)
16. Janzen, B.F., Teather, R.J.: Is 60 fps better than 30? the impact of frame rate and latency on moving target selection. In: CHI'14 Extended Abstracts on Human Factors in Computing Systems, pp. 1477–1482 (2014)
17. Zhu, J.-Y., Park, T., Isola, P., Efros, A.A.: Unpaired image-to-image translation using cycle-consistent adversarial networks. In: Proceedings of the IEEE International Conference on Computer Vision, pp. 2223–2232 (2017)
18. Kuehne, H., Jhuang, H., Garrote, E., Poggio, T., Serre, T.: HMDB: a large video database for human motion recognition. In: 2011 International Conference on Computer Vision, pp. 2556–2563. IEEE, Nov 2011
19. Rakwatin, P., Takeuchi, W., Yasuoka, Y.: Stripe noise reduction in modis data by combining histogram matching with facet filter. IEEE Trans. Geosci. Remote Sens. **45**(6), 1844–1856 (2007)

20. Van Erven, T., Harremos, P.: R´enyi divergence and kullback-leibler divergence. IEEE Trans. Inf. Theory **60**(7), 3797–3820 (2014)
21. Mori, C., Gohshi, S.: Real-time non-linear noise reduction algorithm for video. ICETE **1**, 487–493 (2018)
22. Wang, Z., Bovik, A.C., Lu, L.: Wavelet-based foveated image quality measurement for region of interest image coding. In: Proceedings 2001 International Conference on Image Processing (Cat. No. 01CH37205), vol. 2, pp. 89–92. IEEE (2001)

Exploration of Wavelets for Pre-processing of Speech Signals

Mohit Kumar Yadav, Vikrant Bhateja, and Monalisa Singh

Abstract Speech signal carries much more information than spoken words and speaker's information. Noise present in the speech signal is an obstacle in the path of analysing the speech signal precisely. Therefore, speech signal denoising has an important role in the speech signal analysis. In this paper, Wavelets are explored for pre-processing and suppression of noise from the noisy speech signals. The main aim of the proposed methodology involves, the pre-processing of the speech signals by using the combination of Discrete Wavelet Transform (DWT) and hard thresholding and to evaluate the result to find the optimal wavelet family with there optimal order followed by the suitable level of decomposition. This paper involves the exploration of the Wavelet Family and an improved Signal to Noise Ratio (SNR) values of the acquired speech signal which is shown in the result analysis.

Keywords Speech signals · Discrete wavelet transform (DWT) · Wavelet transform (WT) · Hard thresholding · Signal to noise ratio (SNR)

1 Introduction

Communication is the necessity of all human beings. Speech is one of the most important communication sources. It is the most natural, intuitive and fastest means of interaction among humans [1]. The speech signal contains the background noise and the distorted speech signal. The noise present in the speech signal causes the degradation in the quality of the speech. Over the last few decades, the pre-processing of speech signal which involves the removal of noise from speech has been an area of interest for the researchers [2]. Therefore, to remove a noise from the speech is very important as it may affect the further processing of the speech signal. So, filtering the noise from the acquired speech signal plays an important role in the speech signal analysis. As the frequency domain based signal processing can be designed easily

M. K. Yadav · V. Bhateja (✉) · M. Singh
Department of Electronics and Communication Engineering, Shri Ramswaroop Memorial Group of Professional Colleges (SRMGPC), Lucknow, Uttar Pradesh 226028, India

Dr. A.P.J. Abdul, Kalam Technical University, Lucknow, Uttar Pradesh, India

V. Bhateja et al. (eds.), *Computer Communication, Networking and IoT*, Lecture Notes in Networks and Systems 197, https://doi.org/10.1007/978-981-16-0980-0_44

and most of the noise suppression methods require the use of Short Term Fourier Transform (STFT) but in today's scenario Wavelet Transform (WT) is gaining much importance as it is simple and efficient method for speech signal de-noising. Wavelet transform can analyze the signal and can select the information present in it that other signal de-noising techniques lack [2]. Therefore, lot of study has been done by the researchers for the suppression of noise in speech signals. Aggarwal et al. [2] introduced an approach of multi-resolution analysis using WT and found that the modified universal threshold gives better results of denoising. Bhowmick et al. [3] proposed a method of Voiced Speech Probability Detection Wavelet Decomposition (VSPDWD) and compared it with seven different techniques. It was found that the VSPDWD technique gave an improved SNR values at all levels of decomposition. Akkar et al. [4] in their study made a comparison between different thresholding techniques. It was found that the square wavelet thresholding gave better results than the traditional thresholding techniques. Babichev et al. [5] introduced a de-noising method that is the Emperical Mode and the Wavelet Decomposition techniques. It was found that there was a relative change in the values of the Shannon Entropy used for the quality criterion. This indicates that the technique used was effective. Hadi et al. [6] in their study made a comparison between different threshold selection rule. It was found that the sawtooth wavelet thresholding gave better results than the traditional thresholding techniques.

In this paper, a combination of Discrete Wavelet transform and Hard Thresholding technique for noise reduction has been proposed. Wavelet Transform provides a multi-resolution and is a better technique than Fourier Transform and STFT. This paper consists of following sections: Section 2 gives the description of the pre-processing of noisy speech signals using wavelets. Section 3 consists of experimental setup, Sect. 4 consists of exploration of suitable Wavelet Family for speech signal analysis and Sect. 5 consists of simulation results. Section 6 consists of the conclusion of the analysis done in the proposed work.

2 Pre-processing of Noisy Speech Signals Using Wavelets

2.1 Discrete Wavelet Transform

For suppressing the noise present in the speech signal, Discrete Wavelet Transform has been used. It involves the decomposition of the speech signal in the time frequency domain. The noise present in the speech signal cannot be easily removed by using the Kalman or Chebyshev filters. Therefore, it can be removed by applying the wavelet transform [7–9]. A wavelet is a wave like oscillation that begins at zero, increases and then again goes to zero [1, 10]. The scaled and shifted version of fundamental or mother wavelet Ψ is elucidated below [3]:

$$\Psi_{\tau,\beta}(t) = \beta^{-1/2}\Psi\left(\frac{t-\tau}{\beta}\right) \tag{1}$$

where β is the scaling parameter and τ is the translation parameter. The noisy speech signal $s(t)$ is decomposed into sub-bands through DWT into approximation and detailed coefficients. The detailed coefficient or the higher frequency component $D(p, k)$ has been elucidated below [3]:

$$D(p,k) = 2^{-p/2}\sum\nolimits_n s(n)\Psi^*(2^{-p}n - k) \tag{2}$$

where *p, k and n are integers and* $\psi^*(t)$ *is the complex conjugate of* $\psi(t)$. The approximation coefficients or the lower frequency component $A(p, k)$, has been elucidated below [3]:

$$A(p,k) = 2^{-p/2}\sum\nolimits_n s(n)\phi^*(2^{-p}n - k) \tag{3}$$

where *$*(n)$ is the complex conjugate of the scaling function(n)*. When DWT is applied to the noisy signal $s(t)$ at different level then the speech signal decomposes to approximation and detailed coefficients [3, 11]. The detailed coefficients are obtained by filtering the high frequency component present in the noisy speech signals through high pass filter and the approximation coefficients are obtained by filtering the low frequency component present in the noisy speech signals through low pass filter. The reconstruction of the original speech signal is done by applying the Inverse Discrete Wavelet Transform (IDWT) to the filtered speech signal which is formed by combining the detailed and the approximation coefficients from the last level of decomposition to the first level. Figures 1 and 2, shows the two level wavelet decomposition and reconstruction process in which s is the noisy speech signal, cA1 and cD1 are the first level approximation and detailed coefficients and cA2 and cD2 are the second level approximation and detailed coefficients [2].

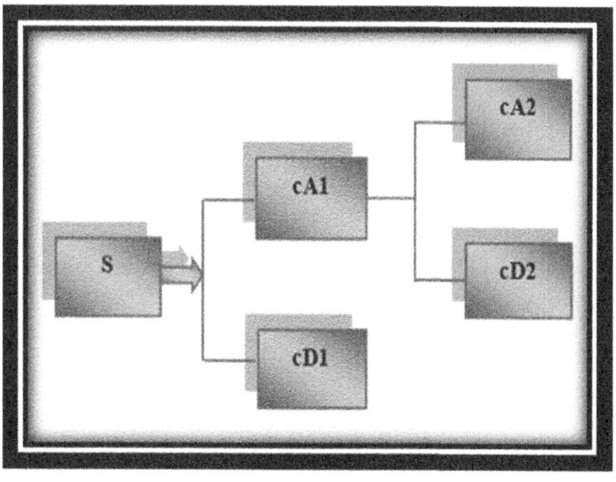

Fig. 1 Wavelet decomposition [2]

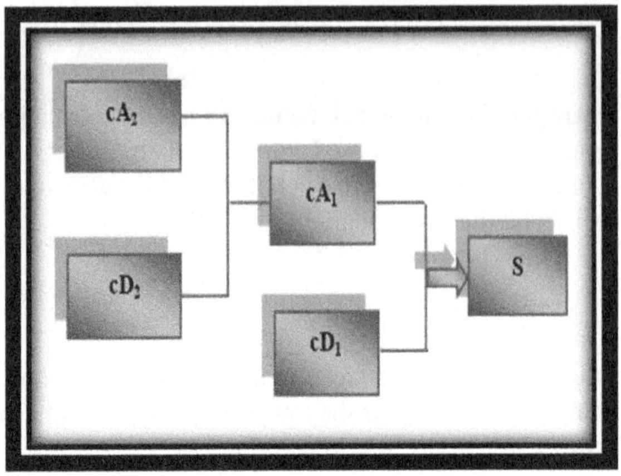

Fig. 2 Wavelet reconstruction [2]

2.2 *Universal Thresholding Based Filtering Method for Pre-processing of Speech Signals*

The noise present in the speech signal is a major issue in the speech signal analysis. In the proposed work, a universal thresholding based filtering technique using DWT has been proposed. The higher frequency components acquired through DWT is having a residual noise that cannot be removed by applying the simple filtering process [9, 12, 13].

2.2.1 Threshold Selection

The universal threshold value T can be evaluated by the equation elucidated below [3]:

$$T = \sigma\sqrt{2\ln(L)} \tag{4}$$

where L is the noisy speech signal sample. The standard deviation σ can be evaluated as [5]:

$$\sigma = \frac{\mathrm{MAD}(|D(n)|)}{0.6745} \tag{5}$$

where MAD is the Median Absolute Deviation and $D[n]$ is the detailed coefficient of noisy speech signal.

2.2.2 Threshold Function

The universal threshold function, hard thresholding has been used in the proposed work. The calculation formula for hard ($H_{m,n}$) threshold function is given below [1]:

$$H_{m,n} = \begin{cases} \omega_{m,n} \left|\omega_{m,n}\right| \geq \mu \\ 0 \left|\omega_{m,n}\right| < \mu \end{cases}, \tag{6}$$

where $\omega_{m,n}$ is the wavelet decomposition coefficient of the noisy speech signal and μ is the threshold value. The threshold value μ is placed to zero and if the value of the coefficients is more than the threshold value then all the coefficients are threshold and this is known as Hard Thresholding.

3 Experimental Setup

The proposed method is performed and evaluated on NOIZEUS database [14]. The database accommodates 30 IEEE sentences contaminated with eight different noises at different SNRs. The noise is added to the sentences from the AURORA database [15] that includes train, babble, car, exhibition hall, restaurant, street, airport and train-station noise. In this experiment the noise from the noisy speech signal is removed using Discrete Wavelet Transform technique. In the proposed methodology the noisy speech signal is decomposed into Approximation and Detailed coefficients by using different types of wavelets like Daubechies, Coiflets, Symlets and Haar wavelet. The detail coefficients are difficult to remove through filters. Therefore, noise suppression in the noisy speech signals is done through hard thresholding [16–18]. The evaluation of the proposed work is done by calculating the SNR by using the mathematical expression as elucidated in Eq. (7) [3, 11]:

$$\text{SNR} = 10 \log 10 \left(\frac{\sum_{m=1}^{L} s^2(m)}{\sum_{m=1}^{L} |s(m) - \hat{s}(m)|^2} \right) \tag{7}$$

where L is the sample size for the filtered speech signal, $s(m)$ is the noisy speech signal and $\hat{s}(m)$ is the clean speech signal.

4 Exploration of Suitable Wavelet Family for Speech Signal Analysis

In the proposed work different Wavelet families has been explored for the suppression of noise from speech signals. From the different family used which are Daubechies,

Table 1 Comparative analysis of SNR values for different wavelet families

Wavelet family	SNR (0 dB) Noisy signal = 2.7767 dB	SNR (5 dB) Noisy signal = 5.7318 dB	SNR (10 dB) Noisy signal = 10.0106 dB
Daubechies	3.7698	6.6212	11.4593
Symlets	3.7627	6.6698	11.3870
Coiflets	3.8136	6.6626	11.5248
Haar	3.0689	5.8379	9.2828

Table 2 Comparative analysis of SNR values for different order of Coiflet wavelet

Order of Coiflet wavelet	SNR (0 dB) Noisy signal = 2.7767 dB	SNR (5 dB) Noisy signal = 5.7318 dB	SNR (10 dB) Noisy signal = 10.0106 dB
coif1	3.3330	6.1603	10.3155
coif2	3.5644	6.4282	10.9109
coif3	3.7176	6.4762	10.9817
coif4	3.7345	6.6274	11.3477
coif5	3.8136	6.6626	11.5248

Symlets, Coiflets and Haar wavelet, Coiflets tends to give the optimal SNR value. The comparative analysis of SNR values for different wavelet family is shown in Table 1. The noise in the signal decreases from 0 dB to 10 dB and the SNR value of the respective signal tends to increase. The order 5 of the Coiflet wavelet gives the optimal SNR value. The comparative analysis of SNR values for different order of Coiflet Wavelet are shown in Table 2. Here it is examined that as the order of the Wavelet family increases, the SNR value increases. The comparative analysis of SNR values for different level of decomposition of noisy speech signal is shown in Table 3. Here it is analyzed that the SNR improves to a certain level of decomposition and then it stops as the sample number decreases in lower sub-bands. So, based on the explored parameters the simulation of result has been done.

Table 3 Comparative analysis of SNR values for different level of decomposition

Level of decomposition	SNR (0 dB) Noisy signal = 2.7767 dB	SNR (5 dB) Noisy signal = 5.7318 dB	SNR (10 dB) Noisy signal = 10.0106 dB
Level 1	3.6323	6.4968	11.4626
Level 2	3.7274	6.5847	11.5248
Level 3	3.8136	6.6626	11.3051
Level 4	3.7429	6.6096	10.8951
Level 5	3.7326	6.5798	10.8382

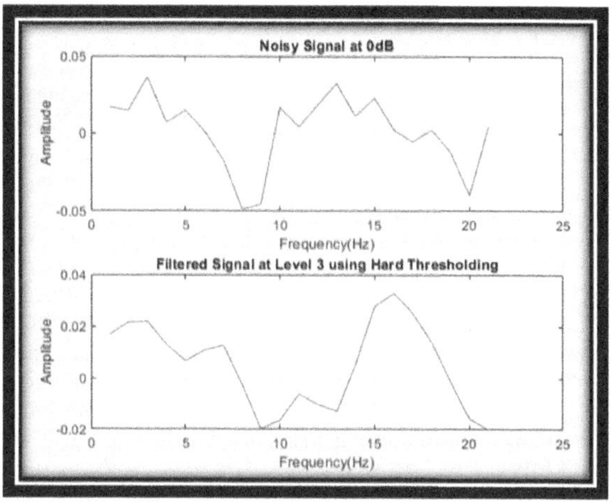

Fig. 3 Noisy and the filtered speech signal. SNR (noisy signal) = 2.7767 dB. SNR (filtered signal) = 3.7698 dB

5 Simulation Results

The speech signals contain the noise which is important to remove as it causes the difficulty in the further processing of the signal. Figure 3 shows the noisy and filtered speech signal. The decomposition of the noisy speech signal is done through DWT at various levels along with a different wavelet family. The hard thresholding is applied by calculating the threshold value T to the coefficients obtained through the decomposition of the noisy speech signal [19, 20]. From the exploration of suitable wavelet family the result is simulated. Therefore, an improved result of pre-processing of speech signal by using the combination of DWT and Hard Thresholding has been obtained. The reconstruction of the noisy speech signal is done through IDWT. The important information present in the reconstructed speech signal is not lost [21, 22].

6 Conclusion

In the proposed work, the pre-processing of the noisy speech signal through the combination of DWT and hard thresholding has been done and the Wavelet Family has been explored to obtain the improved result. Comparative analysis for the best wavelet family, order of the wavelet and the best level of decomposition has been obtained and based on the explored parameters the simulation of the result for the suppression of noisy speech signal has been done.

References

1. Zhong, X., Dai, Y., Dai, Y., Jin, T.: Study on processing of wavelet speech denoising in speech recognition system. Int. J. Speech Technol. **21**, 563–569 (2018)
2. Aggarwal, R., Singh, J.K., Gupta, V.K., Rathore, S., Tiwari, M., Khare, A.: Noise reduction of speech signal using wavelet transform with modified universal threshold. Int. J. Comput. Appl. **20**, 14–19 (2011)
3. Bhowmick, A., Chandra, M.: Speech enhancement using voiced speech probability based wavelet decomposition. Comput. Electr. Eng. **62**, 706–718 (2017)
4. Akkar, H.A.R., Hadi, W.A.H., Al-Dosari, I.H.: A squared-Chebyshev wavelet thresholding based 1D signal compression. Defence Technol. **15**, 426–431 (2019)
5. Babichev, S., Mikhalyov, O.: A hybrid model of 1-D signal adaptive filter based on the complex use of Huang transform and wavelet analysis. Int. J. Intell. Syst. Appl. 1–8 (2019)
6. Akkar, H.A.R., Hadi, W.A.H., Al-Dosari, I.H.: Implementation of sawtooth wavelet thresholding for noise cancellation in one dimensional signal. Int. J. Nanoelectron. Mater. **12**, 67–74 (2019)
7. Bhateja, V., Urooj, S., Verma, R., Mehrotra, R.: A novel approach for suppression of powerline interference and impulse noise in ECG signals. In: Proc. of IMPACT-2013. pp. 103–107. Aligarh, India (2013)
8. Chieng, T.M., Hau, Y.W., Bin Omar, Z., Lim, C.W.: Qualitative and quantitative performance comparison of ECG noise reduction and signal enhancement method based on various digital filter designs and discrete wavelet transform. Int. J. Comput. Digital Syst. **9**, 534–544 (2020)
9. Bhateja, V., Srivastava, A., Tiwari, D.K.: An approach for the preprocessing of EMG signals using canonical correlation analysis. Smart Comput. Inform. **78**, 201–208 (2017)
10. Jakati, J.S., Shridhar, S.K.: Efficient speech de-noising algorithm using multi-level discrete wavelet transform and thresholding. Int. J. Emerg. Trends Eng. Res. **8**, 2472–2480 (2020)
11. Taquee, A., Bhateja, V., Shankar, A., Srivastava, A.: Combination of wavelets and hard thresholding for analysis of cough signals. In: 2018 Second World Conference on Smart Trends in Systems, Security and Sustainability (WorldS4). pp. 266–270. IEEE Press, London (2018)
12. Wang, K., Su, G., Liu, L., Wang, S.: Wavelet packet analysis for speaker-independent emotion recognition. Neurocomputing **398**, 257–264 (2020)
13. Venkateswarlu, S.C., Karthik, A., Kumar, D.N.: Performance on speech enhancement objective quality measures using hybrid wavelet thresholding. Int. J. Eng. Adv. Technol. (IJEAT) **8**, 3523–3533 (2019)
14. NOIZEUS: A Noisy Speech Corpus for Evaluation of Speech Enhancement Algorithms. https://ecs.utdallas.edu/loizou/speech/noizeus/
15. The AURORA Experimental Framework For The Performance Evaluation of Speech Recognition Systems Under Noisy Conditions. https://www.isca-speech.org/archive_open/asr2000/asr0_181.html
16. Verma, R., Mehrotra, R., Bhateja, V.: A new morphological filtering algorithm for preprocessing of electrocardiographic signals. In: Proc. (SPRINGER) of the Fourth International Conference on Signal and Image Processing (ICSIP 2012). pp. 193–201. Coimbatore, India (2012)
17. Bhateja, V., Devi, S.: Combination of wavelet analysis and non-linear enhancement function for computer aided detection of microcalcifications. In: Proc. of International Conference on Biomedical Engineering and Assistive Technologies (BEATs-2010), p. 44. Jalandhar, India (2010)
18. Alabbasi, H.A., Jalil, A.M., Hasan, F.S.: Adaptive wavelet thresholding with robust hybrid features for text-independent speaker identification system. Int. J. Electr. Comput. Eng. **10**, 2088–8708 (2020)
19. Mishra, A., Bhateja, V., Gupta, A., Mishra, A.: Noise removal in EEG signals using SWT–ICA combinational approach. Smart Intell. Comput. Appl. **105**, 217–224 (2018)
20. Srivastava, A., Bhateja, V., Shankar, A., Taquee, A.: On analysis of suitable wavelet family for processing of cough signals. Front. Intell. Comput. Theory Appl. **1014**, 194–200 (2019)

21. Bhateja, V., Taquee, A., Sharma D.K.: Pre-processing and classification of cough sounds in noisy environment using SVM. In: Proc. of 2019 4th International Conference on Information Systems and Computer Networks (ISCON), pp. 822–826. Mathura, India (2019)
22. Vani, H.Y., Anusuya, M.A.: Improving speech recognition using bionic wavelet features. AIMS Electron. Electr. Eng. **4**, 200–215 (2020)

First Experiment to Human Behavior Modeling

T. Edwin Fabricio Lozada, N. Freddy Patricio Baño, Wladimir L. Tenecota, and Jorge A. Benitez

Abstract A situation has been modeled from its components, activities of high semantic level, and these through events of actions of low semantic level. The following describes the modeling of more complex situations to model and is that in this type of modeling, the participation of the fotograma is observed with greater clarity. The proposal is to develop a tool that allows modeling human behavior. The results that may be possible to create a visual tool that can obtain valid conclusions of human behavior.

Keywords Visual tools · Ontology · Human behavior

1 Introduction

The evaluation of the sequences of the activities completed by people needs temporal restrictions. The description of the scene is done by means of the detection of low-level semantic activities and generating a geometric description of the scene (a priori knowledge). Using fuzzy temporal management, fuzzy concepts can be determined that allow for the establishment of relationships between objects in the setting to be studied, for example, one person follows another (Predicate), if they have similar orientation directions or if they are going to the same place, and of course person A

T. Edwin Fabricio Lozada (✉) · N. Freddy Patricio Baño
Carrera de Software, Universidad Regional Autónoma de Los Andes—UNIANDES, Ambato, Ecuador
e-mail: ua.edwinlozada@uniandes.edu.ec

N. Freddy Patricio Baño
e-mail: ua.freddybn@uniandes.edu.ec

W. L. Tenecota
Facultad de Ciencias Humanas de la Educación y Desarrollo Social, Universidad Indoamérica, Ambato, Ecuador
e-mail: wladimirtenecota@uni.edu.ec

J. A. Benitez
Departamento de Ciencias Juridicas, Universidad Técnica Particular de Loja, Loja, Ecuador
e-mail: jabenitezxx@utpl.edu.ec

V. Bhateja et al. (eds.), *Computer Communication, Networking and IoT*, Lecture Notes in Networks and Systems 197, https://doi.org/10.1007/978-981-16-0980-0_45

is behind person B. In this case, verifying the actions of person A and of person B, we can find that there are measurements to be made about the distance or orientation, so between person A and B, there is a difference of distance or orientation called Dif1 and between person B and A (second substantiation), there is a difference of distance or orientation called Dif2 [1], person A has the same direction as person B and is probably following person B. In this work, we use inferences rules to identify Backpack_Theft to form an example of how the mechanisms of classification can be used to identify human behavior. The rules do not consider the real and absolute position of the object in the scene, but rather they analyze the position of an object form the point of view of the camera [2–4] present a work related to the video-vigilance in airports used to manage internal traffic. The problem arises when the video-vigilance produces videos of areas with a high level of traffic, where each image presents a set of moving objects, and it is necessary to obtain the best trajectory to follow, for this, fuzzy logic is used, with the goal of predicting the best routes, based on the heuristic set that describes the characteristics of mobile objects and the scenes associated with them. This type of system is based of a set of rules previously given by an expert or by a simulation of the system in the best condition, for which one of the biggest problems is to find the rules that allow a stable exit from the system in high traffic conditions, in visibility problems (occlusion, rain, etc.) and the management of time so that the transit of moving objects is coordinated. Modeling of situations of high semantic level requires methodological tools that facilitate the task. And that is also a fundamental principle of the framework of semantic management that we propose in this work, facilitating the modeling and reuse of semantic tools. Hence, we have developed a graphic management tool that makes use of the logical capacity of VERL. To facilitate the explanation of the developed software tool, an example is presented below in which a low-level semantic event is graphically modeled, and high-level semantic activities are composed. For Backpack_Theft, the results show that the situation can be inferred. The inference of this situation is based on advances in artificial vision algorithms [5–7]. The problem with this kind of solution is that temporal variables increase during the study, and there is no analysis of dependence between them, as they are treated separately; thus, the temporal treatment of state sequences is not allowed. Examples of these variables (velocity of the walking passer-by, location respecting the place and the placement of the cameras, etc.) can be observed in [8]. In this work, we describe the case studies, in which the expert knows the situation and can detail it. In each of them, it is about modeling the situations of high semantic level Backpack_theft taking advantage of model potentialities. These situations are directly related to the Loiter situation that occurs when a person visits the same places several times at different times, as if walking aimlessly. The situation Backpack_theft also has among its components to loiter, because a person regularly, before committing a theft, wanders around the place. Next, we will detail the modeling of the situations of interest. Since Backpack_theft is common for the case studies, we start modeling it in the inference.

2 Methodology

Model an example of modeling activities that lead to situations of high semantic level. In the framework, we have enough tools that allow us to achieve this goal. The modeling language to be used is VERL, and we use again the data of the backpack to describe in detail the modeling of a composite activity of high semantic level. To use the methodology:

- Detect the event
- Use the rules to infer the event in the ontology
- Infer the new event with the ontology
- Experiment with other events and proposal new components.

3 Experimentation

In this section, we describe the case studies, in which the expert knows the situation and can detail it. Backpack_theft taking advantage of SH's potentialities. These situations are directly related to the Loiter situation that occurs when a person visits the same places several times at different times, as if walking aimlessly. Next, we will detail the modeling of the situations of interest. The task of the knowledge engineer is to translate these definitions to VERL. For this, we use the SH tools:

```
Primitive_Event Backpack_theft(p,z1,z2,t)
  PhysicalObjects:
(p = VERL:Person,z1 = BA:Building_Component, z2 =
BA:Building_Component,t = Time:Period)
  Components:
(Changes_Zone(p,z1,z2,t))
  constraint
  VERL: Repeat(BA:BuildingComponent),t(4)).
```

Code 1
Backpack modeled in VERL (Table 1)

Rule 1
Rule 1 infers the presence activity, which occurs when a person entered the place that was being monitored by the camera. The following rule verifies that the person is in the monitored area (presence), and who repeatedly visits some places. The rule gets the number of times (? countVisits) that a person visits the same place (? in_Places). Also, compare if this number of visits can be considered normal (? NormalVisit). It can be infer that if? countVisits is greater than? normalVisit (Table 2).

Table 1 Code1, generate Rule 1 that allows the inference of the situation

cam_Detects_Person(?camDetects)		∧
camera(?camDetects,?videoCamera)	∧	
at_Instant(?camDetects,?atInst)	∧	
person_At(?camDetects,?Person)		∧
is_In(?videoCamera,?place)		∧
swrlx:createOWLThing(?Presence, "Presence",?atInst)		
->		
presence(?Presence)		∧
person_At(?Presence, ?Person)		∧
at_Instant(?Presence, ?atInst)		∧
in_Place(?Presence, ?place)		

Table 2 Rule 1. Inference of the presence activity from cam_Detects_Person

Loitering	∧	
time:during(?atInst, ?monitoring_time)		∧
swrlx:createOWLThing(?loitering, "Backpack", ?atInst1)		
->		
loitering(?loitering)		∧
person_At(?Backpack, ?Person)		∧
in_place(?loitering,?inPlace)		∧
at_instant(?loitering, ?atInst1)		

4 Conclusions

Our proposal is make a methodology to identify an alarm human behavior, Back-Pack_Theft. The results are not exactly same to human behavior, and for this reason, we select to fuzzy logic algorithm to representing the alarm behavior. The fuzzy logic is a mechanism to generate alarms in different places. We described two experiment to representing a suspicious person and theft. The rules showed that our inference generate alarms. The next studies, we will make a mechanism that obtains a minimal error and experiment in different places and with more persons.

References

1. Fernández, C., González, J.: Ontology for semantic integration in a cognitive surveillance system. Seman. Multimedia, 4816:260–263 (2007) (Springer, Ed.)
2. Albusac, J.: Vigilancia Inteligente: Modelado de Entornos Reales e Interpretación de Conductas para la Seguridad. Tesis de Master, Castilla-La Mancha, España (2008)

3. Gomez, A.H., Arias, T.S., Torres, P., Sanchez, J., Hernandez, V.: Emotions analysis techniques: their application in the identification of criteria for selecting suitable Open Educational Resources. In: International Conference on Interactive Collaborative and Blended Learning, Mexico (2015)
4. Molina, J., Perez, O.: Appying Fuzzy Logic in Video Surveillance Systems. Universidad Carlos III, Madrid (2005)
5. Williem, A., Vamsi, M., Boles, K., Wageeh, W.: Detecting uncommon trajectories. Digital Image Comput. Tech. Appl. 398–404 (2008)
6. Masoud, O., Papanikolopoulos, N.: A novel method for tracking and counting pedestrians in realtime using a single camera. IEEE Trans. Veh. Technol. 1267–1278 (2001)
7. Isard, M., Blake, A.: Condensation-conditional density propagatin for visual tracking. Int. J. Comput. Vision 5–28 (2008)
8. Barake, B., Saddik, A.: Towards and intelligent tele-surveillance system for public transport areas: fuzzy Logic based camera control. IEEE Xplore 87–90 (2008)

Methodology for the Selection of Technological Tools: Second Round

Freddy G. Salazar Carrillo, T. Edwin Fabricio Lozada, N. Freddy Patricio Baño, and Jorge A. Benitez

Abstract The research work describes the processes executed by the author of the research work from the statement of the problem to the generation of a product to positively satisfy the problem through theoretical research, and the implementation of methodology for the selection of technological tools in the process of foreign language learning in order to provide a solution to the deficiency in foreign language learning caused by the selection of ICT tools. The proposal of the investigative work is carried out through a set of phases which together in a systematic way make up the solution to the detected problem.

Keywords Language · Methodology · Tools

1 Introduction

ICTs are developed for use in classrooms as complete instruments that, together with traditional materials, facilitate the teaching and learning process as individual language courses that users administer independently and individually. Taking this into account, it is necessary to mention that there has been a massive development of applications and programs used to learn languages, more specifically, English [1]. These applications present their users with interactive and innovative materials

F. G. Salazar Carrillo
Centro de Posgrados, Universidad Técnica de Ambato, Ambato, Ecuador
e-mail: fg.salazar@uta.edu.ec

T. Edwin Fabricio Lozada (✉) · N. Freddy Patricio Baño
Carrera de Software, Universidad Regional Autónoma de Los Andes–UNIANDES, Ambato, Ecuador
e-mail: ua.edwinlozada@uniandes.edu.ec

N. Freddy Patricio Baño
e-mail: freddy.bano@institutotraversari.edu.ec

J. A. Benitez
Departamento de Jurisprudencia, Universidad Tecnica Particular de Loja, Loja, Ecuador
e-mail: jabenitezxx@utpl.edu.ec

V. Bhateja et al. (eds.), *Computer Communication, Networking and IoT*, Lecture Notes in Networks and Systems 197, https://doi.org/10.1007/978-981-16-0980-0_46

related to specific topics and organized in units, creating a complete language course and including from most elementary to the most advanced content. The use of these language applications provides students with opportunities to learn English on mobile devices, regardless of place and time. In addition, the interface, content organization and versatility of these programs allow users to manage their learning process, obtain a record of their progress and acquire the English language independently. For Petr Škuta [2], the design principles are means to increase the motivation to learn, this makes interesting the fact that part of the mechanical play helps to achieve more than a single principle of play, while others are strictly for one purpose, to demonstrate the actual use of gamification, the work will analyze a learning environment that takes advantage of the gamification concept and uses it to motivate, and the Duolingo Web site was chosen for an analysis in which the gamification design and game mechanics are elaborated in the examples of the application thus obtaining the attention of those who use this application. For his part, Rodríguez Atencio [3] makes reference to the effects of techno-culture in the educational field, its effects, transformations and experimentations, and it is also stated that urgently a change due to the rapidity of the transformation of technology in relation to intellectual and collective life to propose pedagogical approaches to achieve participation, equity and recognition of the digital native.

In the research carried out by Marulanda et al. [4] on the information society, argues that it has been generated a re-configuring effect of the possibilities opened by ICT, also refers to the information society as a form of social organization in which processing and transmission of information become the primary sources; these technological conditions are born, by different institutional, organizational and human factors. ICTs and their applications generate in society new ways of informing, interacting, having fun and learning, thus, asserting that the best technology is not the one that is avant-garde, but the best one adapts to the needs where it is applied. Thus, currently, the focus groups in most cases are young people since they are the ones who currently use and appropriate technology the most. For Castillo, Bueno [5], "since the beginning of the new millennium, European and national regulations call for education in a competence framework that takes into account digital skills to train citizens of a future that seems dominated by new technologies." That is why currently, it seeks to change said work methodology trying to raise awareness in the effective use of technology, but always depending on the technological endowment, always taking care that the technology does not transform into the traditional way and one step would be through training and research with teachers and students. According to Hernández Rodríguez and Bautista Maldonado [6], the incorporation of ICT in classrooms is an issue with a large number of factors to consider, including national policies and laws, as well as the infrastructure that a school requires, and the competencies that the teacher must be able to incorporate the use of technology in an appropriate way and make the necessary adjustments and correct curriculum. In the publication made by Castells [7] on the Internet, it is proposed as an information and communication technology that allows society to interact as a whole, knowing the conceptualization of the Internet from the point of view of several authors, it can be mentioned that the Internet is a great trunk of information resources considered as a network of computer networks capable of communicating with each other. For this reason, technology is much more than a technology, and it is also a means of communication, interaction and social organization. When the Internet was considered "novelty," society considered it interesting, although in a minority, for other people considered Internet users it was considered as the peak of its evolution. But this has now changed dramatically to capture a record for the planet's population in terms of Internet use, according to the researcher, he presents the following report referring to the most developed societies, which in the societies in our contextual framework where the penetration rates indicate it will be around 75% or 80%. As can be seen since the appearance of the Internet, the rate of use grew rapidly where most of the population has accessed the Internet at least once. For Hernández Martín and Martín de Arriba [8] given the growing number of digital classrooms for educational services, it slowly generates a change of vision of educational institutions and therefore of all the members participating in it: teachers, students and authorities. The tools of the Web are currently being used in different social or educational contexts, which is why it is also sought that methodological approaches take a clearly collaborative environment. All this produces the birth of a new teaching trend with its respective collaborative learning methodologies mediated by ICT, regardless of the educational level where it is desired to implement, since nowadays educational innovation is necessary in different areas. In this job, the Duolingo, Memrise, Bussu, Babbel tools are easy to use, they have a version for the Web and for mobile devices, they also present a user-friendly interface; however, these characteristics identified in the analyzed applications are not enough to determine the appropriate tool for the foreign language learning process. In this work, we propose a methodology for selecting the appropriate ICT tools to support foreign language learning and implementing a methodology for selecting tools for learning.

2 Methodology

For the application of the proposed methodology for the selection of technological tools in the process of learning the foreign language-English with ICT, a set of phases that are part of a cycle for the acceptance of a technological tool for foreign language learning is proposed—English.

The GITICIE methodology (management and integration of information and communication technologies in educational institutions): The methodology presented by Soto Jiménez et al. [9] is based on seven ICT axes that are: ICT pedagogical use axis, ICT training axis, content management axis, strategy and leadership axis, strengthening and accompaniment axis, evaluation and monitoring axis, infrastructure axis and raises five moments such as they are: ICT orientations, characterizing ICTs, planning of ICT improvements, execution and monitoring, assessment and adjustment of actions that begin with raising awareness about the benefits of integrating ICTs, subsequently developing a characterization of the institutional needs, which underpin the planning of ICT improvements, designed and implemented by the educational communities, in charge of monitoring, evaluating and adjusting actions repeatedly until optimal results are achieved.

The analysis comprises a process and a separation which is applied according to the field in which it is applied, this process consists of investigating and comparing the decomposition of a whole into its parts.

Step 1 The teacher must search for ICT tools for learning the foreign language-English.

Step 2 The teacher must select at least two ICT tools for foreign language learning-English.

Step 3 The teacher must ensure that the selected tools allow experimentation with them for free.

3 Experimentation

Experimentation is a questionable process in which one must work with variables that lead to a systematic and complex process, and the researcher checks if a theory is real, for this the variables that participate in the process are subjected to experiments. In this phase, the teacher must perform the validation of each previously selected tool, by applying the ICT tools characterization matrix.

4 Conclusions

The success of the application of the methodology (see Table 1) for the selection of suitable technological tools to support the foreign language-English learning process, proposed in the present work; it requires the commitment and joint work of authorities and especially of students and teachers willing to be trained in the use of ICT tools aimed at promoting learning to achieve the educational objectives set.

Table 1 Methodological application matrix

Features	Excellent	Very good	Okay	Regular
Fidelity	The person in charge of the site or company that manages it is recognized, it is also recognized that the information is verifiable and true	The person in charge of the site or company that manages it is recognized, and the information is verifiable and true	The person in charge of the site or company that manages it is recognized, and the information is not fully verifiable and true	The person in charge of the site or the company that manages it is unknown, and the information is not verifiable
Upgrade	The update date is indicated and is less than 1 year	The update date is indicated and is less than 2 years	The update date is indicated and is less than 3 years	The update date is not indicated
Navigability	The structure is very neat and clear, and it greatly facilitates the orientation for use	The structure is neat and if it facilitates its use	The structure is not very organized and does not facilitate its use	The structure or is clear and tends to make it difficult to use
Organization	The content organization follows a very clear and logical sequence, easy to understand what needs to be done	The contents follow a sequence	Most of the contents follow sequence, although some do not	The contents have no sequence
Content selection	The contents are relevant and significant for the level and characteristics of the students	The contents contribute to the level of the students	Most of the contents contribute to the level of the students	Most content does not contribute to the level and characteristics of the students
Readability	Combination of the font, colors, images, graphics and text make the use of resources appropriate, attractive and easy	The combination of colors, images, graphics and text helps proper use	The combination of colors, images, graphics and text confuses the proper use	The combination of colors, images, graphics and text does not aid proper use
Usability	The software is easy to use	The software is easy to learn to use	The software is difficult to learn to use	The software is not easy to use

(continued)

Table 1 (continued)

Features	Excellent	Very good	Okay	Regular
Utility	The software is 100% adaptable to its class	The software is adaptable to its class, but with certain constraints	The software can be used in class, but not constantly	The software can be used occasionally in class
Intention of use	Would use the software in your class	I would try to use the software in class	I would use it a few times	Would use the software in your class occasionally
Group management	The software allows to have control and monitoring of the work group	The software allows to have control of the work group	The software allows you to monitor the work group	The software does not allow to have control and monitoring of the work group
Acknowledgments (gamification)	The software allows to give recognitions at the end of activities	The software allows to give recognitions in the majority of activities	The software allows to give recognitions in few activities	The software does not allow to give acknowledgments at the end of activities
Competitiveness	The software allows you to create educational competitiveness	The software in certain activities allows to generate competitiveness	The software offers very few possibilities to generate competitiveness	The software does not allow to create educational competitiveness
Means	The software presents videos, images, audios and text	The software presents some videos, images, audios and text	The software features few videos, images, audios and text	The software presents few videos, images, audios and text
Tracing	The software allows monitoring of academic progress and student participation	The software allows monitoring of academic progress and student participation in certain activities	The software allows monitoring of academic progress and student participation in a few activities	The software does not allow tracking of academic progress and student participation

References

1. Martínez Gómez, M.: Autonomous learning through the use of mobile applications: Duolingo, Babbel and Busuu, Universitad Jaume (2016)
2. Petr Škuta, K.K.: The inclusion of gamification elements. In: International Scientific Conference on Distance Learning in Applied Informatics, pp. 421–429 (2016)
3. Rodríguez Atencio, M.: Rethinking pedagogical work from techno-culture. Saber **25**(3) (2013)
4. Marulanda, C.E., Giraldo, J., López, M.: Access and Use of Information and Communication Technologies (ICTs) in Learning. The Case of the Pre-university Youth in Caldas, Colombia. Colombia (2014)

5. Castillo Bueno, C.M.: The inclusion of information and communication technologies (ICT) in the school methodology. Universidad de Cantabria (2016)
6. Hernández Rodríguez, H., Bautista Maldonado, S.: ICT in the Mexican educational system. Electron. J. Technol. Educ. Soc. (2017)
7. Castells, M.: Internet Y La Sociedad Red (2013)
8. Hernández Martín, A., Martín de Arriba, J.: Conceptions of Non-University Teachers on Collaborative Learning with ICT. UNED (2017)
9. Soto Jiménez, J.C., Franco Enzuncho, M.L., Giraldo Cardozo, J.C.: Development of a methodology to integrate ICT (Information and communication technologies) in the EIs (Educational Institutions) of Montería. J. Inst. Educ. Stud. Universidad del North (2014)

Karger's Randomness of Graphs to Detect XSS Attacks on IoT Devices

E. Arul and A. Punidha

Abstract Cross-webpage scripting (otherwise called XSS) is a web security helplessness that enables an aggressor to bargain the associations that clients have with a defenseless application. It enables an assailant to go around a similar inception approach, which is intended to isolate various sites from one another. Cross-site scripting vulnerabilities typically enable an aggressor to take on the appearance of an injured individual client, to do any activities that the client can perform, and to get to any of the client's information. On the off chance that the injured individual client includes advantaged access inside the application, at that point, the aggressor may have the option to deal with the majority of the application's usefulness and data. Cross-website scripting works by controlling a powerless site so it returns malignant JavaScript to clients. At the point when the malignant code executes inside an injured individual's program, the aggressor can completely bargain their cooperation with the application. In this paper, the proposed Karger's randomness of graphs to detect cross-site attacks based on the max-stream min-cut on DOM while IoT devices exchanging information through webpages. The resultant algorithm shows effective and improves the detection percentage over the cross-site attack. The proposed algorithm finds true positive of 97.69% and false positive of 0.03%.

Keywords XSS · Karger's randomness · Max-stream · Min-cut · Scripting vulnerabilities · JavaScript · DOM · IoT

E. Arul (✉)
Department of Information Technology, Coimbatore Institute of Technology, Coimbatore, Tamil Nadu, India

A. Punidha
Department of Computer Science and Engineering, Coimbatore Institute of Technology, Coimbatore, Tamil Nadu, India

V. Bhateja et al. (eds.), *Computer Communication, Networking and IoT*, Lecture Notes in Networks and Systems 197, https://doi.org/10.1007/978-981-16-0980-0_47

1 Introduction

A cyber-attack is a pernicious endeavor to adventure, harm, as well as addition unapproved access to sites using IoT devices, PC frameworks, or systems. Cyber-attacks are frequently brought about by malware, which is the umbrella term used to portray programming made for malignant purposes. You might be most acquainted with PC malware, for example, Trojan infections and Spyware, which can be utilized to recover delicate information from a PC or even assume responsibility for the framework [1]. Correspondingly, site malware exploits sites for their assets, traffic, or guest data. You have likely found out about prominent cyber-attacks in the features after a noteworthy information break, for example, the Equifax rupture in 2018. While enormous scale assaults on big business associations are the most generally advertised, independent companies can be similarly as defenseless and focused by cybercriminals. Truth be told, about 60% of private companies have detailed being a casualty of a cyber-attack [2].

XSS vulnerabilities are seen as less risky than for instance SQL injection vulnerabilities. Outcomes of the capacity to execute JavaScript on a website page may not appear to be desperate from the outset [3]. Most Internet browsers run JavaScript in a firmly controlled condition. JavaScript has restricted access to the client's working framework and the client's documents [4]. Be that as it may, JavaScript can in any case be risky whenever abused as a major aspect of malignant substance: Pernicious JavaScript approaches every one of the items that the remainder of the page approaches. This incorporates access to the client's treats. Treats are frequently used to store session tokens. In the event that an aggressor can get a client's session treat, they can mimic that client, perform activities in the interest of the client, and access the client's touchy information [5].

JavaScript can peruse the program DOM and make subjective alterations to it. Fortunately, this is just conceivable inside the page where JavaScript is running. JavaScript can utilize the XMLHttpRequest article to send HTTP demands with subjective substance to discretionary goals. JavaScript in present day programs can utilize HTML5 APIs [4]. For instance, it can access the client's geolocation, webcam, amplifier, and even explicit documents from the client's record framework. The vast majority of these APIs require client pick in, yet the assailant can utilize social building to circumvent that constraint [6].

The abovementioned, in mix with social designing, enable culprits to draw off cutting edge assaults including treat robbery, planting Trojans, keylogging, phishing, and data fraud. XSS vulnerabilities give the ideal ground to raise assaults to progressively genuine ones. Cross-site scripting can likewise be utilized related to different kinds of assaults, for instance, cross-site request forgery (CSRF) [1, 5]. There are a few sorts of cross-site scripting assaults: put away/constant XSS, reflected/non-relentless XSS, and DOM-based XSS. You can peruse increasingly about them in an article titled types of XSS. There are two phases to a normal XSS assault [7].

To run noxious JavaScript code in an injured individual's program, an assailant should initially figure out how to infuse malignant code (payload) into a page that the unfortunate casualty visits. From that point forward, the injured individual must visit the website page with the malevolent code. On the off chance that the assault is aimed at specific unfortunate casualties, the aggressor can utilize social designing or potentially phishing to send a pernicious URL to the person in question.

2 Related Work

A. Abusnaina analyzes the use of CFG-based features and image recognition systems is commonly discussed in the field of IoT threat identification [8]. The key purpose of this analysis is to analyze the power of certain prototypes toward opponent training. They also developed two solutions to IoT tech, which rejects the development of off-shelf solutions and GEA techniques. In opposing way of training, we analyze eight separate methods of learning against the adverse, in order to oblige the paradigm to misclassify. The GEA method attempts to maintain the purpose and convenience of the opponent experiment by keeping with a friendly test carefully in a harmful experiment. Complex tests are carried out with the aim of testing the efficacy of the suggested system, which illustrate that technologies of off-shelf attacks can obtain an error sensitivity of 99%. In comparison, they find that certain IoT security patches can be misclassified as harmless in the GEA methodology. The results of the studies show the need for other comprehensive analysis techniques toward opponent training, even apps which, unlike CFG, really are not simple to handle. As the methodology questioned in this research is commonly also in other implementations utilizing graphs, its ramifications are very general.

A. Kumar's research has focused on the pervasiveness of cloud computing [9]. After the 2016 activate-based network attacks that affected connected devices, he has created a bunch of new ransomware with leaking programming language from Mirai and attacking IoT devices, such as Reaper, memory loss, Masuta and several more. Rather than opening Tcp outlets like ransomware, these ransomware use program bugs that render it tougher to interrupt utilizing established packet sniffers approaches. The whole study introduces EDIMA, the decentralized lightweight approach that will be used throughout inspecting/inundating or even during an intrusion to identify IoT ransomware system operations in large-scale networks (e.g., ISP, business channels). EDIMA uses a congestion control deep learning model, a vector repository for stream communication functionality, a protocol as well as an additional sub-sampling unit stream. They measure EDIMA's success in identification by tests in development kits and discuss the data.

Zhongliu Zhuo sites fingerprinting assaults can uncover the beneficiary in unknown systems and cause a potential risk to clients' protection. Past examinations center more around recognizing singular pages [10]. They likewise disregard the hyper-interface progress data, since it instigates extra "clamor" to group the unique page. In any case, it is a typical situation that clients surf a site by clicking hyper-connects on the site page. In this paper, we propose a site demonstrating technique dependent on profile hidden Markov model (PHMM) which is broadly utilized in bioinformatics for DNA sequencing examination [11, 12]. Our procedure expressly represents conceivable hyper-interface advances made by clients when fingerprinting an objective site, and along these lines can work in a more practical condition than existing techniques. Utilizing SSH and Shadowsocks, author gathers different datasets and direct broad assessments.

Likewise demonstrate that our methodology could work both in site page and site recognizable proof in a shut world setting [6]. The trial results exhibit that our site fingerprinting is more precise and powerful than existing methods. A site fingerprinting assault in light of the PHMM that would deanonymize the collector [4]. Author actualized the proposed technique on two unknown systems: Shadowsocks and SSH. To apply the succession investigation strategies from bioinformatics to display the site page and site fingerprints, we understood two conceivable types of arrangement portrayal techniques (stream and parcel arrangement) just as one symbolization conspire [13].

3 Delineation of Karger's Randomness of Graphs to Detect XSS Attacks on IoT Devices

Complex graph systems, generally regardless of whether an exceptionally single edge or wire is disengaged, an enormous bit of the system can be down a circumstance definitely slow down the process of finding cross side attacks [14]. There are by and large bunches DOM construction from hyper test markup (HTML), extensible markup language (XML), unevenly appropriated portions, inside systems. While managing detachment of these portions or "bunches" proficiently, search for edges between them with the end goal that the quantities of edges are minimum. It can apply a similar contention to different fields, for example, "DOM segmentation." On the off chance that you need to break a DOM into sections to find the unwanted details cross-site attacks like adware attached, most straightforward path is to break joins with low fondness, for example, least number of cuts.

```
Complex Pseudo code
Input to on-demand anomaly detection method: data
Application data set with R features output-an array of R classification models of
each webpage's
For each feature where i=0,1,2....n to perform for webpage each
 If(.js||.vb||.xss)
   Parse(preprocess)
  Unzip(.class||tag->all available DOM types)
If(.class)
create (node) .class tag file
      identify all statements
      convert into array of tables
for each application maintain separate row in database table
     go to Karger's pseudo code graph frame
  if
      Invalid application malicious node found
else
   file cleaned, load webpage and exist
end for each
          return.
```

Considering a webpage graph work is the accumulation of vertices, edges, and faces that make up an object, breaking pictures joins with low partiality can give you separate unmistakable DOM leafs [35]. As you see, there are some genuine min-cut issues. To manage them a customary calculation is "max-stream min-cut." It is deterministic, which means when apply this algorithm, DOM gets the correct min-cut accuracy more. While approaching DOM cut, considering "max-stream min-cut" other side assurance accompanies a cost is high: it is extremely moderate.

In the runtime, the best case scenario is

$$\text{bestcase} = O(\text{number of vertices} * \text{number of edges squared})$$

so about a request for n^3 time multifaceted nature.

There are numerous sharp algorithms, which are quick with a runtime of $O(n^2)$, yet none of them, even predominant ones, as to think, is as richly planned as Karger's calculation [1], which use "arbitrary testing" in a diagram webpage DOM construction and cut issue. Just like a Monte Carlo calculation, Karger's algorithm offers an answer with a certain (yet little) likelihood with much quicker approach than max-stream min-cut calculation over the dense system like webpage's. In this way, the state we have an undirected diagram $G = (V, E)$. So as to cut an associated chart G into two associated sub-graphs in as few cuts as could be expected under the circumstances, the algorithm essentially utilizes withdrawal of an edge. To actualize it, one should pull together two vertices on different sides of the edge [15].

Execute the algorithm in Karger's pseudocode as pursues:

```
Let G=(V,E)
 While |V|>2:
          Pick an edge e from E uniformly at random
          Merge or Contract e in a single vertex
          Remove all self–loops
   Return the only cut left with 2 final vertices.
```

In a webpage DOM with n vertices, a solitary keep running of Karger's calculation gives the outcome with a triumph likelihood of 1/(n pick 2) and a runtime of $O(n^2)$. Not awful, however, one should be fortunate to discover the min-cut edge in the principal run [16]. The likelihood of accomplishment can be expanded by doing numerous runs. On the off chance that we run Karger's calculation n^2 times, the likelihood of disappointment that do not get a base cut is $1/e$. A dependable it is not dependent on the number of vertices. In any case of any number of pages with child's nodes, on the off chance that run [17].

Karger's calculation $n^2 \ln(n)$ times, this is viewed as the ideal worth, at that point $P(\text{fail}) \leq 1/n$, and $P(\text{success}) \geq 1 - 1/n$.

Which means as you get greater cut in the DOM, the likelihood of accomplishment of Karger's calculation shows signs of improvement [18]. In this way, in a chart with 50 hubs, in the event that run the calculation around multiple times, the likelihood of accomplishment is 98%. Such a decent achievement pace of 98% accompanies an expense, however, the running time is polynomial in n (the quantity of vertices) and m (the quantity of edges), which is entirely moderate: $O(n^2m)$. A somewhat adjusted rendition, Karger-Stein calculation, has double the achievement likelihood as the first calculation of finding DOM cutting to find unwanted elements in webpages [19].

4 Experimental Results and Comparison

Style class with script file and malicious class tag file affects other class tag file by making unrelated activities like sending messages to all webpages opened in the browser or by keeping separate cookie to attach, changing the permission setting of our browser without asking for permission like saving our user login information of net banking, user accounts like Gmail, Facebook, etc., making call to any recent contacts, searching our browsing history and store activity log unknowingly. Due to this activity mobile performance degrades and results in unauthorized access of our personal by attacker and results in breaching of our personal data. Direct testing for reflected and put away XSS typically includes presenting some straightforward one of a kind info, (e.g., a short alphanumeric string) into each passage point in the application; recognizing each area where the submitted information is returned in HTTP reactions; and testing every area exclusively to decide if reasonably created information can be utilized to execute subjective JavaScript [20].

Testing by applying Karger's randomness of graphs on extracted tuples from dataset. DOM-based XSS emerging from URL parameters includes a comparable procedure: setting some straightforward interesting contribution to the parameter,

Table 1 Permission and types

User activity	Authorization
Date activity	READ_CALENDAR,WRITE CALENDAR
User information	READ_ CONTACTS
	WRITE_ CONTACTS
	GET_ ACCOUNTS
Internal mobile operation	READ_memory_STATE
	PROCESS_STATE
	PROCESS_OUTGOING_PORTS

utilizing the program's designer apparatuses to scan the DOM for this information, and testing every area to decide if it is exploitable. Be that as it may, different sorts of DOM XSS are more diligently to distinguish. To discover DOM-based vulnerabilities in non-URL-based info, (e.g., document. Cookie) or non-HTML-based sinks (like setTimeout), there is not a viable alternative for investigating JavaScript code, which can be amazingly by better performance max-stream min-cut.

There is also a chance to fully disable our network port, unlimited usage of Internet, publishing our gallery images, videos, documents and private information in public network, etc. So our proposed method maintains activity log1 which is in like Table 1 and initially contains information about application when installing in our mobile. Later on if there is any updating in application that is also maintained as activity log2 and so on. There is an chance that attacker can add malicious code in software updating also. If there is any deviation between activity log1 and log2, we can easily identify our application is affected by malicious code. For analysis purpose various malicious scripts (JavaScript, VB Scirpt, stored cross-site scripting) are taken and test against with some existing work and result in Table 2.

Table 2 Comparisons of proposed Karger's randomness of graphs with other algorithms

Comparisons	Number of malware APK are detected	TP ratio (%)	FP detected	FP ratio (%)
A. Abusnaina	635	91.63	75	0.07
A. Kumar	609	87.87	95	0.09
Proposed Karger's -randomness of graphs	677	97.69	32	0.03

Total number of malware script file taken for analysis: 693
Total number of normal file taken for analysis: 970

5 Conclusion and Future Work

IoT machines or personal programs have been permitted to use the proposed method of identifying unauthorized scripts on websites. It gathers up-to-date features from android web server archives. It overcomes the disadvantages of static and dynamic cross-site attack detection by measuring the features of each webpage by very reliable and frequency of occurrence. It also checks the required state before writing or uploading requests from websites, whether or not they are currently requested. The results indicated a high true positive (97.69%) and less false positive (0.03%) approach to the approaches in use. Using Karger's randomness max-stream min-cut over the server hybrid data center, IoT webpages would like to introduce security for cloud data storage access in future.

References

1. Zhang, T., Antunes, H., Aggarwal, S.: Defending connected vehicles against malware: challenges and a solution framework. IEEE Internet Things J. **1**(1), 10–21 (2014)
2. Suo, X., Han, X., Gao, Y.: Research on the application of honeypot technology in intrusion detection system. In: IEEE Workshop on Advanced Research and Technology in Industry Applications (WARTIA) ON Ottawa, pp. 1030–1032 (2014)
3. Arul, E., Manikandan, V.: Malware detection using higher order statistical parameters. In: Communications in Computer and Information Science, vol. 679. Springer Singapore, pp 42–56. https://doi.org/https://doi.org/10.1007/978-981-10-3274-5_4
4. https://contagiodump.blogspot.in/2010/11/links-and-resourcesfor-malware-samples.html
5. Yin, M.P.P., Suzuki, S., Yoshioka, K., et al.: IoTPOT: analyzing the rise of IoT compromises. In: Usenix Conference on Offensive Technologies. USENIX Association, 9 (2015)
6. http://link.springer.com/chapter/https://doi.org/10.1007/978-981-10-3274-5_4
7. https://tuts4you.com/download.php?list.89
8. Abusnaina, A., Khormali, A., Alasmary, H., Park, J., Anwar, A., Mohaisen, A.: Adversarial learning attacks on graph-based IoT malware detection systems. In: 2019 IEEE 39th International Conference on Distributed Computing Systems (ICDCS), Dallas, TX, USA, pp. 1296–1305 (2019). https://doi.org/https://doi.org/10.1109/ICDCS.2019.00130
9. Kumar, A., Lim, T.J.: EDIMA: early detection of IoT malware network activity using machine learning techniques. In: 2019 IEEE 5th World Forum on Internet of Things (WF-IoT), Limerick, Ireland, pp. 289–294 (2019). https://doi.org/https://doi.org/10.1109/WF-IoT.2019.8767194
10. Li, Z.-Y., Tao, R., Cai, Z.-H., Zhang, H.: A web page malicious code detect approach based on script execution. In: 2009 Fifth International Conference on Natural Computation, pp. 308–312 (2009). https://doi.org/https://doi.org/10.1109/ICNC.2009.363
11. Zhuo, Z., Zhang, Y., Zhang, Z.-l., Zhang, X., Zhang, J.: Website fingerprinting attack on anonymity networks based on profile hidden markov model. IEEE Trans. Inf. Forens. Secur. **13**(5), 1081–1095 (2018). https://doi.org/https://doi.org/10.1109/TIFS.2017.2762825
12. Hallman, R., Bryan, J., Palavicini, G., et al.: IoDDoS—The internet of distributed denial of service attacks (2017)
13. Paik, I.: Improved Malicious Code Classification Considering Sequence by Machine Learning, pp. 1–2. IEEE ISCE (2014) 1569953651
14. Lee, C., Shen, C., Sahin, G., Choi, K.: A novel and scalable communication-history-based Knapsack authentication framework for IEEE 802.11 networks. In: Communications and Network Security, pp. 44–52. IEEE (2015)

15. Churi, P.P., Bondre, S., Gavankar, N.: Honey patterns: recognizing pattern based attacks on websites. Int. J. Comput. Appl. **161**(9), 8–11 (2017)
16. Li, D.: Internet of Things (IoT) Attacks. https://www.infosecwriters.com/Papers/VEllsworth_IoT.pdf
17. Arul, E.: Deep nonlinear regression least squares polynomial fit to detect malicious attack on IoT devices. J. Amb. Intell. Hum. Comput. (2020). https://doi.org/10.1007/s12652-020-02075-y
18. Arul, E.: Hypervisor injection attack using X-cross API calls (HI-API attack). J. Amb. Intell. Hum. Comput. (2020). https://doi.org/10.1007/s12652-020-01837-y
19. Dongxia, L., Yongbo, Z.: An intrusion detection system based on honeypot technology. In: 2012 International Conference on Computer Science and Electronics Engineering (ICCSEE), Vol. 1, pp. 451–454. IEEE (2012)
20. Chakraborty, N.: Intrusion detection system and intrusion prevention system: a comparative study. Int. J. Comput. Bus. Res. (IJCBR) **4**(2), 1–8 (2013)

Security Attacks on RFID and their Countermeasures

Anuj Kumar Singh and B. D. K. Patro

Abstract Radio frequency identification (RFID) has become very popular technique in the modern computing environment due to its ability to track moving objects. Moreover, with the growth of Internet of things, the usage of RFID has increased largely. But, due to the less computational capacity and wireless nature of communication of RFID many attacks have been devised by the attackers to compromise the security of RFID. This paper explores the attacks which can be attempted over the RFID system and also recommends the possible countermeasures to thwart those attacks. Specifically, the attacks on RFID have been divided into physical attacks, channel attacks, and system attacks. The physical attacks on RFID have been further categorized as disabling, cloning, tampering, and reverse engineering. Snooping, skimming, jamming, eavesdropping, and replay attack are the attacks that come under channel attacks. And system attacks includes password cracking, spoofing, tracking, and denial of service. The brief explanation of each specific attack along with the possible defenses has also been presented in this paper.

Keywords RFID · Security attacks · Countermeasures

1 Introduction

In the present computing scenario, automation of tasks and process has become fundamental need since it reduces errors and saves time at the same time resulting huge benefits in productivity. Automatic identification and data capture (AIDC) refers to the methods and techniques that automatically identifies objects, gather the required data from them, and enter the gathered data to the computer system without involving a human. Many AIDC techniques have been developed from time

---

A. K. Singh (✉)
Amity University Haryana, Gurugram, India
e-mail: aksingh@ggn.amity.edu

B. D. K. Patro
Rajkiya Engineering College, Kannauj, India

V. Bhateja et al. (eds.), *Computer Communication, Networking and IoT*, Lecture Notes in Networks and Systems 197, https://doi.org/10.1007/978-981-16-0980-0_49

to time including smart cards, bar codes, magnetic stripes, radio frequency identification (RFID), biometrics, etc. But RFID has now become most prominent out of these techniques because it is has the ability to track moving objects. Khattab et al. [1] have pointed out that against parameters of readability by machine, data density, readability by people, range, effect of moisture, cost, reading speed, and effect of sight distraction, RFID performs best in comparison with the other AIDC techniques.

1.1 RFID Communication Scenario

The two major components of a RFID communication system are RFID tags also called as transponders and the RFID receiver also known as transceiver. The other components of an RFID system are the antenna, the server, and the database. The RFID communication scenario which is based on context, capture, and control has been shown in Fig. 1. Context refers to the environment in which RFID is integrated, capture refers to the selection of appropriate types of equipment (tags and readers) to gather data about the processes and environment, and control is concerned with real-time control of different parts of the RFID system including readers, middleware, user interfaces, and database [2]. The RFID tag contains a unique identification number stored in its ROM, and it provides the required information to the RFID receiver when the receiver interrogates the tag.

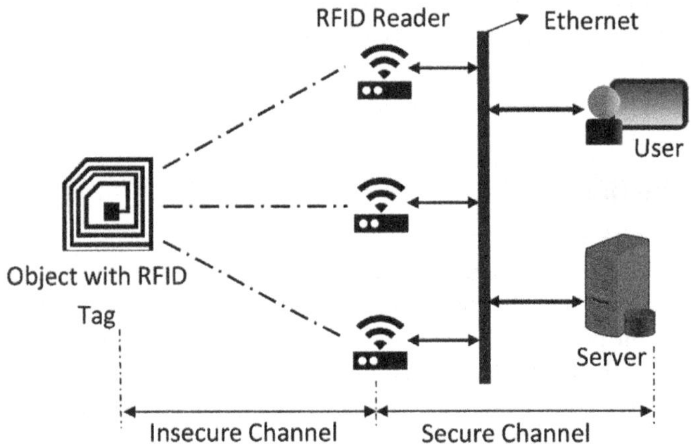

Fig. 1 RFID communication scenario

1.2 RFID Security Requirements

RFID suffers from the constraints of limited computing capacity, limited power, and unreliable nature of communication. These constraints enforce the implementation of many security features. Tan and Wu [3] have figured out that RFID systems must be protected from skimming, illicit tracking, and unauthorized access. These three security features can be ensured by implementing mutual authentication and data confidentiality. Knospe and Pohl [4] have concluded that availability, anonymity, authenticity, confidentiality, and integrity are the security attributes that must be implemented in an RFID system. Dinarvand and Barati [5] pointed out that scalability and forward security are also necessary to make RFID systems secure. The necessary security requirements for an RFID system are briefly explained below.

Data Confidentiality—It is the assurance that the information being transmitted is kept secret, and only authorized entities can access the secret information.

Mutual Authentication—The tag and the reader both must authenticate each other before transfer of data takes place.

Data Integrity—It is the assurance that the data in transit has not been altered or modified by an opponent.

Non-Repudiation—This property ensures that the tag, and the reader cannot deny after sending or receiving the data.

Availability—It ensures that the security protocol for the tag, and the server can be executed all the time and the de-synchronization of private data between the server and the tag should not be able to avert the execution of the security protocol [6].

Tag Anonymity—Original identity of the RFID tag, location, and other confidential information is transmitted securely.

Forward Secrecy—It ensures that an opponent cannot deduce the previous transmitted tag information using the current tag data.

Scalability—The security functionalities of the protocol should not be affected, even if the number of tags in the system is increased.

2 Security Attacks on RFID

Mitrokotsa et al. [7] performed an analysis on the vulnerabilities existing in the RFID systems and broadly classified the attacks on RFID as single layer attacks and multilayer attacks. According to them, the single layer attacks can be further classified as physical layer attacks, network-transport layer attacks, application layer attacks, and strategic layer attacks while multilayer attacks include covert channels, denial of service, traffic analysis, crypto attacks, side channel attacks, and replay attacks. Li et al. [8] classified the threats for RFID into three classes, namely physical threats, channel threats, and system threats. Habibi and Aref [9] performed an analysis over the attacks made on RFID authentication protocols and figured out

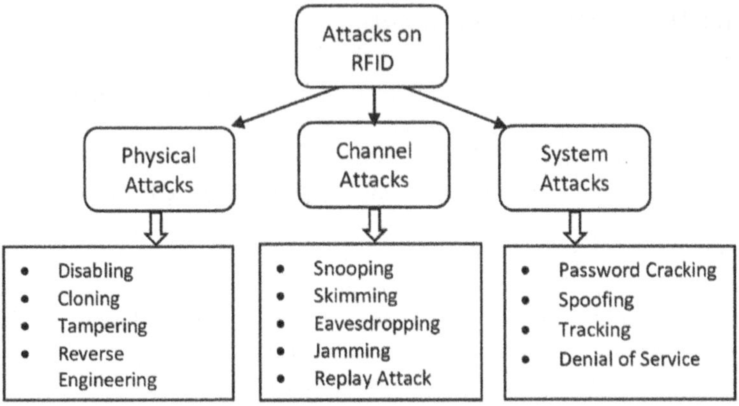

Fig. 2 Classification of attacks on RFID

that these protocols are vulnerable to de-synchronization attack and attacks on traceability. Pande and Unuakhalu [10] carried a study on low level attacks on RFID which can be classified as sniffing, tracking, spoofing, replay attack, and denial of service. Kulkarni et al. [11] classified the attacks on RFID as jamming, eavesdropping, replay attack, tag deactivation, spoofing, cloning, and man-in-the-middle attack. In [12], Guizani stressed that the major attacks on RFID can be attempted over three security properties, namely confidentiality, integrity, and authentication. Kannauf [13] divided the attacks on RFID into three classes—reader attacks, attacks on tags, and attacks on network protocols. Khattab et al. [14] specified that the attacks on RFID can be divided into physical attacks, channel attacks, and system attacks. According to them, the physical attacks on RFID include tag disabling, tag cloning, tag disabling, and reverse engineering while channel attacks include eavesdropping, snooping, skimming, replay attack, relay attack, and electromagnetic interference. In addition to this, the system attacks include counterfeiting and spoofing, tracking, password decoding, and denial of service attack. The taxonomy of attacks on RFID as mentioned by Khattab et al. has been shown in Fig. 2.

3 Physical Attacks on RFID and their Countermeasures

The physical attacks which can be attempted over RFID tags have been briefly explained in this subsection. In addition to this, the countermeasures against different physical attacks have also been discussed. The summary of physical attacks on RFID and their countermeasures has been publicized in Table 1.

Table 1 Physical attacks on RFID and their countermeasures

Specific physical attack	Elucidation	Countermeasures
Tag disabling	Disabling the temporarily or permanently	MAC with secret key
Tag cloning	Creating the duplicate tag for intrusion	Cryptographic fingerprint, MAC with secret key
Tampering	Modifying or removing the tag	Tamper-resistant case, building secure zone
Reverse engineering	Obtaining technical and functional details of the tag	Cryptographic fingerprint, strong encryption

3.1 *Tag Disabling*

In tag disabling, an opponent modifies the tag temporarily or permanently in a way that it cannot be recognized by the reader or back-end server. Tag disabling can be prevented by implementing the MAC using a shared secret key between the tag and the reader or the server. A challenge is generated by the first party, and the response is sent back by the second party using the secret key.

3.2 *Tag Cloning*

In tag cloning, the attacker steals the identity of a legitimate tag and use it to create the duplicate tag. The attacker then uses this tag to penetrate into the system or network. The two possible countermeasures for tag cloning are cryptographic fingerprint and MAC using a secret key [15]. Cryptographic fingerprint for the tag can be created by generating the hash code of the tag identity and other related parameters. In this case, if the attacker clone the tag, then the hash code for the duplicate tag will be different. The MAC with shared secret key can also be generated by the tag so that it can be identified and recognized by the reader or the server.

3.3 *Tampering*

In tampering, the tag can be modified or removed by an attacker if he can access it. The first countermeasure to avoid tampering is building a secure zone around the tag such that an attacker cannot access it [16]. The other alternative is to keep the tag in a tamper-resistant case so that when an attacker tries to tamper the case he cannot perform it since this act of tampering is detected by the system.

3.4 *Reverse Engineering*

In reverse engineering, the detailed functioning of the tag and its internal structure are exploited either to clone the tag or to intercept the communication between the tag and the reader. The first way to counter-reverse engineering is to implement strong authentication algorithm and cryptographic mechanism. The second way to defend against reverse engineering is to use cryptographic fingerprint [17].

4 Channel Attacks on RFID and their Countermeasures

The attacks which can be made on the channel of communication in RFID system and their countermeasures have been explained in this subsection. Since the communication between the tag and the reader or server is wireless in nature, there is more probability that the security of this channel is compromised. The summary of channel attacks on RFID and their countermeasures has been shown in Table 2.

4.1 *Snooping*

This attack involves illegitimate reading the tag identity and the other data stored on the tag without the knowledge of the tag owner. Snooping can be prevented by implement strong authentication and confidentiality. This is a two-level approach where the data stored on the tag must encrypted and the second party trying to access the tag's data must be authenticated.

Table 2 Channel attacks on RFID and their countermeasures

Specific channel attack	Elucidation	Countermeasures
Snooping	Illegitimate reading of the tag	Strong authentication and encryption
Skimming	Masquerade or clone the tag by creating a fake document	Hash lock, MAC with secret key
Eavesdropping	Obtaining secret information by listening to the channel	Strong encryption
Jamming	Electromagnetic jamming the communication between the tag and the reader	Polarization of antenna
Replay attack	Record the message and replays it later	OTP, nonce, hash code, dynamic update

4.2 Skimming

In skimming, the attacker monitors the communication between the tag and the reader, then he creates a fake document and uses this either to clone the tag or to masquerade it. The first approach to counterskimming is using the shared secret key to create the MAC for the challenge sent by the reader. The second approach to defend against skimming is to implement hash lock where a metaID is used by the tag instead of its actual identity [18]. The metaID is the hash code of the actual identity of the tag which can be verified by the reader or the server.

4.3 Eavesdropping

This attacks involves listening to the channel between the tag and the reader to obtain confidential information. Eavesdropping can be either passive or active. In passive eavesdropping, the attacker only listens to the channel in order to get secret data, whereas in active eavesdropping the adversary masquerades as a legitimate reader and tries to obtain secret information from the legitimate tags. Eavesdropping can be countered by implementing strong encryption algorithms in order to provide adequate confidentiality to the transmitted data.

4.4 Jamming

In jamming, an adversary tries to obstruct the communication between the tag and the reader using radio frequency and electromagnetic signals through a jamming device. Jamming can be prevented by polarization of antenna used in the reader and the tags.

4.5 Replay Attack

In replay attack, an adversary captures the message between the tag and the reader and replays it later to produce an unauthorized effect. The countermeasures of replay attack include time-stamping, dynamic updating the information OTP, and using nonce [14]. Another effective approach for defeating replay attack is to store the hash code of the message at the receiving side. If the message is new its hash code will not be found at the receiving side but if the message is replayed, then its hash code will be found at the receiving side and therefore it will be discarded.

5 System Attacks on RFID and their Countermeasures

System attacks launched by the attackers intends to target either the tag or the reader specifically. The aim of these attacks is to compromise the security of a particular device. In this subsection, password cracking attack, spoofing attack, tracking attack, and denial-of-service attack, and their countermeasures have been explained. The summary of system attacks on RFID and their countermeasures have been demonstrated in Table 3.

5.1 Password Cracking

Since in RFID many security functionalities can be implemented using strong encryption algorithm, the primary focus of the attackers is on breaking the ciphers and deducing the key. Once the key is obtained, the attacker can read the messages in the communication. To defeat the cracking of the ciphertext, the system must use the strong encryption algorithm and long size keys. The efficient and secure encryption algorithm like AES-128 can be used for RFID systems.

5.2 Spoofing

In spoofing attack, the adversary replaces the identity of the tag with its own identity. Once this is done, the adversary can get the secret information about the system. Spoofing attacks can be avoided by implementing strong authentication techniques [19].

Table 3 System attacks on RFID and their countermeasures

Specific system attack	Elucidation	Countermeasures
Password cracking	Obtaining the key to break the cipher	Strong encryption
Spoofing	Replacing the tag's identity with attacker's identity	Strong authentication
Tracking	Obtaining the location information of the tag	Strong encryption
Denial of service	Targeting a device to block it	Physical unclonable function, strong authentication

5.3 *Tracking*

In tracking, the attacker tries to obtain the location information of the tag. The tag contains the identification and location information which is required by the reader and the server whenever the tag interacts with the system. The location information can lead to the tag traceability. The tracking of the tag can be prevented by implementing the strong encryption which encrypts the communication between the tag and the reader.

5.4 *Denial of Service (DoS)*

The DoS attack involves targeting a particular device or machine and makes it to perform huge amount of fake computations. In this case, when a legitimate device wants to interact with the targeted device it cannot do so because the targeted device is unable to respond the request of the legitimate device. The first solution in countering the DoS attack is implementation of strong authentication mechanism. The second and more effective approach to counter DoS attack is the use of physical unclonable function (PUF). The advantage of using PUF is that it provides unclonability, consumes less power, and is unpredictable [20].

6 Conclusion

RFID has become an integral component of many computing applications including object identification, record capturing, animal tagging, transportation, inventory, and many more. Security and privacy issues in RFID must be resolved to prevent these applications from different attacks and threats. This paper has explored the security requirements of RFID which must be implemented to make the RFID system secure. Furthermore, the detailed classification of attacks on RFID has been given systematically so that appropriate security solutions can be designed to secure RFID systems form these attacks. The content presented in this paper has high significance for the researchers and professionals interested in the area of securing RFID systems.

References

1. Khattab, A., Jeddi, Z., Amini, E., Bayoumi, M.: Introduction to RFID. In: RFID Security. Analog Circuits and Signal Processing, pp. 3–26. Springer, Cham (2017)
2. Chetouane, F.: An overview on rfid technology instruction and application. IFAC Pap OnLine **48**(3), 382–387 (2015)

3. Tan, C.C., Wu, J.: Security in RFID networks and communications. In: Chen, L., Ji, J., Zhang, Z. (eds.) Wireless Network Security, pp. 247–267. Springer, Berlin (2013)
4. Knospe, H., Pohl, H.: RFID security. Inf. Secur. Tech. Rep. **9**(4), 39–50 (2004)
5. Dinarvand, N., Barati, H.: An efficient and secure RFID authentication protocol using elliptic curve cryptography. Wirel. Netw. **25**(1), 415–428 (2019)
6. Zhang, Z., Qi, Q., Kumar, N., Chilamkurti, N., Jeong, H.Y.: A secure authentication scheme with anonymity for session initiation protocol using elliptic curve cryptography. Multimedia Tools Appl. **74**(10), 3477–3488 (2014)
7. Mitrokotsa, A., Rieback, M.R., Tanenbaum, A.S.: Classifying RFID attacks and defenses. Inf. Syst. Front **12**, 491–505 (2010)
8. Li, H., Chen, Y., He, Z.: The survey of RFID attacks and defenses. In: 2012 8th International Conference on Wireless Communications, Networking and Mobile Computing, pp. 1–4, Shanghai (2012)
9. Habibi, H.M., Aref, M.R.: Attacks on recent RFID authentication protocols. J. Sig. Process. Syst. **79**, 1–13 (2013)
10. Pande, S., Unuakhalu, M.F.: Privacy and security challenges of RFID. In: Proceedings of the Information Systems Educators Conference, pp. 1–10. San Antonio, Texas (2013)
11. Kulkarni, G., Shelke, R., Sutar, R., Mohite, S.: RFID security issues & challenges. In: International Conference on Electronics and Communication Systems (ICECS), pp. 1–4. Coimbatore (2014)
12. Guizani, S.: Security applications challenges of RFID technology and possible countermeasures. In: International Conference on Computing, Management and Telecommunications (ComManTel), pp. 291–297, Da Nang (2014)
13. Kannouf, N., Douzi, Y., Benabdellah, M., Azizi, A.: Security on RFID technology. In: International Conference on Cloud Technologies and Applications (CloudTech), pp. 1–5, Marrakech (2015)
14. Khattab, A., Jeddi, Z., Amini, E., Bayoum, M.: RFID security threats and basic solutions. In: Ismail, M., Sawan, M. (eds.) RFID Security. Analog Circuits and Signal Processing, pp. 27–41, Springer, AG (2017)
15. Bu, K., Weng, M., Zheng, Y., Xiao, B., Liu, X.: You Can Clone but you cannot hide: a survey of clone prevention and detection for RFID. IEEE Commun. Surv. Tutor. **19**(3), 1682–1700 (2017)
16. Barenghi, A., Breveglieri, L., Koren, I., Naccache, D.: Fault injection attacks on cryptographic devices: theory, practice, and countermeasures. Proc. IEEE **100**(11), 3056–3076 (2012)
17. Bokslag, W.: Reverse engineering of RFID devices. CoRR abs/1507.02196, pp. 1–14 (2015)
18. Haver, T.: Security and privacy in RFID applications. Master's thesis. Norwegian University of Science and Technology, Trondheim, Norway (2006)
19. Khemissa, H., Tandjaoui, D., Bouzefrane, S.: An ultra-lightweight authentication scheme for heterogeneous wireless sensor networks in the context of internet of things. In: Bouzefrane, S., Banerjee, S., Sailhan, F., Boumerdassi, S., Renault, E. (eds.) Mobile, Secure, and Programmable Networking. MSPN 2017. Lecture Notes in Computer Science, vol. 10566, pp. 49–62, Springer, Berlin (2017)
20. Gassend, B., Clarke, D., Dijk, M.V., Devadas, S.: Silicon physical random functions. In: Proceedings of the 9th ACM Conference on Computer and Communications Security, pp. 148–160. ACM (2002)

Analytic and Simulative Modeling of RoF System with Intensity Modulation Including Fiber Dispersion Effect

Rajeev, Shelly Garg, and Vishal Jain

Abstract Radio over fiber (RoF) network is integration of two networks named as wireless and optical fiber networks. The conversion of the light signal from electric signal at the transmitter is the most challenging task in the radio over fiber system. Radio over fiber transmission of double sideband (DSB) optical signal undergoes frequency-dependent signal fading due to fiber dispersion. In this paper, fiber dispersion effect on RF signal amplitude is theoretically analyzed while using intensity modulation. Numerical and simulation result indicates that the variation of relative RF signal amplitude with respect to depth of modulation and length of the fiber is sinusoidal in nature. Analysis shows the dependence of RF signal amplitude on the modulation index, radio frequency of RF signal, and the length of the optical fiber. It is seen that the RF signal fundamental amplitude which is free from the harmonic sidebands, suffers from the dispersion induced fading only but the variation of the RF signal harmonic amplitude which include the harmonic sidebands with second order as well as the harmonic sidebands with fourth order, has the cyclic property with respect to fading. This cyclic variation depends on modulation index and is due to the harmonic sidebands that are generated through the nonlinear response of the intensity modulation.

Keywords Optical fiber communication · Electro-optic modulation non linear dispersion · Intensity modulation · Direct detection (IM-DD) · Optical modulation

1 Introduction

A radio over fiber-based network has the capability to enhance both the coverage and efficiency of wireless network. Since its benefit is provide signal diversity to increase the coverage and capacity of communication. In this method, modulation

Rajeev · V. Jain (✉)
Panipat Institute of Engineering and Technology, Samalkha, Panipat, India
e-mail: vishaljain.ece@piet.co.in

S. Garg
Greater Noida Institute of Technology, Greater Noida, India

V. Bhateja et al. (eds.), *Computer Communication, Networking and IoT*, Lecture Notes in Networks and Systems 197, https://doi.org/10.1007/978-981-16-0980-0_50

of radio frequency subcarrier signal occurs onto an optical carrier signal for transmission over a fiber network. For RoF network, optical fibers are considered due to their large number of advantages like resistant to EMI, lightweight, very high bandwidth, small cross section. To access high-speed integrated Internet applications, present wireless systems needs to provide very high data transmission capacities beyond the current standards of wireless network. The use of fiber links to distribute RF signal from a central BS to multiple RAUs is the basis of RoF technology. In [1] described that radio over fiber system provides a simple architecture for the transmission of wireless signals in high-capacity wireless networks in the form of optical signals. The radio over fiber technology has been developed since from last thirty years, with many problems which are already solved and with many more to be solved. In previous years [2], large amount of research has been carried out to combine the radio frequency and optical fiber technologies for the transmission of millimeter-wave (mm-wave) signals. This gives birth of new technology named as radio over fiber (RoF) technology. This system has a great capability to provide reliable, economic, and very high-capacity mobile/wireless communication for the future applications of broadband, interactive, and multimedia wireless services. ROF technology is expected [3] to become the backbone of the future high-capacity 5G wireless systems. The achievable system performance and the associated cost make the most important parameter of ROF system. The remote access units (RAUs) are very simple in architecture as they only have to perform only optical to electric conversion and amplification of electric signal. Nonlinearities occur in Radio over fiber component resulting in ultimate degradation of the system. In [4], it is suggested that in radio over fiber network, it is advantageous to enhance the value of optical modulation in favor of high signal-to-noise ratio. However, signal-distortion compensation capability by chromatic dispersion is restricted when the receiver noise is high or the modulation index is low.

In this paper, we study the impact of fiber dispersion on propagation of QAM signals both numerically and by simulation using RoF. The nonlinear effect of modulator and detector will be considered along with fiber dispersion [5].

1.1 Theoretical Model for Direct Modulation with Effect of Fiber Dispersion

Consider the propagation of optical double sideband signal in which carrier and both upper and lower sidebands are propagated in the optical fiber [6]. Here we can write the complex amplitude of optical field as $f(L, T)$ where L is the optical fiber length and T is the time scale. The Fourier series of light field at $L = 0$, which is direct modulated by a sinusoidal radio frequency signal whose angular frequency is Ω_0 can be described as

$$f(0, T) = \sqrt{E} \sum_{m=0}^{\infty} a_m(n) \cos(m\Omega_o T) \tag{1}$$

where E is the optical signal power, n is the modulation depth, m is the order of harmonics and $a_m(n)$ is the Fourier series coefficient of mth harmonic. The Fourier transform of $f(0, T)$ given in equation can be written as $F(0, \Omega)$ given below:

$$F(0, \Omega) = 2\pi \sqrt{E} \sum_{m=0}^{\infty} a_m(n) \frac{\delta(\Omega - k\Omega_0) + \delta(\Omega + k\Omega_0)}{2} \tag{2}$$

where $\delta(x)$ is the dirac delta function.

The complex envelope amplitude of optical field for a particular propagation distance, i.e., L, obeys the following differential equation in a purely dispersive fiber

$$\frac{\partial f(L, T)}{\partial L} = j\beta_2 \frac{\partial^2 f(L, T)}{\partial T^2} \tag{3}$$

where β_2 is the second-order dispersion of transmission fiber [7]. The corresponding differential equation for its Fourier transform can be written as

$$\frac{\partial F(L, \Omega)}{\partial L} = i \frac{\beta_2 \Omega^2}{2} F(L, \Omega) \tag{4}$$

whose analytical solution is

$$F(L, \Omega) = F(0, \Omega) \exp\left(-\frac{j\beta_2 \Omega^2 L}{2}\right) \tag{5}$$

By using (2) and (5) and by using inverse Fourier transform, we get

$$f(L, T) = \frac{1}{2\pi} \int_{-\infty}^{\infty} F(L, \Omega) \exp(j\Omega T) \mathrm{d}\Omega \tag{6}$$

$$f(L, T) = \sqrt{E} \sum_{m=0}^{\infty} a_m(n) \cos(m\Omega o T) \exp\left(-\frac{jm^2 \beta_2 \Omega^2 L}{2}\right) \tag{7}$$

The RF signal obtained from the output of photo detector is proportional to

$$\frac{|f(L, T)|^2}{E} = a_0^2(n) + \sum_{m=1}^{\infty} a_m^2(n) \cos^2(m\Omega_1 T)$$
$$+ 2 \sum_{k \neq m, 1 \leq m < k} a_m(n) a_k(n) \cos(m\Omega o T) \cos(k\Omega o T) \cos\left(m^2 - k^2\right)\varphi$$

$$\equiv \sum_{p=0}^{\infty} u_p(n, \varphi) \cos(p\Omega o T) \quad (8)$$

where $\varphi = \frac{\beta_2 \Omega^2 L}{2}$ is called as fading parameter.

Where $u_p(n, \varphi)$ is the strength of the nth harmonic of the propagated radio frequency signal which is function of n (modulation depth). It is proportional to n if n is small and for large depth n, it varies non linearly with n. If we consider only the fundamental signal, (8) can be expressed as

$$u_1(n, \varphi) = 2a_0(n)a_1(n) \cos\varphi + \sum_{k=1}^{\infty} a_k(n)a_{k+1}(n) \cos(2m + 1)\varphi \quad (9)$$

In (9), if we consider the harmonic side bands up to second order as they are dominant then the approximation of $u_1(n, \varphi)$ can be expressed as

$$u_1(n, \varphi) = 2a_0(n)a_1(n) \cos\varphi \times \left\{1 + \frac{a_2(n)}{a_0(n)}(2\cos 2\varphi - 1\right\} \quad (10)$$

Here 2φ dependence in Eq. (10) occurs from the beating of fundamental and second-order harmonic sidebands. If we remove this, then we will have dispersion induced signal fading [8, 9].

These issues will clearly explained with the help of ideal intensity modulator with direct detection by taking into consideration the modulator and detector nonlinearities which arc as:

(1) Intensity modulation is not the same as amplitude modulation. For this, it uses two considerations:

   (a) The amplitude ratio of carrier amplitude and the fundamental modulation sidebands varies in accordance with modulation index.
   (b) Beat signals between the fundamental and higher-order harmonics lie in between the frequency range of the propagating signal.

(2) The direct detection is a nonlinear square law detection in which detector gives the response by detecting power of the optical signal not the amplitude of received signal.

Intensity modulation with direct detection is the simplest method for transmitting RF signal over the optical fiber, where the intensity of light carrier is changed according to the amplitude of the RF signal propagated. It consists of application of the radio frequency signal to the operating voltage of a laser diode. Here the radio frequency signal is superimposed on a DC operating voltage so that the intensity of light output of laser varies proportionately with the variation of the RF signal [10] (Fig. 1).

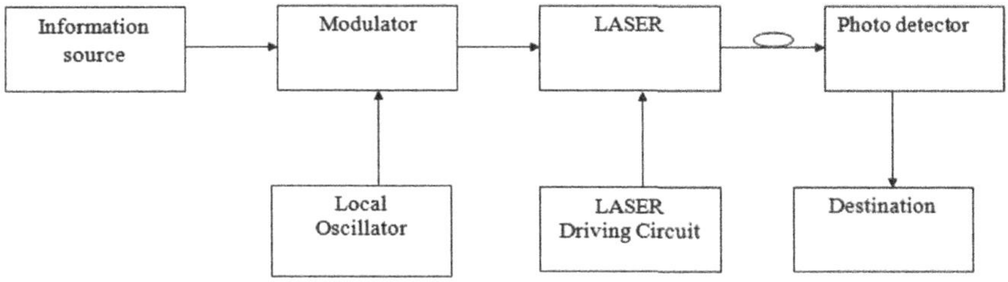

Fig. 1 Block diagram of RoF transmission using direct intensity modulator

2 Result and Discussion

Numerical simulation has been done to study the effect of fiber dispersion on RF signal amplitude using MATLAB. RF amplitude variation is also analyzed with the change in fiber length, modulation depth, and radio frequency by using the ideal intensity modulation.

Variation of relative fundamental amplitude of the transmitted RF signal of frequency 10 GHz with the optical fiber length for three different modulation indices has been plotted in Fig. 2a. Here the optical fiber length varies from 0 to 50 km. For mod index = 0.2, relative fundamental amplitude of RF signal oscillates between −0.12 and +0.12, whereas for mod index = 0.4, it fluctuates between −0.25 and +0.25. The RF amplitude increases with the increase in modulation depth.

Variation of relative fundamental amplitude of the transmitted RF signal of frequency 20 GHz with the optical fiber length for three different modulation indices has been plotted in Fig. 2b. Here the optical fiber length varies from 0 to 50 km. For mod index = 0.4, relative fundamental amplitude of RF signal oscillates between −0.15 and +0.15, whereas for mod index = 0.8, it fluctuates between −0.25 and +0.25. The RF amplitude increases with the increase in modulation depth. Figure 2c shows the sinusoidal variation of relative harmonic amplitude with the change in optical fiber length which varies from 0 to 50 km but for two different radio frequencies one is 10 GHz and other is 20 GHz at modulation index 0.5. Figure 2d shows the sinusoidal variation of relative harmonic amplitude with the change in optical fiber length which varies from 0 to 50 km but for two different radio frequencies one is 10 GHz and other is 20 GHz at modulation index 0.7.

Simulation result indicates that the variation of relative RF signal amplitude with respect to depth of modulation and length of the fiber is sinusoidal in nature. It is seen that the RF signal fundamental amplitude which is free from the harmonic sidebands suffers from the dispersion induced fading only but the variation of the RF signal harmonic amplitude which include the harmonic sidebands upto the second order as well as the harmonic sidebands upto the fourth order has the cyclic property with respect to fading.

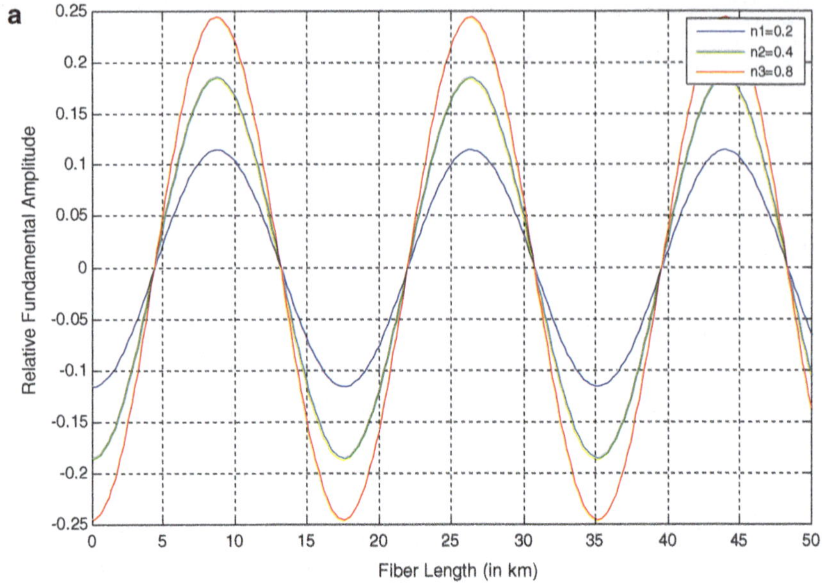

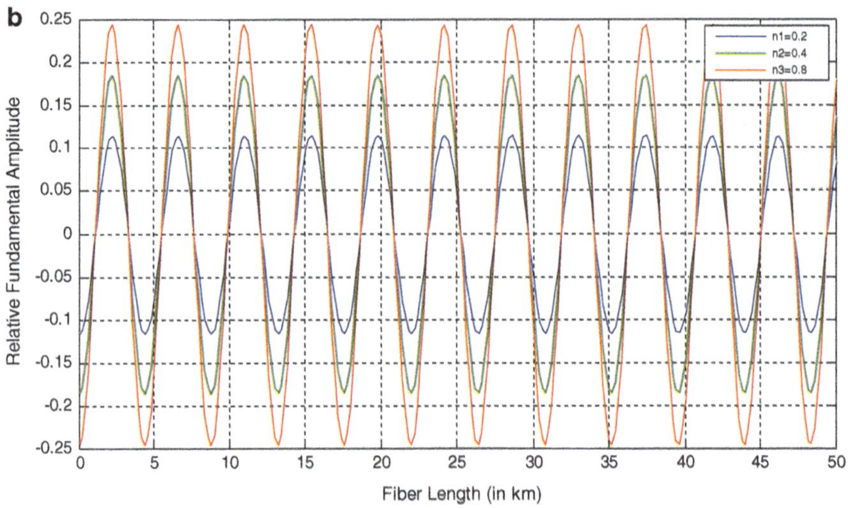

Fig. 2 **a** Relative fundamental amplitude versus length at RF frequency 10 GHz. **b** Relative fundamental amplitude versus length at RF frequency 20 GHz. **c** Relative harmonic amplitude versus length for $n = 0.5$. **d** Relative harmonic amplitude versus length for $n = 0.7$

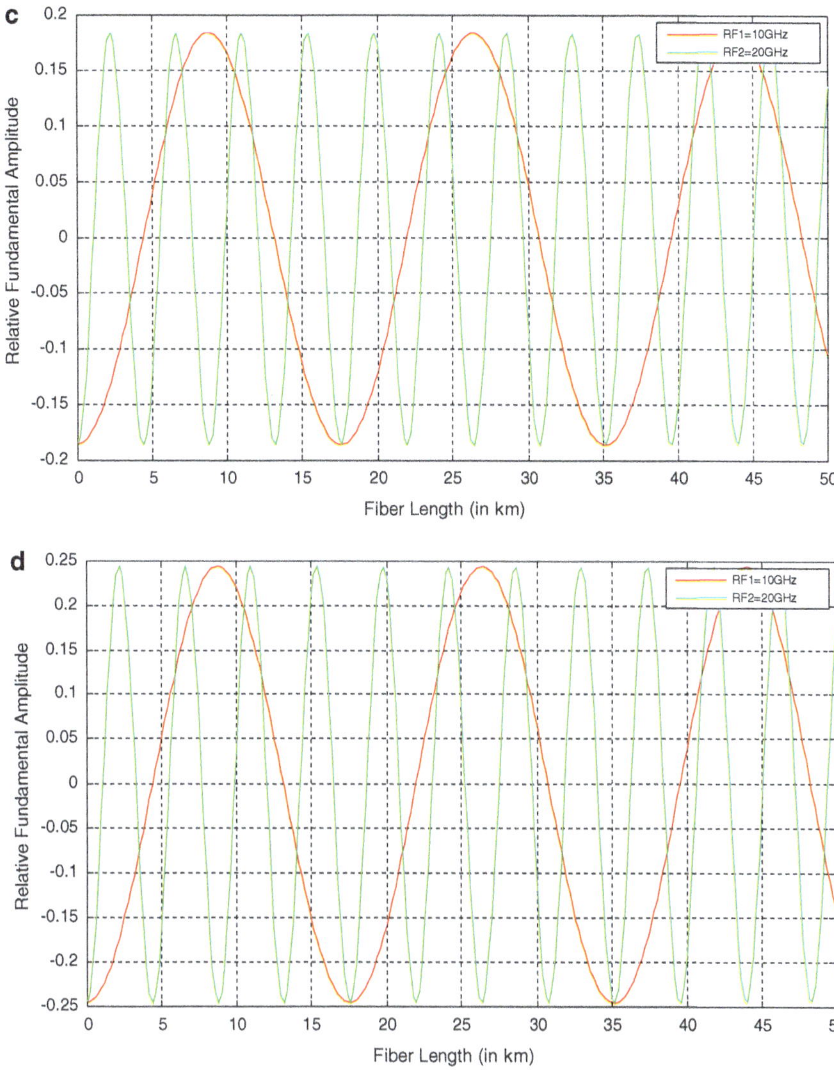

Fig. 2 (continued)

3 Conclusion

Analysis on relative harmonic amplitude of RF signal has been carried out to find the effect of fiber dispersion using ideal intensity modulation. It shows that the RF amplitude variation with respect to the modulation depth and length of fiber is sinusoidal due to dispersion effect. This sinusoidal nature depends on the modulation index and is due to harmonic sidebands which are produced due to nonlinear characteristics of the intensity modulation.

References

1. Lim, C., Tian, Y., Ranaweera, C., Nirmalathas, T.A., Wong, E., Lee, K.-L.: Evolution of radio-over-fiber technology. J. Lightw. Technol. **37**(6), 1647–1656 (2018)
2. Beas, J., Castanon, G., Aldaya, I., Zavala, A., Campuzano, G.: Millimeter-wave frequency radio over fiber system: a survey. IEEE Commun. Surv. Tutor. **15**(4), 1593–1619 (2013)
3. Thomas, V.A., El-Hajjar, M., Hanzo, L.: Performance improvement and cost reduction techniques for radio over fiber communications. IEEE Commun. Surv. Tutorial. **17**(2), 627–670 (2015)
4. Endo, S., Sampath, K.I.A., Maeda, J.: Chromatic dispersion-based modulation distortion compensation for analog radio-over-fiber: performance improvement in OFDM transmission. IEEE Photonics J. **36**(24), 5963–5968 (2018)
5. Maeda, J., Kato, T., Ebisawa, S.: Effect of fiber dispersion on subcarrier QAM signal in radio-over-fiber transmission. J. Lightwave Technol. **30**(16), 2625–2632 (2012)
6. Elwan, H.H., Khayatzadeh, R., Poette, J., Cabon, B.: Impact of relative intensity noise on 60-GHz radio-over-fiber wireless transmission systems. J. Lightw. Technol. **34**(20), 4751–4757 (2016)
7. Huang, Y.-R., Chiu, P.-C., Chyi, J.-I., Sun, C.-K.: A study on the fiber dispersion effect for the generation of quasi-sinusoidal terahertz modulations on optical pulses. J. Lightw. Technol. **33**(23), 4899–4907 (2016)
8. Maeda, J., Kimura, K., Ebisawa, S.: Experimental study on variation of signal amplitude in radio-over-fiber transmission induced by harmonics of modulation sidebands through fiber dispersion. J. Lightw. Technol. **32**(20), 3536–3544 (2014)
9. Khwandah, S.A., Cosmas, J.P., Glover, I.A., Lazaridis, P.I., Prasad, N.R., Zaharis, Z.D.: Direct and external intensity modulation in OFDM RoF links. IEEE Photonics J. **7**(4), 1–10 (2015)
10. Cabon, B., Khayatzadeh, R., Elwan, H.H., Poette, J.: Impact of amplitude noise in millimeter-wave radio-over-fiber systems. J. Lightw. Technol. **33**(13), 2913–2919 (2015)

Residual Energy and Throughput Enhancement in Underwater Sensor Network Routing Using Backward Forwarding

Swati Gupta and Niraj Pratap Singh

Abstract Underwater sensor networks (UWSNs) have been an area of continuous research because of the human quest on ocean exploration. But the study of underwater sensor networks faces a lot of challenges due to small bandwidth, limited energy resources, long propagation delay and high deployment cost. These limitations severely hamper the performance of underwater networks especially in delay critical applications. The communication strategies and sensing too tend to differ with underwater environment. Routing is one of the major performance measures to be considered in designing an underwater sensor network. Network throughput and lifetime are the major factors connected with effective routing. In this paper, a simulated performance analysis is performed using the proposed routing algorithm in a moving node environment with an energy saving and an improved throughput using a combination of clustering and backward propagation mechanism.

Keywords Underwater wireless sensor networks · Energy efficiency · Throughput · Routing protocol · Network lifetime

1 Introduction

Due to its broad and various applications, underwater sensor networks have sparked interest in science and business communities. Some of the major applications of UWSN are underwater environmental monitoring, mine exploration, disaster detection and prevention, oceanographic data collection and applications for tactical surveillance [1]. The nodes deployed in underwater collect information from severe aquatic setting and send it to offshore stations. Sensor nodes deployed underwater have acoustic modems since they work in aquatic environment only, but the sink

S. Gupta (✉) · N. P. Singh
National Institute of Technology Kurukshetra, Kurukshetra, India

N. P. Singh
e-mail: npsingh@nitkkr.ac.in

S. Gupta
P.I.E.T., Panipat, India

V. Bhateja et al. (eds.), *Computer Communication, Networking and IoT*, Lecture Notes in Networks and Systems 197, https://doi.org/10.1007/978-981-16-0980-0_51

nodes have both acoustic as well as radio modems because they have to receive messages from deployed underwater sensors and had to convey this information through radio links to offshore base stations [2].

UWSNs suffer from a lot of drawbacks. The propagation speed of acoustic signals is 1500 m/s which is quite low as compared to that of radio signals speed which is 3×10^8 m/s. Also due to the mobility of nodes, no fixed topology can be considered. Acoustic communication offers low bandwidth resulting into low data rates. Moreover, sensor nodes cannot be recharged, and battery replacement is not feasible. Delay and Doppler spread cause inter-symbol interference (ISI) in underwater signals [3]. Prominently, two communication architectures have been projected for underwater networks: two dimensional and three dimensional [4]. Static two-dimensional UWSNs can be applied to ocean bottom monitoring. They are made up of sensor nodes which are anchored to ocean floor and are connected to one or more underwater sinks by wireless acoustic links. The underwater sinks pass on information from the marine floor network to a surface station. Characteristic applications are ecological monitoring or monitoring of marine plates in tectonics. Three-dimensional networks of self-governing underwater vehicles include fixed part made up of anchored sensors and mobile part made up of self-governing vehicles. Some of the applications are oceanography, environmental monitoring and underwater resources study.

Underwater routing faces many difficulties owing to the peculiar characteristics of underwater acoustic environment [5–9]. Underwater network communication experiences significant loss of signals [10] and complex multi-path effects in the underwater acoustic system, resulting in high rate of communication errors [11]. Additionally, the delay in propagation in underwater acoustic channels is nearly five-order longer than in terrestrial radio channels [12, 13]. Communication signals suffer from severe Doppler distortions caused by the transceiver motion, or by changing environments such as surface waves, internal turbulence and sound velocity fluctuations due to the very low sound rate in water [22]. Therefore, routing protocols should have lightweight signalling and the ability to adapt to the highly dynamic nature of the links in order to get effective end-to-end packet transmission over UWA links.

2 Related Work

UWSN is complex due to water currents, and thus terrestrial routing protocols are not appropriate for underwater sensor networks. Routing protocols for UWSNs are divided into two major categories—geographic information-based and geographic information-free routing. Geographic routing uses sensor node location information to route between source and destination [5]. It is further divided as receiver based and sender based and more further as location information-based routing or depth-based routing. Some of the receiver-based protocols are DBR, WDFAD DBR, VAPR, DSD DBR and H2 DAB, and sender-based protocols are VAR, Hydrocast and ARP.

Depth-based routing (DBR) works on routing decisions which are based on the depth of sensor nodes [4]. This information about sensor depth is present in the packet header. After the packet is received at a node, this node compares its depth with the information contained in the receiving message. The packet will be transmitted further only if the receiving node's depth is lower or else it gets dropped. WDFAD-DBR is another protocol which is also based on depth but it works on the principle of dividing the forwarding area into a primary fixed area and two auxiliary areas. These auxiliary areas can be extended based on the density of nodes. It is a two-hop protocol but the main disadvantage of this protocol is that the fixed area can limit the routing area and receiving of acknowledgement messages in response to every transmitted packet causes wastage of network resource. In [14], VAPR is proposed which also considers the depth information for the data packets forwarding. By exchanging beacon messages, VAPR predicts all the information about the path to the destination in advance. HydroCast [15] is a routing protocol based on a load. It operates in two parts, with a greedy routing algorithm based on pressure and a lower depth-first method of local recovery. Recovery paths are established a prior in HydroCast, by making use of the depth properties of deployed nodes. But the discovery and maintenance of the recovery route will incur high overhead, especially when the path is very long. A virtual pipe from source to destination used for transmitting data is built in the vector-based forwarding protocol. For data forwarding, only the nodes that are in the pipe are picked. But, selecting forwarders inside the pipe over and over again does not balance energy consumption. Network life span is high because of the imbalanced energy consumption. In [3], a hop-by-hop vector-based forwarding protocol (HH-VBF) is provided to address this problem and improve the performance of VBF. In HH-VBF, each node makes the virtual pipe, because of which the path of the vector varies according to the location of the node. Compared to VBF, the energy consumption gets balanced. However, the hop-by-hop virtual pipe method produces large overhead communication. Compared to that in VBF, the overhead is higher. Another protocol adaptive hop-by-hop forwarding-based vector is given in [12]. This is based on the fixed distance of a virtual routing pipeline. When a packet is received at a node, it is initially tested for its distance from the forwarder to see if it is within a predefined threshold.

By using holding time, it minimizes unnecessary broadcasts. A node with more number of neighbours has a high holding time as compared with the nodes having less number of neighbours [16, 17]. Though this improves the rate of packet collision, but its increases end-to-end delay. The proposed protocol uses the information about the location of the neighbouring nodes for deciding the set of next hop forwarder neighbours. For avoiding unnecessary transmissions, the higher depth nodes are prohibited from the forwarding process [18]. It results in improvement of residual energy of the network. Also, the packet delivery ratio is improved by avoidance of unnecessary transmissions and collisions [16]. While it is assumed that the sinks are stationary once deployed, the sensor nodes follow the random-walk mobility pattern [19]. Each sensor node randomly selects a direction and moves to the new position with a random speed between minimal and maximal speed, which are 1 m/s and 5 m/s, respectively, unless specified otherwise. A performance analysis will be carried out between the proposed protocol and the HH-VBF protocol, using a simulation environment, analysing the metrics: PDR, residual energy and lifetime of nodes. In HH-VBF, a large number of virtual pipes are created with even larger communication overhead, but in the proposed protocol, if forwarder with lower depth is not possible, then backward recovery method is adopted for successful delivery of packet.

3 System Model

This section discusses in detail the network model which gives the network architecture, propagation model, acoustic channel speed, noise and mobility of sensors. An UWSN has a 3D network topology consisting of anchored, relay and sink nodes. Sink nodes are placed on the water surface as the final destinations of all packets in a multi-sink network model [20]. They are fitted with both underwater and land communications acoustic and radio modems, respectively. Anchored nodes are positioned at the bottom of the ocean to sense, collect and transmit information to one of the sinks using the relay nodes deployed at different depths between them [6]. Upon receipt, sink nodes can send the collected information via satellite to the monitoring centre for further analysis. Here it detailed the channel model used to calculate signal attenuation and total noise loss in underwater acoustics.

At a given frequency f, Thorps channel model calculates absorption loss $\alpha(f)$ [2] as follows:

$$10\log\alpha(f) = \frac{0.11f^2}{(1+f^2)} + \frac{44f^2}{4*100+f} + 2.75*104f^2 + 0.003 \quad f \geq 0.4 \tag{1}$$

$$= 0.002 + 0.11\left(\frac{f}{1+f}\right) + 0.011f \quad f < 0.4 \tag{2}$$

Here $\alpha(f)$ is in dB/km and f in kHz. We determine the value of α by using the value of absorption loss as follows:

$$\alpha = \frac{10^{\alpha(f)}}{10} \tag{3}$$

likewise, all tunable parameters are given in dBreμPa. The total attenuation $A(l, f)$ can be computed by combining absorption loss and spreading loss, as follows:

$$10\log(A(l, f)) = k * 10\log(l) + l * 10\log(\alpha(f)) \tag{4}$$

where the first term corresponds to spreading loss and the second term to the absorption loss [7]. The spreading coefficient k defines the geometry of the signal propagation in underwater acoustics (i.e. $k = 1$ for cylindrical, $k = 2$ for spherical and $k = 1.5$ for practical spreading).

The ambient noise in underwater networks is composed of turbulence noise, shipping noise, wind noise and thermal noise. The power spectral density of the four noise components in dBreμPa is given by the following formulae as a function of frequency in kHz

$$10\log(N_t(f)) = 17 - 30\log(f) \tag{5}$$

$$10\log(N_s(f)) = 40 + 20(s - 0.5) + 26\log(f) - 60\log(f + 0.03) \tag{6}$$

$$10\log(N_w(f)) = 50 + 7.5w^{1/2} + 20\log(f) - 40\log(f + 0.4) \tag{7}$$

$$10\log(N_{th}(f)) = -15 + 20\log(f) \tag{8}$$

where N_t, N_s, N_w and N_{th} represent turbulence noise, shipping noise, wind noise and thermal noise. The overall noise power spectral density can be calculated as follows:

$$N(f) = N_t(f) + N_s(f) + N_w(f) + N_{th}(f) \tag{9}$$

There are different mobility models capable of recording the physical motion of the sensor nodes. Meandering current mobility (MCM) [21] model is the best model for evaluating the efficiency of routing protocols under the continuous changes in topology. Find a sensor in coordinates x and y. By using the stream function Ψ, the sensor trajectory can be determined as follows:

$$\dot{x} = \frac{\partial\Psi(x, y, t)}{\partial y} \tag{10}$$

$$\dot{y} = \frac{\partial\Psi(x, y, t)}{\partial x} \tag{11}$$

where $\dot{x}$ is the zonal (eastward) and $\dot{y}$ is the meridional (northward) drift of the sensor in time.

4 Simulation Results

In the proposed model in Fig. 1, 100 sensor nodes are deployed in a three-dimensional space of length = 500 m, width = 500 m and height = 500 m along with four sinks. The simulation is carried out in NS2. The proposed routing algorithm works on a cluster-based approach where the cluster members are based on one hop distance from the sender node. The interesting fact is that here the cluster forwarding is only in the forward direction and no backward transmission is considered normally. But if a packet gets stuck up where it could not find next hop member using the greedy

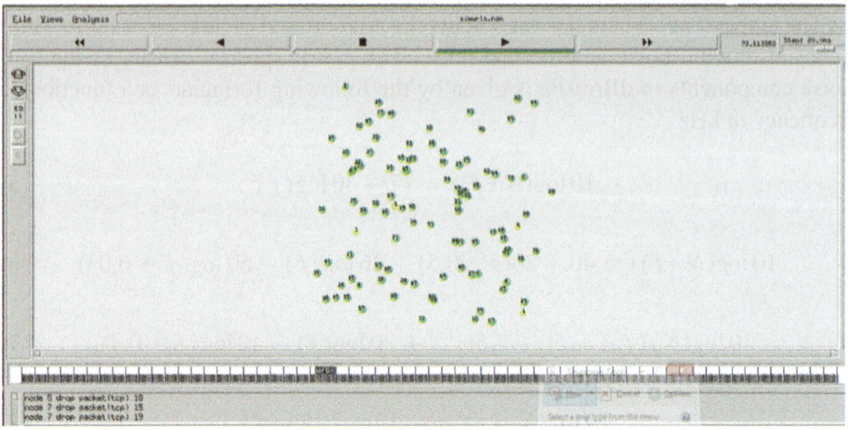

Fig. 1 Network model with mobile nodes simulated in NS2

forwarding strategy, the backward path is taken so as to ensure the delivery of the packet to the onshore sink. Also, if there are two neighbours with the similar hop count, the node with the least hop count for the backup route is taken as the final forwarder. The routing table at each node includes the next hop neighbours, their node IDs, their distance in terms of hop count from the sink and also the residual energy of each neighbour in the list. The forwarding set also takes into consideration the residual energy of the neighbour nodes, and the nodes with maximum residual energy are taken as next hop forwarder nodes. Each node which receives the packet correctly verifies itself as a next hop neighbour. Three performance metrics of packet delivery ratio, residual energy and network throughput are compared.

The protocol proposed has a higher packet delivery ratio because it works both on the concept of one hop neighbour selection through clustering method and also takes into account the residual energy of the nodes. As the network comes into operation, the nodes start dying early in the reference protocol because it does not work on the concept of residual energy, whereas considering residual energy as one of the major affecting factors in forwarder selection, the proposed protocol is able to achieve a better PDR. Figure 2 shows the packet delivery ratio of both the protocols, and Fig. 3 represents the residual energy of the network in case of both the protocols.

Energy consumption for the proposed protocol is also improved because of the energy balancing being adopted due to the selection of forwarder nodes based on their residual energies. The comparison is shown in Fig. 4, wherein in HH-VBR, residual energy is not considered as a parameter for forwarder selection.

Because of the long propagation delays in acoustic communication, intricate protocols with large overheads and heavy signalling can drop the network throughput. The proposed protocol does not include signalling overheads and is very simple to implement. It provides high network throughput as compared to HH-VBR as can be seen from the simulation results in Fig. 5.

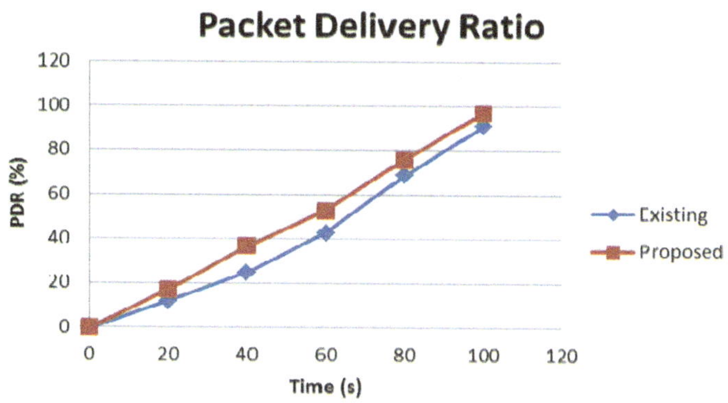

Fig. 2 Packet delivery ratio

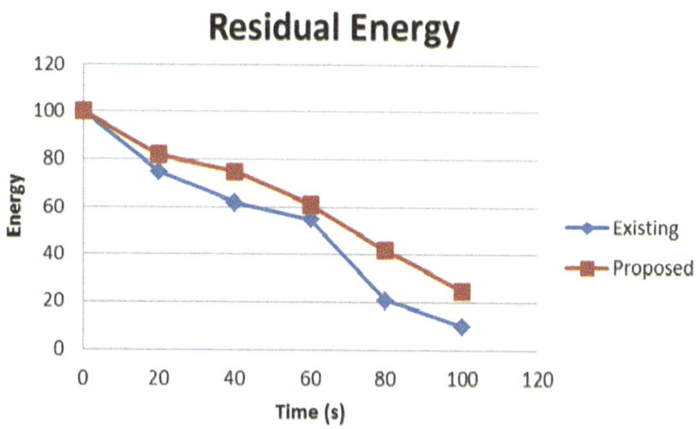

Fig. 3 Residual energy

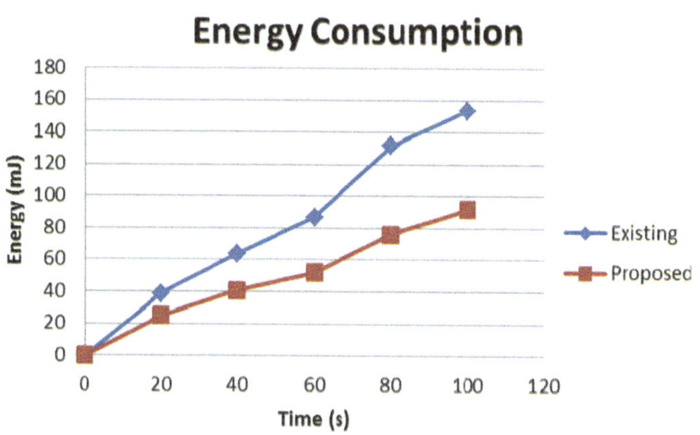

Fig. 4 Energy consumption comparison

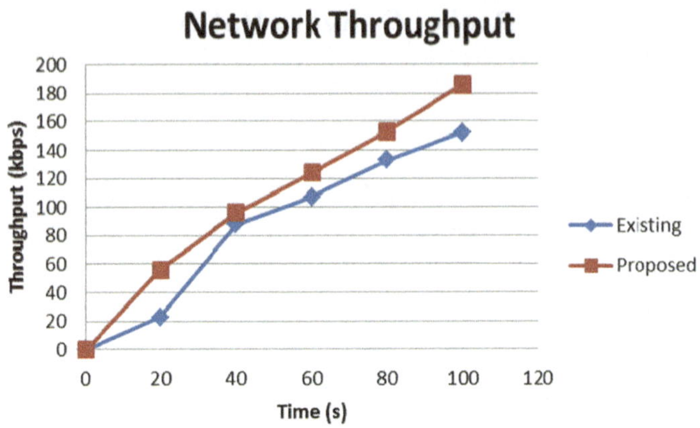

Fig. 5 Network throughput comparison

5 Conclusion

The proposed protocol outperforms the base protocol in terms of residual energy, throughput and also packet delivery ratio to some extent. The protocol presented also leads to energy balancing in the network by giving an equal opportunity to the nodes with higher residual energy in packet forwarding. Also, since it does not include signalling overhead, it helps in optimizing the packet length for maximization of throughput efficiency.

References

1. Akyildiz, I.F., Pompili, D., Melodia, T.: Challenges for efficient communication in underwater acoustic sensor networks. SIGBED Rev. **1**(2), 3–8 (2004)
2. Akyildiz, I.F., Pompili, D., Melodia, T.: Underwater acoustic sensor networks: research challenges. Ad Hoc Netw. **3**(3), 257–279 (2005)
3. Xie, P., Cui, J.-H., Lao, L.: VBF: vector based forwarding protocol for underwater sensor networks. In: Networking 2006, Lecture Notes in Computer Science, vol. 3976, pp. 1216–1221. Springer, Berlin, Heidelberg (2006)
4. Yan, H., Shi, Z.J., Cui, J.-H.: DBR: depth-based routing for underwater sensor networks. In: International Conference Resolution Networks, pp. 72–86. Springer, Berlin, Germany (2008)
5. Noh, Y., Lee, U., Wang, P., Choi, B.S.C., Gerla, M.: VAPR: void aware pressure routing for underwater sensor networks. IEEE Trans. Mob. Comput. **12**(5), 895–908 (2013)
6. Souiki, S., Feham, M., Feham, M., Labraoui, N.: Geographic routing protocols for underwater wireless sensor networks: a survey. Int. J. Wirel. Mob. Netw. (IJWMN) **6**(1), 69–87 (2014)
7. Cayirpunar, O., Kadioglu-Urtis, E., Tavli, B.: Optimal base station mobility patterns for wireless sensor network lifetime maximization. IEEE Sens. J. **15**(11), 6592–6603 (2015)
8. Yu, H., Yao, N., Wang, T., Li, G., Gao, Z., Tan, G.: WDFAD-DBR: weighting depth and forwarding area division DBR routing protocol for UASNs. Ad Hoc Netw. **37**, 25–282 (2016)

9. Menon, V.G., Jogi Priya, P.M., Joe Prathap, P.M.: Analyzing the behavior and performance of greedy perimeter stateless routing protocol in highly dynamic mobile ad hoc networks. Life Sci. J. **10**(2), 1601–1605 (2013)
10. Nicolaou, N., See, A., Xie, P., Cui, J.H., Maggiorini, D.: Improving the robustness of location-based routing for underwater sensor networks. In: OCEANS 2007 Europe, pp. 1–6, Aberdeen (2007)
11. Ismail, N.S.N., Hussein, L.A., Ariffin, S.H.S.: Analyzing the performance of acoustic channel in underwater wireless sensor network (UWSN). In: Proceedings of the Asia Modelling Symposium 2010: 4th International Conference on Mathematical Modelling and Computer Simulation, iAMS 2010, pp. 550–555, Malaysia (2010)
12. Javaid, N., Shah, M., Ahmad, A., Imran, M., Khan, M.I., Vasilakos, A.V.: An enhanced energy balanced data transmission protocol for underwater acoustic sensor networks. Sensors **16**, 487 (2016)
13. Wadud, Z., Hussain, S., Javaid, N., Bouk, S.H., Alrajeh, N., Alabed, M.S., Guizani, N.: An energy scaled and expanded vector-based forwarding scheme for industrial underwater acoustic sensor networks with sink mobility. Sensors **17**(10), 2251 (2017)
14. Noh, Y., Lee, U., Wang, P., Choi, B.S.C., Gerla, M.: VAPR: void-aware pressure routing for underwater sensor networks. IEEE Trans. Mob. Comput. **12**(5), 895–908 (2013)
15. Noh, Y., et al.: HydroCast: pressure routing for underwater sensor networks. IEEE Trans. Veh. Technol. **65**(1), 333–347 (2016)
16. Jouhari, M., Ibrahimi, K., Tembine, H., Ben-Othman, J.: Underwater wireless sensor networks: a survey on enabling technologies, localization protocols, and internet of underwater things. IEEE Access **7**, 96879–96899 (2019)
17. Latif, K., Javaid, N., Ullah, I., Kaleem, Z., Abbas Malik, Z., Nguyen, L.D.: DIEER: delay-intolerant energy-efficient routing with sink mobility in underwater wireless sensor networks. Sensors (2020)
18. Xi, Z., et al.: Research on underwater wireless sensor network and MAC protocol and location algorithm. IEEE Access **7**, 56606–56616 (2019)
19. Feng, P., Qin, D., Ji, P., et al.: Improved energy-balanced algorithm for underwater wireless sensor network based on depth threshold and energy level partition. J Wirel. Comput. Netw. 228 (2019)
20. Coutinho, R.W.L., A. Boukerche, L.F.M. Vieira, A.A.F. Loureiro, GEDAR: Geographic and opportunistic routing protocol with depth adjustment for mobile underwater sensor networks. In: 2014 IEEE, International Conference on Communications (ICC), pp. 251–256. Sydney, NSW (2014)
21. Javaid, N., Ahmed, F., Wadud, Z., Alrajeh, N., Alabed, M.S., Ilahi, M.: Two hop adaptive vector based quality forwarding for void hole avoidance in underwater WSNs. Sensors **17**(8), 1762 (2017)
22. Kim, S., Yoo, Y.: Practical multiple user system using heterogeneous frequency modulation for high data rate in underwater sensor network. Wirel. Pers. Commun. **108**, 1393–1416 (2019)

Multifactor Authentication for Smart Grid Cyber Security

Spandana Shashikumar and B. S. Manohar

Abstract Deployments of smart grids have made the operations of power systems more efficient by using distributed computing power grid—interoperation schemes. The development and evolution of information and communication technologies (ICT) in traditional power systems further increase the improvement of cyber-physical systems (CPS) on the smart grid. IoT-based smart grid systems are critical infrastructures, and they also have complex architectures and include critical device communication systems that can lead to national security deficits, disruption of public order, loss of life, or large-scale economic damage when confidentiality, integrity, or communication availability is broken down owing to cyber threats. There is ample research scope in the industry, government, and academia to strengthen smart security. Authentication is one of the very critical requirements in the design of cyber security. In this paper, multifactor authentication using RSA securID has been proposed to ensure that only approved users can access sensitive grid information with RSA's proven, self-risk engine.

Keywords Smart grids · Cyber-attacks · Cyber security · Multifactor authentication · RSA SecurID

1 Introduction

The introduction of smart grid has encouraged the convergence of traditional power with renewable energy, meeting the increasing demand for electricity, and reducing environmental hazards as well. Besides, the smart grid provides a multi-directional flow of information between two or more units in the network to achieve a revolutionized power industry, thereby providing sufficient data from metering to substations, networks, transmission and generations, improving efficiency, resilience and efficient control, and monitoring of assets and services [1]. As per the National Institute

S. Shashikumar (✉) · B. S. Manohar
Department of ECE, JSS Academy of Technical Education, Bangalore, Karnataka, India

B. S. Manohar
e-mail: manoharbs@jssateb.ac.in

V. Bhateja et al. (eds.), *Computer Communication, Networking and IoT*, Lecture Notes in Networks and Systems 197, https://doi.org/10.1007/978-981-16-0980-0_52

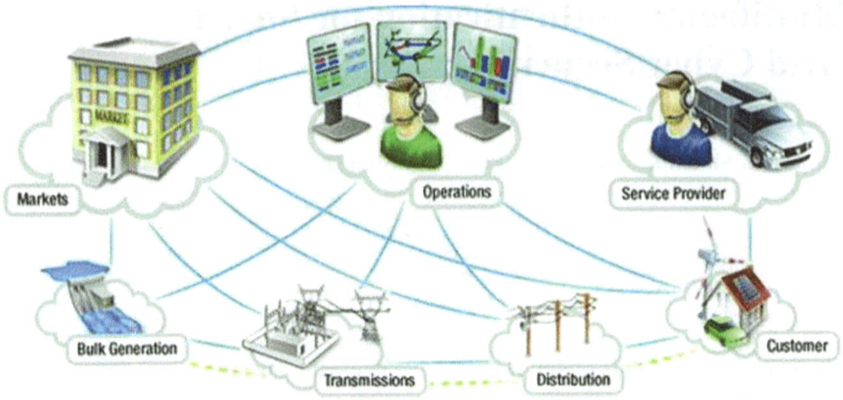

Fig. 1 Smart grid model

of Standards (NIST), the conceptual model of the smart grid and communications systems link modules of the power system are shown (see Fig. 1). There are seven logical domains in the smart grid: markets, operations, service provider, transmission and distribution and bulk production, and customer [2]. The first three deal with the storage and control of data while the last four deal with the transfer of power and energy through the smart grid. Stable contact ties connect up these domains.

The smart grid is set to transform a centralized, producer-controlled network into a decentralized, consumer-interactive network supported by fine-grained monitoring. Consumers, for example, react to price signals (i.e., supply) with the help of smart meters to achieve active load management. On the monitoring side, old metering data recorded hourly or monthly is replaced by a smart metering [3].

They comprise numerous modules and properties, such as generations, distribution, consumers, substations, field systems, communication systems, networking, units for measuring the phase, relays protection, smart electronic equipment, remote controls, human–machine interface, home appliances, computer controls, circuit breakers, protocol, logging servers, data concentrators, tap changers, smart meters, etc. All of these elements are plugged in smart grid for running, tracking, regulating, and measuring the power flow.

Smart grid provides new features for the collection, communication, and sharing of energy-related data, and these innovations, in turn, poses privacy concerns. The introduction of smart grid has increased the value of information technology (IT) and telecommunications infrastructures to ensure the efficiency and safety of the electricity sector [4]. Cyber security must counter not only malicious attacks by disgruntled workers, industrial espionage officers, and terrorists, but also unintended information technology contributions due to user mistakes, equipment failures etc.

Smart grids are cyber-physical networks that fundamentally alter the substantive difference between physical power and network infrastructure, with the Internet providing the framework for services to assimilate information, implement the strategy, and also to connect with customer devices. The smart grid system is vulnerable to cyber-attack, which is different from the current grid. Since smart grid environments require two-way data transfer, sufficient power supply and network operation are conducted automatically by gathering specific data [5]. The transmission network smart grid serves as a carrier for enormous consumption data from a broad range of connected devices to manage dynamic energy loads. In the sense of energy storage and distribution in the smart grid, information infrastructure is crucial, and cyber-attacks on the same will affect the broader economy. Due to the existence of multiple entry points, the possibility of increased attack periphery lies with smart grids. The cyber threat landscape for smart grids is expanding significantly affecting all interconnected components in some way or the other across generation, transmission, and distribution channels [6].

These attackers can cause a wide variety of attacks, divided into three main categories—component-wise, protocol-wise, and topology-wise. Component-wise attacks target field components including the remote terminal unit (RTU). RTUs have typically been used by engineers to remotely control and troubleshoot smart grid systems. This remote access mechanism could be subject to an attack that allows malicious users to take control of the devices and issue problematic states, such as shutting down the devices. Protocol-wise attacks threaten the communication protocol itself using techniques such as reverse engineering and false data injection [7]. Topology-wise attacks exploit the topology of the smart grid by initiating a denial-of-service (DoS) attack that prevents operators from getting a full view of the power system that triggers incorrect decision making.

Research and development challenges on cyber security as identified by NIST include

- Safety at equipment level,
- Key management and cryptographic,
- Networking security issues, and
- Security at the system level [8].

This paper explains authentication methods for security threats in the smart grid as this authentication method enables devising new security measures and strategies to thwart such attacks to make the grid more robust and resilient.

2 Potential Risks for Smart Grid Security

The NISTIR 7628 study outlines a few instances of the potential smart grid risks:

- Significantly higher ambiguity increases the risk of unintended errors and potential attackers.
- Networks that more frequently link to other networks introduce security weaknesses which can now span multiple smart grid domains and increase the potential for cascading failures.

- More connectivity allows enhanced opportunities for "denial-of-service" attacks, malicious code inclusion (in firmware/software) or jeopardized hardware, and possible attacks and intrusions associated with it.
- As the number of system nodes increases, the number of entry points and paths potentially exploited by potential adversaries could also increase.
- Enormous data collection and two-way information flows could expand the potential for data confidentiality compromises and customer privacy violations [9].

2.1 Objectives of Cyber Security in Smart Grids

The reliability of the power grid operated by power networks has power quality as the primary and information confidentiality as the secondary requirement. Confidentiality of consumer information is of utmost importance for privacy considerations in the billing process.

Power *AVAILABILITY* is important security objective. Time latency aligned to availability varies for each section of the power system, such as protective relaying has $\leq$4 ms, seconds for substation, meter reading has hours/days, and long-term data collection for power quality issues takes days/weeks/months.

INTEGRITY includes assurance for operation, assurance like authorized modification of data, data source authentication, and data quality [10].

CONFIDENTIALITY is the least important to the stability of the power grid. However, it is becoming more relevant, especially with the growing availability of online consumer details.

3 Authentication as a Cyber Security Requirement for Smart Grids

Modeling a smart grid cyber infrastructure is important so that it is susceptible to any invasion of the cyber workspace. It can only be done if the evaluation is made of the possible entry points that cause an adversary to enter the network (Fig. 2) [5].

Passwords are a weak method of single-factor authentication, as access is only protected by one factor with a user-selected confidential word or phrase [11]. When the wrong person finds this password, the entire system's protection will be compromised. Multifactor authentication offers greater security by allowing two or more unrepeatable factors to verify the identity of a user [12].

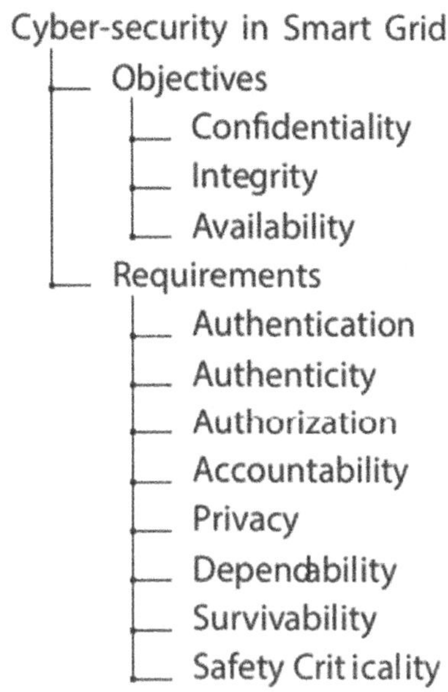

Fig. 2 Security objectives and requirements

4 Authentication with RSA SecurID Tokens

A two-factor authentication is an authentication method, which requires the presentation of two different kinds of evidence for someone who they say they are. The authentication is based on two factors—something the person knows (a password or PIN) and something the person has (an authenticator) [13–15]. Two-factor authentication decreases the probability that an attacker can impersonate a user. It is the basic principle on which the RSA SecurID solution is built [15, 16]. SecurID system identifies and authenticates authorized users on designated devices, while denying access to unauthorized attempts to provide an enterprise-wide authentication that protects the most valuable resources and information of the organization [12].

The RSA SecurID authentication process involves the following components as shown in the (see Fig. 3)

- SecurID authenticators provides hardware or software tokens which generate a user's unique credentials for one-time authentication.
- Authentication agents are built on client computers and trigger authentication requests to the authentication manager.
- Authentication manager is the multifactor authentication program that centrally verifies authentication requests and administers the authentication policies of enterprise networks.

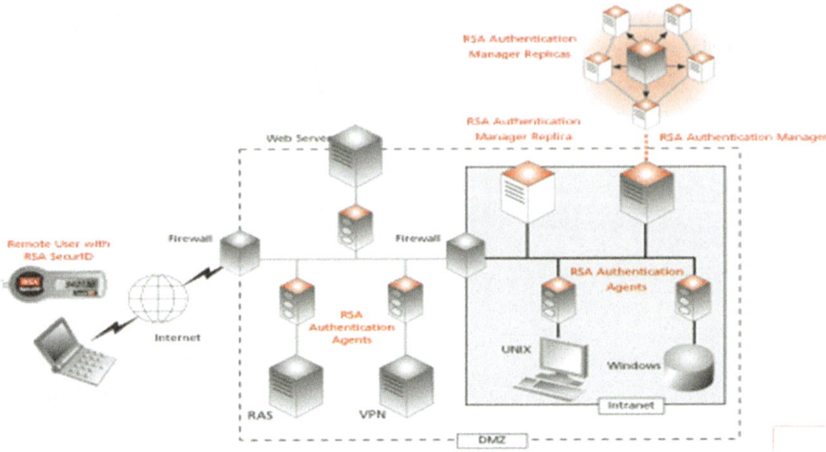

Fig. 3 Authentication manager architecture

RSA Authentication Manager Architecture

Authentication manager is the server component of security client/server software designed to protect access to selected networks as shown (see Fig. 3). The software can reside on several platforms such as UNIX, LINUX, or Windows and validate authentication requests from strategically placed agents [17].

The manager deployment consists of an authentication server and its associated database, a primary instance. This basic structure processes requests from authentication agents and stores and manages users. Multiple replica servers are sharing the function of authentication requests with each replica instance containing a copy of the primary instance database contents. Authentication manager software is expandable, so it can authenticate large number of users. It is compatible with the network, remote access, wireless and virtual private networks, Internet, and application products. Authentication manager provides an administrative dashboard, which has a consolidated view of a particular user, what tokens they have been assigned, what groups they belong to, what protected resources they can access, and what authentication activity they have performed [18]. There is also a self-service web portal that end users can use for common token management tasks, such as the ability to reset personal identification number or move from hardware to software-based tokens or vice-versa.

Primary Instance The authentication manager primary instance as shown (see Fig. 4) is the first component installed in a deployment. The authentication manager primary instance includes an embedded version of Oracle database. Additional server nodes to a primary instance can be added to improve performance. The primary instance function is to allow the administration of data stored in the internal database by authentication manager to perform tasks such as importing and assigning

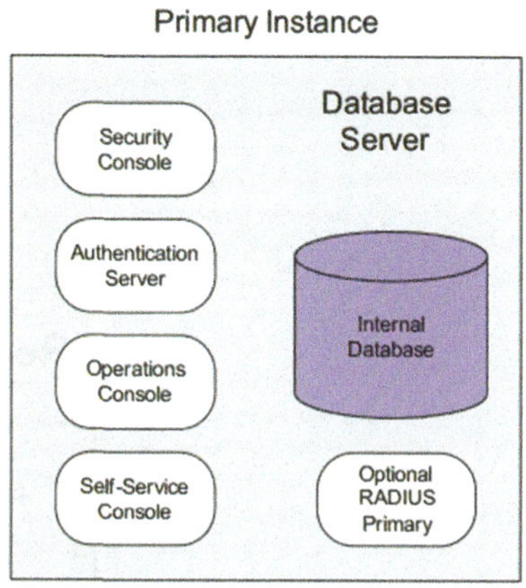

Fig. 4 Primary instance of AM

SecurID tokens, allowing risk-based authentication (RBA), adding lightweight directory access protocol, identity sources, configuring self-service, generating replica packages, and generating configuration files and node secrets for agent [17].

Primary server instance provides administration and coordinates the work of up to fifteen replica server instances. The replica server contains its own dedicated database synchronized to the primary.

Replica Instance The overall structure of a replica instance is identical to the primary instance as shown (see Fig. 5). The replica instance provides additional load support for authentications. Transactions are temporarily stored in the database of the replica instance and are periodically sent to the primary. Replicas support disaster recovery by including a real-time copy of the primary database contents.

RSA SecurID Authentication Agents

Authentication manager authenticates a user's identity based upon the user's supplied passcode. It is the job of the authentication agent to obtain a user's passcode and communicate with the server to verify that passcode. Authentication agents are installed at the point where the user has prompted for authentication and challenges a user for SecurID credentials. Initially, agents are configured with a primary server instance address. Upon communication with the primary server, they obtain information about other servers with replicas. Subsequently, agents contact primary or replica servers as needed for authentication requests. Authentication agents can be installed on different platforms, such as windows agent, UNIX-based PAM agent, and web agent for the web server.

Authentication agents securely perform the following steps (see Fig. 6):

Intercept all attempts at entry, such as attempts to log in to or a URL entry. Determine if SecurID protects the requested resource.

- If the requested resource is not protected, the agent either ignores the request or in the case of a custom agent, takes whatever action is appropriate, such as writing an audit message in the UNIX Syslog or the Windows event log.

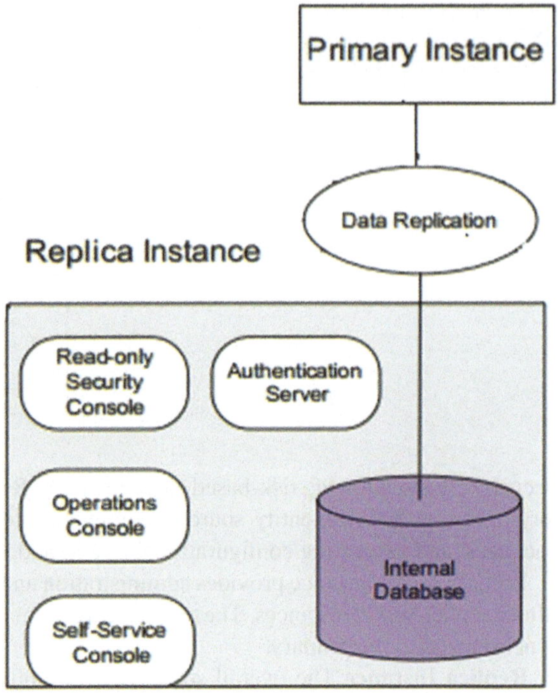

Fig. 5 Replica instance of AM

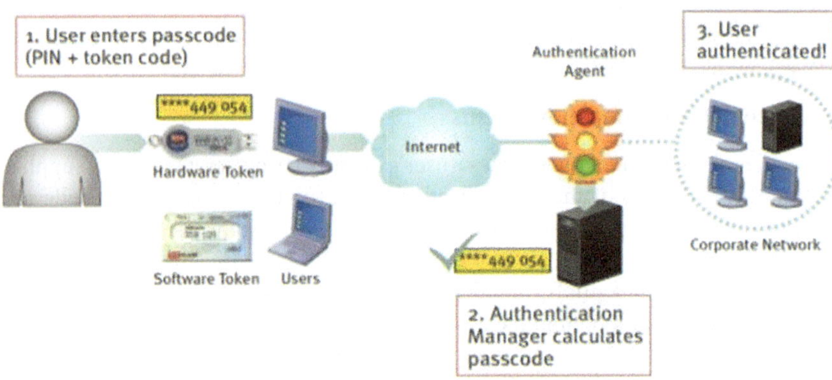

Fig. 6 Two-factor authentication mechanism

- SecurID protects the requested resource while the agent continues the authentication process. Determine the user name of the person logging on, so that the authentication manager will verify that the token code originates from that person's registered authentication device, where replicas are deployed, lock the user name to prevent replay attacks by requesting the user's passcode. Combine the passcode with data only known to the agent and its associated authentication manager in the realm and deliver the combined data to a server for validation.

- If the request is approved, the agent permits access to the protected resource. If the access is denied, the agent prevents the user from logging on to the protected resource and takes any other appropriate actions.

SecurID Authenticators

SecurID tokens are of two kinds, hardware tokens and software tokens.

Hardware tokens produce token codes with an embedded clock and a factory-encoded random key. Software tokens require an application that is relevant to the platform of the intended user, such as a specific operating system on smart phones, computers, or tablets. Users get symmetric token key from the app by scanning a QR code or importing an attachment from the email. The software token apps create token codes on the specified device and provide the same passcode serviceability as hardware token does.

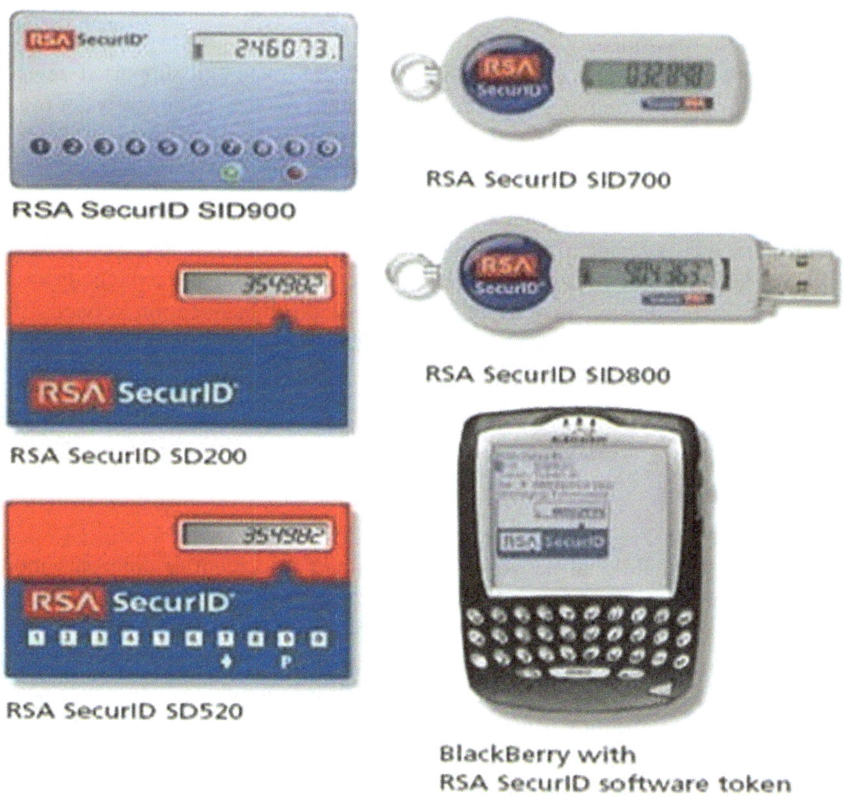

Fig. 7 Hard and soft token authenticators

SecurID authenticators (see Fig. 7) offer token codes which allow users to authenticate and access secure resources from authentication manager and cloud authentication service. A token code is a pseudorandom number, typically with a length of six digits [9]. Token code is dependent on time, which switches at regular intervals. A user enters a personal identification number (PIN) and the number displayed on the token code to access the protected services. The combination of the two is termed a passcode.

Multifactor Authentication

Multifactor authentication is a two-factor authentication extension. The integration of the RSA authentication manager with the cloud authentication service offers an opportunity to extend the resources that has to be protected, and users are provided with authentication methods [11]. Users can authenticate with approve, device biometrics, or authenticate token code after installing the RSA SecurID authentication app on a supported specified device. Authentication factors for a multifactor system may include:

1. Something the person should be knowing (Personal Identification Number, password, passphrase)
2. Something the person should have (a hardware token and supported devices such as laptop, computer, or mobile phone used for generating software token code)
3. Something the person is going to do (specific actions or a behaviour pattern).

For RSA authentication manager, RSA RADIUS can be used to directly authenticate users who attempt to access network resources through RADIUS-enabled devices. Hence, RADIUS deployment can expand the authentication methods available to users by moving to RADIUS for the cloud authentication service. The cloud authentication service responds to the RADIUS server, which responds to the RADIUS clients that requests it.

5 Conclusion

Compared with traditional network security, smart grid security confronts more obstacles. This is because it needs both power and IT security. The smart grid is a next-decade intelligent electrical grid that optimizes energy efficiency by using IT technology. Nonetheless, there is a high likelihood of specific security threats in such a smart grid environment, including data disclosure and data piracy, which occur in two-way communication. Smart grid security approaches have to be inherently cognizant for device performance. In particular, these technologies must integrate seamlessly into its service and output with minimal disruption to its operation and performance. This trade-off between security and efficiency is pronounced in RSA SecurID.

In this article, multifactor authentication methodology has been proposed for smart grids cyber security using minimal IT resources to ensure that only approved users can access sensitive grid information. SecurID access provides a wide range of authentication methods including modern multifactor mobile authenticators such as one-time passwords, push alerts, SMS and biometrics, as well as conventional hard and soft tokens for encrypted access to all applications, whether they reside in the site or in the cloud. RSA SecurID suite aims to help companies and organizations to mitigate the risk of identification and ensure compliance without impeding productivity of users.

References

1. Otuoze, A.O., Mustafa, M.W., Larik, R.M.: Smart grids security challenges: classification by sources of threats. J. Electr. Syst. Inf. Technol. **5**(3), 468–483 (2018)
2. Wang, W., Lu, Z.: Cyber security in the smart grid: survey and challenges. Comput. Netw. **57**(5), 1344–1371 (2013)
3. Khurana, H., Hadley, M., Lu, N., Frincke, D.A.: Smart-grid security issues. IEEE Secur. Priv. **8**(1), 81–85 (2010)
4. Rawat, D.B., Bajracharya, C.: Cyber security for smart grid systems: status, challenges and perspectives. In: SoutheastCon 2015, pp. 1–6. IEEE (2015)
5. Gunduz, M.Z., Das, R.: Cyber-security on smart grid: threats and potential solutions. Comput. Netw. **169**, 107094 (2020)
6. Pandey, R.K., Misra, M.: Cyber security threats—smart grid infrastructure. In: 2016 National power systems conference (NPSC), pp. 1–6. IEEE (2016)
7. Aloul, F., Al-Ali, A.R., Al-Dalky, R., Al-Mardini, M., El-Hajj, W.: Smart grid security: threats, vulnerabilities and solutions. Int. J. Smart Grid Clean Energy **1**(1), 1–6 (2012)
8. Sanjab, A., Saad, W., Guvenc, I., Sarwat, A., Biswas, S.: Smart grid security: threats, challenges, and solutions (2016). arXiv preprint arXiv:1606.06992
9. https://community.rsa.com/community/products/securid
10. Mishra, P.K., Tiwari, P.: Cyber security in smart grid. In: First international conference on advances in SCIENCE HUB (ICASH 2020) Proceedings (Special Edition)
11. McAfee Case Study—Securing the cloud with strong two-factor authentication through McAfee one time password
12. Florencio, D., Herley, C.: A large-scale study of web password habits. In: Proceedings of the 16th international conference on the World Wide Web, ACM Digital Library, pp. 657–666 (2007)
13. El Mrabet, Z., Kaabouch, N., El Ghazi, H., El Ghazi, H.: Cyber-security in smart grid: survey and challenges. Comput. Electr. Eng. **67**, 469–482 (2018)
14. Dmitrienko, A., Liebchen, C., Rossow, C., Sadeghi, A.-R.: On the (In) security of mobile two-factor authentication. in: Lecture Notes in Computer Science, pp. 365–383 (2015)
15. Kale, R., Gore, N., Kavita, Jadhav, N., Shinde, S.: Review paper on two factor authentication using mobile phone. Int. J. Innovative Res. Stud. **2**(5), 164–170 (2015)
16. Vaithyasubramanian, S., Christy, A., Lalitha, D.: Two factor authentication for secured login using array password engender by Petri net. Procedia Comput. Sci. Elsevier (2016)
17. https://searchsecurity.techtarget.com/definition/multifactor-authentication-MFA
18. Adeoye, O.S.: Evaluating the performance of two-factor authentication solution in the banking sector. Int. J. Comput. Sci. Issues (IJCSI) **9**(4), 457 (2012)

Super Resolution Techniques for COVID-19 Chest X-ray Images

Seema S. Bhat, M. C. Hanumantharaju, and V. N. Manjunath Aradhya

Abstract COVID-19 is a respiratory infectious disease caused by a novel coronavirus-II. COVID-19 has affected a few millions of people across the globe with alarming death toll rate. Chest X-rays are widely available modality for immediate care in diagnosing COVID-19. X-ray images inherently possess significant amount of noise, and image resolution depends on radiation dosage. Diagnosis of COVID-19 chest X-rays would be more precise with high-resolution images. The aim of the current work is to explore selected super-resolution techniques for COVID-19 chest X-ray images. The samples are taken from COVID-19 chest X-ray public dataset. We conducted the experiments on COVID-19 chest X-ray public dataset. The super-resolution methods produced high resolution chest X-ray images the clear view of the pulmonary opacities in chest x-ray images. Evaluation metrics used in the experiments demonstrated improvement in overall quality enhancement of the chest X-ray images.

Keywords COVID-19 · Chest X-ray images · Deep learning · Super resolution

1 Introduction

COVID-19 is an infectious disease caused by severe acute respiratory syndrome coronavirus 2. COVID-19 cases were first emerged in Wuhan, China, in late December 2019 and now affected most of the countries across the globe [1]. COVID-19 is

S. S. Bhat (✉)
Department of ISE, Dayananda Sagar College of Engineering, Bengaluru, India
e-mail: seemasbhat@dayanandasagar.edu

M. C. Hanumantharaju
Department of Electronics and Communication, BMS Institute of Technology and Management, Bengaluru, India
e-mail: mchanumantharaju@bmsit.in

V. N. M. Aradhya
Department of Computer Applications, JSS Science and Technology University, Mysuru, India
e-mail: aradhya@jssstuniv.in

V. Bhateja et al. (eds.), *Computer Communication, Networking and IoT*, Lecture Notes in Networks and Systems 197, https://doi.org/10.1007/978-981-16-0980-0_53

regarded as general respiratory infections in radiology studies. Chest CT and X-ray widely used imaging modalities. Though CT produces better contrast, X-ray is the most available imaging modality for immediate care in diagnosing COVID-19. COVID-19 medical images are examined for ground glass patterned areas which are dense opacities obscuring vessels and bronchial walls in lungs. The opacities can damage both the lungs at an early stage of COVID-19. These opacities could be visualized in the posterior segments of lower lobes with a peripheral and subpleural distribution. Bilateral lower zone consolidation is one more symptom of COVID-19, usually appears after 10–12 days of onset of the disease. These symptoms were observed in one of the COVID-19 chest X-ray study [2], and the study has claimed 69% of sensitivity in detecting COVID-19 symptoms with baseline chest X-ray images. X-ray images have low resolution (LR) and a significant amount of noise. This is mainly due to radiation levels that are minimized to maintain patient safety [3]. High-resolution (HR) X-ray images may increase sensitivity and specificity in detection and diagnosis of the disease under investigation. Post processing of X-ray images would generate high-resolution images without high dose of radiation and with less cost.

In order to take the advantage of super resolution methods, this work explores some of the well-known techniques and analyzes them on COVID-19 chest X-ray images. The study involves experimenting with basic interpolation-based super resolution method to state-of-the-art deep learning-based super resolution methods. The main contribution of this work is preserving and enhancing finer details in chest X-ray images through super resolution techniques. The hazy pulmonary opacities representing COVID-19 infections can be more clearly viewed with sharper chest X-rays. Image quality metrics used in the study reassured significant improvement in the resolution of test images.

2 Materials and Methods

2.1 Dataset Collection

The current work explores selected dictionary-based super resolution methods. HR images used for dictionary generation and training the network are derived from T91 dataset from Kaggle [5]. T91 datasets are designated for dictionary-based super resolution experiments. Training of convolution neural network (CNN)-based super resolution method uses BSD-200 standard HR images [6]. Efficiency of the trained dictionaries and CNN was verified against to Set 5 and Set 14 standard test images. Selected samples of chest X-ray images from public dataset of COVID-19 images [7] are considered for testing purpose. The dataset exclusively has 123 frontal view chest X-ray images. There are 244 total samples: inclusive of COVID-19 and other viral/bacterial pneumonias in posterior-anterior (PA) or vice-versa (AP) view of chest CT and X-ray images. It also has 38 samples of AP supine view chest images.

Though, public COVID-19 chest x-ray images are available, they are inclusive of non-COVID-19, pneumonia, and other chest X-ray images. Also there are no ground truth and standard HR chest X-ray images specified. Hence, in the current work, known standard datasets are utilized for training purposes [5, 6].

2.2 *Super Resolution Techniques for Medical Image Enhancement.*

High resolution in medical imaging is receiving much attention for accurate disease detection and diagnosis [4]. Pulmonary diseases often have hazy structures. Finer images can be obtained with higher dose of radiation or expensive imaging set up. The commonly used chest X-rays are inherently low resolved, but available for immediate care in COVID-19 like pulmonary disease. Super resolution-based enhancement would be one of the easier ways to get high-quality chest X-rays with minimal radiation dosage to the patients [3]. These super resolutions are a class of techniques that enhance noisy LR images using set of LR images [3, 4]. Most of the super resolution methods were basically experimented with natural images. Eventually extended their proficiencies in medical images. The current work briefs about the methods which are applicable for X-ray like LR images. The efficiency of these methods is verified with standard test images.

Interpolation-based super resolution. Image interpolation is the conventional method for upscaling image with poor quality. Nearest neighbor, bilinear, and bicubic interpolations are the widely used linear interpolation methods that does transformation of pixel coordinates to resize, rotate, or shear of the input image [8]. Nearest neighbor is the basic method which operates one pixel at a time, closet one to the pixel to be interpolated. It takes least execution time and simply makes each pixel bigger. Bilinear considers 2×2 neighborhood of known pixel values surrounding the unknown pixel values. Interpolated pixel will have weighted average of these 4 pixels. Bicubic considers 4×4 neighborhood of known pixel values, and more weightage will be given to the least distanced pixels. Bicubic interpolation is commonly used in cameras and other imaging applications. Interpolation methods often yield blurry images due to over smoothening. Aliasing and halo around the edges in upscaled images may not give a unique solution [9, 10].

Image reconstruction-based super resolution. Super resolution reconstruction (SRR) uses HR reference image to improve the resolution of the LR image. They are basically unsupervised machine learning methods which does not need a training database. On local means (NLM), SRR is one such method, which assumes that image content can be repeated within some neighborhood of the image itself [11]. Each HR pixel's value is calculated by weighted average of all pixels in its neighborhood by following equation:

$$\hat{x}[k,l] = \frac{\sum_{(i,j)\in N(k,l)} w[k,l,i,j]y[i,j]}{\sum_{(i,j)\in N(k,l)} w[k,l,i,j]} \quad (1)$$

where $N(k, l)$ represents the neighborhood of the pixel (k, l) and the term $w[k, l, i, j]$ is the weight for the (i, j)th neighbor pixel, input pixel $y[k, l]$, and output pixel is $\hat{x}[k, l]$. Weights and neighborhood sizes are defined by regularization terms using regression of varying orders and/or image priors. The method is faster than any supervised learning reconstruction as there no need to have training dataset. But the choice of optimal neighborhood size and regularization parameters is a challenge. The methods often produce outliers and artifacts.

Learning-based super resolution. With advance in image processing, local image information is insufficient to predict the lost high-resolution details. Therefore, learning-based super resolution used spatial neighborhood information for the high-resolution pixel. Learning-based methods involve creation of dictionaries. Dictionaries are usually comprising of low and high-resolution patch pairs. Mapping of low to high-resolution patches is done by exploiting neighborhood relationships with the help of probabilistic models. Learning-based SR is classified into neighborhood-based methods [12–14] and sparsity-based methods [15].

To take an example, anchored neighborhood regression (ANR) is one such learning-based SR method [13]. It uses global regression and works on fixed size patches in a LR and HR patch pair dictionary. By searching for nearest neighbors in a low-resolution dictionary, a number of corresponding high-resolution candidates are retrieved. This method involves grouping the dictionary atoms into neighborhoods based on K nearest neighbors. Nearest neighbors grouped based on similarity measure between input and output patch and precomputed projection matrix. The neighborhood size and searching time for similar patterns in training set are the two controlling factors of NE-based super resolution.

In order to reduce the computation time, instead of working directly with the image LR-HR patch pairs, sparsity-based learning methods use compact representation of patch pairs to capture the co-occurrence prior [15]. The local sparse model uses a global reconstruction model to obtain a global optimum. Optimal sparse dictionary creation can still be a challenge. Example-based super resolution methods may consume more time to reconstruct HR versions with limited real-time medical data.

Deep learning-based super resolution. Deep learning is a branch of artificial neural network based on representation learning [16]. Convolutional neural networks (CNN) are used in first ever super resolution technique based on deep learning with 2–4 CNN layers. Several deep learning methods are available for super resolution [16, 17]. The current work uses one of the well-known deep learning-based super resolution methods for upscaling COVID-19 chest X-rays [18]. Rapid and accurate image super resolution (RAISR) is faster deep learning method [18]. It classifies input image patches according to the patch's angle, strength, and coherence and then learns maps from LR image to HR image among the clustered patches. Thus, all the RAISR filters are trained according to edge features. The filters are applied selectively to each pixel of the LR image, to recreate details that are of comparable quality to the HR image.

X-ray images suffer from inherent system noise and non-Gaussian quantum noise. Conventional enhancement methods and learning-based approaches make use of hand-crafted image priors to remove artifacts. Deep learning-based super resolution methods extract image features and learn the mapping function automatically. They learn mapping automatically through convolution operations and enhance details.

3 Results

The test images are taken from COVID-19 chest X-ray public dataset that are shown in Fig. 1(first row) [7]. Figure 1a, b (matrix size: 365 × 435 pixels) is posterior-anterior (PA) view of the patient taken on first day and a follow up test on eighth day. Figure 1c (matrix size: 702 × 621 pixels) image is COVID-19 positive patient's chest X-ray from the same dataset. Medical image super resolution methods escalate a lot of uncertainty in their accuracy. Hence, original test images are down-sampled by ×2 factor to generate LR input images, and super resolution techniques were

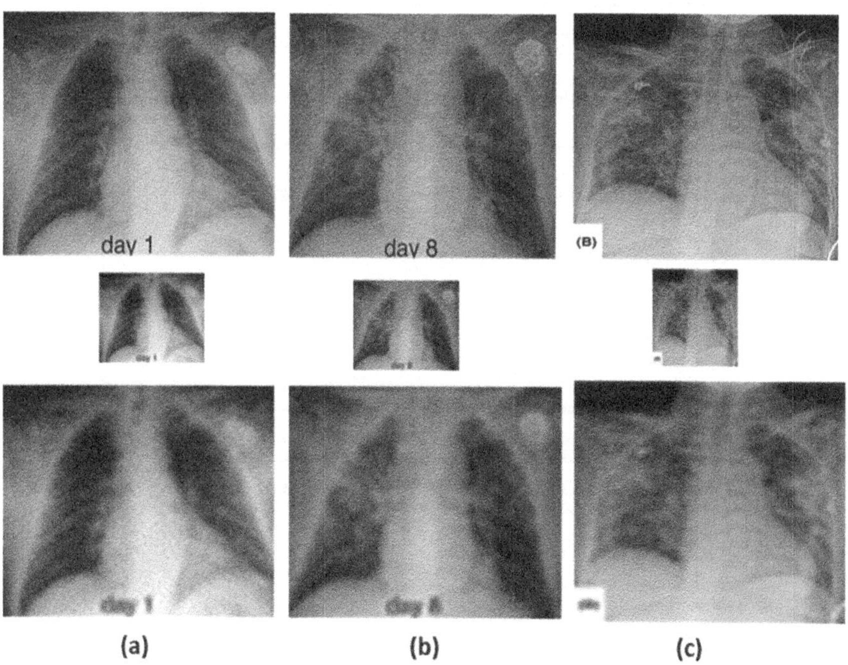

Fig. 1 First row: **a** Sample 1, **b** Sample 2, **c** Sample 3, COVID-19 chest X-ray test images [7], second row: LR test images, down-sampled with ×2 factor, third row: zoomed version of LR test images

applied on down-sampled version. Down-sampled version of test images is shown in Fig. 1 (second row, third shows the zoomed version of LR images). Then quality of the reconstructed images was compared with original image to ensure details are preserved.

Experimenting with bicubic interpolation was straight forward, and input images were upscaled at ×4 factor. Although the approach is simple and fast, it resulted in blurred edges and unclear textures in the super-resolved HR image. The objective and subjective measures of interpolation method demonstrate the same (shown in Figs. 2a and 4). There was a significant amount of details lost while upscaling the input image (shown in Fig. 3a).

Learning-based super resolution methods work in two phases: training and testing. HR images [6] were down-sampled by bicubic interpolation to generate HR–LR image pairs in the training phase. A 3 × 3 HR–LR patch with a two-pixel overlap was used in both the training and testing phases. Dictionary size of 1024 with ×3 upscaling factor was chosen for neighborhood and sparse-based super resolution methods [13].

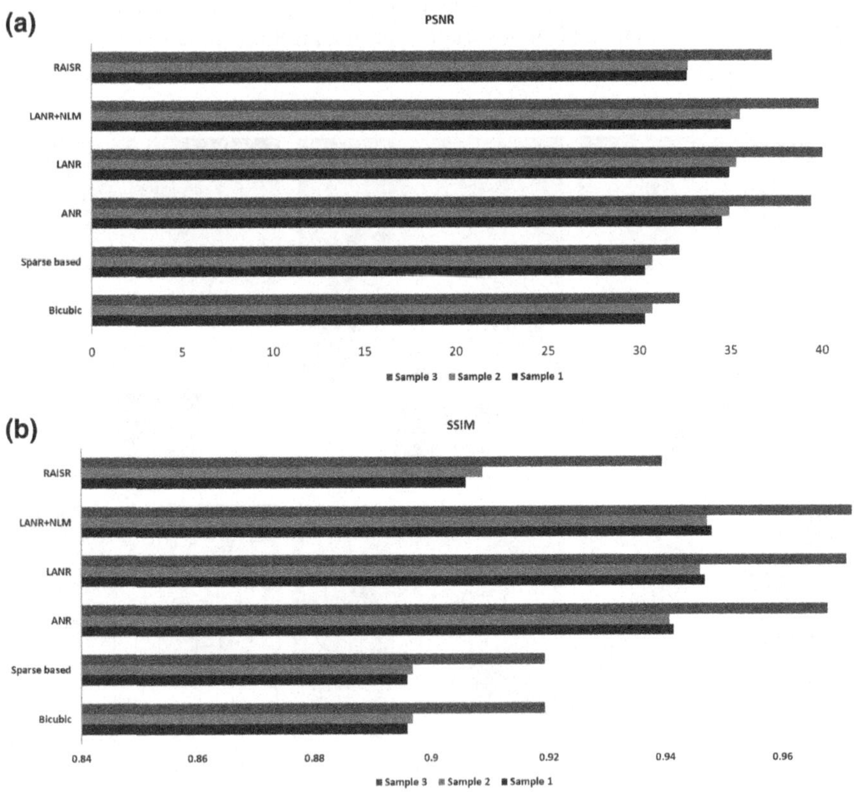

Fig. 2 **a, b** PSNRs and SSIMs of selected super resolution schemes and figure demonstrate learning-based super resolution methods achieve better quality metrics compared to interpolation-based methods

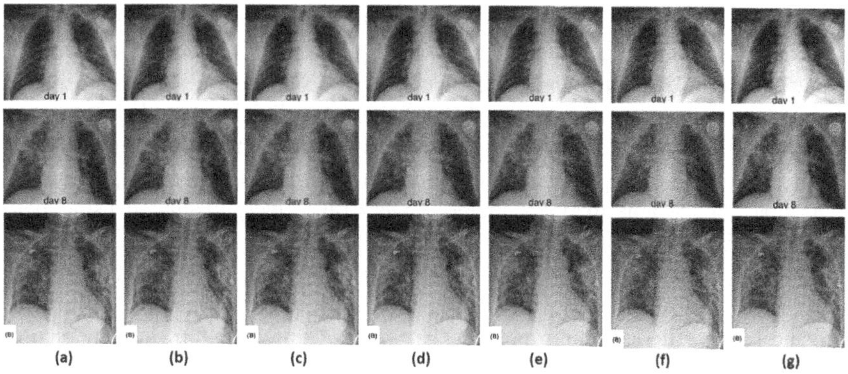

Fig. 3 High-resolution COVID-19 chest X-ray images using selected super resolution methods (First row: sample 1, second row: sample 2, third row: sample 3): **a** Original image [16] **b** Bicubic interpolation [9] **c** Sparse representation [15] **d** ANR [13] **e** LANR [14] **f** NLM + LANR [12] **g** RAISR[18]

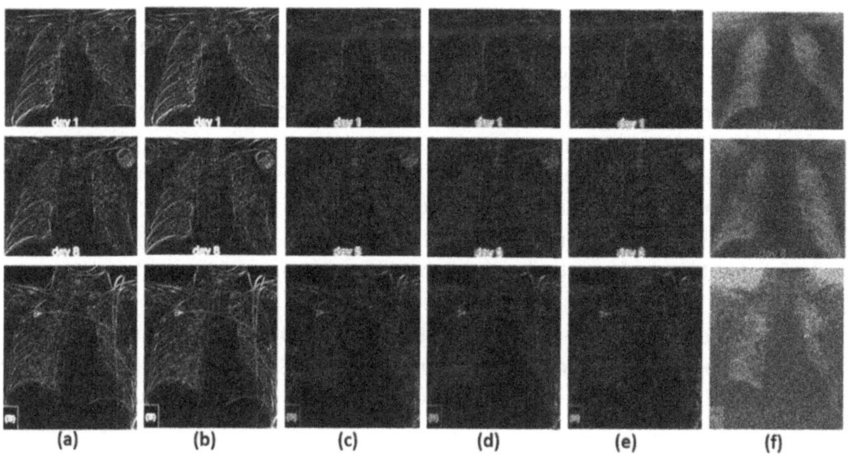

Fig. 4 Image residuals obtained by subtracting COVID-19 test image, and their high-resolution version obtained using selected super resolution schemes (First row: sample 1, second row: sample 2, third row: sample 3): **a** Bicubic interpolation [6] **b** Sparse representation [15] **c** ANR [13] **d** LANR [14] **e** NLM + LANR [12] **f** RAISR[18]

Neighborhood size $K = 200$, regularization parameter $\lambda_1 = 0.00001$ and $\lambda_2 = 0.005$ and weight decay power parameter α was set to 11 in ANR, LANR and LANR + NLM-based super resolution methods. These tuning parameters were selected based on the literature [13]. Sparse-based and neighborhood-based super resolution methods yielded better PSNR and SSIM values compared to bicubic interpolation method (as shown in Fig. 2a, b). Subtraction images in Fig. 4b, c, d, e illustrate high-resolution details are well preserved during upscaling.

Deep learning-based RAISR method used BSD200 dataset for training [6, 18]. The training outputs three files: filter, Qfactor_str, and Qfactor_coh, corresponding to patch's angle, strength, and coherence which are needed in the testing phase. These three files assist edge-based mapping filters to upscale each pixel in the test image. Upscaling factor $\times 2$, $\times 3$ or $\times 4$ can be set during training and testing. RAISR obtained sharper SR output without compromising high-frequency components.

All the above-mentioned SR methods were objectively evaluated using peak signal-to-noise ratio (PSNR) [19] and structural similarity (SSIM) [20]. Higher the measures better the restoration. The image quality metrics of the selected super resolution techniques were tested with COVID-19 chest X-ray image and were compared with the bicubic up-sampling method. The mean $\pm$ SD PSNR values of bicubic and sparse based, ANR, LANR, NLM and RAISR were 31.06 ± 1.00 dB, 36.26 ± 2.72 dB, 36.73 ± 2.83, 36.77 ± 2.64 and 33.13 ± 2 dB, respectively. The mean $\pm$ SD SSIM of bicubic and sparse based, ANR, LANR, NLM and RAISR were 0.9040 ± 0.012, 0.9499 ± 0.012, 0.9545 ± 0.014, 0.9556 ± 0.014 and 0.9180 ± 0.018. PSNR and SSIM scores of learning-based super resolution methods are better than the interpolation-based SR. Figure 2a, b represents PSNRs and SSIMs of interpolation-based, sparse-based, ANR and NLM-based super resolution schemes, respectively. There is a substantial enhancement in deep learning-based RAISR method, but there is a significant reduction in SSIM and PSNR scores. This observation recommends the need of using new evaluation metrics that can evaluate the SR image enhancement techniques unlike the present metrics.

Figures 3 captures high-resolution images obtained using selected super resolution schemes for three different LR test images (shown in Fig. 1). Test images were trained with non- medical images for all learning-based super resolution schemes [5, 6]. The methods can offer significantly improve chest X-ray image quality with prominently visible details.

Bicubic interpolation upscaled LR input image with factor $\times 4$. But the method yielded poor objective and subjective scores. Visual assessment confirmed that the learning-based SR methods restored most of the image details close to the original. It was also noticed that, the neighborhood embedding methods produce ringing artifact at the boundaries and low frequency regions were over smoothed. Deep learning-based RAISR method produced finer lung structures compared to other tested methods. Sample 1 and sample 2 were follow up images of the COVID-19 patient with a gap of 7 days. More opacities can be observed with increase in infection after a week. Darker contrast represents air and hazy structures indicates infection. These opacities are more near to lobes, pleural space and chest walls with gray contrast.

Figure 4 shows image residuals obtained by subtracting original images by super-resolved images. The residuals in subtracted images visualize the components lost from the original image. Prominent details were lost while upscaling for bicubic and sparse-based super resolution. ANR and NLM-based methods performed better as they were able to preserve high-frequency components while upscaling. RAISR produced little puzzling image residual, and there is complete restoration of prominent details. But low-resolution areas are over smoothed.

Image residuals in Fig. 4 represent how much details were able to restore in each of the super resolution methods explored in the current study. However, this study has few limitations dictionary-based and CNN-based SR methods were trained with non-medical images. Additional study is required to find the better network configuration, and a good number of training images is concerned. Super resolution experiments done in this study may achieve better result with 3D chest CT and high-resolution chest X-ray images in their training phase.

4 Discussion and Conclusion

In this study, selected super resolution methods available in the literature were applied to increase the resolution of COVID-19 chest X-ray images. The experimental results obtained using clinical COVID-19 chest X-ray images demonstrated, that reconstruction-based and learning-based schemes significantly outperforms conventional interpolation method. Analysis has been done on a limited data and obtained preliminary results. The enhancement schemes are purely exploratory, and clinical trials/medical validations have not been done on the approach.

References

1. Zhu, N., Zhang, D., Wang, W., Li, X., Yang, B., Song, J., Niu, P.: A novel coronavirus from patients with pneumonia in China, 2019. New England J. Med. (2020)
2. Wong, H.Y.F., Lam, H.Y.S., Fong, A.H.T., Leung, S.T., Chin, T.W.Y., Lo, C.S.Y., Lee, E.Y.P.: Frequency and distribution of chest radiographic findings in COVID-19 positive patients. Radiology, 201160 (2020)
3. Shimizu, M., Kariya, H., Goto, T., Hirano, S., Sakurai, M.: Super-resolution for X-ray images. In: 4th IEEE global conference on consumer electronics, pp. 246–247. IEEE Press (2015)
4. Robinson, M.D., Chiu, S.J., Toth, C.A., Izatt, J.A., Lo, J.Y., Farsiu, S.: New applications of super-resolution in medical imaging. In: Super-resolution imaging, pp. 401–430. CRC Press (2017)
5. Kaggle: Your Machine Learning and Data Science Community. https://www.kaggle.com/datasets
6. Martin, D., Fowlkes, C., Tal, D., Malik, J.: A database of human segmented natural images and its application to evaluating segmentation algorithms and measuring ecological statistics. In: 8th IEEE international conference on computer vision, vol. 2, pp. 416–423. IEEE Press (2001)
7. Cohen, J.P., Morrison, P., Dao, L.: COVID-19 image data collection (2020). arXiv preprint arXiv:2003.11597
8. Zhou, F., Yang, W., Liao, Q.: Interpolation-based image super-resolution using multisurface fitting. IEEE Trans. Image Process. **21**(7), 3312–3318 (2012)
9. Gilman, A., Bailey, D. G., Marsland, S.R.: Interpolation models for image super-resolution. In: 4th IEEE International Symposium on Electronic Design, Test and Applications, pp. 55–60. IEEE Press (2008)
10. Hung, K.W., Siu, W.C.: Fast image interpolation using the bilateral filter. IET Image Proc. **6**(7), 877–890 (2012)

11. Yong-Rim, K., Yong-Jin, K.: Novel super-resolution method based on high order nonlocal-means (2015). arXiv preprint arXiv:1503.04253
12. Roweis, S.T., Saul, L.K.: Nonlinear dimensionality reduction by locally linear embedding. Science **290**(5500), 2323–2326 (2000)
13. Timofte, R., De Smet, V., Van Gool, L.: Anchored neighborhood regression for fast example-based super-resolution. In: IEEE International Conference on Computer Vision, pp. 1920–1927. IEEE Press (2013)
14. Jiang, J., Ma, X., Chen, C., Lu, T., Wang, Z., Ma, J.: Single image super-resolution via locally regularized anchored neighborhood regression and nonlocal means. IEEE Trans. Multimedia **19**(1), 15–26. IEEE Press (2016)
15. Yang, J., Wright, J., Huang, T.S., Ma, Y.: Image super-resolution via sparse representation. IEEE Trans. Image Process. **19**(11), 2861–2873 (2010)
16. Wang, Z., Chen, J., Hoi, S.C.: Deep learning for image super-resolution: a survey. IEEE Trans. Pattern Anal. Mach. Intell. (2020)
17. Greenspan, H.: Super-resolution in medical imaging. Comput. J. **52**(1), 43–63 (2009)
18. Romano, Y., Isidoro, J., Milanfar, P.: RAISR: rapid and accurate image super resolution. IEEE Trans. Comput. Imaging **3**(1), 110–125 (2016)
19. Huynh-Thu, Q., Ghanbari, M.: Scope of validity of PSNR in image/video quality assessment. Electron. Lett. **44**(13), pp. 800–801 (2008)
20. Wang, Z., Bovik, A.C., Sheikh, H.R., Simoncelli, E.P.: Image quality assessment: from error visibility to structural similarity. IEEE Trans. Image Process. **13**(4), 600–612 (2004)

Author Index

V. Bhateja et al. (eds.), *Computer Communication, Networking and IoT*, Lecture Notes in Networks and Systems 197, https://doi.org/10.1007/978-981-16-0980-0

The manufacturer's authorised representative in the EU is Springer Nature Customer Service Centre GmbH, Europaplatz 3, 69115 Heidelberg, Germany. If you have any concerns regarding our products, please contact ProductSafety@springernature.com

Printed and bound by CPI Group (UK) Ltd, Croydon, CR0 4YY

30/09/2026

02225292-0015